Mechanics of Biological Tissue

Preface

This book contains a collection of papers that were presented at the IUTAM Symposium on *Mechanics of Biological Tissue*, which was held in Graz, Austria, from June 27 to July 2, 2004. The setting of Graz was very appropriate for the symposium since it is the city where such illustrious scientists as *Johannes Kepler, Ernst Mach, Ludwig Boltzmann, Erwin Schrödinger* and *Otto Kratky* spent parts of their lives, while the cultural life of Graz provided ample opportunity for complementing the scientific proceedings. Graz has an historic centre, which is one of the best preserved old town centres in Europe, and which was added to the UNESCO world cultural heritage list in 1999.

The symposium brought together 96 participants from universities, research centres and clinics in 19 countries. There were 42 oral presentations, including 7 keynote lectures, and 15 poster presentations. The keynote lectures were given by P.B. Canham (University of Western Ontario, Canada), S.C. Cowin (City University of New York, USA), K. Hayashi (Osaka University, Japan), J.D. Humphrey (Texas A&M University, USA), P.J. Hunter (University of Auckland, New Zealand), R.S. Lakes (University of Wisconsin, USA), and P.D. Richardson (Brown University, USA).

The mechanics of biological tissues, both soft and hard tissues, is a multidisciplinary area of research. Several prestigious universities have established new focuses on this rapidly expanding field, which involves researchers in engineering, physics, biology, medicine and applied mathematics. A common goal is to analyze and tabulate, to search for causal connections and to make predictions on the basis of abstraction and general principles. Therefore, we need to ask fundamental questions and to create new technologies for solving problems related to the mechanics of biological tissues; in other words, tissue biomechanics offers a positive heuristic. Accordingly, the symposium aimed to provide a forum for discussion, for comparison of different methods and approaches in the field and for the unification of these approaches, to bring together young researchers and the world's leading scientists working in the field, and to stimulate the study of challenging new topics in the mechanics of biological tissue. Major objectives of the symposium were to survey the current state-of-the-art and to point to important new directions, with an emphasis on the need to combine the diverse disciplines involved.

Part of the aim was to provide a deeper understanding of tissue structure and function by considering the mechanics of both soft and hard tissues at the molecular, cellular, tissue and organ levels, to better understand inter-relations between the mechanical and biological processes such as growth, remodelling, and repair, and how mechanical information is processed and programmed by the cells (mechanobiology). This involves experimental, microstructural, continuum mechanical and computational approaches, and a key ingredient is the modelling of the mechanical behaviour of tissues, the models being used in simulating therapeutic and diagnostic procedures such as balloon angioplasty and stenting. Because of the complexity of the material properties and the geometries encountered in applications, implementation of the biomechanical models in numerical codes is vital. An efficient numerical tool is one of the prerequisites for the design and development of soft and hard tissue prostheses. Thus, numerical aspects were also well represented.

The symposium included presentations on

- *Tissue growth and remodelling*

  This topic of major interest included discussion of general continuum theories of growth, remodelling and adaptation, with particular reference to mass change and collagen fibre orientation development, and applications of those theories to arterial, tendon and cartilage growth and to bone healing.

- *Micromechanics, cells and matrix*

  This area covered measurements of the mechanical properties of cells, the engineering of cell systems, constitutive and computational modelling of cells and cell-substrate interactions, and micro-to-macro modelling considerations in the context of the mechanics of nacre.

- *Arteries in health and disease*

  Here a major concern was the determination of properties of arterial tissue, both in its healthy state and when subject to disease, such as atherosclerosis, or in the presence of aneurysms. This included experimental analysis of residual stress, modelling the constitutive properties of arterial walls and the computational analysis of the effect of stenting on the arterial wall. The effect of collagen fibre orientation distributions in saccular aneurysms was examined, as was the interaction between blood flow and aneurysm development.

- *Biological tissues*

  Soft tissues other than arterial tissue were the focus of several presentations. These included discussion of the musculo-skeletal system, heart valves, ligaments, intervertebral discs, the uterus and vocal fold tissues, with experimental, modelling and computational perspectives represented.

- *Image-based analysis*

  The presentations in this area illustrated that imaging techniques have great potential for the analysis of tissue properties and pathologies and for guiding the design of engineered tissue constructs.

Bearing in mind that *statements that are not open to revision are not scientific*, the symposium provided scientific impetus and new ideas for future research directions and also some cultural impressions and new friends. The organizers would like to thank the members of the Scientific Committee for their support in helping to make the symposium such a success. They are S.C. Cowin (New York, USA), J. Engelbrecht (Tallinn, Estonia), J.D. Humphrey (College Station, USA), J.M. Huyghe (Eindhoven, The Netherlands), P.J. Prendergast (Dublin, Ireland), and A. Rachev (Sofia, Bulgaria).

This volume has been made possible by the considerable efforts of all the authors, and we are very grateful for their contributions. Special thanks also go to Bettina Strametz for her great help with the detailed organization of the symposium and with the production of the symposium literature. Finally, we would like to thank Dieter Merkle of Springer-Verlag, Heidelberg, for his encouragement to publish this volume.

Stockholm, July 2005

Gerhard A. Holzapfel, Chairman (Graz, Austria)
Ray W. Ogden, Co-Chairman (Glasgow, UK)

Contents

Part I Tissue Growth and Remodelling

Towards a Theory of Vascular Growth and Remodeling
J.D. Humphrey ... 3

Complementary Roles of Theoretical Modeling and Computer-controlled Experimentation in Vascular Growth and Remodeling
R.L. Gleason, E. Wilson, J.D. Humphrey ... 17

On the Modeling of Growth and Adaptation
S.C. Cowin ... 29

Growth in Soft Biological Tissue and Residual Stress Development
A. Guillou, R.W. Ogden ... 47

Characterization and Modeling of Growth and Remodeling in Tendon and Soft Tissue Constructs
E.M. Arruda, S.C. Calve, K. Garikipati, K. Grosh, H. Narayanan ... 63

Modeling and Simulation of Remodeling in Soft Biological Tissues
E. Kuhl, A. Menzel, K. Garikipati, E. M. Arruda, K. Grosh ... 77

Anisotropic Remodelling of Biological Tissues
A. Menzel ... 91

A Mechanobiological Formulation of Bone Healing
M. Doblaré, J.M. García-Aznar, M.J. Gómez-Benito ... 105

Continuum Models of Growth with Emphasis on Articular Cartilage
S.M. Klisch 119

Part II Micromechanics, Cells and Matrix

Tensile Properties and Local Stiffness of Cells
K. Hayashi 137

Microfluid-dynamics in Three-dimensional Engineered Cell Systems
F. Boschetti, M.T. Raimondi, F. Migliavacca, M. Cioffi, R. Pietrabissa . 153

Nonlinear Constitutive Models for Cochlear Outer Hair Cells
K. Grosh, N. Deo 165

Prediction of Changes in Cell-substrate Contact under Cyclic Substrate Deformation Using Cohesive Zone Modelling
J.P. McGarry, B.P. Murphy, P.E. McHugh 177

Micromechanics and Macromechanics of the Tensile Deformation of Nacre
H.J. Qi, B.J.F. Bruet, J.S. Palmer, C. Ortiz, M.C. Boyce 189

Part III Arteries in Health and Disease

Mechanical Properties of Atherosclerotic Tissues
P.D. Richardson 207

Towards a Computational Methodology for Optimizing Angioplasty Treatments with Stenting
G.A. Holzapfel, M. Stadler, Th.C. Gasser 225

Computational Modeling of Stented Arteries: Considerations for Evolving Stent Designs
M.R. Moreno, J. Bedoya, C. Meyer, J.E. Moore Jr. 241

Simulation of In-stent Restenosis for the Design of Cardiovascular Stents
C. Lally, P.J. Prendergast 255

Material Property Alterations with Early Atheroma in an Animal Model
K.B. Chandran, H. Kim, D.D. McPherson 269

Microscopic Analysis of Residual Stress and Strain in the Aortic Media Considering Anisotropy of Smooth Muscle Layer
T. Matsumoto, T. Furukawa, K. Nagayama 283

Parameter Identification in Arteries Using Constraints
J. Stålhand, A. Klarbring 295

Collagen Organization and Biomechanics of the Arteries and Aneurysms of the Human Brain
P.B. Canham, R.M. Korol, H.M. Finlay, R.R. Hammond, D.W. Holdsworth, G.G. Ferguson, A.R. Lucas 307

Image-based Simulation of Blood Flow and Arterial Wall Interaction for Cerebral Aneurysms
M. Oshima, R. Torii, T. Takagi 323

Part IV Biological Tissues

A Framework for Soft Tissue and Musculo-skeletal Modelling: Clinical Uses and Future Challenges
J.W. Fernandez, H. Schmid, P.J. Hunter 339

Invariant Formulation for Dispersed Transverse Isotropy in Tissues of the Aortic Outflow Tract
D.R. Einstein, A.D. Freed, I. Vesely 355

Mathematical Modelling of Cardiac Mechanoenergetics
J. Engelbrecht, M. Vendelin 369

Creep and Relaxation in Ligament: Theory, Methods and Experiment
A.L. Oza, R. Vanderby, R.S. Lakes 379

Viscoelastic Constitutive Law Based on the Time Scale of the Mechanical Phenomena
D.P. Pioletti 399

A Coupled FE Analysis of the Intervertebral Disc Based on a Multiphasic TPM Formulation
W. Ehlers, B. Markert, N. Karajan, A. Acartürk 405

Is the Free Energy of Hydrogel the Sum of Elastic Energy and Ionic Energy?
R.W. Roos, J.M. Huyghe, F.P.T. Baaijens 421

In Vivo Experiments to Characterize the Mechanical Behavior of the Human Uterine Cervix
E. Mazza, A. Nava, M. Bauer, R. Winter, M. Bajka, G.A. Holzapfel . . . 431

Viscoelastic Response of Vocal Fold Tissues and Scaffolds at High Frequencies
R.J. Clifton, X. Jia, T. Jiao, C. Bull, M.S. Hahn . . . 445

An Alternative Fabric-based Yield and Failure Criterion for Trabecular Bone
Ph. Zysset, L. Rincón . . . 457

Part V Image-based Analysis

Functional Micro-imaging at the Interface of Bone Mechanics and Biology
R. Müller, A. Nazarian, P. Schneider, M. Stauber, P. Thurner, G.H. van Lenthe, R. Voide . . . 473

Strain Measurement Using Deformable Image Registration
J.A. Weiss, A.I. Veress, G.T. Gullberg, N.S. Phatak, Q. Sun, D. Parker, R.D. Rabbitt . . . 489

Image-based Hierarchical Analysis and Design of Tissue Engineering Scaffolds
S.J. Hollister, C.Y. Lin, C.Y. Lin . . . 503

Index . . . 517

Part I

Tissue Growth and Remodelling

Towards a Theory of Vascular Growth and Remodeling

J.D. Humphrey

Texas A&M University, Department of Biomedical Engineering and
M.E. DeBakey Institute, USA
jhumphrey@tamu.edu

Diverse reports over the last three decades reveal the ubiquitous role of mechanical factors in controlling many aspects of vascular growth and remodeling. Examples include vascular adaptations during normal development, those due to altered loads ranging from exercise to a microgravity environment, those due to interventional procedures such as stenting or bypass surgery, and those incurred during the development of diseases ranging from hypertension to atherosclerosis or the formation of aneurysms. Concurrently, there has been an increased realization that mathematical modeling must play a central role in our attempts to understand and better predict vascular growth and remodeling. Models will help guide the identification and interpretation of revealing experiments and will allow an efficient exploration of the consequences of competing hypotheses.

1 Introduction

Normal arterial development as well as functional adaptations, responses to injury, and many disease processes appear to occur via similar means – a cell mediated turnover of individual wall constituents at different rates, to different extents, and in different biomechanical states. It is because of the diverse repertoire of mechanosensitive cellular activities (e.g., migration, proliferation, apoptosis, synthesis and degradation of matrix, and production of adhesion molecules, vasoactive molecules, growth factors, matrix metalloproteinases, and cytokines), however, that there are so many different manifestations of arterial growth and remodeling (G&R). For example, the lumen of an artery enlarges in response to sustained increases in flow, the thickness of the arterial wall increases in response to sustained increases in pressure, the unloaded length of an artery increases in response to sustained increases in axial load, the inner portion of the wall experiences neointimal hyperplasia in cases of balloon-induced damage to the endothelium and sub-intimal elastin, and so forth. There is a pressing need, therefore, for a general theory of vascular growth and remodeling that will allow us to predict better these diverse manifestations based on a small set of fundamental concepts and principles as well as knowledge of the altered mechanical loading. The goal of this chapter is to review briefly some of the recent research in vascular G&R and to suggest some avenues for future work based in part on new hypotheses that we have identified.

2 Histo-mechanical Background

Growth and remodeling is fundamental to all blood vessels – arteries, arterioles, capillaries, venules, and veins – yet our focus herein will be on large arteries. Perusal of the literature reveals an interesting evolution in our understanding of the mechanics of large arteries (see the review in Humphrey (1995)). Prior to the late 1960s, most stress analyses were inappropriately based on the linearized theory of elasticity. From the late 1960s to the early 1980s, however, attention shifted to the nonlinear behavior of passive arteries under finite deformations. Key contributors included I. Mirsky, R.P. Vito, and H. Demiray. During this same period, investigators began to account for the anisotropic behavior as well, key contributors being R.N. Vaishnav, B.R. Simon, Y.C. Fung, and W. von Maltzahn. Figure 1 shows a typical computation of the transmural distribution of Cauchy stress in a large artery based on a finite strain anisotropic model. In particular, note the strong gradient in stress, particularly for the circumferential component. This finding led some investigators to suggest that the high stress in the inner region of the wall was a risk factor for atherosclerosis, an insidious disease of the inner layer of the wall. Indeed, this suggestion was consistent with parallel developments in hemodynamics, which also began in the late 1960s, that implicated excessive wall shear stresses in atherosclerosis.

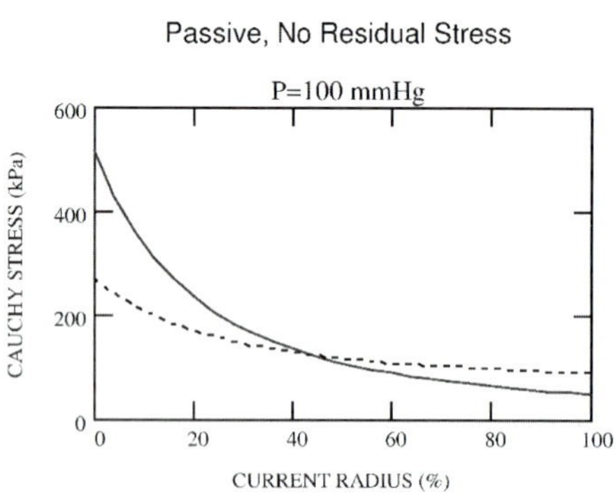

Fig. 1. Predicted Cauchy stress versus pressurized radius, passive behavior. The solid line is circumferential and the dashed line is axial stress.

In 1983, however, Y.C. Fung and R.N. Vaishnav independently reported that excised, unloaded arterial segments are not stress-free. Rather, such segments are residually stressed, which is revealed by their 'springing open' when cut radially. Inclusion of this residual stress (values of which appear to be on the order of 3 kPa) in the stress analysis by Chuong and Fung (1986) and later by Takamizawa and Hayashi (1987) suggested that the transmural stress field was much more uniform than first thought (Fig. 2). This finding is important for two reasons: first, it showed the importance of accounting for the nonlinearities, for superposition of the residual and load-induced stress fields would not significantly reduce the previously predicted gradient in stress (cf. Fig. 1), and second, our community began thinking teleologically about the potential advantages of a uniform stress field. Indeed, when Rachev and Hayashi (1999) later proposed a model for smooth muscle activation, we found that this further reduced the gradient in the stress field (Fig. 3). In other words, out of complexity (i.e., modeling the nonlinear, anisotropic, active-passive behavior

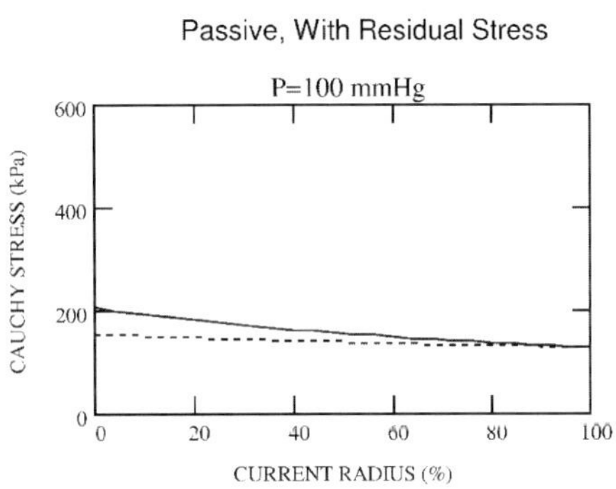

Fig. 2. Predicted Cauchy stress versus radius, including residual stresses. The solid line is circumferential and the dashed line is axial stress.

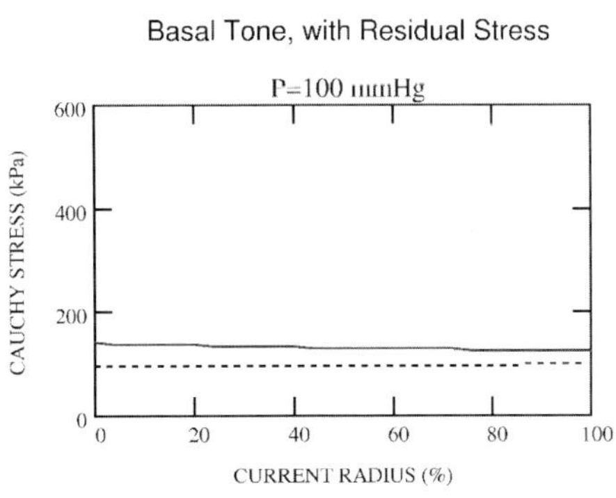

Fig. 3. Predicted Cauchy stress versus radius, including the effects of residual stress and muscle activation.

over finite deformations while accounting for residual stress) came simplicity (a predicted nearly uniform transmural stress field). It seems, therefore, that smooth muscle cells residing in the inner part of the wall may experience the same mechanical environment as the cells residing in the outer part of the wall, which is teleologically favorable. If this is true, it highlights the importance of a fundamental question: *What rules govern the development of a structurally complex artery so that the resulting stress field in maturity is uniform and optimal?* In other words, as suggested by Aristotle (384-322 B.C.), 'Here and elsewhere we shall not obtain the best insights into things until we actually see them growing from the beginning.' This same sentiment is advocated in two highly recommended papers: Taber (1995) and Cowin (2000).

Whether via vasculogenesis (i.e., *de novo* formation) or angiogenesis (formation from pre-existing vessels), arteries develop in a similar way (e.g., see Hungerford and Little (1999)). Precursor cells are recruited to form a primitive tube-like endothelial structure. These endothelial cells produce an abluminal basement membrane, initially consisting largely of the adhesion molecule fibronectin, and they send chemotactic signals to angioblasts that recruit these cells to form a layer around the endothelial tube; these angioblasts differentiate into smooth muscle cells of the synthetic phenotype, which in turn synthesize numerous extracellular matrix proteins that contribute to the structural integrity of the wall. These matrix proteins include the fibrillins and fibulins as well as tropoelastin and eventually various collagens; note, too, that it appears that the tropoelastin is chemotactic for the recruitment and organization of additional smooth muscle cells. As a result, the layered wall begins to exhibit a structure that is characteristic of a mature artery (Fig. 4). That is, during this developmental process, the basement membrane matures by consisting more of collagen IV and laminin, and the extracellular matrix matures by cross-linking the elastin (which thereby becomes one of the most stable proteins in the body, with a long half-life), elaborating the fibrillar collagens I and III, and finally changing the phenotype of the smooth muscle towards contractile. Whereas the endothelium and basement membrane constitute the so-called intima (inner layer), the elastin, smooth muscle, much of the collagens, and the ground substance (i.e., glycosaminoglycans) constitute the so-called media (middle layer). The ground substance plays many important roles, including sequestering significant water and serving as a repository for a host of growth factors, inhibitors of matrix metalloproteinases, etc. Finally, the adventitia (outer layer) consists largely of collagen I with admixed elastin, fibroblasts, and nerves, and in large arteries, a network of blood vessels called the vasa vasorum.

Again, we emphasize the remarkable observation that it is out of this complex, but well orchestrated, sequence of events that a simple, uniform stress field appears to arise in normalcy in maturity. Indeed, as suggested many years ago by Clark and Glagov (1985), it appears that the increased thickening of the wall during development correlates well with increases in blood pressure such that the mean circumferential wall stress tends to remain

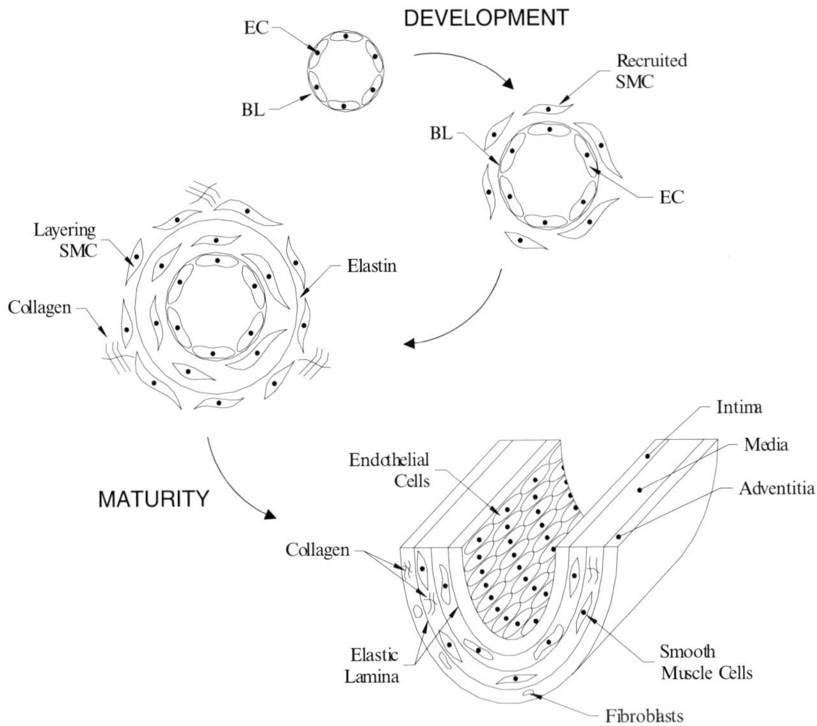

Fig. 4. Schema of the development of an elastic artery.

nearly constant. That is, the value of the uniform stress appears to be a preferred, or target, value.

Although developmental vascular biology has long been studied, it has only been through recent advances in molecular and cell biology that we have begun to appreciate the importance of the sequence of the many changes in the developing matrix. Whether for the purpose of guiding tissue engineering or motivating mathematical modeling of vascular G&R, an increasingly better understanding of vascular development promises to provide great insight; thus, this area of research must be given a high priority.

3 Need for New Hypotheses

To say that we seek a biomechanical theory for arterial growth and remodeling necessarily implies that we seek to formulate constitutive/evolution relations that describe the properties of the arterial wall and how they change in response to altered loads. That is, we may tacitly assume that the arterial wall obeys the fundamental postulates of mechanics (e.g., balance of mass, momentum, and energy), thereby rendering the requisite equations of motion and methods of solution (e.g., finite element codes) well known. As is often the case in biomechanics, therefore, it is the constitutive relation that is the most pressing need (cf. Fung (1995)). The first step in any constitutive formulation is delineation of the general characteristics of the behavior, under conditions of interest, so we can establish an appropriate theoretical framework.

Amongst other things, studying arterial development reveals that growth and remodeling processes rely on the organization of individual structurally significant constituents in different configurations and at different rates. In other words, from the perspective of mechanics, regardless of whether we consider homogeneous or heterogeneous properties, the arterial wall is materially non-uniform, consisting of multiple structurally important constituents that contribute to the overall load bearing capability. This observation suggests that we should not seek to formulate a theory for materially uniform bodies (i.e., a standard continuum theory, which remains useful in analyses such as those presented in Figs 1–3). Rather, the continuum theory of mixtures (see Section 130 in Truesdell and Noll (1965)) is probably a better framework on which to build. We must realize, however, that despite its generality in principle, there are numerous practical limitations with regard to implementing a full mixture theory. First, the theory of mixtures requires constitutive relations for both the individual constituents (e.g., hyperelastic descriptors for the stress responses) and their interactions (e.g., momentum exchanges between constituents). The latter, in particular, are not well known for soft tissues, whether it be interactions between networks of collagen and elastin during deformation or between migrating cells and matrix. Second, the theory of mixtures requires that one satisfy momentum equations for individual constituents, which typically requires specification of boundary conditions for partial tractions. These, too, are currently unknown. Hence, we (Humphrey and Rajagopal (2002)) suggest that it would be best to exploit the full mixture equations for mass balance (i.e., to model fully the production and removal of individual constituents) but to enforce only a single momentum balance equation for the mixture via a standard rule-of-mixtures formulation for the stress response. Moreover, we suggest that it is reasonable to assume, on average, that individual constituents deform together in the current configuration despite each constituent possibly having a separate natural (i.e., stress-free) configuration, which may evolve separately. The non-existence of a single 'natural configuration' in soft tissues was recognized early on by Fung (1973), though not for individual constituents, and the importance of the possible evolution of stress-free configurations was noted by Tozeren and Skalak (1988). Finally, the idea of modeling the stress response via a rule of mixtures, with constituents having individual natural configurations, was suggested long ago by Brankov et al. (1975), though they did not follow up on this idea within the context of G&R, and it is consistent with the ideas of microstructural relations postulated by Lanir (1983). Hence, albeit not planned, in hindsight it appears that our suggestion to combine a full mixture theory for mass balance, a rule-of-mixtures relation for the stress response in the classical momentum balance equation, the concept of a constrained mixture, and evolving natural configurations represents a melding of numerous earlier ideas, some of which came to our attention only recently.

Our constrained mixture theory of growth and remodeling is sufficiently general for addressing three-dimensional initial-boundary value problems (Humphrey and Rajagopal (2004)), yet the inherent complexities in such problems may render it difficult to build intuition from the computational results. Rather, it seems prudent to pursue solutions to simpler (well-studied) problems that will allow us initially to build intuition, to test various competing hypotheses, and to motivate immediate experimental needs. In a way, this is consistent with the inductive approach of biology: to infer general insights from many different individual observations (or, in our case, individual computations). Towards this end, we (Gleason et al. (2004)) proposed a 2D constrained mixture theory applicable to sustained, but modest, alterations in pressure, flow, or axial loads that act on straight, thin-walled cylindrical arteries consisting of three dominant structural constituents: elastin, collagen, and smooth muscle. By modest altered loads, we mean those for which an initial vasoactive response (dilatation or constriction) maintains or restores the wall shear stress near its target value. By dominant structural constituents, we mean locally homogenized responses that include the smaller contributions of other constituents such as the fibrillins or proteoglycans.

Briefly, we assume that altered loads change the biomechanical state, or configuration, in which previously existing (i.e., original) constituents reside, thereby perturbing them from their homeostatic values of stress or strain. As a result, our primary hypothesis, in non-pathological situations, is that over time these original constituents are replaced with new constituents that have new natural configurations but otherwise the same material properties (e.g., collagen is collagen); moreover, we hypothesize that the rate of turnover tends to increase with the degree to which the load is initially altered from normal. Hence, within the context of a rule-of-mixtures, we not only assume that multiple constituents co-exist mathematically at a point (i.e., within a homogenized neighborhood), constituents produced at different times, having different natural configurations, can also co-exist spatially and temporally. Whereas we allow new constituents to have new natural configurations, they are necessarily deposited within and organized into an extant matrix that is stressed. Hence, there is a need to prescribe a 'deposition' stress or stretch for the new constituents, which in turn will dictate the individual natural configuration. One candidate hypothesis is that the deposition stress or strain is the homeostatic value for that constituent. In other words, regardless of the state of stress or strain of the extant material, new constituents are always organized within the wall at a preferred (homeostatic) value. Predicted consequences of these basic hypotheses (with illustrative assumptions for the individual stress response functions and kinetics of turnover) are presented in

Gleason et al. (2004) and Gleason and Humphrey (2005, 2004) for the cases of modest alterations in flow, pressure, and axial load, respectively. In each case, the predicted results agree qualitatively with the majority of the reports in the literature, thus suggesting that our basic hypotheses are tenable, at least to first order. See these papers for details on the formulation as well as particular illustrative results.

Figure 5 illustrates the basic concept of a 2D constrained mixture model of an artery. In particular, β_0 and β_t represent experimentally accessible configurations for the mixture (i.e., the intact thin-walled vessel) in the absence of applied loads and under physiologic loading, respectively. Shown, too, are the associated individual configurations for the three primary structural constituents. Each constituent has the same configuration in the loaded configuration (i.e., constituents are constrained to deform together), but possibly different configurations when unloaded. In particular, we assume that the elastin will be stretched more in the physiologic configuration than will be the collagen, which is to say that the elastic lamella has a smaller unloaded radius. This assumption is motivated by observations that elastin is deposited earlier in development (when the vessel is smaller) and that it is biologically very stable, with a half-life perhaps on the order of the life span of the organism. Conversely, collagen is deposited later in development and turned-over throughout life. An interesting consequence of these potentially different natural configurations is that the intact unloaded artery need not be stress-free; rather, the elastin could be in tension, trying to return to its smaller natural configuration, whereas the collagen (and likely associated glycosaminoglycans) could be compression, trying to reach its larger natural configuration.

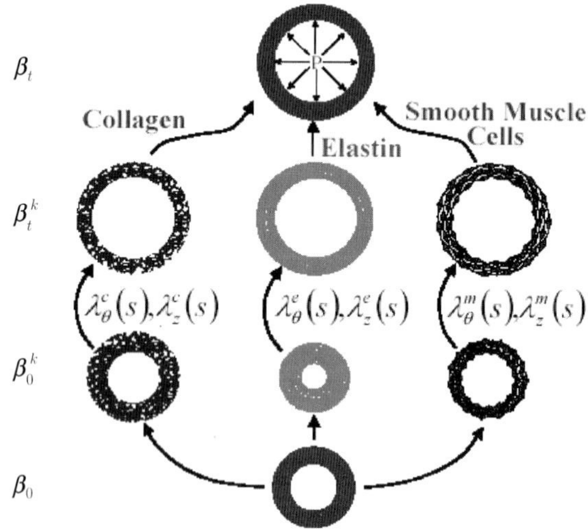

Fig. 5. Schema of the configurations assumed by the intact vessel as well as those assumed by the three individual structurally significant constituents.

The co-existence of self-equilibrating constituent stresses is actually supported by the experiments of Zeller and Skalak (1998) as well as those by Greenwald et al. (1997). Hence, whereas the total (i.e., mixture) stress is necessarily zero in an unloaded, thin-walled cylinder, the constituent stresses need not be zero (of course, the artery never experiences the unloaded configuration *in vivo*). Indeed, it is easy to imagine that gradients in non-zero constituent stresses in an unloaded thick-walled cylinder could give rise to the aforementioned residual stresses of Fung and Vaishnav. That is, a mixture theory description of growth and remodeling during development may help explain 'how' residual stresses arise in thick-walled arteries. We have not examined this question, however.

Finally, I wish to highlight two provocative consequences of our simple simulations. First, recalling that elastin is thought to be very stable into maturity (but prior to the onset of 'aging'), we compared predictions wherein all three structurally significant constituents turned over versus predictions wherein only the collagen and smooth muscle turned over. In each case, the basic finding was that in response to a modest change in pressure, flow, or axial stretch, the model artery was able to adapt perfectly when all constituents turned over completely. That is, in these cases, the mixture stresses were returned to homeostatic values and the vessel changed its lumen, wall thickness, or axial length as expected for an ideal adaptation. In contrast, when the elastin was not allowed to turnover, the stresses were predicted to return towards, but not exactly to, homeostatic values, and the predicted changes in geometry tended towards, but not exactly to, optimal values. In other words, predicted adaptations were sub-optimal in the presence of a constraint against the turnover of elastin. Indeed, a careful examination of data in the literature suggests that this is often the case in maturity (e.g., the wall will thicken in response to an enlarged lumen induced by an increased flow, but the mean circumferential stress does not return to its homeostatic value). Conversely, it appears reasonable to assume that the continued 'adaptations' during development (as a vessel expands, thickens, and lengthens) are optimal, which in fact is consistent with our predictions in the case of complete turnover of all constituents, which would be expected more in development. In other words, it appears that a vessel can conserve its preferred mechanical state despite significant growth when all constituents are free to turnover ideally. Again, there is clearly a need for more data on the developmental vascular biology to explore this and similar situations.

For the simple cases we considered (e.g., a single, modest alteration in loading), we were able to assume a general first order kinetic model for turnover wherein the rate parameter was an arbitrary function of differences between current and homeostatic constituent stresses. In these cases, the rate of change of mass production and removal correlated nearly linearly with a 'measurable' difference between the current and homeostatic circumferential mixture stress, that is Pa/h, where P is the transmural pressure, a the arterial radius, and h the thickness. This correlation is consistent with that which is assumed *a priori* in many models of growth and remodeling kinetics (e.g., Taber (1998); Rachev (2000)). In Gleason and Humphrey (2004), however, we argue that first order kinetics do not mimic well the data in the literature on changes in the rates of turnover of cells or constituents. That is, whereas a first order model necessarily predicts the highest rate of turnover at the instant of the alteration in loading (wherein the current stresses will differ the most from homeostatic provided G&R restore the stresses toward homeostatic), all data suggest that there is initially a slow increase in the rate of turnover followed by a peak and then a decrease back toward baseline. Indeed, such a trend is intuitive: the cells cannot respond instantaneously, so there must be sufficient time for the appropriate gene transcription and translation. Not knowing what kinetics actually controls such processes, we then used a gamma distribution function to prescribe phenomenologically the rate of turnover of cells and matrix. Although the predicted time-course of changes in the mixture stress-difference and vessel geometry differed little from that predicted by the first order kinetics, the predicted correlation between stress-differences and rates of turnover were much more complex, indeed non-monotonic. We submit that this finding suggests a word of caution when attempting to model growth and remodeling – one of the primary needs is evolution equations for mass production and removal in terms of convenient metrics of the altered biomechanical state. We must be careful not to overly simplify such relations, particularly when trying to find correlations directly from data.

4 Discussion and Future Needs

Until recently, much of our thinking with regard to soft tissue G&R has stemmed from three seminal ideas: the concept of global optimization (Murray (1926)), the use of reaction-diffusion equations to describe the distribution of 'morphogens' (Turing (1952)), and the concept of kinematic growth (Skalak (1981)). Of particular relevance herein, the idea of kinematic growth was critical in bringing the modeling of G&R within the purview of nonlinear continuum mechanics. Indeed, the subsequent extension of Skalak's idea by Rodriguez et al. (1994) provided a fertile framework for the development of many models of arterial G&R (e.g., Taber and Eggers (1996), Taber (1998), Rachev et al. (1996, 1998, 2000), Rachev (1997, 2000), Fridez et al. (2001)) as well as models for other tissues (e.g., Epstein and Maugin (2000), Ambrosi and Mollica (2002), Lubarda and Hoger (2002), Quiligotti (2002), Garikipati et al. (2004)). Although the concept of kinematic growth has enabled many reasonable predictions, we submit that it focuses primarily on the consequences of G&R, not the means by which G&R occurs. That is, whereas kinematic growth relies on prescribing evolution relations for the expansion of fictitious stress-free elements, which need not change in a compatible fashion and thereby can induce residual stresses in materially uniform bodies, all G&R in soft tissues necessarily occurs in stressed configurations. Moreover, G&R necessarily occurs via the production and removal of individual constituents. Hence, consistent with the suggestion of Fung (1995), it appears best to seek evolution equations for the production and removal of individual constituents, which is ever-present but the rates of which appear to change dramatically in response to altered mechanical stimuli. To date, our group has investigated only very simple situations, but it is hoped that they provide both motivation and guidance for extending such ideas to incorporate new data and to describe more realistic circumstances. Towards this end, the reader is referred to the paper by Watton et al. (2004), which couples a rule-of-mixtures approach with a prescribed degradation of elastin to study the development of both axisymmetric and non-axisymmetric abdominal aortic aneurysms.

In conclusion, although the focus herein was on arterial growth and remodeling, similar mechanisms are likely operative in the arterioles, capillaries, venules, and veins – indeed, probably in most soft tissues – despite some important differences. For example, G&R in the microcirculation appears to be governed by a strong coupling between the blood pressure and wall shear stress (Pries et al. (1995)). It seems prudent, therefore, to seek a unified theory of vascular growth and remodeling while being mindful of potential differences in the specific constitutive (including evolution) equations that govern a specific type of vessel under specific circumstances. Indeed, there is likely much to learn by studying in parallel G&R for many different types of vessels and under many different alterations in applied loads.

Acknowledgements. This research was supported, in part, by grants from the NSF (BES-0084644) and NIH (R01 HL-64372). I also acknowledge illuminating discussions with many colleagues at Texas A&M University, including Drs. G. Davis, T. Fossum, R. Gleason, L. Kuo, L. Martinez-Lemus, G. Meininger, J. Moore, K.R. Rajagopal, and E. Wilson, as well as Dr. L. Taber at Washington University in St. Louis.

References

Ambrosi, D., and Mollica, F. (2002). On the mechanics of a growing tumor. *Int. J. Eng. Sci.* 40:1297–1316.

Brankov, G., Rachev, A., and Stoychev, S. (1975). A composite model of large blood vessels. *Mech. Biol. Solids (Bulgaria)* 71–78.

Chuong, C. J., and Fung, Y. C. (1986). On residual stress in arteries. *J. Biomech. Eng.* 108:189–192.

Clark, J. M., and Glagov, S. (1985). Transmural organization of the arterial media. The lamellar unit revisited. *Arteriosclerosis* 5:19–34.

Cowin, S. C. (2000). How is a tissue built? *J. Biomech. Eng.* 122:553–569.

Epstein, M., and Maugin, G. A. (2000). Thermodynamics of volumetric growth in uniform bodies. *Int. J. Plasticity* 16:951–978.

Fridez, P., Rachev, A., J. Meister, J., Hayashi, K., and Stergiopulos, N. (2001). Model of geometrical and smooth muscle tone adaptation of carotid artery subject to step change in pressure. *Am. J. Physiol.* 280:H2752–H2760.

Fung, Y. C. (1973). Biorheology of soft tissues. *Biorheology* 10:139–155.

Fung, Y. C. (1995). Stress, strain, growth, and remodeling of living organisms. *Z. Angew. Math. Phys.* 46:469–482.

Garikipati, K., Arruda, E. M., Grosh, K., Narayanan, H., and Calve, S. (2004). A continuum treatment of growth in biological tissue: the coupling of mass transport and mechanics. *J. Mech. Phys. Solids* 52:1595–1625.

Gleason, R. L., and Humphrey, J. D. (2004). A mixture model of arterial growth and remodeling in hypertension: Altered muscle tone and tissue turnover. *J. Vasc. Res.* 41:352–363.

Gleason, R. L., and Humphrey, J. D. (2005). Effects of a sustained extension on arterial growth and remodeling: a theoretical study. *J. Biomech.* 38: 1255–1261.

Gleason, R. L., Taber, L. A., and Humphrey, J. D. (2004). A 2-D model for flow-induced alterations in the geometry, structure, and properties of carotid arteries. *J. Biomech. Eng.* 126:371–381.

Greenwald, S. E., Moore, Jr., J. E., Rachev, A., Kane, T. P. C., and Meister, J.-J. (1997). Experimental investigation of the distribution of residual strains in the artery wall. *J. Biomech. Eng.* 119:438–444.

Humphrey, J. D., and Rajagopal, K. R. (2002). A constrained mixture model for growth and remodeling of soft tissues. *Math. Model. Meth. Appl. Sci.* 12:407–430.

Humphrey, J. D., and Rajagopal, K. R. (2004). A constrained mixture model for arterial adaptations to a sustained step change in blood flow. *Biomech. Model. Mechanobio.* 2:109–126.

Humphrey, J. D. (1995). Mechanics of the arterial wall: Review and directions. *Crit. Rev. Biomed. Eng.* 23:1–162.

Hungerford, J. E., and Little, C. D. (1999). Developmental biology of the vascular smooth muscle cell: Building a multilayered vessel wall. *J. Vasc. Res.* 36:2–27.

Lanir, Y. (1983). Constitutive equations for fibrous connective tissues. *J. Biomech.* 16:1–12.

Lubarda, V. A., and Hoger, A. (2002). On the mechanics of solids with a growing mass. *Int. J. Solids Structures* 39:4627–4664.

Murray, C. D. (1926). The physiological principle of minimum work. I. the vascular system and the cost of blood volume. *Proc. Nat. Acad. Sci. USA* 12:207–214.

Pries, A. R., Secomb, T. W., and Gaehtgens, P. (1995). Design principles of vascular beds. *Circ. Res.* 77:1017–1023.

Quiligotti, S. (2002). On bulk growth mechanics of solid-fluid mixtures. *J. Theoret. Appl. Mech.* 28/29:277–288.

Rachev, A., and Hayashi, K. (1999). Theoretical study of the effects of vascular smooth muscle contraction on strain and stress distributions in arteries. *Ann. Biomed. Eng.* 27:459–468.

Rachev, A., Stergiopulos, N., and Meister, J.-J. (1996). Theoretical study of dynamics of arterial wall remodeling in response to changes in blood pressure. *J. Biomech.* 29:635–642.

Rachev, A., Stergiopulos, N., and Meister, J.-J. (1998). A model for geometric and mechanical adaptation of arteries to sustained hypertension. *J. Biomech. Eng.* 120:9–17.

Rachev, A., Manoach, E., Berry, J., and Moore Jr., J. E. (2000). A model of stress-induced geometrical remodeling of vessel segments adjacent to stents and artery/graft anastomoses. *J. Theoret. Biol.* 206:429–443.

Rachev, A. (1997). Theoretical study of the effect of stress-dependent remodeling on arterial geometry under hypertensive conditions. *J. Biomech.* 30:819–827.

Rachev, A. (2000). A model of arterial adaptation to alterations in blood flow. *J. Elasticity* 61:83–111.

Rodriguez, E. K., Hoger, A., and McCulloch, A. D. (1994). Stress-dependent finite growth in soft elastic tissues. *J. Biomech.* 27:455–467.

Skalak, R. (1981). Growth as a finite displacement field. In Carlson, D. E., and Shield, R. T., eds., *Proceedings of the IUTAM Symposium on Finite Elasticity, 1981.* The Hague: Martinus Nijhoff Publishers.

Taber, L. A., and Eggers, D. W. (1996). Theoretical study of stress-modulated growth in the aorta. *J. Theoret. Biol.* 180:343–357.

Taber, L. A. (1995). Biomechanics of growth, remodelling, and morphognesis. *Appl. Mech. Rev.* 48:487–545.

Taber, L. A. (1998). A model for aortic growth based on fluid shear and fiber stress. *J. Biomech. Eng.* 120:348–354.

Takamizawa, K., and Hayashi, K. (1987). Strain energy density function and uniform strain hypothesis for arterial mechanics. *J. Biomech.* 20:7–17.

Tozeren, A., and Skalak, R. (1988). Interaction of stress and growth in a fibrous tissue. *J. Theoret. Biol.* 130:337–350.

Truesdell, C., and Noll, W. (1965). The nonlinear field theories of mechanics. In Flügge, S., ed., *Handbuch der Physik*. Berlin: Springer-Verlag.

Turing, A. (1952). The chemical basis of morphogenesis. *Philos. Trans. R. Soc. Lond.* B237:37–72.

Watton, P. N., Hill, N. A., and Heil, M. (2004). A mathematical model for the growth of the abdominal aortic aneurysm. *Biomech. Model. Mechanobio.* 3:98–113.

Zeller, P. J., and Skalak, T. C. (1998). Contribution of individual structural components in determining the zero-stress state in small arteries. *J. Vasc. Res.* 35:8–17.

Complementary Roles of Theoretical Modeling and Computer-controlled Experimentation in Vascular Growth and Remodeling

R.L. Gleason, E. Wilson, J.D. Humphrey

Texas A&M University, Departments of Biomedical Engineering and Medical Physiology, USA
rudy77@neo.tamu.edu, emilyw@tamu.edu, jhumphrey@tamu.edu

It is now unquestioned that cell-mediated vascular growth and remodeling (G&R) plays a key role in many physiological and pathophysiological processes, as well as in the success (or failure) of many clinical interventions. Despite an explosion of information on these processes, from intra-cellular, cellular, cell-matrix, and tissue levels, attempts at integrating these data into a predictive model are still in their infancy. We present a fundamentally different approach to quantify G&R by modeling the tissue as a constrained mixture. This approach, which can incorporate the underlying mechanisms of G&R (e.g., production, removal, and remodeling of individual constituents), can be used to simulate the effects of imposed biological constraints (e.g., changing the rate of elastin turnover), to test competing hypotheses, and to perform parametric studies that guide both experimental design and data interpretation.

Motivated by this theoretical framework, we designed and built a novel computer-controlled organ culture device with biomechanical testing capabilities to study vascular adaptations in small caliber (50–5,000 μm) blood vessels. With this device we are able to accurately controlling mean vessel pressure, mean flow rates, pulse pressure magnitude and frequency (up to 15 Hz), and axial load, while maintaining an adequate biochemical and physiological tissue culture environment. In addition to controlling the loading, this device enables both intermittent biaxial (pressure-diameter and axial load-length) and functional tests to quantify adaptations in mechanical behavior and vaso-reactivity to agonists by the same vessel at different time-points throughout G&R.

Herein we emphasize the synergistic relationship between developing a mathematical theory that can be used to motivate experiments and interpret data and performing theoretically-motivated experiments that can be used to further improve the model and test the underlying hypotheses of specific simulations.

1 Introduction

It is well known that many tissues grow and remodel in response to an altered mechanical environment. In the vasculature, for example, arteries exposed to increased blood pressure tend to thicken, those exposed to increased flow tend to dilate, and those exposed to increased axial extension tend to lengthen. These adaptations correlate well with the stresses in the tissue. Kamiya and Togawa (1980) suggested that mean 'wall shear stress was the predominant factor in initiating and controlling the adaptive response [enlargement of the lumen]' in cases of sustained increases in flow. Clark and Glagov (1985) found that the mean circumferential stress correlates well with aortic thickening during normal development whereas Jackson et al. (2002) suggest that the axial stress may also be regulated via G&R mechanisms. Recall, therefore, that equations for the mean wall shear stress (τ_w), mean circumferential stress (σ_θ) and mean axial stress (σ_z), and typical homeostatic values, are

$$\tau_w = \frac{4\mu Q}{\pi a^3} \approx 1.5\,\text{Pa}, \quad \sigma_\theta = \frac{Pa}{h} \approx 150\,\text{kPa}, \quad \sigma_z = \frac{f}{\pi h(2a+h)} \approx 125\,\text{kPa}, \tag{1}$$

respectively, where Q is the luminal flow, P is the transmural pressure, f is the axial force, a is the luminal radius, h is the thickness of the vessel wall, and μ is the viscosity of blood.

In addition, it appears that not only do the mean stresses correlate well with vascular growth and remodeling (G&R), so too do the local stresses. Stress analyses performed based on mechanical data from healthy arteries suggest that, when one includes material non-linearities, anisotropy, finite deformations, residual stress, and a basal smooth muscle contribution, the distribution of stress across the arterial wall under physiological loading is nearly uniform. Non-physiologic loading, however, often results in non-uniform stress distributions across the wall; for example, supraphysiologic pressures may produce a greater increase in stress at locations near the inner versus the outer wall. Matsumoto and Hayashi (1996) illustrated well that not only do blood vessels thicken in response to sustained increases in pressure, but also different layers thicken to different degrees, with the innermost layers thickening more early on than outer layers. These results, and many others, suggest that the vascular endothelial, smooth muscle, and fibroblast cells sense and adapt their local mechanical environment so as to achieve a preferred homeostatic state.

In this paper, we describe the parallel and complementary approaches of developing a theoretical framework capable of testing competing hypotheses regarding the underlying mechanisms of G&R and devising an experimental approach to test theoretical predictions that support or oppose underlying hypotheses. In particular, we outline our approach toward developing a fundamentally new mathematical framework for vascular G&R and describe a novel experimental system for subjecting isolated ($\sim$500 μm diameter) arteries to well controlled loads for multiple days.

2 Motivating a Theory and Experiments

A mechanical signal can be transduced to a cellular response in many ways. Increased load on focal adhesion sites (cell-matrix level), for example, can alter cell-signaling pathways (sub-cellular level), ultimately leading to altered gene expression. Altered gene expression, of course, can elaborate itself in many different ways, including an altered expression of growth factors, apoptotic signals, and other cell cycle regulating signals, as well as the production of vasoactive molecules, extracellular matrix (ECM) degrading proteins, ECM proteins, ECM cross-linking proteins, chemotaxic molecules, and cytoskeletal proteins, amongst many others. The net effect of these (and other) mechanotransduction pathways, at the cell and tissue level, include increases in cell and matrix turnover, local growth (or atrophy), structural and functional remodeling of existing cells, remodeling of matrix, and altered cell-matrix and matrix-matrix interactions, all (presumably) aimed towards evolving the local mechanical environment from an 'undesirable' condition to a 'desirable' condition. Mathematically modeling G&R thus requires that one track local balances or imbalances in the continual production and removal of individual constituents (e.g., smooth muscle cells, collagen, elastin, etc.), the mechanical state in which the constituents are formed, how these constituents are organized, and how the organization and interaction of constituents evolves (Humphrey and Rajagopal (2002)). Experimentally testing this model requires one to observe adaptations at the sub-cellular, cellular, cell-matrix, and tissue levels.

Although normal growth and remodeling tends to be an active process that seeks to optimize structure and function locally with respect to some set of parameters, in comparison to processes during development there appear to be constraints on this optimization process during maturity. For example, Clark and Glagov (1985) observed that aortic thickening during development is achieved by adding 'musculo-elastic fascicles', that is the layers of smooth muscle and collagen interspersed between elastic lamina. In contrast, Matsumoto and Hayashi (1996) observed that mature blood vessels thickened by adding material to existing musculo-elastic fascicles. Thus, although both of these vessels aim to thicken, the mature vessel appears to have a more limited 'set of tools' to meet this goal; we will refer to these limitations as biological constraints.

3 Theoretical Framework

The overall aim for developing a theoretical framework for G&R should be to allow for the incorporation of many diverse observations, whether by molecular biologists, cell biologists, geneticists, biochemists, physiologists, clinicians, biomechanists, or others; indeed, it should allow for the inclusion of observations that are made at many different length scales. Moreover, such a framework should allow one to test competing hypotheses and to impose biological constraints on the underlying biological processes.

Towards this end, our theoretical framework is based on a constrained mixture theory. Thus, we allow multiple, load-bearing constituents to coexist at any point within the tissue (Fig. 1), but because the material is a constrained mixture, each constituent follows the same motion as the mixture. We also allow the tissue to be materially non-uniform by allowing the natural configuration of one constituent (the configuration wherein the stress of that constituent is zero) to be different from the natural configuration of another constituent. Thus, within the mixture at its reference configuration (which we take as the global stress free configuration), one constituent may be in tension whereas another may be in compression; the reference configuration of the mixture is that configuration in which these tensile and compressive loads balance. Herein, we impose an additional assumption that the total stress behavior can be described by a simple rule of mixtures approximation, namely

$$\mathbf{t} = \sum_k \phi^k \mathbf{t}^k,$$

where $\mathbf{t}$ is the total stress at any point, ϕ^k is the mass fraction of constituent k, and $\mathbf{t}^k$ is the stress response function for constituent k.

To illustrate this approach, we modeled thin walled arteries (e.g., mouse carotid) mechanically as a membrane consisting of three structurally significant constituents (elastin, collagen, and smooth muscle) and imposed a single,

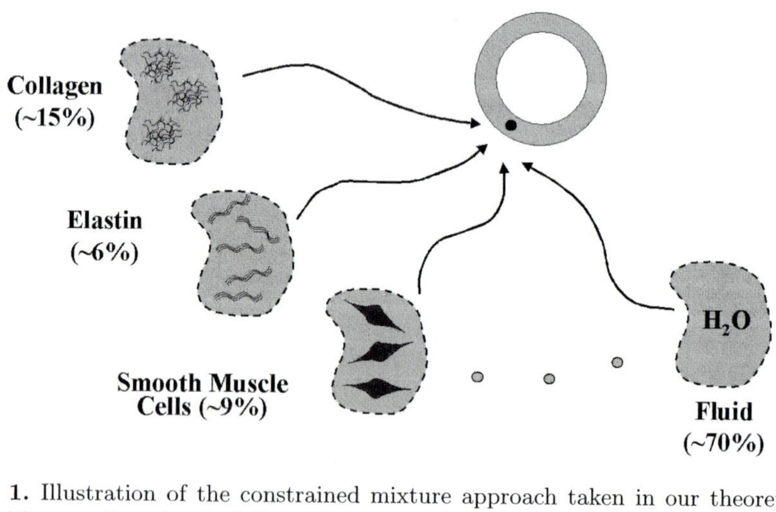

Fig. 1. Illustration of the constrained mixture approach taken in our theoretical model for growth and remodeling. We imagine that multiple constituents can coexist at each point, but they are constrained to move together in response to applied loads. Here, the mass fractions (percent) are given for a typical carotid artery.

moderate step change in blood flow (Gleason et al. (2004b)), blood pressure (Gleason and Humphrey (2004)), or axial stretch (Gleason and Humphrey (2005)). The findings were based on several hypotheses with regard to the fundamental means by which blood vessels adapt. First, we proposed that in response to altered loading (whether it be flow, pressure, axial load, or combinations thereof), the vessel will experience an immediate 'passive' elastic response (e.g., distention in response to increased pressure), a subsequent short-term adaptation via a vasoactive control of the lumen (which occurs in minutes to hours and seeks to restore τ_w toward normal values) that may result in a new 'vasoaltered' configuration, and if the altered mechanical loading persists, a long-term adaptation involving turnover and remodeling (i.e. coordinated production and removal) of cells and extracellular matrix in the vasoaltered configuration. This long-term response occurs over days to weeks (Langille et al. (1989)). Notice from eq. (1) that if the luminal flow increases from $Q(s=0)=Q_o$ to $Q(s>0)=\varepsilon\, Q_o$, then, in order to restore wall shear stress, the luminal radius must increase from $a(s=0)=a_o$ to $a(s>0)\rightarrow \varepsilon^{(1/3)}a_o$; here, s is G&R time with $s=0$ denoting the time that altered loading was imposed and the subscript o denotes an initial value prior to altered loading.

Second, we hypothesized that, as a new constituent is produced, it is deposited at a homeostatic stretch, denoted $\lambda_i^k|_h$, where $i=\theta$ or z denotes the two in-plane directions. Thus, in this 2D setting, if we know the configuration in which a constituent was produced and the stretch at which it is was laid-down, then the natural configuration for that constituent (defined here as the unloaded radius (A^k) and axial length (L^k) for 'constituent' k) can be determined as,

$$(A^k, L^k) = \left(\frac{a_p}{\lambda_\theta^k|_h}, \frac{l_p}{\lambda_z^k|_h} \right), \tag{2}$$

where a_p and l_p are the loaded luminal radius and axial length of the (mixture) vessel when the constituent k was produced. Combining this hypothesis with our first hypothesis, that the vessel remains in a constant (mean) vasoaltered configuration (i.e. a_p and l_p are constant for all $s>0$), then we see that all new material (e.g., new collagen, denoted $k=nc$) is produced at a single a_p and l_p and all new collagen will possess a single set of stress-free lengths, which we denote as (A^{nc}, L^{nc}). In contrast, the original collagen (denoted $k=oc$) was produced in the original, pre-vasoregulated configuration. Thus, even though original and new material were laid down with the same homeostatic stretches, since they were produced in different states, they will possess a different natural configuration; that is, $(A^{oc}, L^{oc}) \neq (A^{nc}, L^{nc})$ in general. In this simple case, our rule-of-mixtures equation can be written as

$$\begin{aligned}\sigma_i = {} & \phi^{oe}(s)\hat{\sigma}_i^e(\lambda_\theta^{oe}, \lambda_z^{oe}) + \phi^{ne}(s)\hat{\sigma}_i^e(\lambda_\theta^{ne}, \lambda_z^{ne}) \\ & + \phi^{oe}(s)\hat{\sigma}_i^c(\lambda_\theta^{oc}, \lambda_z^{oc}) + \phi^{nc}(s)\hat{\sigma}_i^c(\lambda_\theta^{nc}, \lambda_z^{nc}) \\ & + \phi^{om}(s)\hat{\sigma}_i^m(\lambda_\theta^{om}, \lambda_z^{om}) + \phi^{nm}(s)\hat{\sigma}_i^m(\lambda_\theta^{nm}, \lambda_z^{nm}),\end{aligned} \tag{3}$$

where σ_i is the mean stress, with $i = \theta$ or z denoting the two in-plane directions, $\hat{\sigma}_i^j(\lambda_\theta^k, \lambda_z^k)$ denotes the constitutive behavior of constituent $j = e, c$, or m, denoting elastin, collagen, or muscle. Notice that we assume that original material and new material exhibit the same material behavior (albeit evaluated at different stretches); this need not be the case, in general. Each constituent is modeled as elastic and thus, in a 2D setting, can be described as a function of the two in-plane stretches of that material, λ_θ^k and λ_z^k, where $k = oe, ne, oc, nc, om$, or nm denotes original or new elastin, collagen, or smooth muscle, respectively. The simulation is completed by prescribing evolution equations for the rates of production and removal (i.e. evolution of mass fractions) and stress response functions for the individual materials (elastin, collagen, smooth muscle); the reader is encouraged to consult Gleason et al. (2004b) for details on the evolution and stress response functions used below.

Let us consider employing this theoretical framework, with the above-mentioned assumptions, to impose a single biological constraint. In Gleason et al. (2004b) we imposed a 33% increase in blood flow and simulated the evolution of biomechanical behavior and vessel geometry for a model artery in which all constituents are allowed to turnover (Complete Turnover) and an artery in which muscle and collagen are allowed to turnover completely, but elastin turnover is negligible (No Elastin Turnover). These simulations were motivated by the observation by Clark and Glagov (1985) that vessels in development are capable of producing new musculo-elastic fascicles very similar in appearance to existing material, whereas that made by Matsumoto and Hayashi (1996) suggested that mature vessels produce additional collagen and muscle to thicken, with negligible elastin production. Notice that for both complete turnover and no elastin turnover, the total *in vivo* stresses increase upon vasodilation (Fig. 2). As constituent turnover proceeds, the total stress decreases monotonically in both cases (consistent with the assumed kinetics). However, whereas the *in vivo* stress is completely restored at long times for the case of complete turnover, it is not restored to initial values for the case no elastin turnover. Here $s = s_h(> 1)$ denotes the time in which a new steady state (i.e. existing constituents are replaced with new constituents with identical properties and reference lengths) is achieved. Notice, too, that for the case of complete turnover, the normalized unloaded radius (that is, the current unloaded radius divided by the initial unloaded radius, with $A(s_h) = \varepsilon^{(1/3)} A_o$) equals the normalized loaded radius (with $a(s_h) = \varepsilon^{(1/3)} a_o$). Thus, for the case of complete turnover, the circumferential stretch $(\lambda_\theta(s)/A(s))$ of the mixture in the new steady state (at time $s = s_h$) equaled the initial value. For the case of no elastin turnover, however, $A(s_h)/A(0) < a(s_h)/a(0)$ and so the circumferential stretch $\lambda_\theta(s_h) > \lambda_\theta(0)$. Finally, whereas $L(s_h) = L(0)$ for the complete turnover case (thus, $\lambda_z(s_h) = \lambda_z(0)$), for the case of no elastin turnover $L(s_h) > L(0)$ and $\lambda_z(s_h) < \lambda_z(0)$.

Our illustrative results were not performed to mimic specific experimental results or clinical observations, but rather to demonstrate the ability of our theoretical framework to test competing hypotheses and to impose biological

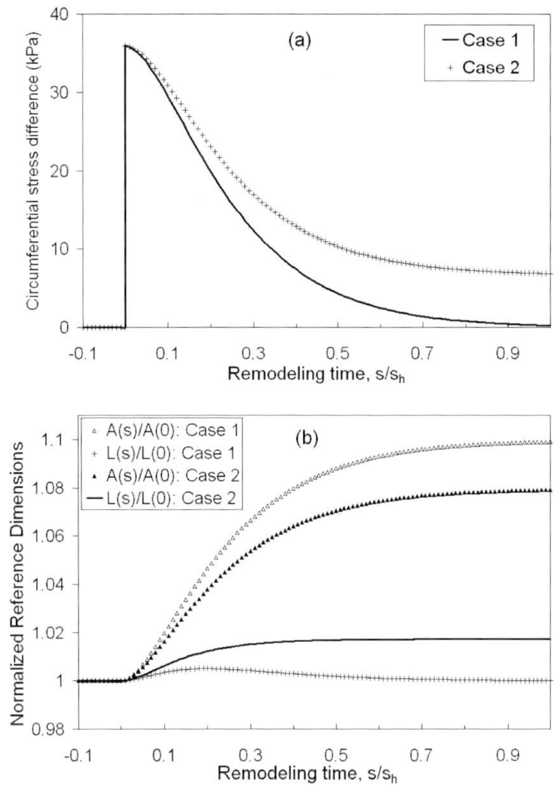

Fig. 2. Simulation results for the evolution of the difference between the current total circumferential stress $\sigma_\theta(s)$ and its original homeostatic value $\sigma_\theta(0)$ (panel a) and the unloaded radius and length of the mixture (panel b) for an abrupt, 33% increase in flow for complete turnover of all constituents (Case 1) and no elastin turnover (Case 2).

constraints. Nevertheless, it is provocative that consistent with our prediction, many have observed that, whereas wall shear is restored, σ_θ is not restored and the unloaded length often increases with flow-induced G&R. Herein we invoked a very simple hypothesis regarding how G&R proceeds; that is, replacing existing constituents with new constituents with identical material properties, but new natural configurations. Clearly, there are many other mechanisms of vascular G&R that we did not include in these simulations (e.g., evolving constituent behaviors, constituent-to-constituent interactions, etc.) Nevertheless, the general theoretical framework is structured to be able to include such complexities as data become available. Unfortunately, such data are currently lacking. Towards this end, we have developed a novel experimental model, described below.

4 A New Experimental Model

Although much of our understanding of vascular mechanotransduction has come from studies using cell cultures or *in vivo* animal models, the recent success of perfused organ culture systems offers an exciting alternative approach (e.g., Chestler et al. (1998); Matsumoto et al. (1999)). Every experimental model has particular advantages, but organ culture systems offer the potential for a more tightly controlled mechanical and biochemical environment compared to *in vivo* models and they can better maintain vessel geometry, structure, and cell-extracellular matrix attachments than cell culture techniques. In addition, an organ culture model isolates the cell-mediated mechanisms of vascular adaptation by removing neural- and hormonal-induced mechanisms. If combined with multiaxial biomechanical testing systems (e.g., Humphrey et al. (1993)), organ culture offers the opportunity to perform multiaxial *in vitro* mechanical and functional testing, at multiple time-points of adaptation by the same vessel, that provides data sufficient to perform stress analyses.

We developed a computer-controlled organ culture and biomechanical testing device designed for small caliber (50 μm to 5 mm) vessels (see Gleason et al. (2004a) for additional details). The blood vessel is bathed and perfused with a cell culture medium that is maintained at physiological temperature and pH (Fig. 3). A control program was written in LabView to independently maintain luminal flow, transmural pressure, and axial force. In addition, the independent control of the magnitude (0 to 40 mmHg) and frequency (0 to 15 Hz) of the pulse pressure can be accomplished manually. A 5-gram force transducer is used to measure the axial force and computer-controlled actuators allow feedback control of the axial force. Notice three flow loops in the system: a pulse flow loop, luminal flow loop, and adventitial flow loop (Fig. 4).

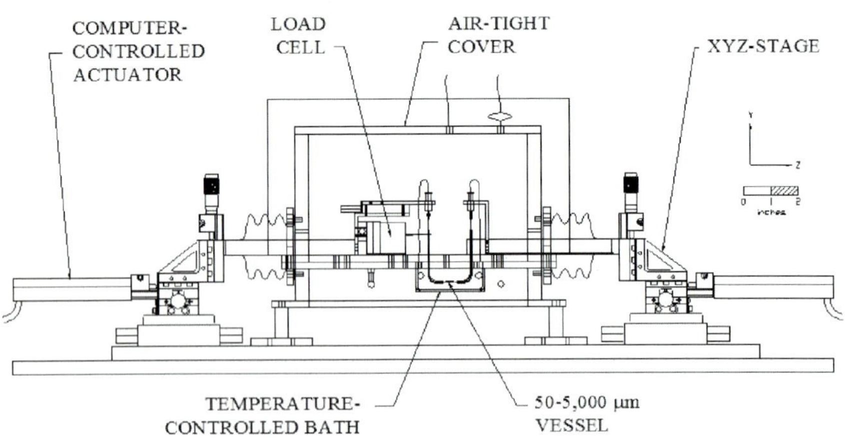

Fig. 3. Profile view of the organ culture and biomechanical testing device.

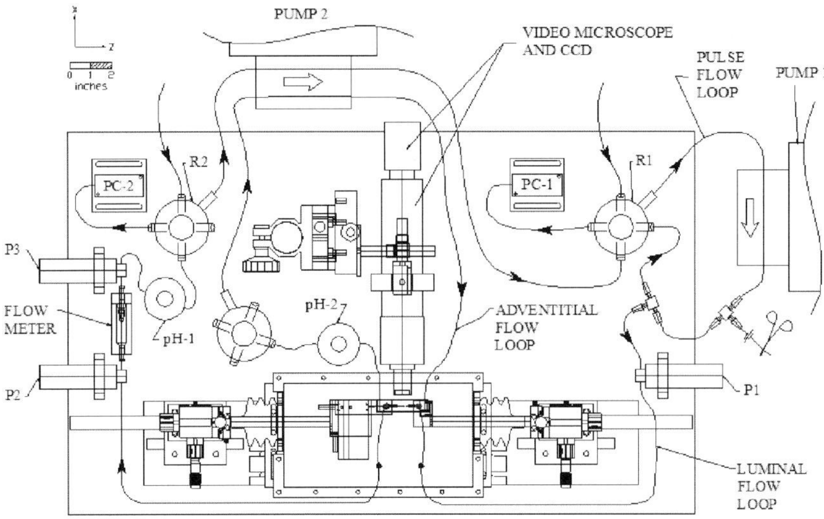

Fig. 4. Plan view of the organ culture and biomechanical testing device. P1, P2, and P3 are pressure transducers, PC-1 and PC-2 are pressure controllers, R1 and R2 are 125 mL reservoirs (Erlenmeyer flasks), PUMP 1 and PUMP 2 are peristaltic pumps and pH-1 and pH-2 are pH in-line meters. Note, too, that pressure drop (P2 − P3) due to flow through a small gauge needle (FLOW METER) is calibrated to measure the mean flow rate.

The pulse flow loop recirculates media from the proximal reservoir (R1) via a peristaltic pump (PUMP 1) that generates pulse magnitudes greater that 40 mmHg. Luminal flow is drawn off from the pulse flow loop, perfuses the vessel and is discharged to a distal reservoir (R2); this loop is completed by pumping (PUMP 2) media from R2 back to R1. The mean flow and pressure are controlled by independently controlling the pressure in R1 and R2 via separate pressure controllers (PC-1 and PC-2). By increasing the pressure in both reservoirs equally, the mean pressure can be controlled, and by increasing the pressure gradient between the proximal and distal reservoirs, the mean flow rate can be controlled. Proximal and distal pressure transducers (P1 and P2, respectively) are used to measure the mean pressure and a custom flow meter is used to measure the flow. The flow meter consists of a small gauge needle placed between the distal pressure transducer and a third pressure transducer (P3); the pressure drop across the needle can be accurately calibrated to the mean flow rate. Finally, a video microscope and CCD interface with the computer via an image acquisition board to allow real-time tracking of the vessel diameter (or the tracking of markers placed on the vessel).

Because our theoretical framework admits observations that are made at many different length scales, we were motivated to include measurements at different scales in our experimental design. At the tissue level, of course, we have designed the system to collect data sufficient for stress analyses (Fig. 5) and to observe functional adaptations (Fig. 6). We have chosen first to study mouse common carotid arteries. These vessels are well suited for *ex vivo* and *in vitro* tests because they are easy to excise with little dissection-induced damage, they are long, straight, and cylindrical with no branches, and the contra-lateral vessel provides a natural control for each animal. Our animal selection was motivated by the availability of genetic 'knockout' models, wherein specific genes have been suppressed. Thus, with this experimental model (as with our theoretical framework), one can impose very specific biological constraints at the genetic level; that is, we can study how a blood vessel will adapt to a specific perturbation in loading with or without the ability to express a certain gene. In addition to these *in vitro* observations, post-experimental analyses can be performed at the genetic level (e.g., gene arrays, reverse transcriptase polymerase chain reaction (RT-PCR)), molecular level (e.g., electrophoresis, western blotting), and the cell-cell and cell-matrix level (e.g., histology, immunohistochemistry). Taken together, these data can be used to test the underlying hypotheses in our theoretical model.

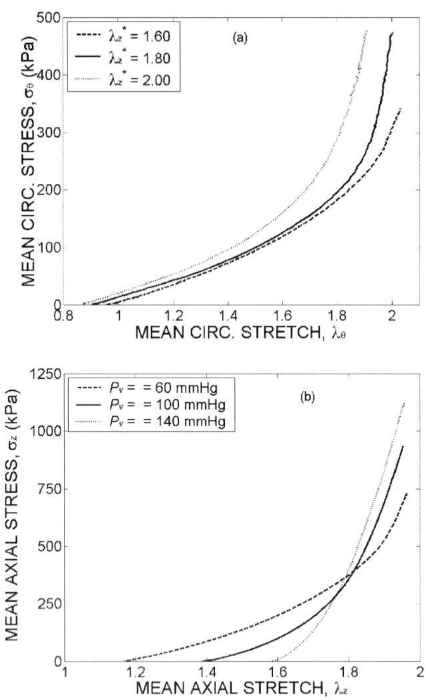

Fig. 5. Mean circumferential (panel a) and axial (panel b) stress-stretch curves for a native mouse carotid artery.

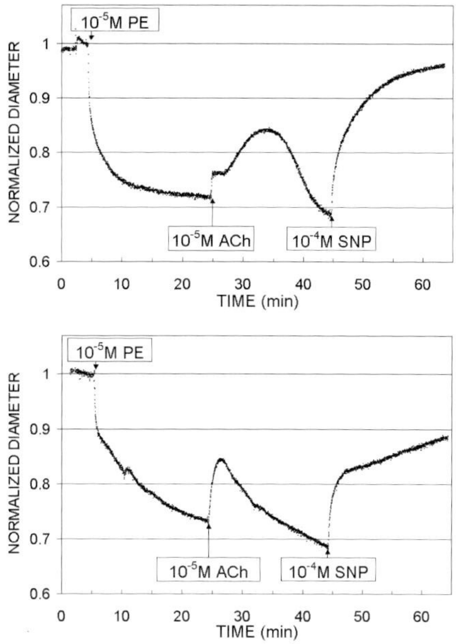

Fig. 6. Functional testing of a freshly-isolated (panel a) and 4-day cultured (panel b) mouse carotid artery. Functional testing consisted of sequential administration of 10^{-5} M phenylephrine (PE), 10^{-5} M acetylcholine (ACh), and 10^{-4} M sodium nitropruiside (SNP).

5 Closure

We emphasize that there is a pressing need for synergistic developments of a mathematical theory that can be used to motivate experiments and interpret data and development of novel experimental capabilities that can be used to feed back into the model to test underlying hypotheses of specific simulations. Our theory demanded tissue-level data sufficient to perform stress analyses, track evolving configurations, and observe changes in functional behavior at multiple time points during G&R; we have designed a device to capture this data. In parallel, our theory also demands information at the cellular and molecular levels, for example, regarding production and removal of individual constituents and adapting constituent-to-constituent interactions that can be correlated to changes at the tissue level; these data can be obtain via standard techniques in cell biology (for experiments run for different durations). It is only by such multi-scale approaches that integrate theory and experiment that one may hope to understand tissue G&R in sufficient detail to provide the biomedical community with a model having predictive capabilities.

Acknowledgements. This research was supported, in part, by grants from the NSF (BES-0084644) and NIH (R01 HL-64372, R01 HL-58856).

References

Chestler, N. C., Conklin, B. S., C. Han, H., and Ku, D. K. (1998). Simplified ex vivo artery culture techniques for porcine arteries. *J. Vasc. Invest.* 4: 123–127.

Clark, J. M., and Glagov, S. (1985). Transmural organization of the arterial media. *Arteriosclerosis* 5:19–34.

Gleason, R. L., and Humphrey, J. D. (2004). A mixture model of arterial growth and remodeling in hypertension: Altered muscle tone and tissue turnover. *J. Vasc. Res.* 41:352–363.

Gleason, R. L., and Humphrey, J. D. (2005). Effect of a sustained extension on arterial growth and remodeling: a theoretical study. *J. Biomech.* 38: 1255–1261.

Gleason, R. L., Gray, S., Wilson, E., and Humphrey, J. D. (2004a). A multiaxial computer-controlled organ culture and biomechanical testing device for mouse carotid arteries. *J. Biomech. Eng.* 126(6):787–795.

Gleason, R. L., Taber, L. A., and Humphrey, J. D. (2004b). A 2-D model for flow-induced alterations in the geometry, structure, and properties of carotid arteries. *J. Biomech. Eng.* 126:371–381.

Humphrey, J. D., and Rajagopal, K. R. (2002). A constrained mixture model for growth and remodeling of soft tissues. *Math. Model. Meth. Appl. Sci.* 12:407–430.

Humphrey, J. D., Kang, T., Sakarda, P., and Anjanappa, M. (1993). Computer-aided vascular experimentation: A new electromechanical test system. *Ann. Biomed. Eng.* 21:33–43.

Jackson, Z. S., Gotlieb, A. I., and Langille, B. L. (2002). Wall tissue remodeling regulates longitudinal tension in arteries. *Circ. Res.* 90:918–925.

Kamiya, A., and Togawa, T. (1980). Adaptive regulation of wall shear stress to flow change in the canine carotid artery. *Am. J. Physiol.* 239:H14–H21.

Langille, B. L., Bendeck, M. P., and Keeley, F. W. (1989). Adaptations of carotid arteries of young and mature rabbits to reduced carotid blood flow. *Am. J. Physiol.* 256:H931–H939.

Matsumoto, T., and Hayashi, K. (1996). Stress and strain distribution in hypertensive and normotensive rat aorta considering residual strain. *J. Biomech. Eng.* 118:62–73.

Matsumoto, T., Okumura, E., Miura, Y., and Sato, M. (1999). Mechanical and dimensional adaptation of rabbit carotid artery in vitro. *Med. Biol. Eng. Comput.* 37:252–256.

On the Modeling of Growth and Adaptation

S.C. Cowin

The School of Engineering of The City College and The Graduate School of The City University of New York, The New York Center for Biomedical Engineering and the Departments of Biomedical and Mechanical Engineering, USA
scowin@earthlink.net

The growth and adaptation of a tissue depends upon certain features in the history of its mechanical environment as well as its genetic makeup. The mechanical environment influences the tissue's developing morphology, the process of simply increasing the size of existing morphological structures, and the formation of the proteins of which the tissue is constructed. The developmental growth and remodeling of most structural tissues is enhanced by the use of those tissues and retarded by their disuse. The mechanical or mathematical modeling of tissue growth and development continuum mechanical models is reviewed.

1 Introduction

A tissue is a composite material whose constituents, and therefore structure, are continually changing due to growth and response of the tissue to its physical and chemical environment. A tissue's physiochemical environment includes the environment it is presently experiencing and the recent history of that environment. A tissue itself is a collection of cells and extracellular matrices that perform specialized functions. The extracellular matrix (ECM) consists of fibers (for example, the proteins collagen and elastin) and a ground substance (for example, proteoglycans).

Growth is the process of gradual increase in the net volume of a tissue, but it may also include some resorption. A distinction is made between *appositional* growth and *interstitial* growth; that is the difference between growth on the surface and growth within a volume of tissue. Appositional growth occurs, for example, when a tree adds another layer to the outside of its trunk and interstitial growth occurs within the cavities or spaces in the tissues of animals or plants. Hard tissues, bone and teeth, grow by apposition and soft tissues grow interstitially. In this work the primary concern is with the influence of the recent history of mechanical loading on growth rate although growth rate may also be slowed or accelerated by hormones, vitamins, bioelectrical factors, surgical intervention, imbalances between deposition and resorption and other factors.

Morphogenesis refers to the processes that are responsible for producing the complex shapes of adults from the simple ball of cells that derives from division of the fertilized egg. Morphogenetic events include pattern and template formation in tissue development. Morphogenetic processes interact with the growth and remodeling processes and often all three process types occur simultaneously.

The subject of this contribution is the mathematical modeling of tissue growth and remodeling. A section on modeling considerations is followed by four sections on continuum models, the first on kinematic models of growth, the second on bone remodeling, the third on soft tissue growth and remodeling and the fourth on heart and joint growth, remodeling and morphogenesis.

2 Some Modeling Considerations

Models to describe tissue growth and remodeling may be classified as phenomenological or mechanistic. The phenomenological models attempt to simulate cause and effect (e.g., changed mechanical loading leading to changed tissue architecture) without a consideration of the intermediary mechanical and biological mechanisms involved. Phenomenological models allow for conveniently testing the consequences of different hypotheses about tissue growth and adaptation. This approach is often useful for eliminating some assumptions that don't match experimental or clinical results and observations (e.g., only compressive static loading leads to a particular tissue formation) or stimulate further investigations (e.g., strain rates and spatial gradients may regulate adaptation). The theory of adaptive elasticity for bone, reviewed and critiqued in Section 4 is a phenomenological model. The basis concepts underlying phenomenological models for soft tissue growth and adaptation will be described in Section 5.

Mechanistic models, on the other hand, start instead with parameters (e.g., cell activities and microenvironment) that are linked to portions of the biological processes involved in tissue maintenance, turnover, and repair. These models, currently less developed than some of the phenomenological models because they are more complex, may lead to successfully linking mechanical and biological causes and effects. These models offer the promise of not only extending the descriptive and predictive capabilities of phenomenological models, but may offer insights into manipulation of the tissue response, and development of pharmacological therapeutic agents. A mechanistic model for bone adaptation is described by Cowin and Moss (2000). A mechanistic model that relates the effect of mechanical load applied to a whole bone to the bone fluid flow around the cells buried in the bone and, most significantly, to the bone adaptation process has been presented in a series of papers over the last decade by Cowin, Weinbaum and associates: Cowin et al. (1991), Weinbaum et al. (1994), Cowin et al. (1995), Zhang et al. (1998), Cowin (1999), Wang et al. (1999, 2000), You et al. (2001), Smit et al. (2002), Wang et al. (2003). Mechanistic bone models are not reviewed here; aspects of these models are reviewed in Cowin and Moss (2000) and Cowin (2002).

In the remainder of this contribution attention will be focused on continuum models for tissue growth and remodeling, but in this paragraph a significant caveat concerning the possibility of accomplishing this objective with continuum models, or purely continuum models, will be considered. The question of the adequacy of continuum models for this task has been discussed before, for example, by Humphrey (2001) who raises the question of whether '*... continuum measures such as stress and strain be applied to present the effect of mechanical loading on the development and remodeling of tissue*'. The problem is that most continuum models assume smooth bijective mappings of the reference state of an object onto a subsequently deformed state of the same object; bijective means one-to-one and onto. If one examines activities at the cellular level during tissue growth and remodeling the conditions for mappings of this type are not satisfied. Cells move around like guests circulating at a cocktail party, they replicate themselves with some ease and they produce new material for the tissue of which they are a part. If one takes a picture of the reference state of a tissue at the cellular level and then examines a picture of a subsequent state of the tissue one sees regions that were once neighbors now separated by other regions, that there now exist regions that did not exist before and there are regions that existed before that do not exist now. To imagine this, suppose that two jigsaw puzzles were made from the picture of the reference state of a tissue and from the picture of the subsequent state. In a continuum model the puzzle pieces would still fit together in both puzzles but the two puzzles could be different sizes and shapes; the puzzle associated with the deformed state would appear locally as if its picture were stretched like a rubber sheet before the puzzle was cut into pieces. The pieces of the actual cell level puzzles will not be compatible for the two states of the tissue. A piece-by-piece comparison of the cell level puzzle pieces of the two puzzles will uncover missing and additional pieces and pieces that will not fit together. Clearly the real cell level mapping will not be a continuum one-to-one mapping.

Cellular automata models that do effectively consider the complexities of these cellular level non-bijective mappings have been successfully formulated for several situations in which straining of the tissue between subsequent states of the tissue was not considered (and was not an issue). One example of such cellular automata models is given by the growth of epithelial tissue layers (Walker et al. (2004)) and another by the models of cell-sorting that occur during development. The differential adhesion hypothesis (Steinberg (1962, 1964)) suggests that combinations of different cell types behave as immiscible liquids (for example, oil and water) because of surface tension, but with the cell-to-cell adhesion forces playing the role of molecule-molecule attraction forces. In immiscible liquids a liquid with a higher surface tension will form droplets enclosed by another liquid with lower surface tension. In combinations of different cell types, cells with higher surface tension will band more tightly together, forcing the cells with a lower surface tension to the outside. This sorting has been effectively modeled with cellular automata (Mombach et al. (1995), Ramsey et al. (1996), Beysens et al. (2000), Alber et al. (2002)). There appears to be a great deal of potential for modeling the mechanics of biological tissue development using cellular automata if a method can found for the inclusion of deformation measures such as strain. The use of cellular automata models at the cellular level and continuum models at hierarchical scale levels is also a possibility.

In the creation of phenomenological mathematical models for tissue growth and remodeling, the modeler must select a parameter to represent the effect of the mechanical loading on the development and remodeling of the tissue. In a mechanistic model such a parameter would arise from biophysical considerations. Do biological tissues sense the continuum variables of stress, strain, strain rate or some other parameter of the mechanical loading? This is equivalent to asking if a tissue has baroreceptors (stress receptors) or stretch receptors (strain receptors) or another type of receptors. One answer to the question is that tissues sense strain or stretch or strain rate or stretch rate and not stress, Cowin (1984). The reason that tissues sense strain or strain rate and not stress is that strain is a primary, directly measurable, physical quantity whereas stress is not. Stress is an abstract concept, a creation of man that can be measured only indirectly. Baroreceptors are described as 'a spray-type nerve ending lying in the walls of arteries that are stimulated when stretched' (cf., e.g., Guyton (1976)). Clearly, baroreceptors sense stretch. From this perspective, the terminology 'baroreceptor' is a misnomer; they are stretch receptors.

Strain is excluded as a possible growth stimulus for soft tissue, Rodriguez et al. (1994), because of the difficulty in defining a reference configuration for its measurement in a growing material. The question is one of gage length. If the gage length is changing due to the addition of new tissue, how can one measure strain? Several studies have considered this question of the appropriate stimulus (Taber (1995), Omens (1998)) and some have developed models in which the growth of soft tissue is dependent upon applied stress (Rodriguez et al. (1994), Taber et al. (1992), Lin and Taber (1995), Taber and Eggers (1996), Taber (1998c)). The substantial and important structure of these studies is independent of the fact that it is assumed that the growth of soft tissue is dependent upon applied stress; the alternative structure necessary to convert the stimulus from applied stress to loading rate of deformation or loading strain is easily constructed. The exact stimulus for growth is presently unknown (in bone there is strong evidence to suggest that it is strain rate). However, the argument sometimes presented is that strain requires a reference configuration and stress does not. The problem is that reference configuration for strain will change in time due to tissue growth and remodeling. There is a concern that this is really a problem and, if it is a problem, it can be addressed in another way.

First, is this really the problem? It is not possible that the gage length used for the sensing of strain does not change as the tissue grows, as noted by Cowin (1996). For example, the cell may sense the strain itself and not change its length on the remodeling time scale t. Numerous cell culture studies have shown that the key connective tissue building cells begin to respond to strains above about 0.5% with various signals that indicate the cells are aware that they are being excited. Generally cells in tissues at different stages of growth are the same size. Cells probably do not change their length on the time scale t as the tissue grows; the cells can however stiffen or relax their cytoskeleton on a time scale that is greater than the loading time scale τ and less than the remodeling time scale t. The tissue either adds or subtracts cells or ECM to achieve the volume change; it is unlikely that it changes the (sensor) cell size. One cell does not likely determine the response of a tissue; it is more likely that the response of the tissue is determined by averaging the response of numerous cells by intercellular communication. Thus, it appears that it is possible that the gage length used for the sensing of strain does not change as the tissue grows. However this observation does not entirely solve the problem because it leaves open the question of how to construct a model with such a microstructural gage length.

Second, if it is a problem, can it be addressed in another way? Appropriate measures of the rate of deformation or strain upon which the growth stretch may depend may be constructed if the two-time scales in living tissue mechanics are recognized. These time scales differ by many orders of magnitude Cowin (1996). Let T_r denote the time that biological processes take to complete significant growth (or remodeling) associated with a mechanical loading, and let T_L denote the characteristic period of mechanical loading. Rough estimates of these numbers are two weeks and one second, respectively; thus T_L/T_r is a small number, of the order 10^{-6}. In order to keep these time scales separate let τ denote time on the loading time scale and t denote time on the remodeling time scale. Measures of tissue strain may be constructed using two facts, the fact that the tissue is strained on the time scale τ and the fact that the gage length changes on the time scale t, a million times longer.

The loading rate of deformation of an object is an instantaneous kinematic measure independent of reference configuration or gage length, and thus does not have the disadvantage that might exclude strain as a growth stimulus. The rate of deformation is suggested as a remodeling stimulus for endothelial and bone cells in the sense that the shear stress due to fluid motion over these epithelial cells is a stimulus for their activity and shear stress is proportional to the rate of deformation of a fluid. It is an open question as to how much of the rate of deformation is actually transferred to the epithelial cells. The cell layer is compliant and the undulations of the cell contour are observed in laminar shear flows in which the cells are attached to a supporting plate. Adaptive remodeling of the vasculature has been observed so as to maintain nearly constant endothelial fluid shear stress (rate of deformation to the cell) throughout the entire arterial side of the circulation (Kamiya et al. (1984)). Explicit forms of appropriate or plausible measures of loading rate of deformation or loading strain upon which the growth stretch rate are presented in Cowin (1996); see also Luo et al. (1995). The main point is that the recognition of two time scales in the growth process, the loading time scale and the growth (or remodeling) time scale, scales that differ by many orders of magnitude, permit the consideration of strain and rate of deformation as growth stimuli in finite growth models of soft tissues.

3 Continuum Kinematic Models of Growth

The growth and development of hard and soft tissues are kinematically different. The growth of hard tissues is appositional, that is to say the growth is by deposition and resorption at a surface. The growth of soft tissues is interstitial volumetric growth. Thus an increase in a hard tissue occurs by adding material on an existing surface while soft tissues grow by adding material internally. The kinematics of growth of an organism was a primary interest of D'Arcy Thompson and the subject is well illustrated in his famous book (D'Arcy Thompson (1942)). The classical ideas of growth kinematics were recast in the mathematics of contemporary continuum mechanics in Skalak et al. (1982). This included continuum kinematics models for volumetrically and surface distributed growth by deposition or resorption, the growth of horns and spiral shells (a favorite of D'Arcy Thompson), allometric growth and the initial idea for a kinematic model in which simultaneously occurring growth and deformation are considered as a sequence of two mappings, one representing stress free growth and the other representing the deformation of the tissue due to forces acting upon the tissue. This last topic will be discussed at the beginning of Section 5. However, the other topics on the modeling of the kinematics of growth of an organism are not reviewed here because the review of Skalak et al. (1982), with its refinements (Skalak et al. (1996, 1997)), is still current and readable.

4 Continuum Models for Bone Remodeling

The first continuum model for bone adaptation to mechanical loading was the (phenomenological) theory of adaptive elasticity due to Cowin and Hegedus (1976), Hegedus and Cowin (1976) and Cowin and Nachlinger (1978). The objective of these papers was the formulation of a model for the understanding and prediction of the strain-controlled remodeling properties of normal living bone. The strain adapting properties of living bone are represented by a strain-controlled chemical reaction that transfers mass, momentum, entropy and energy to and from the porous elastic solid. The addition of mass to the porous solid modifies its porosity. Bone adaptation to environmental strain is a collective phrase for the continual processes of growth, reinforcement and resorption that occur in living bone. The resulting theory describes an elastic material that adapts its structure to applied loading, hence the term *adaptive elasticity*. Seen from the perspective of a quarter of a century, there were major successes in the 1976 theory of adaptive elasticity and there were things that should be done differently if the theory is to be renovated. The largest success was the development of a thermodynamic open-system model of tissue adaptation.

This model has been followed, or assumed as a starting point, by most subsequent models of tissue adaptation (a survey of these models appears in Hart (2001)). The essence of the model was the assumption that the load adapting properties of living bone can be modeled by a chemically reacting porous medium in which the rate of reaction is strain controlled. The porous medium has two components: a porous elastic solid representing the matrix structure of bone including the bone cells and a perfusant that represents the extracellular fluid and the blood plasma which flow through the matrix structure. A schematic diagram of this model is shown in Fig. 1. The fact that living bone is encased in a living organism is reflected in the model by setting the porous structure in a bath of the perfusant. The perfusant bath is assumed to be an isothermal heat reservoir, an assumption that appears to be easily justified by common knowledge concerning living organisms. The mechanical load is applied directly to the porous structure across the walls of the perfusant bath, as illustrated in Fig. 1. The system consisting of the porous structure and its perfusant bath is considered to be closed with respect to mass, heat energy, and entropy transfer, but open with respect to momentum transfer from loading. The system consisting of only the porous structure without its entrained perfusant is open with respect to momentum transfer as well as

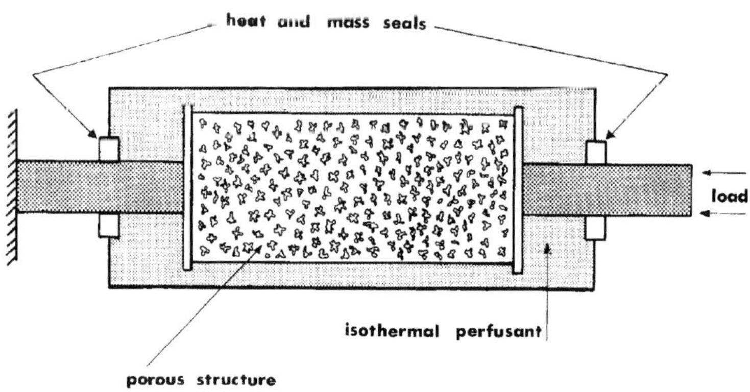

Fig. 1. A schematic diagram of the conceptual model for bone adaptation to mechanical loading. From Cowin and Hegedus (1976).

mass, energy, and entropy transfer. The bone matrix was considered to be the control system since the mechanical properties of the bone matrix alone determine the mechanical properties of the bone.

Balance and constitutive equations were developed only for the bone matrix. The perfusant was accounted for only insofar as it transfers mass, momentum, energy, or entropy to the bone matrix. The rate at which mass, momentum, energy and entropy transfers occur was assumed to depend on the local strain and the other independent variables. One of the strongest assumptions made was the assumption that the internal energy of the mass added to the porous tissue was equal to the internal energy of the porous tissue at the site where the mass is added. In retrospect this is a fairly easy assumption to justify if one recalls how bone tissue is deposited and if the bone adaptation is viewed on two time scales, as discussed in Section 5 above.

New bone deposition begins with a deposit of osteoid on an existing bone surface. The osteoid is of much lower modulus than the existing bone. Over a time period of a month the osteoid will become mineralized and its modulus will increase approximately linearly with the mineralization. Since the mechanical loading time scale is much, much shorter than the adaptation time scale the osteoid/bone will undergo many million cycles of mechanical loading in the course of the adaptation tissue deposition time period. It follows that the internal energy can be gradually increased with the mineralization in the tissue deposition time period so that, when viewed from the long-term adaptation time scale, it will be equivalent to having the new mass deposited at the same internal energy as the existing mass in the porous bone structure. Other features common to most of the later models of internal adaptation (described by Hart (2001)) include the formulation of the theory in terms of a constitutive relation for the mass supply term introduced by the open system formulation of the mass balance equation; the porosity of the medium could be increased or decreased. The increased or decreased porosity is then assumed to influence the elastic constants through their constitutive dependence upon porosity.

There are changes recommended if the theory of adaptive elasticity is to be reformulated. A major change would be the introduction of two time scales discussed in Section 5 above. The reformulated theory of adaptive elasticity should allow the mass supply and the free energy to depend upon the remodeling time and the history of deformation; thus, functions of deformation gradients in the original theory should be replaced by functionals of the history of reference solid volume fraction and deformation gradients. Such strain history dependent relationships are familiar from non-linear viscoelasticity. The reformulated theory would then consist of the usual equations for an elastic object and equations which change on the long time scale, very slowly adding or subtracting mass and free energy and, thereby, changing the elastic constants. Perhaps the point most in need of revision in the theory of adaptive elasticity is the stimulus for remodeling. The model pointed to time-averaged strain and strain-energy as the first measures of stimulus. The experimental literature suggests, however, that it must be a stimulus that involves the rate-of-strain or the rate-of-loading. It has been shown that a moderate periodic strain applied to living bone is more effective in causing remodeling than a static strain of the same magnitude. It has also been shown that a constant load applied with fixed springs to isolated living bone caused it to resorb as if there were no load placed on the bone (Lanyon and Rubin (1984)). In other words, the living bone showed no effect from the constant, non-time varying loading. Further evidence that the effective stimulus was strain rate is provided in O'Connor et al. (1982), and Rubin and Lanyon (1984). The mechanistic models (see Section 5) that relate the effect of mechanical load driven bone fluid flow around the cells buried in bone to the bone adaptation process strongly suggest strain rate as a strong mechanical stimulus (Cowin and Moss (2000), Weinbaum et al. (1994)).

The theory of adaptive elasticity has been recently reviewed and critiqued in Cowin (2003). This model and other phenomenological models for cortical and cancellous bone adaptation have recently been reviewed in Hart (2001). These reviews are still current.

5 Continuum Models for Soft Tissue Growth and Remodeling

5.1 The Separability Hypothesis

In the final section of the Skalak et al. (1996) paper on the analytical description of growth, discussed in Section 6, a kinematic model is described in which simultaneously occurring growth and deformation are considered as a sequence of two mappings, one representing stress-free growth and the other representing the deformations of the tissue due to forces acting upon the tissue. This may be the first statement of the separability hypothesis that simultaneously occurring growth and deformation may be decomposed into a growth deformation and an elastic deformation. The validity of this hypothesis is easy to imagine in a thought experiment involving a whole bone without residual stress. One simply measures the size of the bone in a fixed, unstressed, sit uation periodically over a significant growth period. Since the measurement is done in the unstressed situation, and there is no residual stress, the measurement represents the growth of the bone. The hypothesis is not valid if there is an inelastic deformation of the bone since there would be an Ilizarov lengthening procedure.

The hypothesis was extended (Rodriguez et al. (1994)) to a general three-dimensional theory of mechanically modulated volumetric growth for soft incompressible biological tissues. The mapping composition idea described in Skalak et al. (1982) was rendered in Rodriguez et al. (1994) as a composition of deformation gradient mappings. The overall growth deformation is represented by a mapping, denoted by $\mathbf{F}_{eg}$, of the initial configuration of the object $B(t_0)$ into the instantaneous configuration $B'(t_1)$. The decomposition of the mapping suggested by Rodriguez et al. (1994) is represented by

$$\mathbf{F}_{eg} = \mathbf{F}_e\mathbf{G}, \tag{1}$$

where $\mathbf{G}$ is a symmetric tensor representing the growth deformation gradient and $\mathbf{F}_e$ represents an elastic deformation necessary to maintain overall compatibility of the mapping $\mathbf{F}_{eg}$. The representation (1) follows from the fact that any deformation gradient can be decomposed into a product of deformation gradients as long as there is a definition of one of the elements of the decomposition that provides a method to calculate the other element of the decomposition. That is the case because of our ability to measure growth (Bookstein (1978)). The symmetry of $\mathbf{G}$ follows from the fact that any deformation gradient may be decomposed into the product of an orthogonal tensor representing the rigid object rotation and a symmetric positive tensor representing the shape changing deformation. When growth is measured any effect of rigid object rotation is easily removed, hence $\mathbf{G}$ is necessarily symmetric, as pointed out by Skalak et al. (1982). The mapping decomposition (1) is illustrated in Fig. 2. Note that there is no mechanical loading applied to any of the objects in Fig. 2, but it is possible for the final configuration $B'(t_1)$ to have residual stresses due to incompatibilities induced by growth.

This decomposition (1) is analogous to the decomposition of the deformation gradient used in plasticity theory to separate finite elastic and plastic deformations since the 1960's (Kröner (1960), Lee (1969), Casay and Naghdi (1981)) and in the literature on polymeric swelling to separate elastic and

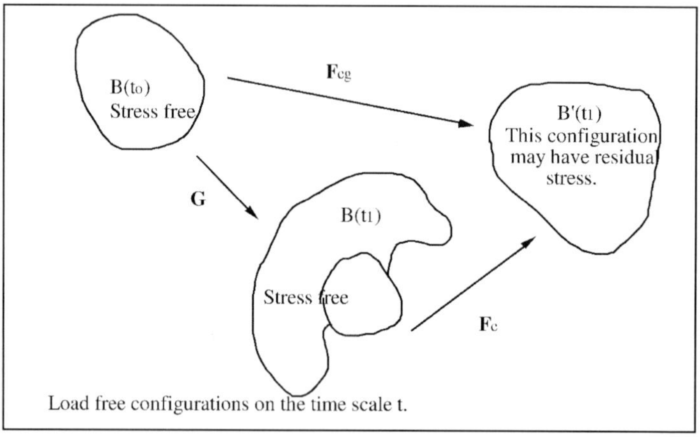

Fig. 2. An illustration of the relationship between the various (unloaded) configurations considered and the deformation gradients that represent mappings between these configurations. This diagram is conceptually identical with Fig. 1 of Rodriguez et al. (1994). Adapted from Cowin (1996).

swelling deformations since the 1940's (Boyce (2003)). Each biological situation in which the assumption (1) is to be applied must be carefully scrutinized to evaluate the justification of the separation of the growth and deformation effects. The separability hypothesis is an example of the reductionism problem discussed in Section 2. In the case where the growth is well mapped so that the growth tensor $\mathbf{G}$ is known, there should be no difficulty. However, if there is any morphological development event occurring that cannot be represented by the growth tensor $\mathbf{G}$, the decomposition (1) is not possible. This restriction is due to the definition of growth; if we were able to describe morphological events mathematically, the growth tensor could be replaced by a growth and morphogenesis tensor and the predictive capacity of the model increased.

Models of tissue growth in the cardiovascular system employing these ideas have been described by Taber and coworkers (Lin and Taber (1995), Taber and Eggers (1996), Taber (1998c)), Rodríguez et al. (2003) and Rodríguez (2003).

5.2 The Loading Time Scale and the Growth Time Scale

The recognition of two time scales in the growth process, the loading time scale and the growth (or remodeling) time scale, scales that differ by many orders of magnitude, permits the consideration of strain and rate of deformation as growth stimuli in finite growth models of soft tissues. Explicit forms of appropriate or plausible measures of loading rate of deformation or loading strain upon which the growth stretch rate may depend may be constructed using the two time scales, the loading time scale τ and the remodeling time scale t. An extension of the representation in Fig. 2 to include these two time scales and the two loaded configurations $B''(t_0, \tau)$ and $B''(t_1, \tau)$ is shown in Fig. 3; Fig. 2 is repeated in the box outlined with dashed lines in Fig. 3. The two loaded configurations $B''(t_0, \tau)$ and $B''(t_1, \tau)$ are related to the configurations $B(t_0)$ and $B'(t_1)$, respectively, by the deformation gradients $\mathbf{F}_L(t_0, \tau)$ and $\mathbf{F}_L(t_1, \tau)$, as illustrated in Fig. 3. The growth tensor $\mathbf{G}$ is now a function of t, not τ; thus, $\mathbf{G} = \mathbf{G}(t)$. The deformation gradient associated with the loaded configuration from the unloaded, but not necessarily stress free, configuration is denoted by $\mathbf{F}_L(t, \tau)$. The tissue is mechanically strained on the loading time scale τ and therefore the deformation gradient of loading $\mathbf{F}_L$ is considered to be dependent on both time scales. If the residual elastic deformation $\mathbf{F}_e$ is significant and thought to be a growth stimulus, then rather than employing the loading deformation $\mathbf{F}_L$ as the basis for the independent constitutive variable in the growth stimulus, the basis could be provided by the composition $\mathbf{F}_{eL} = \mathbf{F}_L\mathbf{F}_e$ of the residual elastic deformation $\mathbf{F}_e$ and the loading deformation $\mathbf{F}_L$. This would make the growth stimulus dependent on both the loading and the residual strain:

$$\mathbf{F} = \mathbf{F}_L\mathbf{F}_{eg} = \mathbf{F}_L(t, \tau)\mathbf{F}_e(t, \tau)\mathbf{G}(t). \tag{2}$$

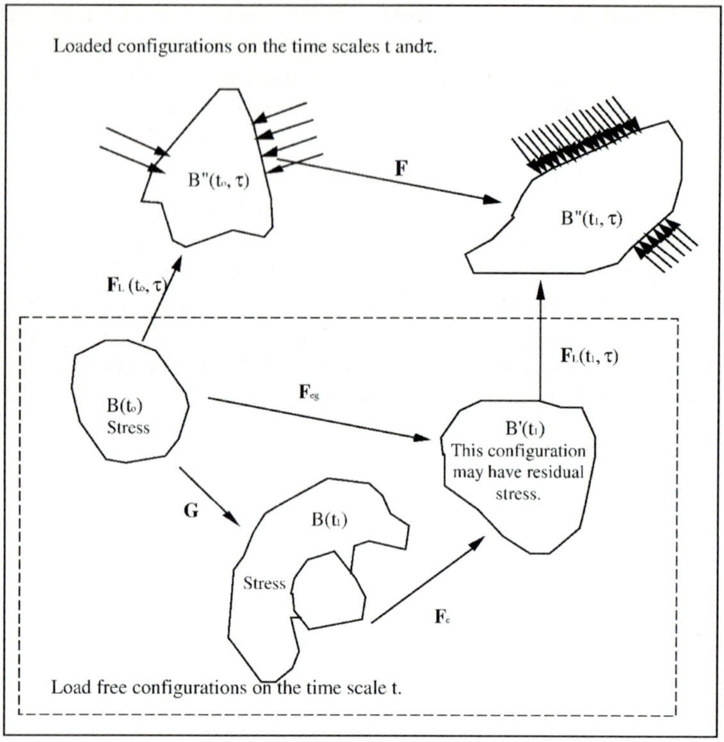

Fig. 3. An illustration of the relationship between the various configurations considered. Broadly there are two classes of configurations, loaded and unloaded. The unloaded ones are within the portion of the diagram outlined with the dashed lines. This portion of the diagram is identical to Fig. 2 above. The loaded configurations are above and outside the dashed lines. The deformation gradient associated with the deformation to the loaded configuration from the unloaded (but not necessarily stress free) configuration is denoted by $\mathbf{F}_L(t, \tau)$. It is shown at times t_0 and t_1. Adapted from Cowin (1996).

5.3 Residual Stresses

The presence of residual stress in biological tissues has been widely reported in the last two decades. Residual stresses in arteries were demonstrated by performing experiments on unloaded arterial rings dissected from blood vessels (Vaishnav and Vossoughi (1983, 1987), Choung and Fung (1986), Liu and Fung (1988)). Two transverse cuts of an artery yield an unloaded arterial ring. If that arterial ring is cut so that it is no longer an approximately circular ring but a sector of a circular ring, the unloaded sector of the arterial ring opens and displays an opening angle indicating that, when it was a complete ring, it had internal stresses. In fact if one keeps cutting the sector of circular ring into smaller pieces, each cut produces a new opening angle indicating the presence of internal stresses in each progressively smaller piece. The presence of residual stress has been established in a number of other tissues, including veins (Xie et al. (1991)), ventricular myocardium (Omens and Fung (1990), Rodriguez et al. (1993)) and the trachea (Han and Fung (1991)). The residual stress in biological organs is thought to be a consequence of non-uniform growth, resorption, and remodeling (Fung (1990), Skalak (1981), Skalak et al. (1982)). Residual stresses in the heart and arteries are thought to minimize the peak stresses experienced by these tissues *in vivo* (Fung and Liu (1989), Liu and Fung (1989), Choung and Fung (1986)) and thus improve their mechanical function. The inclusion of the residual stress in the stress constitutive equations used to describe these soft tissues has been the focus of Hoger and coworkers (Hoger (1997), Johnson and Hoger (1998), Chen and Hoger (2000)).

5.4 General Continuum Models of Growth

A general constitutive theory of the stress-modulated growth of soft tissues was developed by Lubarda and Hoger (2002). This work provides an explicit representation of **G** for various material symmetries, and an incremental formulation for the stress-modulated growth process. A theory of material growth (mass creation and resorption) is presented in Epstein and Maugin (2000). In this work growth is viewed as a local rearrangement of material inhomogeneities.

6 Continuum Models for Heart and Joint Growth, Remodeling and Morphogenesis

Modeling the development of an organ or an organism is an objective that is presently experiencing a pioneering exploration. Such models will necessarily involve the processes of growth, remodeling and morphogenesis. Two examples are briefly described here, the modeling of the embryonic chicken heart (Taber and Perucchio (2001), Taber (1995, 1998a,b)) and a model for articular joint morphogenesis (Heegaard (1999), Heegaard et al. (1999)).

The heart is the first functioning organ in the embryo and it continues to function without interruption although its morphology changes dramatically during development. Development consists of a coordinated, dynamic interaction between genetic and environmental factors that regulate the primary developmental processes of volume change (growth), tissue property change (remodeling), and shape change (morphogenesis). A mathematical model for the embryonic heart development in the chicken as it transforms from a single tube into a four-chambered pump is given in Taber and Perucchio (2001).

A mathematical model for joint morphogenesis encompassing the hypothesis that the stress distribution created in a functional joint will modulate the growth of the primordial template and lead to the development of congruent articular surfaces has been described by Heegaard and coworkers (1999). In a computational model the morphogenesis of a human finger joint (proximal interphalangeal joint) was simulated for the period between days 55 and 70 of fetal life. It was assumed the biological growth rate was proportional to the chondrocyte density in the growing tissue. Cyclic hydrostatic stress caused by joint motion was assumed to modulate the baseline biological growth, with compression slowing it and tension accelerating it. The model prediction was that the articular surfaces became more congruent and the primordial template exhibited an asymmetric sagittal profile similar to that observed in adult phalangeal bones.

Acknowledgement. The Whitaker Foundation, NIH, NSF and the PSC-CUNY Research Award Program of the City University of New York supported this work. A portion of the material in this contribution is taken from Cowin (2004).

References

Alber, M. S., Kiskowski, M. A., Glazier, J. A., and Jiang, Y. (2002). On cellular automaton approaches to modeling biological cells. In Rosenthal, J., and Gilliam, D. S., eds., *Mathematical Systems Theory in Biology, Communication, and Finance.* New York: Springer-Verlag. IMA Volume 142.

Beysens, D. A., Forgacs, G., and Glazier, J. A. (2000). Cell sorting is analogous to phase ordering in fluids. *Proc. Nat. Acad. Sci. USA* 97:137–145.

Bookstein, F. L. (1978). *The Measurement of Biological Shape and Shape Change*, volume 24 of *Lecture Notes in Biomathematics.* New York: Springer.

Boyce, M. C. (2003). Private communication.

Casay, J., and Naghdi, P. M. (1981). A remark on the use of the decomposition $\mathbf{F} = \mathbf{F}_e\mathbf{F}_p$ in plasticity. *J. Appl. Mech.* 47:672–675.

Chen, Y. C., and Hoger, A. (2000). Constitutive function of elastic materials in finite growth and deformation. *J. Elasticity* 59:175–193.

Choung, C. J., and Fung, Y. C. (1986). Residual stress in arteries. In Schmid-Schoenbein, G. W., Woo, S. L., and Zweifach, B. W., eds., *Frontiers in Biomechanics.* 117–129.

Cowin, S. C., and Hegedus, D. H. (1976). Bone remodelling I: Theory of adaptive elasticity. *J. Elasticity* 6:313–326.

Cowin, S. C., and Moss, M. L. (2000). Mechanosensory mechanisms in bone. In Lanza, R., Langer, R., and Chick, W., eds., *Textbook of Tissue Engineering.* San Diego: Academic Press, 2nd edition. 723–738.

Cowin, S. C., and Nachlinger, R. R. (1978). Bone remodelling III: Uniqueness and stability in adaptive elasticity theory. *J. Elasticity* 8:285–295.

Cowin, S. C., Moss-Salentijn, L., and Moss, M. L. (1991). Candidates for the mechanosensory system in bone. *J. Biomech. Eng.* 113:191–197.

Cowin, S. C., Weinbaum, S., and Zeng, Y. (1995). A case for bone canaliculi as the anatomical site of strain generated potentials. *J. Biomech.* 28: 1281–1297.

Cowin, S. C. (1984). Modeling of the stress adaptation process in bone. *Calcif. Tissue Int.* 36:S99–104.

Cowin, S. C. (1996). Strain or deformation rate dependent finite growth in soft tissues. *J. Biomech.* 29:647–649.

Cowin, S. C. (1999). Bone poroelasticity. survey article. *J. Biomech.* 32: 217–238.

Cowin, S. C. (2002). Mechanosensation and fluid transport in living bone. *J. Musculoskelet. Neuronal. Interact.* 2:256–260.

Cowin, S. C. (2003). Adaptive elasticity: A review and critique of a bone tissue adaptation model. *Engng Trans.* 51:1–79.

Cowin, S. C. (2004). Tissue growth and remodeling. *Ann. Rev. Biomed. Eng.* 6:77–107.

D'Arcy Thompson, W. (1942). *On Growth and Form.* Cambridge: Cambridge Univ. Press.

Epstein, E., and Maugin, G. A. (2000). Thermomechanics of volumetric growth in uniform bodies. *Int. J. Plasticity* 16:951–978.

Fung, Y. C., and Liu, S. Q. (1989). Relationship between hypertension, hypertrophy, and opening angle of zero-stress state of arteries following aortic constriction. *J. Biomech. Eng.* 111:325–335.

Fung, Y. C. (1990). *Biomechanics: Motion, Flow, Stress, and Growth.* Springer-Verlag.

Guyton, A. C. (1976). *Medical Physiology.* Philadelphia: W. B. Saunders.

Han, H. C., and Fung, Y. C. (1991). Residual strains in porcine and canine tracheas. *J. Biomech.* 24:307–315.

Hart, R. L. (2001). Bone modeling and remodeling: Theories and computation. In Cowin, S. C., ed., *Bone Mechanics Handbook.* Boca Raton, Florida: CRC Press.

Heegaard, J. H., Beaupre, G. S., and Carter, D. R. (1999). Mechanically modulated cartilage growth may regulate joint surface morphogenesis. *J. Orthop. Res.* 17:509–517.

Heegaard, J. H. (1999). Dynamics of joint morphogenesis. In Petersen, P., and Bendsøe, M. P., eds., *IUTAM Symposium on Synthesis in Bio-Solid Mechanics.* Dordrecht: Kluwer.

Hegedus, D. M., and Cowin, S. C. (1976). Bone remodeling, II: Small strain adaptive elasticity. *J. Elasticity* 6:337–352.

Hoger, A. (1997). Virtual configurations and constitutive equations for residually stressed bodies with material symmetry. *J. Elasticity* 48:125–144.

Humphrey, J. D. (2001). Stress, strain and mechanotransduction in cells. *J. Biomech. Eng.* 123:638–641.

Johnson, B. E., and Hoger, A. (1998). The use of strain energy to quantify the effect of residual stress on mechanical behavior. *Math. Mech. Solids* 4:447–470.

Kamiya, A., Bukhari, R., and Togawa, T. (1984). Adaptive regulation of wall shear stress optimizing vascular tree function. *Bull. Math. Biol.* 46:127–173.

Kröner, E. (1960). Allgemeine Kontinuumstheorie der Versetzungen und Eigenspannungen. *Arch. Rat. Mech. Anal.* 4:273–334.

Lanyon, L. E., and Rubin, C. T. (1984). Static vs dynamic loads as an influence on bone remodelling. *J. Biomech.* 17:897–905.

Lee, E. H. (1969). Elastic-plastic deformation at finite strains. *J. Appl. Mech.* 36:1–8.

Lin, I. E., and Taber, L. A. (1995). A model for stress-induced growth in the developing heart. *J. Biomech. Eng.* 117:343–349.

Liu, S. Q., and Fung, Y. C. (1988). Zero-stress states of arteries. *J. Biomech. Eng.* 110:82–84.

Liu, S. Q., and Fung, Y. C. (1989). Change of residual strains in arteries due to hypertrophy caused by aortic constriction. *Circ. Res.* 65:1340–1349.

Lubarda, V. A., and Hoger, A. (2002). On the mechanics of solids with a growing mass. *Int. J. Solids Structures* 39:4627–4664.

Luo, G. M., Cowin, S. C., Sadegh, A. M., and Arramon, Y. (1995). Implementation of strain rate as a bone remodeling stimulus. *J. Biomech. Eng.* 117:329–338.

Mombach, J., Glazier, J. A., Raphael, R., and Zajac, M. (1995). Quantitative comparison between differential adhesion models and cell sorting in the presence and absence of fluctuations. *Phys. Rev. Lett.* 75:2244–2247.

O'Connor, J. A., Lanyon, L. E., and MacFie, H. (1982). The influence of strain rate on adaptive bone remodeling. *J. Biomech.* 15:767–781.

Omens, J. H., and Fung, Y. C. (1990). Residual strain in rat left ventricle. *Circ. Res.* 66:37–45.

Omens, J. H. (1998). Stress and strain as regulators of myocardial growth. *Prog. Biophys. Molec. Biol.* 69:559–572.

Ramsey, A., Foty, C., Pfleger, M., Forgacs, G., and Steinberg, M. S. (1996). Surface tensions of embryonic tissues predict their mutual envelopment behavior. *Development* 122:1611–1620.

Rodriguez, E. K., Omens, J. H., Waldman, L. K., and McCulloch, A. D. (1993). Effect of residual stress on transmural sarcomere length distribution in rat left ventricle. *Am. J. Physiol.* 264:H1048–H1056.

Rodriguez, E. K., Hoger, A., and McCulloch, A. D. (1994). Stress-dependent finite growth in soft elastic tissues. *J. Biomech.* 27:455–468.

Rodríguez, J. S., Goicolea, J., García, J. C., and Gabaldón, F. (2003). Finite element models for mechanical simulation of coronary arteries. Preprint from E.T.S.I. Caminos, Canales y Puertos, Depto. Mecanica de Medios Continuos, Universidad Politécnica de Madrid (UPM). Madrid, 28040, Spain.

Rodríguez, J. S. (2003). *Modelos numéricos para mecanica cardiovascular de las paredes arteriales y sus procesos de adaptacíon.* Ph.D. Dissertation, Universidad Politécnica de Madrid, Escuela Técnica Superior de Ingenieros de Caminos, Canales y Puertos.

Rubin, C. T., and Lanyon, L. E. (1984). Regulation of bone formation by applied dynamic loads. *J. Bone Joint Surg.* 66A:397–402.

Skalak, R., Dasgupta, G., Moss, M., Otten, E., Dullemeijer, P., and Vilmann, H. (1982). Analytical description of growth. *J. Theor. Biol.* 94:555–577.

Skalak, R., Zargaryan, S., Jain, R. K., Netti, P. A., and Hoger, A. (1996). Compatibility and the genesis of residual stress by volumetric growth. *J. Math. Biol.* 34:889–914.

Skalak, R., Farrow, D. A., and Hoger, A. (1997). Kinematics of surface growth. *J. Math. Biol.* 35:869–907.

Skalak, R. (1981). Growth as a finite displacement field. In Carlson, D. E., and Shield, R. T., eds., *Proceedings of the IUTAM Symposium on Finite Elasticity, 1981.* The Hague: Martinus Nijhoff Publishers.

Smit, T. H., Huyghe, J. M., and Cowin, S. C. (2002). Estimation of the linear isotropic parameters. *J. Biomech.* 35:829–835.

Steinberg, M. (1962). Mechanism of tissue reconstruction by dissociated cells, II time-course of events. *Science* 137:762–763.

Steinberg, M. (1964). *Cell Membranes in Development.* San Diego: Academic Press.

Taber, L. A., and Eggers, D. W. (1996). Theoretical study of stress-modulated growth in the aorta. *J. Theor. Biol.* 180:343–357.

Taber, L. A., and Perucchio, R. (2001). Modeling heart development. *J. Elasticity* 61:165–197.

Taber, L. A., Keller, B. B., and Clard, E. B. (1992). Cardiac mechanics in the stage-16 chick embryo. *J. Biomech. Eng.* 114:427–434.

Taber, L. A. (1995). Biomechanics of growth, remodelling, and morphognesis. *Appl. Mech. Rev.* 48:487–543.

Taber, L. A. (1998a). Biomechanical growth laws for muscle tissue. *J. Theor. Biol.* 193:201–213.

Taber, L. A. (1998b). Mechanical aspects of heart development. *Prog. Biophys. Molec. Biol.* 69:225–254.

Taber, L. A. (1998c). A model for aortic growth based on fluid shear and fiber stress. *J. Biomech. Eng.* 120:348–354.

Vaishnav, R. N., and Vossoughi, J. (1983). Estimation of residual strains in aortic segments. In Hall, C. W., ed., *Recent Developments in Biomedical Engineering.* New York: Pergamon Press. 330–333.

Vaishnav, R. N., and Vossoughi, J. (1987). Residual stress and strain in aortic segments. *J. Biomech.* 20:235–239.

Walker, D. C., Southgate, J. S., Hill, G., Holcombe, M., Hose, D. R., Wood, S. M., Macneil, S., and Smallwood, R. H. (2004). The epitheliome: modeling the social behavior of cells. *Biosystems* 76:89–100.

Wang, L., Fritton, S. P., Cowin, S. C., and Weinbaum, S. (1999). Fluid pressure relaxation mechanisms in osteonal bone specimens: modeling of an oscillatory bending experiment. *J. Biomech.* 32:663–672.

Wang, L., Cowin, S. C., Weinbaum, S., and Fritton, S. P. (2000). Modeling tracer transport in an osteon under cyclic loading. *Ann. Biomed. Eng.* 28:1200–1209.

Wang, L., Fritton, S. P., Weinbaum, S., and Cowin, S. C. (2003). On bone adaptation due to venous stasis. *J. Biomech.* 36:1439–1451.

Weinbaum, S., Cowin, S. C., and Zeng, Y. (1994). A model for the excitation of ostecytes by mechanical loading-induced bone fluid shear stresses. *J. Biomech.* 27:339–360.

Xie, J. P., Liu, S. Q., Yang, R. F., and Fung, Y. C. (1991). The zero-stress state of rat veins and vena cava. *J. Biomech. Eng.* 113:36–41.

You, L., Cowin, S. C., Schaffler, M., and Weinbaum, S. (2001). A model for strain amplification in the actin cytoskeleton of osteocytes due to fluid drag on pericellular matrix. *J. Biomech.* 34:1375–1386.

Zhang, D., Cowin, S. C., and Weinbaum, S. (1998). Electrical signal transmission in a bone cell network: The influence of a discrete gap junction. *Ann. Biomed. Eng.* 26:644–659.

Growth in Soft Biological Tissue and Residual Stress Development

A. Guillou, R.W. Ogden

University of Glasgow, Department of Mathematics, UK
ag@maths.gla.ac.uk, rwo@maths.gla.ac.uk

In this paper we summarize a general continuum mechanical theory that takes account of growth in materials capable of large deformations, with particular reference to soft biological tissues such as arterial walls. The material response is elastic at any stage of the growth and the stress depends on the growth. Part of the objective is to relate the growth to the development of residual stress. The theory is applied to a circular cylindrical tube subject to extension and internal pressure when the wall thickness changes as a result of persistent high pressure, and the results are illustrated for a simple example.

1 Introduction

The ability of living tissues to grow and remodel in response to altered loads has long been considered an important characteristic in biomechanics. However, in the last few decades most studies have focussed on hard tissues such as bones. It is important to emphasize that a major difference between hard and soft tissues lies in the magnitudes of the deformations that they can sustain. While infinitesimal deformation theories can reasonably be applied to hard tissues, the deformation of soft tissues requires a nonlinear theory that allows for large deformations. One of the first attempts to formulate a mathematical description of growth in soft tissues was that of Skalak et al. (1982). The first formal framework for soft tissue that accounts for volumetric growth only appeared with the work of Rodriguez et al. (1994). Since then, amongst others, Taber (1995), Rachev et al. (1998) and Rachev (2003) have contributed to our understanding of this growth behaviour. See also the paper by Lubarda and Hoger (2002) and references contained therein. In the present paper we adopt the definition used by many researchers that considers growth as a change in mass and geometry. This distinguishes growth from remodelling, which is often regarded as a rearrangement of the microstructure in the tissue and is not considered here.

This paper is divided into two main sections. In Section 2, we focus on a general formulation of the mechanics of growth. Balance equations of mass, linear and angular momentum and energy incorporating volumetric and surface sources are reviewed. General constitutive equations that include the effect of residual stresses are also described. The theory is illustrated with a simple example in Section 3, in which the problem of extension and inflation of an artery is considered. The artery is modelled as a residually stressed mechanically incompressible circular cylindrical tube. Residual stresses prior to growth are calculated on the assumption that the circumferential stress is uniform at typical physiological pressure. Growth in the thickness of the artery wall due to increased blood pressure is then discussed briefly along with the evolution of residual stress. Because of limitations on space detailed derivations of the equations and the results for the specific problem are not given here but will be the subject of a separate paper.

2 Mechanics of Growth

In modelling the mechanical behaviour of soft tissues, two rather different approaches present themselves. The first approach consists of 'reassembling' knowledge of the mechanics of the constituents of the microstructure of the tissue, together with knowledge of their interactions, so as to develop a theoretical framework that would describe the mechanics of the whole tissue. Mixture theories are based on this concept. A second possible approach is based on the macroscopic description of the tissue as a whole, and, in particular, is concerned with how various features of the tissue evolve under changes in the mechanical environment. Such models, referred to as phenomenological models, are based on the notion of cause and effect and have the advantage of not requiring detailed knowledge of the internal structure of the tissue. A phenomenological approach is adopted here. The mechanical balance equations for a growing material are examined for a single-component material rather than for a mixture of different constituents. For detailed background of the relevant framework for the mechanics of open systems and for discussion of balance equations involving a single-component material, we refer to, for example, Holzapfel (2000), Epstein and Maugin (2000) and Kuhl and Steinmann (2003).

2.1 Basic Kinematics

Let a material point be labelled by its position vector $\mathbf{X}$ in some fixed reference configuration, denoted $\mathcal{B}_0$, and let $\mathbf{x}$ be its position in the current configuration, denoted $\mathcal{B}_t$, at time t. The configuration $\mathcal{B}_t$ corresponds to the loaded grown state of the body. Its geometry evolves with time, and the density at $\mathbf{x}$ at time t in this configuration is denoted ρ. If the body is unloaded from the current configuration and allowed to relax it will take up an unloaded configuration, which we denote by $\mathcal{B}_r$. In general, the configuration $\mathcal{B}_r$ will be residually stressed as a result of the growth that has taken place up to time t (hence the subscript r). Continued growth (in the loaded configuration) gives rise to additional residual stress, so that $\mathcal{B}_r$ is expected to evolve during the growth process. The density in $\mathcal{B}_r$ is denoted ρ_r. The relaxation from $\mathcal{B}_t$ to $\mathcal{B}_r$ is assumed to be isochoric and to occur on a short time scale compared with the time scale of growth, so that this response can be regarded as incompressible and elastic. Thus, we have $\rho = \rho_r$.

The reference configuration $\mathcal{B}_0$ is chosen to be the initial unloaded (but possibly residually stressed) configuration. We consider that growth is stimulated when the mechanical loads on the body are altered in such a way as to disturb the homeostatic state of stress. The density at $\mathbf{X}$ in $\mathcal{B}_0$ is denoted ρ_0. We assume that material points of the body at time t existed at the beginning of the growth process so that they can be mapped contiguously from the reference configuration to the current configuration.

The deformation from $\mathcal{B}_0$ to $\mathcal{B}_t$ is described by a one-to-one mapping χ such that $\chi(\mathbf{X}, t) = \mathbf{x}$. The deformation gradient from $\mathcal{B}_0$ to $\mathcal{B}_t$ is then defined by $\mathbf{F} = \operatorname{Grad}\chi$. We adopt the standard notation and convention $J = \det \mathbf{F} > 0$. Let the deformation gradient from the (evolving) relaxed configuration $\mathcal{B}_r$ to $\mathcal{B}_t$ be represented by $\mathbf{F}_e$ (see Fig. 1). It will be convenient for later reference to denote by $\chi^{(r)}$ the specialization of χ to $\mathcal{B}_r$ and to write $\mathbf{X}^{(r)} = \chi^{(r)}(\mathbf{X}, t)$. The deformation gradient $\mathbf{F}_r = \operatorname{Grad}\chi^{(r)}$ from $\mathcal{B}_0$ to $\mathcal{B}_r$ then contains information about the growth as well as information on the 'rearrangement' of the material that ensures geometric compatibility of the growth and is responsible for the appearance of additional residual stresses. Note that $\mathbf{F} = \mathbf{F}_e\mathbf{F}_r$. We emphasize that since $\mathcal{B}_r$ is a residually-stressed

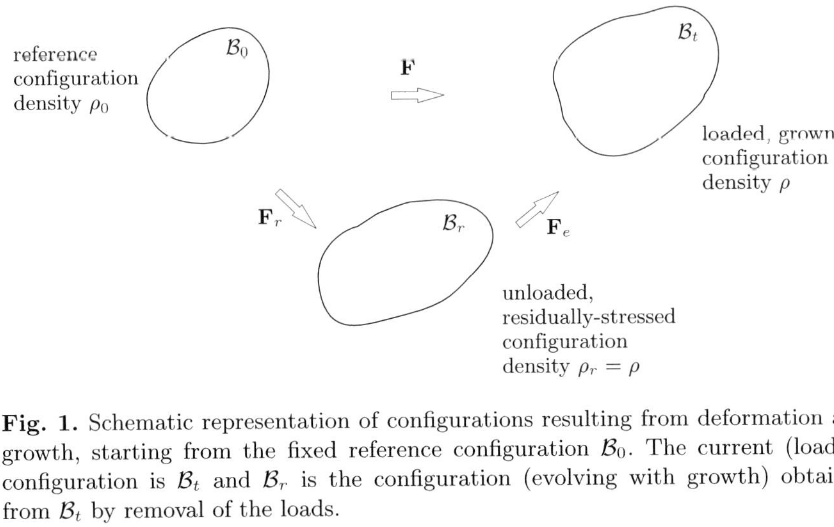

Fig. 1. Schematic representation of configurations resulting from deformation and growth, starting from the fixed reference configuration $\mathcal{B}_0$. The current (loaded) configuration is $\mathcal{B}_t$ and $\mathcal{B}_r$ is the configuration (evolving with growth) obtained from $\mathcal{B}_t$ by removal of the loads.

configuration each of $\mathbf{F}_r$, $\mathbf{F}_e$ and $\mathbf{F}$ is a deformation gradient and no 'fictitious' stress-free configuration is assumed. The approach adopted here is therefore different from that initiated by Rodriguez et al. (1994), which involves a (symmetric) 'growth' tensor that is not in general the gradient of a deformation.

Let $J_r = \det \mathbf{F}_r$. Then, since $\mathbf{F}_e$ is isochoric, we have $J = J_r$. Let $\dot{\mathbf{F}} = \mathbf{LF}$, where $\mathbf{L}$ is the velocity gradient grad $\mathbf{v}$, $\mathbf{v}$ being the material velocity in $\mathcal{B}_t$. Similarly, let $\dot{\mathbf{F}}_r = \mathbf{L}_r\mathbf{F}_r, \dot{\mathbf{F}}_e = \mathbf{L}_e\mathbf{F}_e$. Here and subsequently a superposed dot signifies the material time derivative. Then,

$$\mathbf{L} = \mathbf{L}_e + \mathbf{F}_e\mathbf{L}_r\mathbf{F}_e^{-1}. \tag{1}$$

By incompressibility we have $\mathrm{tr}\,\mathbf{L}_e = 0$, and hence

$$\mathrm{tr}\,\mathbf{L} = \mathrm{tr}\,\mathbf{L}_r, \tag{2}$$

where tr denotes the trace of a second-order tensor.

2.2 Mass Balance

We account for both volumetric sources of mass and surface flux of mass. Let μ denote the bulk mass source per unit mass in $\mathcal{B}_t$ and $\mathbf{m}$ the surface mass influx per unit area of $\partial\mathcal{B}_t$, the boundary of $\mathcal{B}_t$. Then, the Eulerian local form of the mass balance for an open system has the form

$$\dot{\rho} + \rho\,\mathrm{div}\,\mathbf{v} = \rho\mu - \mathrm{div}\,\mathbf{m}, \quad \dot{\rho} + \rho\,\mathrm{tr}\,\mathbf{L}_r = \rho\mu - \mathrm{div}\,\mathbf{m}, \tag{3}$$

the second of which having been obtained from the first by use of eq. (2). Note that in some treatments, such as Epstein and Maugin (2000), the opposite sign convention is used for $\mathbf{m}$, while often the surface flux term is omitted, as in the work of Hoger and co-workers (see, for example, Lubarda and Hoger (2002)). Of course, in the absence of mass sources eq. (3) reduces to the standard continuum conservation of mass equation.

2.3 Momentum Balance Equations

When growth is taken into account the standard continuum linear momentum balance equation is modified. As with mass balance we can consider that there are two contributions to linear momentum production, one volumetric and one surface contribution. Let $\mathbf{b}$ denote the bulk source of momentum production, which incorporates body forces, and also possibly other volumetric sources. The surface source of momentum production consists of a contribution due to surface tractions, and may also include other surface sources. Indeed, in Epstein and Maugin (2000), for example, such terms are included explicitly. Here, for simplicity, it is convenient to combine such sources into one volume and one surface term. The integral form of the linear momentum balance equation is then

$$\frac{\mathrm{d}}{\mathrm{d}t}\int_{\mathcal{B}_t}\rho\mathbf{v}\,\mathrm{d}v=\int_{\mathcal{B}_t}\rho\mathbf{b}\,\mathrm{d}v+\int_{\partial\mathcal{B}_t}\boldsymbol{\sigma}^T\mathbf{n}\,\mathrm{d}a, \tag{4}$$

where $\boldsymbol{\sigma}$ is a stress tensor that embodies all surface sources and may be referred to as the Cauchy stress tensor by analogy with standard continuum mechanics. On use of the mass balance equation and a transport theorem, several equivalent formulations for the Eulerian local form of the linear momentum balance may be obtained. We mention just one of them here, namely

$$\rho\dot{\mathbf{v}}=\rho\bar{\mathbf{b}}+\mathrm{div}\,\bar{\boldsymbol{\sigma}}, \tag{5}$$

where we have defined $\bar{\mathbf{b}}$ and, for consistency of notation, $\bar{\boldsymbol{\sigma}}$ by

$$\rho\bar{\mathbf{b}}=\rho\mathbf{b}-(\rho\mu-\mathrm{div}\,\mathbf{m})\mathbf{v},\quad \bar{\boldsymbol{\sigma}}=\boldsymbol{\sigma}. \tag{6}$$

We refer to $\bar{\mathbf{b}}$ as the 'effective' body force and $\bar{\boldsymbol{\sigma}}$ as the 'effective' Cauchy stress. An alternative form of the equation can be found in, for example, Kuhl and Steinmann (2003).

On the assumption that there are no sources of angular momentum other than those generated by $\mathbf{b}$ and $\boldsymbol{\sigma}$, the counterpart of eq. (4) for the angular momentum balance leads, by standard manipulations, to the conclusion that $\bar{\boldsymbol{\sigma}}$ is symmetric, i.e. $\bar{\boldsymbol{\sigma}}^T=\bar{\boldsymbol{\sigma}}$, just as in standard continuum mechanics without growth. Although in different notation, this is equivalent to the result of Epstein and Maugin (2000).

2.4 Energy Balance and Constitutive Equations

Let u denote the internal energy per unit mass, γ the volumetric growth energy supply (per unit mass), $\mathbf{g}$ the growth energy surface influx vector (per unit area of $\partial\mathcal{B}_t$), h the heat supply (per unit mass) and $\mathbf{q}$ the heat influx vector (per unit area of $\partial\mathcal{B}_t$). Then, by applying standard arguments, slightly modified, the local form of the energy balance equation can be written

$$\rho\,\dot{u}=\mathrm{tr}\,(\bar{\boldsymbol{\sigma}}\mathbf{L})+\rho\,\bar{\gamma}-\mathrm{div}\,\bar{\mathbf{g}}+\rho\,\bar{h}-\mathrm{div}\,\bar{\mathbf{q}}, \tag{7}$$

where the overbars signify 'effective' quantities in a sense similar to that used for $\bar{\mathbf{b}}$ and $\bar{\boldsymbol{\sigma}}$. If thermal terms are ignored then $\bar{h}=0$ and $\bar{\mathbf{q}}=\mathbf{0}$. For the moment, however, we retain the thermal terms.

We now consider the free energy ψ per unit mass, defined by $\psi=u-\theta\eta$, where $\theta\,(>0)$ is the absolute temperature and η denotes the entropy per unit mass. Equation (7) may then be rewritten as

$$\rho\dot{\psi}=\mathrm{tr}\,(\bar{\boldsymbol{\sigma}}\mathbf{L})+\rho\,\bar{\gamma}-\mathrm{div}\,\bar{\mathbf{g}}+\rho\,\bar{h}-\mathrm{div}\,\bar{\mathbf{q}}-\rho\eta\dot{\theta}-\rho\theta\dot{\eta}, \tag{8}$$

where the last four terms represent the contributions from thermal effects. Next, on using a general expression for the rate of change of entropy in $\mathcal{B}_t$ and the second law of thermodynamics, we deduce that

$$\rho\dot{\psi} - \mathrm{tr}\,(\bar{\boldsymbol{\sigma}}\mathbf{L}) - \rho\,\bar{\gamma} + \mathrm{div}\,\bar{\mathbf{g}} + \text{thermal terms} \leq 0. \tag{9}$$

Again, for brevity, details of the derivation are omitted, but we refer to the analysis in, for example, Kuhl and Steinmann (2003) in the present context. The inequality (9) is identical in structure to that arising in the absence of growth, except that the terms in $\bar{\gamma}$ and $\bar{\mathbf{g}}$ are included. Note, however, that it differs from corresponding inequalities in, for example, Epstein and Maugin (2000) and Kuhl and Steinmann (2003), the reasons for which will be discussed elsewhere.

The inequality (9) is quite general and does not include any indication of the variables on which ψ depends. At this point, in order to obtain an explicit constitutive law, we need to specify such variables. There are many possible ways in which this can be done. When the material is deformed in the absence of growth then we assume that its mechanical response is elastic and the stored elastic energy is captured by ψ, which is then essentially a strain-energy function. During growth, on the other hand, externally supplied energy is both expended (non-recoverable) in order to generate the growth and stored (as elastic energy, which is recoverable). The form of ψ should reflect both the recoverable and non-recoverable parts of the energy. In several treatments of the mechanics of growth the existence of a (fictitious) stress-free configuration is assumed and ψ is taken to depend on what in the present notation can be written $\mathbf{F}_e$ (although its interpretation is somewhat different here). This requires that the (evolving) local stress-free configuration is known. Here we do not make this assumption but refer always to the (fixed) initial configuration $\mathcal{B}_0$ and take ψ to depend on $\mathbf{F}$, the total deformation gradient, which includes both elastic and growth contributions. Additionally, we suppose that ψ depends separately on the density ρ and its gradient (this is analogous to thermoelasticity, where initially the temperature gradient is included along with the temperature, but then excluded by application of the Clausius-Duhem inequality). It must also depend on any structural features of $\mathcal{B}_0$, such as preferred directions (associated with collagen fibres, for example). We embody such information by means of the quantity $\mathbf{A}_0$, which may be a vector, a tensor or a collection of such objects. It is not necessary to make this more explicit here. Furthermore, ψ will in general depend on any residual stress within $\mathcal{B}_0$, which we denote by $\boldsymbol{\sigma}_0$ (which may be zero). The symmetry of the material, as generated by both $\mathbf{A}_0$ and $\boldsymbol{\sigma}_0$, must also be reflected in the form of ψ.

For the growth phase it is convenient to use the 'pullback' (or Lagrangian) version of ρ rather than ρ itself. This is denoted ρ_L and defined by $\rho_L = \rho J$ so that the mass balance equation (3) becomes

$$\dot{\rho}_L = J(\rho\mu - \mathrm{div}\,\mathbf{m}). \tag{10}$$

The spatial gradient of ρ_L is denoted $\boldsymbol{\delta} \equiv \text{Grad}\,\rho_L$. Thus, for the considered combination of elasticity and growth, we take ψ to have the dependence

$$\psi = \psi(\mathbf{F}, \rho_L, \boldsymbol{\delta}, \mathbf{A}_0, \boldsymbol{\sigma}_0, \theta), \tag{11}$$

the temperature θ being included for completeness at this stage although we shall not deal with temperature effects subsequently. Since we are considering volumetric growth here and no structural changes (no remodelling) the structural quantities $\mathbf{A}_0$ change only through the effect of $\mathbf{F}$ and no separate evolution law is required for them (although in a more general treatment such a law could be introduced). It follows from eq. (11) and the inequality (9) that, with $\dot{\boldsymbol{\delta}} = \text{Grad}\,\dot{\rho}_L$,

$$\text{tr}\left[\left(\rho\mathbf{F}\frac{\partial\psi}{\partial\mathbf{F}} - \bar{\boldsymbol{\sigma}}\right)\mathbf{L}\right] + \rho\frac{\partial\psi}{\partial\rho_L}\dot{\rho}_L + \rho\frac{\partial\psi}{\partial\boldsymbol{\delta}}\cdot\dot{\boldsymbol{\delta}} - \rho\bar{\gamma} + \text{div}\,\bar{\mathbf{g}}$$
$$+\,\text{thermal terms} \leq 0. \tag{12}$$

For the case of incompressible elastic response the constraint $\text{tr}\,\mathbf{L}_e = 0$ holds, but $\mathbf{L}$ may be treated as arbitrary since $\mathbf{L}_r$, being not on its own constrained by specification of μ and $\mathbf{m}$ in (3), is arbitrary. Thus, during growth, the inequality (12) must hold, in particular, for arbitrary $\mathbf{L}$. This leads to the equation

$$\bar{\boldsymbol{\sigma}} = \rho\mathbf{F}\frac{\partial\psi}{\partial\mathbf{F}}. \tag{13}$$

Note that the expression for the stress $\bar{\boldsymbol{\sigma}}$ does not involve a Lagrange multiplier normally associated with the constraint of incompressibility since growth-induced volume changes are included in the response. In structure, equation (13) has the standard form of conventional (thermo-)elasticity in the absence of growth. We emphasize, however, that the stress depends on the growth and, for suitable specifications of constitutive laws for μ and $\mathbf{m}$, the growth will depend on the stress, so the present formulation provides a two-way coupling. Note that $\dot{\rho}_L$ is not arbitrary since it appears in the mass balance equation (10) and it therefore depends on the specification of μ and $\mathbf{m}$. This leaves the residual (dissipation) inequality

$$\rho\frac{\partial\psi}{\partial\rho_L}\dot{\rho}_L + \rho\frac{\partial\psi}{\partial\boldsymbol{\delta}}\cdot\dot{\boldsymbol{\delta}} - \rho\bar{\gamma} + \text{div}\,\bar{\mathbf{g}} + \text{thermal terms} \leq 0 \tag{14}$$

to be satisfied, which has to be enforced for any choice of constitutive laws for μ and $\mathbf{m}$ that lead to $\dot{\rho}_L$. Analysis of the residual inequality (14) depends on detailed specification of $\bar{\gamma}$ and $\bar{\mathbf{g}}$, which will be discussed elsewhere and need not be considered further here.

2.5 Growth/Elastic Boundary-value Problem

In an equilibrium state in which no growth occurs (a homeostatic state) we can consider $\mathcal{B}_t$ as a configuration that relaxes elastically to $\mathcal{B}_0$ on removal of the applied loads. Suppose now that the boundary conditions in $\mathcal{B}_t$ are changed, resulting in an (instantaneous) elastic change in the geometry of $\mathcal{B}_t$, a change in the stress distribution and a disturbance of the homeostatic state. This instantaneous response, since it involves no growth, is incompressible. The resulting equilibrium of $\mathcal{B}_t$, with stress $\bar{\boldsymbol{\sigma}}$, in the absence of body forces, is described by

$$\operatorname{div}\bar{\boldsymbol{\sigma}} = \mathbf{0}, \quad \bar{\boldsymbol{\sigma}} = \rho_0 \mathbf{F}\frac{\partial \psi_0}{\partial \mathbf{F}} - p_0 \mathbf{I}, \tag{15}$$

where $\psi_0(\mathbf{F}, \mathbf{A}_0, \boldsymbol{\sigma}_0)$ is the specialization of ψ for the absence of growth, $\mathbf{F}$ has only an elastic contribution, $\det \mathbf{F} = 1$, p_0 is the associated Lagrange multiplier and $\mathbf{I}$ is the identity tensor. Here and henceforth thermal terms are omitted. This new stress distribution stimulates the (slow) process of growth and can be considered as an initial condition for the growth.

If we neglect inertia terms (as is appropriate for slow growth), then the equation of motion (5) reduces to

$$\operatorname{div}\bar{\boldsymbol{\sigma}} + \rho\bar{\mathbf{b}} = \mathbf{0}, \tag{16}$$

with $\bar{\boldsymbol{\sigma}}$ given by (13). We note that if there are no *mechanical* body forces then it can be shown that $\rho\bar{\mathbf{b}} = (\operatorname{div}\mathbf{m})\mathbf{v}$, where $\mathbf{v}$ is just the (slow) growth velocity.

During growth equations (10) and (16), together with eq. (13) and constitutive laws for μ and $\mathbf{m}$ (not yet specified), form a coupled system of equations for the determination of $\mathbf{x} = \boldsymbol{\chi}(\mathbf{X}, t)$ and $\rho_L(\mathbf{X}, t)$ subject to given boundary conditions. We consider that growth continues until a new homeostatic equilibrium state is reached, after which the material response is purely elastic unless some further change in the mechanical environment is initiated. At any stage of the growth, however, since growth is a slow process we can consider it to be suspended on the time scale of elastic relaxation if the loads are removed and the body relaxes to the configuration $\mathcal{B}_r$, in which $\mathbf{F}_e = \mathbf{I}$. The (equilibrium) stress in $\mathcal{B}_r$ is given by

$$\operatorname{div}\boldsymbol{\sigma}^{(r)} = \mathbf{0}, \quad \boldsymbol{\sigma}^{(r)} = \rho\mathbf{F}_r\frac{\partial \psi}{\partial \mathbf{F}}(\mathbf{F}_r, \rho_L, \boldsymbol{\delta}, \mathbf{A}_0, \boldsymbol{\sigma}_0) - p^{(r)}\mathbf{I}, \tag{17}$$

where ρ_L and $\boldsymbol{\delta}$ are now known from solution of the growth problem, and the boundary of $\mathcal{B}_r$ is free of traction.

Equations (17) determine $\mathbf{X}^{(r)} = \boldsymbol{\chi}^{(r)}(\mathbf{X}, t)$, with $\mathbf{F}_r = \operatorname{Grad}\mathbf{X}^{(r)}$, and hence $\mathcal{B}_r$, at any stage of the growth process. Since $\mathbf{F}$, and hence $J \equiv \det\mathbf{F}$ is now determined the relaxation response is subject to the constraint $J_r \equiv \det\mathbf{F}_r = J$, and hence the inclusion of the Lagrange multiplier $p^{(r)}$.

Once growth is terminated and a homeostatic state restored the resulting equilibrium solution is determined from eq. (16) with $\bar{\mathbf{b}} = \mathbf{0}$ and $\dot{\rho}_L = 0$.

3 Extension and Inflation of an Artery

In this section we apply the theory of Section 2 to a prototype problem in which the artery is modelled as an incompressible elastic circular cylindrical tube that is residually stressed in some initial (unloaded) reference configuration. We assume that there is a homeostatic state in which no growth is occurring and that this state is associated with an internal pressure P_0, the typical physiological pressure, and an axial extension of the tube corresponding to a uniform axial stretch, denoted λ_z. When the tube is unloaded it relaxes back elastically to the configuration $\mathcal{B}_0$ and the elastic response relative to $\mathcal{B}_0$ is orthotropic. Now suppose that the pressure is increased to P on a time scale that is fast compared with the time scale of growth so that the initial response resulting from the pressure change can be treated as elastic and instantaneous. The homeostatic stress distribution within the wall of the tube is therefore disturbed, and this leads to a biological response that induces growth. The resulting configuration at time t after the initiation of growth is $\mathcal{B}_t$. If at any time during growth the loads are removed then the material relaxes elastically to an unloaded residually stressed configuration $\mathcal{B}_r$, which is in general different from $\mathcal{B}_0$, and evolves as the growth proceeds. The growth continues until a new homeostatic state is established.

We consider separately the initial homeostatic state (and the initial residual stress), the growth phase and the final homeostatic state.

3.1 The Initial Homeostatic State

We describe the geometry of $\mathcal{B}_0$ in terms of cylindrical coordinates (R, Θ, Z) with $A \leq R \leq B$ and $0 \leq Z \leq L$, where A, B and L denote respectively the inner radius, the outer radius and the length of the tube. The (homeostatic) deformed configuration is represented by cylindrical coordinates (r, θ, z) with $a \leq r \leq b$ and $0 \leq z \leq l$. The residual principal Cauchy stress components are denoted σ_{0RR}, $\sigma_{0\Theta\Theta}$ and σ_{0ZZ} in $\mathcal{B}_0$. Because of the symmetry of the problem only the radial component of the equilibrium equation is not satisfied identically, and this has the form

$$\frac{\mathrm{d}\sigma_{0RR}}{\mathrm{d}R} + \frac{1}{R}(\sigma_{0RR} - \sigma_{0\Theta\Theta}) = 0 \quad \text{in } A < R < B, \quad \sigma_{0RR} = 0 \quad \text{on } R = A, B, \tag{18}$$

wherein the associated (zero load) boundary conditions are included.

In the deformed configuration, where the principal stress components are denoted σ_{rr}, $\sigma_{\theta\theta}$ and σ_{zz}, the equilibrium equation is

$$\frac{\mathrm{d}\sigma_{rr}}{\mathrm{d}r} + \frac{1}{r}(\sigma_{rr} - \sigma_{\theta\theta}) = 0 \quad \text{in} \quad a < r < b, \tag{19}$$

and boundary conditions are

$$\sigma_{rr} = -P_0 \quad \text{on} \quad r = a, \qquad \sigma_{rr} = 0 \quad \text{on} \quad r = b. \tag{20}$$

3.2 Residual Stress in the Absence of Growth

Since, in the absence of growth, the deformation is isochoric it is defined by the equations

$$r^2 = a^2 + \lambda_z^{-1}(R^2 - A^2), \quad \theta = \Theta, \quad z = \lambda_z Z, \tag{21}$$

where λ_z is the (uniform) axial stretch. Let λ_r and λ_θ be the principal stretches associated with the radial and azimuthal directions respectively. Then $\lambda_\theta = r/R$ depends on R, while $\lambda_r = \lambda_\theta^{-1}\lambda_z^{-1}$ is given by the incompressibility condition. For the considered deformation the principal directions of stress, deformation and residual stress coincide. Moreover, we consider the material response to be cylindrically orthotropic. It is therefore unnecessary to include either structure tensors or the residual stress explicitly in the free energy function, which can therefore be regarded as a function of $\lambda_r, \lambda_\theta, \lambda_z$ only, and we write $\psi_0(\lambda_r, \lambda_\theta, \lambda_z)$ to represent this. The orthotropy ensures that, unlike for an isotropic material, ψ is not in general a symmetric function of the stretches.

It is convenient now to work in terms of two independent stretches, λ_θ and λ_z, and to introduce the notation $\hat{\psi}_0$ to represent the free energy in this case. This is defined by

$$\hat{\psi}_0(\lambda_\theta, \lambda_z) = \psi_0\left(\lambda_\theta^{-1}\lambda_z^{-1}, \lambda_\theta, \lambda_z\right). \tag{22}$$

Since ψ_0 and the stretches are measured relative to $\mathcal{B}_0$ we may use $\hat{\psi}_0$ to calculate the stress differences in the deformed configuration in the form

$$\sigma_{\theta\theta} - \sigma_{rr} = \rho\lambda_\theta \frac{\partial\hat{\psi}_0}{\partial\lambda_\theta} + \sigma_{0\Theta\Theta} - \sigma_{0RR},$$

$$\sigma_{zz} - \sigma_{rr} = \rho\lambda_z \frac{\partial\hat{\psi}_0}{\partial\lambda_z} + \sigma_{0ZZ} - \sigma_{0RR}, \tag{23}$$

which reduce to the residual stress differences in $\mathcal{B}_0$.

The residual stress distribution throughout the cylindrical tube can be calculated by following the work of Ogden and Schulze-Bauer (2000) and Ogden (2003a,b), where a single-layer tube was considered. As discussed in Ogden and Schulze-Bauer (2000), the literature indicates that an artery in a healthy physiological state maintains a uniform circumferential stress in each of the media and adventitia. By assuming the circumferential stress to be uniform in the current configuration the equilibrium equation (19) can then be integrated and expressions for the pressure P_0 and the stress distribution can be found explicitly. This is the case whether one or more layers are included in the analysis. Here, however, for simplicity we restrict attention to the case of a single layer and we assume that $\sigma_{\theta\theta} = \sigma_0$, where σ_0 is a constant. It follows that

$$a_0 P_0 = \sigma_0(b_0 - a_0), \quad r\sigma_{rr} = -\sigma_0(b_0 - r). \tag{24}$$

Here we have attached a subscript 0 to each of a and b for this special case, so that $a_0 \leq r \leq b_0$.

Furthermore, by using eqs (18), (23) and (24), we obtain an expression for the radial residual stress, namely

$$\sigma_{0RR} = b_0\sigma_0 \int_A^R \frac{\mathrm{d}R}{Rr} - \int_A^R \rho_0 \lambda_\theta \frac{\partial \hat{\psi}_0}{\partial \lambda_\theta} \frac{\mathrm{d}R}{R}. \tag{25}$$

Note that $\sigma_{0\Theta\Theta}$ can then be found by substituting eq. (25) into eq. (18). Explicit expressions are given by Ogden and Schulze-Bauer (2000) and we do not repeat them here.

To illustrate the residual stress distribution we use the free energy proposed by Holzapfel et al. (2000) which, in a general form, is given by

$$\psi_0 = \psi_{0\,\mathrm{iso}} + \psi_{0\,\mathrm{aniso}}, \tag{26}$$

where $\psi_{0\,\mathrm{iso}}$ is the isotropic contribution of the matrix material in which collagen fibres are embedded. It is the collagen fibres that give the material response its anisotropic character, and the associated anisotropic contribution to ψ_0 is denoted $\psi_{0\,\mathrm{aniso}}$. In Holzapfel et al. (2000) the isotropic part was taken as a neo-Hookean material, while the anisotropic part was represented in terms of exponential functions. For full details we refer to the cited paper. At low pressures, the main contribution to ψ_0 is provided by $\psi_{0\,\mathrm{iso}}$, whereas at high pressure $\psi_{0\,\mathrm{aniso}}$ becomes predominant.

Figure 2 shows the resulting radial and circumferential residual stress distributions, plotted as functions of the dimensionless radius R/A, where we have taken $B/A = 1.2$, $\lambda_z = 1.2$ and $P_0 = 13.33\,\mathrm{kPa}$ (which corresponds to the homeostatic pressure, i.e. the mean arterial pressure in a mature healthy human artery). The results are very similar to those obtained by Ogden (2003a) for a different energy function, and also to results based on an 'opening angle' analysis, as discussed in the latter paper. We note that the circumferential stress is compressive in the inner part of the wall and tensile in the outer part, while the radial stress is very small and vanishes at both boundaries. The corresponding axial residual stress is not shown here since we have adopted plane strain conditions and the resulting axial 'residual' stress depends on the radius and not the axial coordinate Z. It is therefore not possible under the considered restrictions to ensure that the axial load vanishes locally on the ends of the tube.

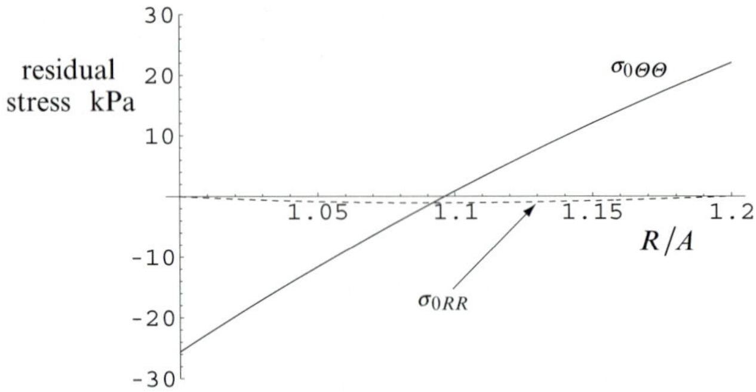

Fig. 2. Residual stress distribution: radial stress (dashed curve); circumferential stress (continuous curve).

3.3 Evolution of Residual Stresses with Growth

Let us now consider that the internal pressure is increased to P, as indicated earlier, i.e. the artery experiences hypertensive pressure, so that growth is initiated. At the start of the growth process, $\mathcal{B}_r$ is identical to the unloaded configuration $\mathcal{B}_0$ associated with no growth and has the residual stress distribution calculated in the last section. As time passes and growth takes place $\mathcal{B}_r$ evolves, and changes in the residual stress distribution are generated.

During growth the relevant governing equation is (16). Let us suppose for simplicity of illustration that $\mathbf{m} = \mathbf{0}$. Then, in its specialization to the considered radially symmetric situation equation (16), on dropping the redundant overbar, has the form (19). The boundary conditions are again (20) except that P_0 is replaced by P. Since we have set $\mathbf{m} = \mathbf{0}$ the mass balance equation in the forms (3) and (10) together reduce to

$$\dot{\rho}_L \equiv J(\dot{\rho} + \rho\,\mathrm{tr}\,\mathbf{L}_r) = \mu\rho_L. \tag{27}$$

Because of the symmetry the radial part of the deformation has the form $r = r(R)$, λ_z is uniform and

$$\lambda_r = r'(R), \quad \lambda_\theta = r/R, \tag{28}$$

where the prime indicates differentiation with respect to R. The free energy is specialized to $\psi(\lambda_r, \lambda_\theta, \lambda_z, \rho_L, \rho_L')$ and the (principal) stresses are given by

$$\sigma_{rr} = \rho\lambda_r\frac{\partial\psi}{\partial\lambda_r}, \quad \sigma_{\theta\theta} = \rho\lambda_\theta\frac{\partial\psi}{\partial\lambda_\theta}, \quad \sigma_{zz} = \rho\lambda_z\frac{\partial\psi}{\partial\lambda_z}, \tag{29}$$

but here the residual stresses are not included separately since they cannot be calculated until after the equations for the total deformation are solved.

The equilibrium equation and boundary conditions together are

$$\frac{\mathrm{d}\sigma_{rr}}{\mathrm{d}r} + \frac{1}{r}(\sigma_{rr} - \sigma_{\theta\theta}) = 0, \quad \sigma_{rr} = -P \text{ on } r = a, \quad \sigma_{rr} = 0 \text{ on } r = b. \tag{30}$$

Combined with (29) and (28), this provides an equation for determination of $r(R)$ subject to the given boundary conditions.

Note that ψ includes dependence on the growth, directly through ρ_L and ρ'_L and indirectly through the total stretches $\lambda_r, \lambda_\theta$ and λ_z. We write the specialization of the decomposition $\mathbf{F} = \mathbf{F}_e\mathbf{F}_r$ for the present situation as

$$\lambda_r = \lambda_r^{(e)}\lambda_r^{(r)}, \quad \lambda_\theta = \lambda_\theta^{(e)}\lambda_\theta^{(r)}, \quad \lambda_z = \lambda_z^{(e)}\lambda_z^{(r)}, \tag{31}$$

where the superscript $^{(e)}$ signifies the elastic part of the deformation, which satisfies the incompressibility condition

$$\lambda_r^{(e)}\lambda_\theta^{(e)}\lambda_z^{(e)} = 1. \tag{32}$$

The residual (growth) part of the deformation is indicated by the superscript $^{(r)}$.

Suppose that R_r and Z_r denote the radial and axial coordinates in the configuration $\mathcal{B}_r$ and that $A_r \le R_r \le B_r$ and $0 \le Z_r \le L_r$. Then, the elastic part of the deformation (from $\mathcal{B}_r$ to $\mathcal{B}_t$) is described by the equations

$$r^2 = a^2 + \lambda_z^{(e)^{-1}}(R_r^2 - A_r^2), \quad z = \lambda_z^{(e)} Z_r. \tag{33}$$

Given that $r(R)$ is determined from solution of the growth equations, eq. (33) determines R_r as a function of A_r, which is itself determined as follows.

At any stage of the growth the equilibrium equation to be satisfied in the relaxed configuration $\mathcal{B}_r$ and the associated boundary conditions are

$$\frac{\mathrm{d}\sigma_{RR}^{(r)}}{\mathrm{d}R_r} + \frac{1}{R_r}(\sigma_{RR}^{(r)} - \sigma_{\Theta\Theta}^{(r)}) = 0 \quad \text{in} \quad A_r \le R_r \le B_r,$$

$$\sigma_{RR}^{(r)} = 0 \quad \text{on} \quad R_r = A_r, B_r, \tag{34}$$

where $\sigma_{RR}^{(r)}$ and $\sigma_{\Theta\Theta}^{(r)}$ denote the radial and circumferential residual stresses in $\mathcal{B}_r$ at time t of the growth process. The specialization of eq. (17) to the present situation, with $\hat{\psi}(\lambda_\theta^{(r)}, \lambda_z^{(r)}, \rho_L, \rho'_L) = \psi(\lambda_r^{(r)}, \lambda_\theta^{(r)}, \lambda_z^{(r)}, \rho_L, \rho'_L)$ is

$$\sigma_{\Theta\Theta}^{(r)} - \sigma_{RR}^{(r)} = \rho\lambda_\theta^{(r)}\frac{\partial\hat{\psi}}{\partial\lambda_\theta^{(r)}}, \tag{35}$$

and integration of eq. (34) subject to the boundary conditions determines A_r from

$$\int_{A_r}^{B_r} \rho\lambda_\theta^{(r)}\frac{\partial\hat{\psi}}{\partial\lambda_\theta^{(r)}}\frac{\mathrm{d}R_r}{R_r} = 0. \tag{36}$$

The configuration $\mathcal{B}_r$ is then completely determined, as is the cross-section residual stress distribution by use of eq. (34). This provides a programme for the evaluation of growth and the resulting development of residual stress in the considered problem.

The driving process for growth in an artery, motivated by experimental findings, is that increased pressure affects the uniformity of the circumferential stress $\sigma_{\theta\theta}$ in the current configuration $\mathcal{B}_t$. As already noted this implies that growth occurs in such a way as to return $\sigma_{\theta\theta}$ to a uniform value. We therefore assume that during growth the circumferential stress in $\mathcal{B}_t$ has the non-uniform, time-dependent form $\sigma_{\theta\theta} = c(r,t)$. This observation provides a basis for the formulation of a growth law. Amongst others, Taber (1995) postulated a growth law based on the special case in which the circumferential stress returns to its initial value. Note that for an artery *in vivo*, the 'normal' value attributed to the uniform circumferential stress in the absence of growth depends on the age and health of the subject, thus suggesting that the final uniform value of the circumferential stress may not be the same as that before growth.

A growth law is provided by specifying μ, or equivalently $\dot{\rho}_L$. This then gives the combination $\dot{\rho} + \rho\,\mathrm{tr}\,\mathbf{L}_r$, which allows changes in density at fixed volume, volume expansion at fixed density, or simultaneous changes in density and volume. Thus, specification of μ does not discriminate between these two effects. An alternative is to specify $\dot{\rho}$ and $\mathrm{tr}\,\mathbf{L}_r$ separately. In fact, the model of Taber (1995) in essence specifies separately the diagonal components of $\mathbf{L}_r$ and hence $\mathrm{tr}\,\mathbf{L}_r$.

A detailed analysis based on the coupled equations discussed above is beyond the scope of the present paper, but will be examined elsewhere. Here we consider a very simplified approach in order to elicit some basic information. Taber considered the growth stretch rates to be given in terms of the difference between the circumferential stress and its homeostatic value, vanishing when the stress returns to that value at the termination of growth. Here we prescribe, instead of the growth stretch rates, the value of μ. And, furthermore, we set it to be proportional to the gradient of the circumferential stress. Thus, we write

$$\dot{\rho}_L = \mu\rho_L, \quad \mu = \alpha\frac{\partial\sigma_{\theta\theta}}{\partial r}, \tag{37}$$

where α is a constant. Thus, growth stops when $\sigma_{\theta\theta}$ becomes uniform.

For simplicity of illustration we now assume that $\sigma_{\theta\theta}$ is linear in r, so that

$$\sigma_{\theta\theta} = c(r,t) = c_0 + c_1(t)r, \tag{38}$$

where c_0 is a constant and c_1 depends on time only. This is not unreasonable as a first approximation for a wall thickness that is not too large. If we choose c_0 to have the uniform value σ_0 in the absence of growth, as given by eq. $(24)_1$, then c_1 can be interpreted as measuring the deviation in the circumferential stress distribution through the wall from its initial uniform value due to the hypertensive pressure P.

Integration of the equilibrium equation with the boundary conditions in eq. (30) gives

$$c_1 = 2P\,\frac{a}{b^2 - a^2} - 2c_0\,\frac{1}{a+b}. \tag{39}$$

Next we note that growth stops when c_1, regarded as a function of time, vanishes. From eq. (39) we deduce that this occurs when

$$\frac{b-a}{a} = \frac{P}{c_0}. \tag{40}$$

In fact, this result can be obtained on the same basis as $(24)_1$ without the above assumptions since it represents the final steady state condition with constant and uniform circumferential stress c_0.

Let $a = a_1\,(> a_0)$ be the internal radius of the artery arising from the (instantaneous) increase in pressure from P_0 to P. Another assumption based on experimental findings is now adopted. When an artery experiences high blood pressure but no change in the blood flow rate, its lumen radius is essentially unchanged (see, for example, Matsumoto and Hayashi (1994)). Thus, if the inner radius a_1 in the current configuration $\mathcal{B}_t$ remains constant during the whole growth process, equation (40), with $a = a_1$, shows how the thickness of the artery has increased as a result of the high blood pressure. In particular, in the special case $c_0 = \sigma_0$ for which the homeostatic circumferential stress is unchanged by the pressure increase, we obtain

$$b - a_1 = \frac{a_1}{a_0}\frac{P}{P_0}(b_0 - a_0). \tag{41}$$

The thickness is then proportional to the pressure and also to the inner radius, but if the uniform circumferential stress is increased as a result of growth then the right-hand side of (41) is scaled by the factor σ_0/c_0 and the thickness increase is proportionately less.

Calculation of the residual stress resulting from the growth requires specification of ψ and solution of (30). Examples of such calculations will be provided in a separate paper.

Acknowledgement. The work of Anna Guillou was supported by the Engineering and Physical Sciences Research Council (UK) and by the University of Glasgow.

References

Epstein, M., and Maugin, G. A. (2000). Thermomechanics of volumetric growth in uniform bodies. *Int. J. Plasticity* 16:951–978.

Holzapfel, G. A., Gasser, T. C., and Ogden, R. W. (2000). A new constitutive framework for arterial wall mechanics and a comparative study of material models. *J. Elasticity* 61:1–48.

Holzapfel, G. A. (2000). *Nonlinear Solid Mechanics. A Continuum Approach for Engineering.* Chichester: John Wiley & Sons.

Kuhl, E., and Steinmann, P. (2003). Mass- and volume-specific views on thermodynamics for open systems. *Proc. R. Soc. Lond. A* 459:2547–2568.

Lubarda, V. A., and Hoger, A. (2002). On the mechanics of solids with a growing mass. *Int. J. Solids Structures* 39:4627–4664.

Matsumoto, T., and Hayashi, K. (1994). Mechanical and dimensional adaptation of rat aorta to hypertension. *J. Biomech. Eng.* 116:278–283.

Ogden, R. W., and Schulze-Bauer, C. A. J. (2000). Phenomenological and structural aspects of the mechanical response of arteries. In Casey, J., and Bao, G., eds., *Mechanics in Biology.* New York: The American Society of Mechanical Engineers (ASME). AMD-Vol. 242/BED-Vol. 46, pp. 125–140.

Ogden, R. W. (2003a). Nonlinear elasticity, anisotropy, material stability and residual stresses in soft tissue. In Holzapfel, G. A., and Ogden, R. W., eds., *Biomechanics of Soft Tissue in Cardiovascular Systems*, 65–108. Wien: Springer-Verlag. CISM Courses and Lectures No. 441, International Centre for Mechanical Sciences.

Ogden, R. W. (2003b). *Nonlinear Elasticity with Application to Material Modelling.* Warsaw: Polish Academy of Sciences. Lecture Notes 6, Institute of Fundamental Technological Research.

Rachev, A., Stergiopulos, N., and Meister, J.-J. (1998). A model for geometric and mechanical adaptation of arteries to sustained hypertension. *J. Biomech. Eng.* 120:9–17.

Rachev, A. (2003). Remodeling of arteries in response to changes in their mechanical environment. In Holzapfel, G. A., and Ogden, R. W., eds., *Biomechanics of Soft Tissue in Cardiovascular Systems*, 221–271. Wien: Springer-Verlag. CISM Courses and Lectures No. 441, International Centre for Mechanical Sciences.

Rodriguez, E. K., Hoger, A., and McCulloch, A. D. (1994). Stress-dependent finite growth in soft elastic tissues. *J. Biomech.* 27:455–467.

Skalak, R., Dasgupta, G., Moss, M., Otten, E., Dullemeijer, P., and Vilmann, H. (1982). Analytical description of growth. *J. Theoret. Biol.* 94:555–577.

Taber, L. A. (1995). Biomechanics of growth, remodelling, and morphognesis. *Appl. Mech. Rev.* 48:487–543.

Characterization and Modeling of Growth and Remodeling in Tendon and Soft Tissue Constructs

E.M. Arruda,[1,2] S.C. Calve,[2] K. Garikipati,[1] K. Grosh,[1,3] H. Narayanan[1]

[1] University of Michigan, Department of Mechanical Engineering, USA
arruda@umich.edu, krishna@umich.edu, grosh@umich.edu, hnarayan@umich.edu
[2] University of Michigan, Macromolecular Science and Engineering Program, USA
scalve@umich.edu
[3] University of Michigan, Department of Biomedical Engineering, USA

Engineered tendon fibroblast and collagen-based soft tissue constructs have been developed to characterize growth and remodeling in soft tissue via *in vitro* studies of the effects of controlled mechanical interventions on collagen content and orientation. Growth and remodeling often occur simultaneously in soft tissue. The present contribution distinguishes these separate biological processes and develops the theoretical frameworks for the formation of constitutive theories of each. Remodeling is considered as a motion in material space that occurs at constant mass whereas growth is a change in concentration due to mass transport and is treated via a system open with respect to mass. Examples illustrating both biological processes are presented.

1 Introduction

Understanding growth and remodeling processes in tendon in health and disease is of great interest because of the role of tendon in transmitting forces produced by muscles during movement. Tendon is also an excellent testbed for studying collagen growth and remodeling processes in soft tissue because it is relatively acellular and avascular. Mature tendon consists largely of type I collagen. Therefore, the mechanical properties of mature tendon can be largely attributed to the parallel collagen fibrils that comprise approximately 75% (Nordin et al. (2001)) of its dry weight. Growth and remodeling experiments on soft tissue are challenging for many reasons, as the data in Fig. 1 demonstrate (Arruda et al. (2005)). The normal tibialis anterior (TA) tendon of a rat is functionally graded along its length. In addition to measuring mechanical properties locally (achieved here via local optical strain measurement), assays for collagen content, cross-linking and cell expression must all be applied locally to formulate relationships between the biochemical variables and

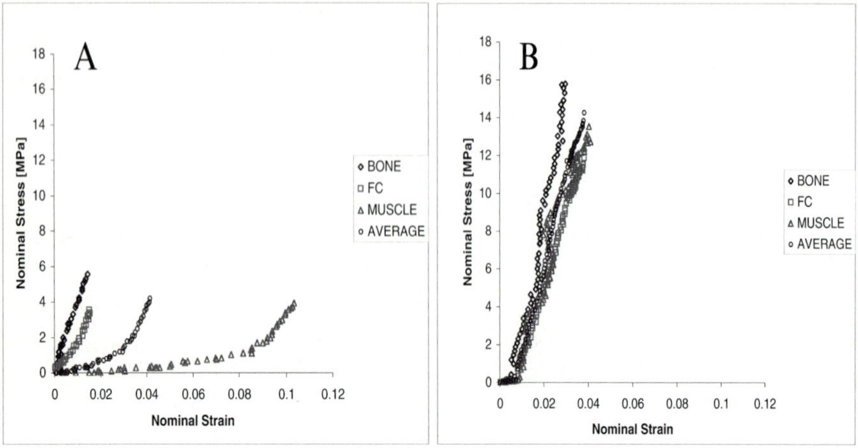

Fig. 1. Mechanical response curve of the mature TA tendon of the rat at three local regions along the length of the tendon: **BONE**, distal end (near the bone); **FC**, fibrocartilage mid-section; and **MUSCLE**, proximal end (near the muscle). The average response **AVERAGE** along the length is also shown. **A** Normal tendon is functionally graded; **B** after five weeks of unloading the previously compliant tendon regions have stiffened and the response is very uniform along its length.

the mechanical properties. Knowledge of *in vivo* mechanical loading histories in tendon is very difficult to obtain and under much less control. Yet it is clear that mechanical loading is critical to collagen growth and remodeling, as can be seen for example in Fig. 1 in which a TA tendon that has been unloaded *in vivo* for a period of time is no longer functionally graded; regions that were relatively compliant under normal loading have undergone significant growth or remodeling to stiffen (Arruda et al. (2005)). The so-called toe region or initially compliant non-linear portion of the normal tendon is truncated as a result of the unloading and the tendon prematurely stiffens with increased tangent stiffness.

To conduct controlled experiments of the effects of mechanical loading on collagen growth and remodeling we have developed *in vitro* models of soft tissue constructs comprised largely of type I collagen and tendon fibroblasts (tenocytes), and functional bioreactors for the application of cyclic mechanical interventions. The tissue engineering methods are described in Section 2 along with preliminary data demonstrating the growth response of the cells in these constructs subjected to static and cyclic mechanical loading protocols. Continuum level mathematical models of growth in soft tissue are currently a very active area of research (Epstein and Maugin (2000), Klisch et al. (2001), Taber and Humphrey (2001), Humphrey and Rajagopal (2002), Kuhl and Steinmann (2002), Sengers et al. (2004)). These models all consider the addition of mass as at least a subset of the phenomenon of growth. Some consider a single species (e.g., collagen) and consider its increasing mass. Others specifically include the transport of the growing species in a system that is open with respect to mass. Some of these formulations also describe a change in structural organization as a growth response. The present formulation, described in Sections 3 and 4, distinguishes between changes in material configurations at constant mass and growth processes. The former, described in Section 3, is treated as a motion in material space and is termed remodeling. The growth treatment in Section 4 describes transport of species such as nutrients and enzymes involved in tissue growth along with the tissue's fluid species, and not the mass transport of the growing extracellular collagen matrix itself. This is in contrast to the previous treatments in which the growing species (e.g., collagen) diffuses, but it is consistent with many physiologically relevant homeostatic tissue growth phenomena found in soft tissue growth.

2 Engineered Tissue Constructs

For growth and remodeling studies, tendon constructs have been engineered *in vitro* using two techniques, self-organization of tenocytes around their autogenous collagen scaffold and tenocyte-induced contraction of fibrin gels. Functional bioreactors were designed and built for mechanical testing of constructs and also cyclic load intervention studies. Preliminary data have been collected that demonstrate cyclic mechanical loading increases collagen content and improves mechanical properties of the engineered constructs.

2.1 Experimental Methods

This section briefly describes the engineered construct methods. See Calve et al. (2004) for the detailed protocols for self-organized tendon constructs and Andrick et al. (2005) for more information on the formation of constructs by fibrin gel contraction.

Self-organized Construct Formation. Primary rat tendon fibroblasts were dissociated from Achilles tendons of Fischer 344 retired breeders and suspended in growth medium containing Ham F-12, 20% fetal bovine serum (FBS) and 1% antibiotic-antimycotic (Invitrogen 10437). The cells were expanded, passaged at approximately 60% confluence and stored in liquid nitrogen until needed. Preparation of 35 mm culture plates involved coating with 1.5 mL of silicone elastomer (SYLGARD), curing the SYLGARD for two weeks, rinsing, then adding natural mouse laminin in Dilbecco's phosphate-buffered saline (DPBS) and allowing the DPBS to evaporate. Two size 0 laminin-coated silk sutures were pinned 12 mm apart into the laminin-coated substrate and the plates were filled with enough growth medium to cover the sutures then sterilized. The growth medium was aspirated and 200,000 cells suspended in 2 mL growth medium were seeded onto each plate and supplemented with ascorbic acid. Media were changed every 2–3 days until cells reached confluency after about 5 days. The medium was changed to differentiation medium (Dulbecco's Modified Eagle Medium containing 7% horse serum and 1% antibiotic-antimycotic) at 2–3 day intervals until constructs formed and were ready for mechanical intervention or testing. Construct formation occurred over a two-week period by the progressive delamination of the cell and autogenous collagen layer beginning at the outer periphery of the culture plate. The constructs formed into cylinders of length 12 mm and attached at their ends to the suture pins. The constructs remained viable for several weeks in culture under static tension due to cell contraction. A formed construct 3 months after plating is shown in Fig. 2.

Contracted Fibrin Gel Formation. Tenocytes were isolated from the Achilles tendon of Sprague Dawley rats using similar protocols as described above for the self-organized constructs. Culture plates were also prepared similarly; however, an oval (5 mm × 15 mm) polydimethylsiloxane (PDMS) mold was pressed into the center of the plates prior to sterilization and cell plating. The growth medium was supplemented with 10 units/mL of thrombin and agitated. Fibrinogen was added and polymerized by the thrombin within 10 minutes. About 15 minutes after gel formation 100,000 tenocytes were seeded on the gel and the culture plates were incubated until ready for mechanical intervention or testing. The growth medium was changed after 2 days to differentiation medium to promote collagen deposition because constructs had completely formed around the oval molds.

Histology. Self-organized constructs were prepared for electron microscopy using histological methods described in Calve et al. (2004). Electron microscopy revealed the constructs contained 60 nm diameter collagen fibers oriented along the main axis of the construct, closely resembling the morphology of neonatal tendon. Histology methods for the fibrin-based constructs may be

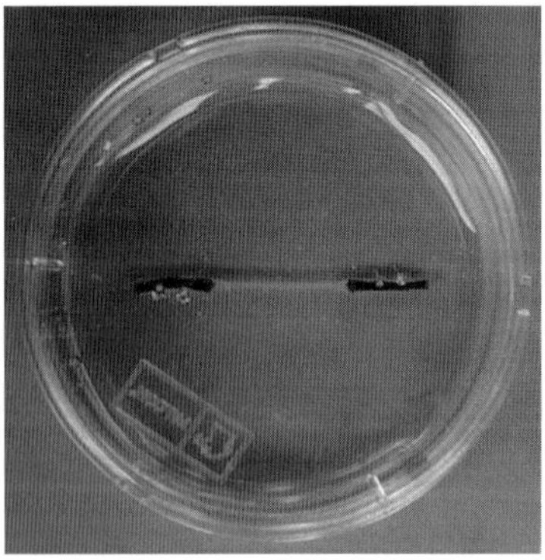

Fig. 2. A 12 mm long self-organized engineered tendon construct in culture and suspended above the SYLGARD substrate via pinned silk sutures at its ends. The construct is shown at three months after cell plating (Calve et al. (2004)).

found in Andrick et al. (2005). Longitudinal sections stained with either Massons trichrome or picosirius red at three weeks after seeding showed that the cells have digested much of the fibrin and deposited and organized a collagen matrix.

Mechanical Testing. Constructs formed by either approach were attached to functional bioreactors for cyclic mechanical loading. The constructs remained in culture and were attached to a stepper motor at one end and held fixed at the other end. The stepper motor was commissioned to cyclically strain the constructs between +10 and −10% at 0.1 Hz for 7 days. The mechanical response curves of the constructs were determined using an optical device for *in situ* strain determination during tensile loading; details may be found in Arruda et al. (2005). Briefly, constructs were kept in a saline solution, their cross-sectional areas were determined by several diameter measurements, then they were attached to stepper motors at either end, one end via an optical force transducer, and marked with India ink spots at several equal intervals along their lengths. Labview software controlled both servomotors as well as a digital video camera attached to a Nikon dissecting microscope. Constructs were loaded at a constant strain rate, without preconditioning until failure and synchronized force and image recordings were collected and converted to nominal stress vs. nominal strain data.

Collagen Content Determination. The hydroxyproline assay of Woessner (1961) was used to determine collagen content in the constructs by assuming the hydroxyproline accounts for 13.8% of the total type I collagen.

2.2 Experimental Results

The tenocytes in the self-organized tendon constructs are mechanically responsive to static tensile stretch. Cells in control constructs held in bioreactors and in culture at the as-formed lengths respond to static tension by depositing collagen. Figure 3 demonstrates that the collagen content in the constructs increases over time as the construct grows. This response is similar to the growth response of tendon *in vivo* as an animal ages. The tenocytes in the engineered constructs also respond to cyclic stretching, as shown in Fig. 3. The collagen content increases as a result of cyclic stretching over that in the control constructs, suggesting these constructs constitute a viable *in vitro* model for characterizing and determining evolution laws for collagen growth in soft tissue. Collagen growth is manifested in the stress-strain response curves of the tendon constructs as improvements in the mechanical properties associated with tendon function. Figure 4 demonstrates that cyclic stretching results in an evolution of the construct response curve towards that of native young tendon as the collagen content is increased, characterized by improved tangent stiffness and strength over those of controls held under static stretch for an equivalent amount of time.

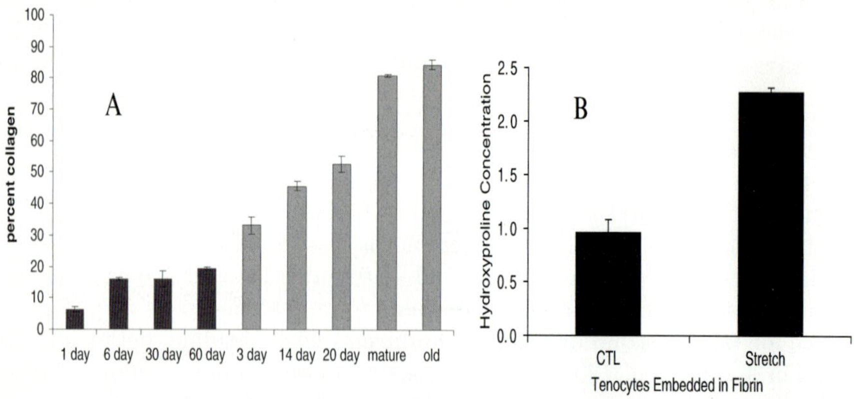

Fig. 3. **A** Tenocytes in self-organized constructs are mechanoresponsive to static stretch. Collagen content vs. time in culture (first four bars) or animal age. **B** Tenocytes in fibrin-based constructs respond to cyclic stretching [Stretch] by depositing a greater than two-fold increase in collagen after one week of cyclic stretching compared to controls [CTL].

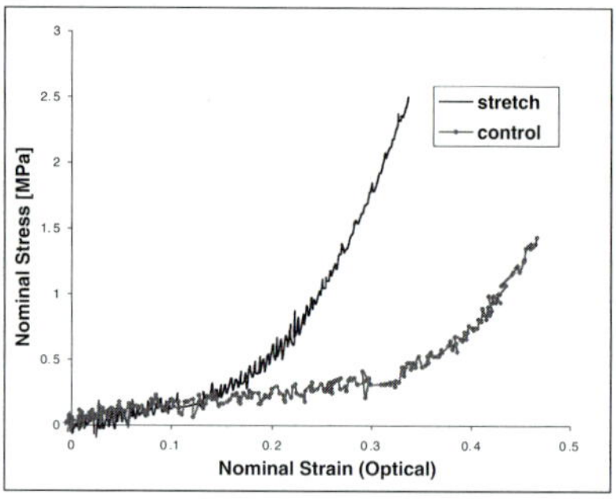

Fig. 4. Stress-strain response curves of self-organized tendon constructs with (solid line) and without (line with symbols) cyclic stretching for approximately one week. The result of cyclic stretching includes an evolution of the mechanical properties of the construct, such as the shape of the initial toe region, tangent modulus and strength, towards those of young native tendon.

3 Mathematical Modeling of Remodeling

The mathematical model of remodeling in biology is treated as a motion in material space or configurational change. Further details of this remodeling framework may be found in Garikipati et al. (2005). The kinematics of remodeling are illustrated in Fig. 5. The deformation gradient of the overall material motion is $\mathbf{F}$ and admits the multiplicative decomposition $\mathbf{F} = \mathbf{F}^*\mathbf{K}^c\mathbf{K}^r$, where $\mathbf{K}^r$ is the tangent map to the preferred remodeled state

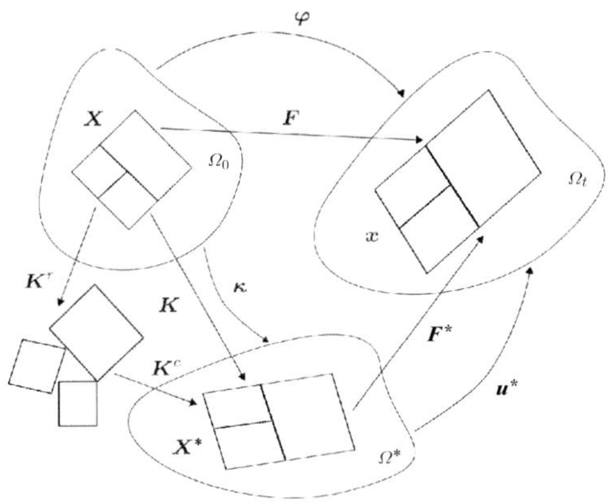

Fig. 5. The kinematics of remodeling (Garikipati et al. (2005)).

that a region would attain if it were free to remodel according to the local cellular mechanotransductive responses. Evolution laws based on experimental results from, e.g., uniformly remodeling tissue would yield $\mathbf{K}^r$. It is incompatible in general because of constraints placed on the region by the surroundings and $\mathbf{K}^c$ represents the compatibility restoring tangent mapping to the remodeled configuration. The deformation gradient $\mathbf{F}^*$ maps to the current configuration.

3.1 Outline of Remodeling Theory

The mechanical theory may be examined by minimizing the potential energy functional in the remodeled configuration Ω_t^* and the total material motion $\boldsymbol{\kappa} + \mathbf{u}^*$, i.e.

$$\Pi[\mathbf{u}^*, \boldsymbol{\kappa}] := \int_{\Omega_t^*} \hat{\psi}^*(\mathbf{F}^*, \mathbf{K}^c, \mathbf{X}^*) \mathrm{d}V^* - \int_{\Omega_t^*} \mathbf{f}^* \cdot (\mathbf{u}^* + \boldsymbol{\kappa}) \mathrm{d}V^* - \int_{\partial\Omega_t^*} \bar{\mathbf{t}}^* \cdot (\mathbf{u}^* + \boldsymbol{\kappa}) \mathrm{d}A^*, \tag{1}$$

where $\boldsymbol{\kappa}$ is the motion of a point in material space (configurational change), $\mathbf{u}^*$ is the displacement, $\psi^* = \hat{\psi}^*(\mathbf{F}^*, \mathbf{K}^c, \mathbf{X}^*)$ is the stored energy function and is assumed to depend on the compatibility restoring motion $\mathbf{K}^c$ in addition to $\mathbf{F}^*$, $\mathbf{f}^*$ is the body force and $\bar{\mathbf{t}}^*$ is the surface traction. Stationarity is assumed with respect to both displacements $\mathbf{u}^*$ and $\boldsymbol{\kappa}$. Standard manipulations using variational calculus lead to the Euler-Lagrange equations that can be localized to the relations in Ω^*. Thus,

$$\mathrm{Div}^* \mathbf{P}^* + \mathbf{f}^* = \mathbf{0}, \tag{2}$$

$$-\mathrm{Div} \left(\psi^* \mathbf{I} - \mathbf{F}^{*T} \mathbf{P}^* + \frac{\partial \psi^*}{\partial \mathbf{K}^c} \mathbf{K}^{cT} \right) + \frac{\partial \psi^*}{\partial \mathbf{X}^*} = \mathbf{0}, \tag{3}$$

and on $\partial\Omega^*$

$$\mathbf{P}^*\mathbf{N}^* = \bar{\mathbf{t}}^*, \tag{4}$$

$$\left(\psi^*\mathbf{I} - \mathbf{F}^{*T}\mathbf{P}^* + \frac{\partial\psi^*}{\partial\mathbf{K}^c}\mathbf{K}^{cT}\right)\mathbf{N}^* = \mathbf{0}, \tag{5}$$

where $\mathbf{P}^*$ is the first Piola-Kirchhoff stress, $\psi^*\mathbf{I} - \mathbf{F}^{*T}\mathbf{P}^* = \mathcal{E}$ is the Eshelby stress and $(\partial\psi^*/\partial\mathbf{K}^c)\mathbf{K}^{cT} = \mathbf{\Sigma}$ is an additional configurational stress arising from the assumption that the configurational change stores energy. The remodeling is subjected to restrictions placed by the dissipation inequality for the mechanical theory, written in terms of the Kirchhoff stress

$$\boldsymbol{\tau} = \det[\mathbf{K}]\frac{\partial\psi^*}{\partial\mathbf{F}^*}\mathbf{F}^{*T}, \tag{6}$$

as

$$\boldsymbol{\tau} : (\dot{\mathbf{F}}\mathbf{F}^{-1}) - \frac{\partial}{\partial t}(\det[\mathbf{K}]\psi^*) \geq 0, \tag{7}$$

where $\mathbf{K} = \mathbf{K}^c\mathbf{K}^r$. This leads to the following reduced dissipation inequality, placing a restriction on the evolution law for $\mathbf{K}^r$ and on the functional form of the extra configurational stress $\mathbf{\Sigma}$:

$$-\det[\mathbf{K}](\mathcal{E} + \mathbf{\Sigma}) : (\dot{\mathbf{K}}^c\mathbf{K}^{c-1}) - \det[\mathbf{K}]\mathcal{E} : (\mathbf{K}^c\dot{\mathbf{K}}^r\mathbf{K}^{-1}) - \det[\mathbf{K}]\frac{\partial\psi^*}{\partial t} \geq 0. \tag{8}$$

3.2 A Remodeling Example

The remodeling theory is illustrated using the anisotropic eight-chain model of Bischoff et al. (2002a,b) to describe a collagen network in a soft tissue such as tendon. Consider the case of a local tendon region initially described by the network in Fig. 6**A**, characterized by the fiber angle θ_0, as shown. If the tendon were stretched uniaxially its true stress vs. true strain response curve would follow the dashed trace in Fig. 6**C** as the collagen fibers orient via an affine transformation that is the tangent map of the incompressible deformation of the tissue. The network is hyperelastic, and in the absence of remodeling, it loads and unloads along the non-linear dashed path to return to the reference configuration of Fig. 6**A** with the fibers returning to their original orientation as the deformation is removed. If instead the tendon region is described by a different fiber direction θ, as shown in Fig. 6**B**, its response curve would follow the solid hyperelastic trace in Fig. 6**C**.

This remodeling example is simplified for purposes of illustration to assume the entire tissue is homogeneous; moreover, it remodels homogeneously and therefore the remodeling is compatible. The kinematics illustrated in Fig. 5 are simplified for this case to $\mathbf{K}^c = \mathbf{I}$ which greatly simplifies the reduced dissipation inequality above. The first term is eliminated because $\dot{\mathbf{K}}^c = \mathbf{0}$ and the third term is also zero because the strain energy density function is assumed uniform throughout the homogeneous tissue. The remaining term in the reduced dissipation inequality is also simplified in this case and the restriction

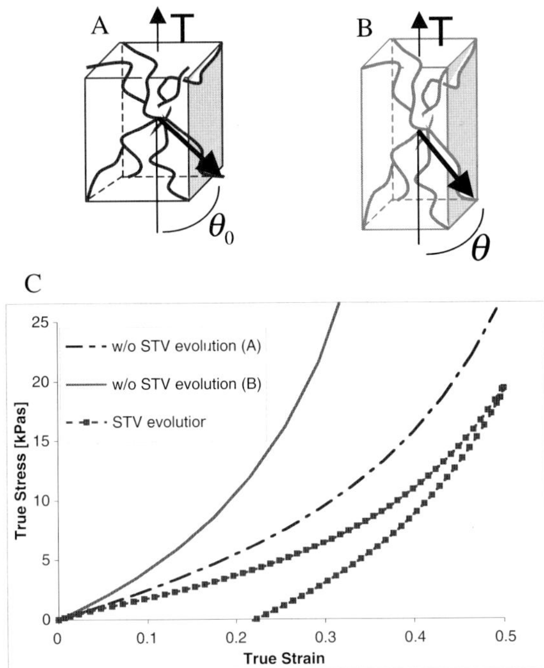

Fig. 6. Numerical simulation of uniform remodeling of an anisotropic network during uniaxial deformation. **A** Hyperelastic network and original reference configuration of network that remodels during deformation; **B** hyperelastic network and remodeled reference configuration of network that has remodeled during uniaxial loading; **C** hyperelastic load-unload response curves for two distinct networks (i.e. **A**, dashed curve, and **B**, solid curve), and an additional response curve (with symbols) of a network that remodels from reference configuration **A** to remodeled reference configuration **B** during uniaxial loading and is fixed in the remodeled configuration during unloading.

placed on the remodeling evolution law, $\dot{\mathbf{K}}^r$, is simplified by the requirement that the fibers in the network remodel towards the principal tensile stretch direction during uniaxial tension loading. We now consider remodeling during uniaxial loading of the network in Fig. 6**A** such that the collagen fibers remodel towards the principal stretch direction in addition to deforming in an affine manner during each loading step. The gradual remodeling evolves the reference configuration of the collagen fibers from that of Fig. 6**A** to that of Fig. 6**B** during the deformation to a true strain of 0.5. At the end of this loading step the tissue is fixed in its evolved configuration and unloaded. The response of the remodeling tissue is shown by the trace with symbols in Fig. 6**C**. The simulation demonstrates that as the remodeling occurs during loading the network stores less strain energy at any given applied deformation level than it does if it deforms without remodeling. Remodeling occurs in this manner to relieve the excess strain energy of the applied deformation in much the same way tissue is hypothesized to remodel in response to increases in loading, such as occurs during exercise. The network is fixed in the remodeled configuration at the end of loading, and no further evolution of the microstructure occurs as the network is unloaded. The change in reference configuration is evidenced by the permanent deformation that remains upon fully unloading.

4 Mathematical Modeling of Growth

The continuum thermodynamics treatment of a system that is open with respect to mass to allow species concentration changes due to mass transport is now considered. The full details of the theory may be found in Garikipati et al. (2004). It is outlined here to demonstrate the differences between the growth and remodeling theories and to illustrate a simple growth example. The kinematical framework assumes local growth by $\mathbf{F}^g$ resulting in an incompatible tissue. Elastic deformation by $\tilde{\mathbf{F}}^e$ restores compatibility, resulting in internal stresses. The applied elastic deformation is $\bar{\mathbf{F}}^e$. The entire mapping admits the multiplicative decomposition $\bar{\mathbf{F}}^e\tilde{\mathbf{F}}^e\mathbf{F}^g$. Multiple species are modeled in the continuum setting. The biological processes by which nutrients are transported and converted to living tissue are described mathematically by sources and sinks. Fluxes are introduced for each species diffusing through the solid phase. In the soft tissue application of tendon the solid phase is type I collagen. Mass balance for a species i is given as $\partial\rho_0^i/\partial t = \Pi^i - \nabla\cdot\mathbf{M}^i$, where ρ_0 is the concentration, Π^i is the source term (or sink term) and $\mathbf{M}^i$ is the mass flux vector of the species. The simplest illustrative example of the theory involves two species, the solid collagen phase and a fluid phase. The latter is assumed to carry nutrients, enzymes, proteases etc. involved in the synthesis and breakdown of the collagen phase, but the two-species approximation assumes these reaction products do not diffuse relative to the fluid phase. Mass balance for the solid phase becomes $\partial\rho_0^s/\partial t = \Pi^s$ and for the fluid phase it is $\partial\rho_0^f/\partial t = -\nabla\cdot\mathbf{M}^f$. Balance of linear and angular momenta and energy are considered for each species and momenta and energy transfer among species are allowed by introducing $\mathbf{q}^i$, the force per unit mass exerted on species i by the other species present, and an interaction energy term $\tilde{e}^i$ transferred to species i by all other species per unit mass of i. The full development of these equations is rather lengthy and the reader is referred to Garikipati et al. (2004) for the details. The entropy production inequality holds for the system as a whole. It is combined with the energy equation and balance laws to derive the reduced entropy inequality for growth processes in the form

$$\sum_i\left(\rho_0^i\frac{\partial e^i}{\partial\mathbf{F}^{ei}} - \mathbf{P}^i(\mathbf{F}^{gi})^T\right):\dot{\mathbf{F}}^{ei} + \sum_i\rho_0^i\left(\frac{\partial e^i}{\partial\eta^i} - \theta\right)\frac{\partial\eta^i}{\partial t}$$

$$+\sum_i\left(\rho_0^i\frac{\partial}{\partial t}(\mathbf{V}+\mathbf{V}^i) - \rho_0^i\mathbf{g} - \nabla\cdot\mathbf{P}^i + \nabla(\mathbf{V}+\mathbf{V}^i)\mathbf{M}^i\right)\cdot\nabla$$

$$+\sum_i \left[\rho_0^i \mathbf{F}^{i-T}(\nabla e^i - \theta\nabla\eta^i)\right]\cdot\nabla + \sum_i \Pi^i\left(e^i + \frac{1}{2}\parallel \mathbf{V}^i \parallel\right)$$

$$+\frac{\nabla\theta\cdot\mathbf{Q}^i}{\theta} + \sum_i \rho_0^i \frac{\partial e^i}{\partial\rho_0^i}\frac{\partial\rho_0^i}{\partial t} - \sum_i \mathbf{P}^i : \left(\nabla\mathbf{V}^i + \mathbf{F}^{ei}\dot{\mathbf{F}}^{gi}\right) \leq 0, \qquad (9)$$

where, for species i, $\mathbf{P}^i$ is the partial first Piola-Kirchhoff stress, e^i is the internal energy per unit mass, η^i is the entropy per unit mass, $\mathbf{V}^i$ is the velocity relative to the solid phase velocity $\mathbf{V}$, and $\mathbf{Q}^i$ is the partial heat flux vector. Also, $\mathbf{g}$ is the body force and θ is the absolute temperature. Constitutive assumptions of the form $e^i = \hat{e}^i\left(\mathbf{F}^{ei}, \eta^i, \rho_0^i\right)$ are made for the thermodynamic variables to obtain the following restrictions on the constitutive relations from the reduced entropy inequality: $\mathbf{P}^i\mathbf{F}^{gi}{}^{\mathbf{T}} = \rho_0^i\partial e^i/\partial\mathbf{F}^{ei}$, hyperelastic constitutive laws for the partial stresses; $\theta = \rho_0^i\partial e^i/\partial\eta^i$, temperature definition; $\mathbf{Q}^i = -\mathbf{K}^i\nabla\theta$, the Fourier's law of heat conduction; and a constitutive relation for the flux as a product of a mobility and a thermodynamic driving force in which the full coupling between mechanics and mass transport emerges. Thus,

$$\mathbf{M}^i = \mathbf{D}^i\cdot\left[\rho_0^i\mathbf{F}^T\mathbf{g} + \mathbf{F}^T\nabla\cdot\mathbf{P}^i - \rho_0^i\left(\nabla e^i - \theta\nabla\eta^i\right)\right]. \qquad (10)$$

Specifically, mass transport is driven by an inertia term, a body force term (e.g., gravity), a stress divergence term such as pressure gradient and a chemical potential gradient. This last term includes strain-gradient dependent diffusion through the gradient in internal energy term and a concentration-gradient dependent diffusion arising from the mixing entropy. This theory has been implemented in the general purpose finite element program FEAP using the anisotropic eight-chain hyperelastic constitutive law of Bischoff et al. (2002a) for the solid collagen phase and a growth law for collagen that prescribes conversion to solid if the fluid concentration exceeds a preset value. A cylindrical construct of the size of the *in vitro* tendon construct models described earlier is simulated. It is assumed to initially contain a uniform solid volume fraction of 0.5. The stress-strain response curve for this construct is first simulated without allowing growth and axially deforming it. The construct is then placed in a fluid bath and the collagen concentration is commissioned to increase locally if the fluid concentration in the construct exceeds its initial value of 0.5. When placed in the fluid bath the construct swells as fluid flows from the high concentration bath to the low concentration construct. Solid is grown because of the rise in fluid concentration; this results in further swelling of the construct. The average final solid concentration is approximately 0.6. After a period of growth the constitutive properties of the construct are again simulated by conducting a tensile test without allowing further growth. The results appear in Fig. 7 as stress on the vertical axis and extension on the horizontal. They may be compared with the experimental results of growth in Fig. 4 to show the theory is capable of capturing the nonlinear constitutive response of the tendon constructs and correctly predicts the qualitative change in tangent modulus with collagen growth. However, further

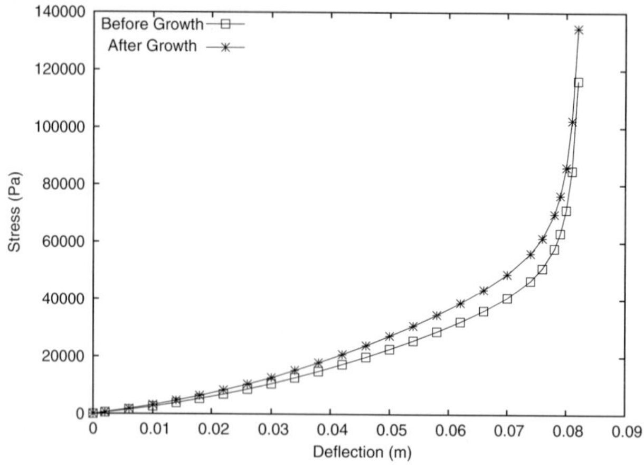

Fig. 7. Results from a finite element simulation of growth in soft tissue. After an increase in collagen concentration from 50% (solid curve) to approximately 60% (dashed curve) the tangent stiffness of the construct is increased, in agreement with experimental results.

experiments of the type introduced here using *in vitro* soft tissue models are needed to provide accurate evolution laws for mathematical growth theories.

5 Conclusions

Separate theories of growth and remodeling have been outlined to illustrate that remodeling may occur at constant mass and is a configurational change whereas growth involves a change in the concentration of species. Engineered tendon constructs were generated for growth and remodeling studies. The constructs demonstrate mechanically responsive cells, grow and remain viable in culture for several weeks. They are excellent *in vitro* models for growth studies.

Acknowledgements. This research was funded in part by DARPA (SPAWAR) contract N66001-02-C-8034, NSF (CMS) contract 9988693, and an LDRD sub-contract from Sandia National Laboratories.

References

Andrick, J. J., Mundy, K., Calve, S. C., Arruda, E. M., and Baar, K. (2005). Uniaxial stretch results in increased collagen in fibrin-based 3D engineered tendon. *J. Appl. Physiol.* submitted.

Arruda, E. M., Mundy, K., Calve, S. C., and Baar, K. (2005). Denervation decreases tendon extensibility and increases tendon stiffness. *J. Physiol.* submitted.

Bischoff, J. E., Arruda, E. M., and Grosh, K. (2002a). A microstructurally based orthotropic hyperelastic constitutive law. *J. Appl. Mech.* 69:570–579.

Bischoff, J. E., Arruda, E. M., and Grosh, K. (2002b). Orthotropic hyperelasticity in terms of an arbitrary molecular chain model. *Tissue Eng.* 10:755–761.

Calve, S. C., Dennis, R. G., Kosnik II, P. E., Baar, K., and Arruda, E. M. (2004). Engineering of functional tendon. *J. Appl. Mech.* 69:199–201.

Epstein, M., and Maugin, G. A. (2000). Thermomechanics of volumetric growth in uniform bodies. *Int. J. Plasticity* 16:951–978.

Garikipati, K., Arruda, E. M., Grosh, K., Narayanan, H., and Calve, S. C. (2004). A continuum treatment of growth in biological tissue: The coupling of mass transport and mechanics. *J. Mech. Phys. Solids* 52:1595–1625.

Garikipati, K., Narayanan, H., Arruda, E. M., Grosh, K., and Calve, S. C. (2005). Material forces in the context of biotissue remodelling. In Steinmann, P., and Maugin, G. A., eds., *Mechanics of Material Forces.* Dordrecht: Kluwer Academic Publishers. E-print available at http://arXiv.org/abs/q-bio.QM/0312002.

Humphrey, J. D., and Rajagopal, K. R. (2002). A constrained mixture model for growth and remodeling of soft tissues. *Math. Model. Meth. Appl. Sci.* 12:407–430.

Klisch, S. M., Van Dyke, T. J., and Hoger, A. (2001). A theory of volumetric growth for compressible elastic biological materials. *Math. Mech. Solids* 6:551–575.

Kuhl, E., and Steinmann, P. (2002). Geometrically nonlinear functional adaptation of biological microstructures. In Mang, H. A., Rammerstorfer, F. G., and Eberhardsteiner, J., eds., *Proceedings of the Fifth World Congress on Computational Mechanics (WCCM V)*, 1–21. Vienna, Austria: International Association for Computational Mechanics.

Nordin, M., Lorenz, T., and Campello, M. (2001). Biomechanics of tendons and ligaments. In Nordin, M., and Frankel, V. H., eds., *Basic Biomechanics of the Musculoskeletal System.* New York: Lippincott Williams and Wilkins. 102–125.

Sengers, B. G., Oomens, C. W. J., and Baaijens, F. P. T. (2004). An integrated finite-element approach to mechanics, transport and biosynthesis in tissue engineering. *J. Biomech. Eng.* 126:82–91.

Taber, L. A., and Humphrey, J. D. (2001). Stress-modulated growth, residual stress, and vascular heterogeneity. *J. Biomech. Eng.* 123:528–535.

Woessner, J. F. (1961). The determination of hydroxyproline in tissue and protein samples containing small proportions of this imino acid. *Arch. Biochem. Biophys.* 93:440–447.

Modeling and Simulation of Remodeling in Soft Biological Tissues

E. Kuhl[1], A. Menzel[1], K. Garikipati[2], E. M. Arruda[2], K. Grosh[2]

[1] University of Kaiserslautern, Chair of Applied Mechanics, Germany
ekuhl@rhrk.uni-kl.de, amenzel@rhrk.uni-kl.de

[2] University of Michigan, Department of Mechanical Engineering, USA
garikipati@engin.umich.edu, arruda@umich.edu, grosh@engin.umich.edu

The present contribution discusses a micromechanically motivated chain network model for soft biological tissues. On the microscale, the model is governed by the statistics of long chain molecules. To simulate the behavior of a single collagen microfibril, a wormlike chain approach is applied. On the macroscale, the network of collagen fibrils is represented by a specific eight-chain unit cell which is characterized through its axis of transverse isotropy. This axis is allowed to reorient itself with respect to a particular biological stimulus, in our case, the direction of maximum principal strain. First, the basic features of the model are illustrated by means of the contour plots of its free energy function in the principal stretch space. Then, the suggested model is applied to the biomechanically relevant boundary value problem of an *in vitro* engineered functional tendon construct.

1 Introduction

Soft biological tissues are typically found to undergo large strains accompanied by nonlinear elastic deformations. Governed by a network of sparsely cross-linked long chain molecules, their constitutive behavior is basically dominated by changes in their configurational entropy rather than in their internal energy. Unlike polymer chains in rubber elasticity, collagen chains in biomaterials show a pronounced orientation which is typically aligned with the predominant loading direction. Due to this directional dependent reinforcement, biomaterials, in particular soft tissues such as muscles, tendons or ligaments, are classical representatives of transversely isotropic materials; see, e.g., the textbooks by Humphrey (2002) or Holzapfel and Ogden (2003).

Soft biological tissues can be characterized within the framework of statistical mechanics of long chain molecules as documented in the classical textbook by Flory (1969); see also Holzapfel (2000). In this approach, each single collagen fibril can be modeled as a so-called wormlike chain with a smoothly varying curvature; see Kratky and Porod (1949). In contrast to the single parameter freely-jointed-chain model applied in rubber elasticity, the wormlike-chain model offers the additional freedom of a second parameter, which accounts for the initial chain stiffness. Traditionally, the wormlike-chain model has been applied to describe the behavior of the DNA double helix; see, e.g., Bustamante et al. (1994) or Marko and Siggia (1995). Only recently, the wormlike-chain approach has been used to simulate the constitutive behavior of a collagen fibril by Bischoff et al. (2002), Garikipati et al. (2004) and Kuhl et al. (2005a,b).

On the macroscopic scale, the behavior of the overall chain network can be described through the effective assembly of a number of chains arranged in a representative unit cell; see, e.g., Treloar (1975). Detailed comparisons of the different chain network approaches can be found in the classical textbook by Treloar (1975) as well as in the excellent articles by Boyce and Arruda (2000) or Miehe et al. (2004). Although the constitutive description of the chain network seems to be well understood for isotropic materials such as rubber, the generalization to anisotropic network structures typically found in soft tissue biomechanics is still an open issue.

It is a well accepted axiom in biomechanics that the biological tissue exhibits a directional strengthening due the local rearrangement of its microstructure that is essentially driven by the mechanical loading situation. In the biomechanical literature, this phenomenon is typically referred to as functional adaptation or remodeling. Micromechanically motivated remodeling theories are based on the rigorous reorientation of collagen fibers; see, e.g., Cowin (1995), Vianello (1996), Sgarra and Vianello (1997) or Driessen et al. (2003). The present work is particularly related to the recent contribution by Menzel (2005), who suggests a reorientation of the material's principal axis with respect to the predominant loading direction; see also Kuhl et al. (2005a,b).

The present contribution is organized as follows. In Section 2, we summarize the basic equations of the transversely isotropic chain network model. Next, in Section 3, we discuss the aspect of remodeling, i.e. the reorientation of the axes of transverse isotropy with respect to the direction of maximum principal strain. In Section 4, we elaborate the free energy function by means of contour plots in the principal stretch space. The simulation of remodeling in an *in vitro* engineered functional tendon construct is presented in Section 5. The manuscript closes with a final discussion in Section 6.

2 Chain Network Models

On the microscale, the model presented in the sequel is based on a particular statistical single chain model for correlated chains, the two parameter wormlike chain. Unlike the classical uncorrelated freely-jointed chain typically applied in rubber elasticity, the wormlike chain is characterized through a smooth curvature whose direction changes randomly but in a continuous manner (see Kratky and Porod (1949) or Flory (1969)). This property is essentially reflected through a second parameter in addition to the contour length L, namely the persistence length A. While the contour length is the end-to-end length of the chain at full extension, the persistence length defines the sum of the average projections of all bonds onto the direction of the first bond. As such, it is a measure of the initial chain stiffness; see Fig. 1, left. The deformation of the chain is then typically defined in terms of the end-to-end distance r, i.e. the length of the vector pointing from one end of the chain to the other. The strain energy ψ^{chn} of a single wormlike chain, given by

$$\psi^{\text{chn}} = \psi_0^{\text{chn}} + \frac{k\,\theta\,L}{4A}\left[2\frac{r^2}{L^2} + \frac{1}{(1 - r/L)} - \frac{r}{L}\right], \tag{1}$$

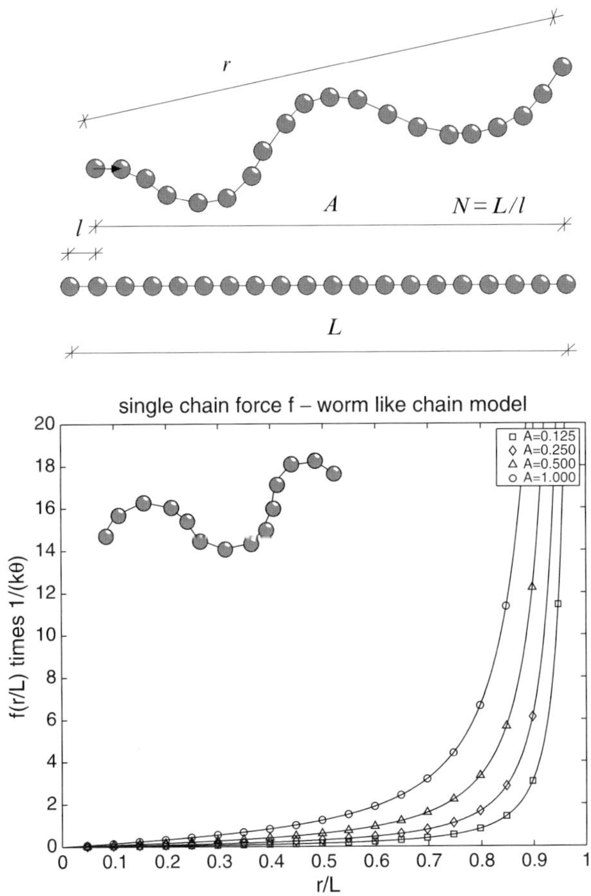

Fig. 1. Single collagen chain – Wormlike-chain model.

can be derived straightforwardly by integrating the force stretch relation

$$f^{\text{chn}} = \frac{k\,\theta}{4A}\left[4\frac{r}{L} + \frac{1}{(1-r/L)^2} - 1\right], \tag{2}$$

as originally suggested for the DNA double helix by Marko and Siggia (1995) and later applied for the collagen fibrils by Bischoff et al. (2002). Therein, $k = 1.38\times10^{-23}$ J/K is the Boltzmann constant, θ is the absolute temperature and ψ_0^{chn} is the value of the chain energy in the unperturbed state. Note that the range of validity of the model is restricted to $0 < r < L$. Typical force stretch relations of a single wormlike chain are illustrated in Fig. 1, right. They essentially reflect the characteristic locking behavior as the end-to-end length r approaches the contour length L. Obviously, the locking response becomes more pronounced as the persistence length A decreases.

On the macroscale, we apply an eight chain representation of the underlying cooperative network structure as originally derived in the context of rubber elasticity by Arruda and Boyce (1993) (see also Boyce and Arruda (2000)) and then applied in the context of biomechanics by Bischoff et al. (2002) and Kuhl et al. (2005a,b). The representative chains are embedded in a transversely isotropic unit cell with initial cell dimensions a and b, as illustrated in Fig. 2. The end-to-end length

$$r = \tfrac{1}{2}\sqrt{I_4(a^2-b^2)+I_1 b^2} \tag{3}$$

can then be expressed in terms of the initial cell dimensions a and b and the first and fourth invariant $I_1 = \boldsymbol{F} : \boldsymbol{F}$ and $I_4 = \bar{\boldsymbol{n}} \cdot \bar{\boldsymbol{n}}$. Therein, $\boldsymbol{F} = \nabla_{\boldsymbol{X}}\varphi$ denotes the deformation gradient related to the deformation map φ and its Jacobian $J = \det \boldsymbol{F}$ introduces the third invariant as $I_3 = J^2 = (\det \boldsymbol{F})^2$. Accordingly, the cell normal in the deformed configuration is given as $\bar{\boldsymbol{n}} = \boldsymbol{F}\cdot\boldsymbol{n}_0$ in terms of the undeformed unit normal $\boldsymbol{n}_0$. The overall

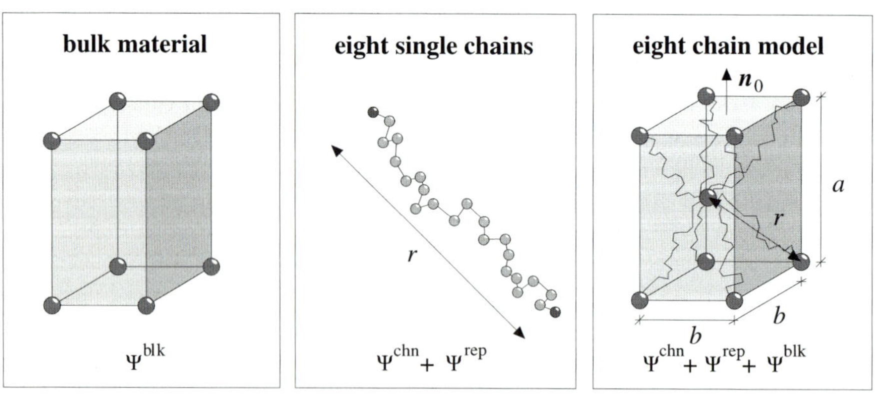

Fig. 2. Kinematics of transversely isotropic eight-chain network model.

energy of a representative unit cell is assumed to consist of three contributions, so that

$$\Psi = \Psi^{\text{blk}} + \Psi^{\text{chn}} + \Psi^{\text{rep}}, \tag{4}$$

where Ψ^{blk} is an isotropic bulk term capturing the effects of the surrounding liquid solvent, $\Psi^{\text{chn}} = \gamma^{\text{chn}}\,\psi^{\text{chn}}$ is a chain term reflecting the assembly of the individual eight-chain energies ψ^{chn}, and Ψ^{rep} is a repulsive term preventing the material from collapsing. These are given by

$$\Psi^{\text{blk}} = \gamma^{\text{blk}} \left[I_1 - 3 + \frac{1}{\beta} \left(I_3^{-\beta} - 1 \right) \right], \tag{5}$$

$$\Psi^{\text{chn}} = \gamma^{\text{chn}} \frac{k\theta L}{4A} \left[2\frac{r^2}{L^2} + \frac{1}{(1-r/L)} - \frac{r}{L} \right], \tag{6}$$

$$\Psi^{\text{rep}} = -\gamma^{\text{chn}} \frac{k\theta}{4A} \left[\frac{1}{L} + \frac{1}{4r_0(1-r_0/L)^2} - \frac{1}{4r_0} \right] \left\{ \ln \left[I_4^{(a^2-b^2)/2} \right] + \tfrac{3}{2} \ln \left(I_1^{b^2} \right) \right\}. \tag{7}$$

The parameter set of the model is thus restricted to the chain density γ^{chn}, the two wormlike-chain parameters, i.e. the persistence length A and the contour length L, the cell dimensions a and b and the two bulk parameters γ^{blk} and β. The free energy Ψ introduced above defines the Kirchhoff stress $\boldsymbol{\tau}$

$$\boldsymbol{\tau} = \boldsymbol{\tau}^{\text{blk}} + \boldsymbol{\tau}^{\text{chn}} + \boldsymbol{\tau}^{\text{rep}} \tag{8}$$

as the push forward of the first Piola-Kirchhoff stress $\boldsymbol{\tau} = \mathrm{d}_{\boldsymbol{F}} \Psi \cdot \boldsymbol{F}^{\text{t}}$. Its individual terms

$$\boldsymbol{\tau}^{\text{blk}} = \gamma^{\text{blk}} \left(2\boldsymbol{F} \cdot \boldsymbol{F}^{\text{t}} - 2 I_3^{-\beta} \boldsymbol{I} \right), \tag{9}$$

$$\boldsymbol{\tau}^{\text{chn}} = \gamma^{\text{chn}} \frac{k\theta}{4A} \left[\frac{1}{L} + \frac{1}{4r(1-r/L)^2} - \frac{1}{4r} \right] \left[\left(a^2 - b^2\right) \boldsymbol{N} + b^2 \boldsymbol{F} \cdot \boldsymbol{F}^{\text{t}} \right], \tag{10}$$

$$\boldsymbol{\tau}^{\text{rep}} = -\gamma^{\text{chn}} \frac{k\theta}{4A} \left[\frac{1}{L} + \frac{1}{4r_0(1-r_0/L)^2} - \frac{1}{4r_0} \right] \left[\frac{1}{I_4} \left(a^2 - b^2\right) \boldsymbol{N} + \frac{3}{I_1} b^2 \boldsymbol{F} \cdot \boldsymbol{F}^{\text{t}} \right], \tag{11}$$

follow straightforwardly from the corresponding energy terms introduced in equations (5)–(7). Here, $\boldsymbol{N} = \bar{\boldsymbol{n}} \otimes \bar{\boldsymbol{n}}$ denotes the push forward $\boldsymbol{N} = \boldsymbol{F} \cdot \boldsymbol{N}_0 \cdot \boldsymbol{F}^{\text{t}}$ of the structural tensor $\boldsymbol{N}_0 = \boldsymbol{n}_0 \otimes \boldsymbol{n}_0$.

Remark 1 (Special cases). Note that the classical isotropic eight-chain model of Arruda and Boyce (1993) and Boyce and Arruda (2000) which was originally introduced in the context of rubber elasticity can be understood as a special case of the present framework. Its undeformed unit cell represents a cube with $a = b$. Accordingly, the end-to-end length $r = \sqrt{I_1}\, a/2$ reduces to a function of the first strain invariant I_1. Another special case of the transversely isotropic chain network model follows from assuming a degenerated unit cell for which the in-plane dimension tends to zero as $b = 0$. All eight chains are thus oriented in a single direction $\boldsymbol{n}_0$ and cross-link network effects are neglected completely. The deformed end-to-end length $r = \sqrt{I_4}\, a/2$ thus degenerates to a mere function of the fourth invariant I_4.

3 Remodeling

In the original isotropic eight-chain model (Arruda and Boyce (1993), Boyce and Arruda (2000)) the representative unit cell is assumed to align its axes instantaneously with the principal strain directions. In the present model, however, the unit cell axis $\boldsymbol{n}_0$ is allowed to gradually align with the eigenvector $\boldsymbol{n}_\lambda^{\max}$ of the right Cauchy-Green tensor $\boldsymbol{F}^{\mathrm{t}} \cdot \boldsymbol{F}$ associated with the maximum eigenvalue $\lambda^{\max} = \max(\lambda_i)$ in the spectral decomposition

$$\boldsymbol{F}^{\mathrm{t}} \cdot \boldsymbol{F} = \lambda_i \, \boldsymbol{n}_\lambda^i \otimes \boldsymbol{n}_\lambda^i, \qquad i = 1, 2, 3. \tag{12}$$

In some biologically relevant cases, we might encounter multiple maximum eigenvalues λ_i. For the sake of clarity, we shall assume that no alignment takes place for multiple maximum eigenvalues. Following the approach suggested recently by Menzel (2005) (see also Kuhl et al. (2005a,b)), we introduce the rotation vector $\boldsymbol{\omega}$ as the scaled vector product of the unit cell orientation $\boldsymbol{n}_0$ and the direction of the maximum principal strain $\boldsymbol{n}_\lambda^{\max}$:

$$\boldsymbol{\omega} = \frac{1}{\tau_\omega} \boldsymbol{n}_0 \times \boldsymbol{n}_\lambda^{\max} = \omega \boldsymbol{n}^\omega. \tag{13}$$

Its decomposition into the unit normal $\boldsymbol{n}^\omega$ and the magnitude ω as

$$\boldsymbol{n}^\omega = \frac{\boldsymbol{n}_0 \times \boldsymbol{n}_\lambda^{\max}}{||\boldsymbol{n}_0 \times \boldsymbol{n}_\lambda^{\max}||}, \qquad \omega = \frac{||\boldsymbol{n}_0 \times \boldsymbol{n}_\lambda^{\max}||}{\tau_\omega} \tag{14}$$

illustrates that the magnitude of the rotation is obviously governed by both the time relaxation parameter τ_ω and the angle between $\boldsymbol{n}_0$ and $\boldsymbol{n}_\lambda^{\max}$. The temporal evolution of the unit cell axis $\boldsymbol{n}_0$ can then be expressed in the abstract form

$$\mathrm{d}_t \, \boldsymbol{n}_0 = \boldsymbol{\omega} \times \boldsymbol{n}_0 = \frac{1}{\tau_\omega} \left[\boldsymbol{n}_\lambda^{\max} - (\boldsymbol{n}_\lambda^{\max} \cdot \boldsymbol{n}_0) \, \boldsymbol{n}_0 \right]. \tag{15}$$

From the above equation, it is obvious that the orthogonality condition $\mathrm{d}_t \boldsymbol{n}_0 \cdot \boldsymbol{n}_0 = 0$ is valid throughout the remodeling history.

Remark 2 (Temporal discretization). For the temporal discretization of the first-order rate equation (15), we suggest an exponential integration scheme. It allows for a closed-form representation of Euler-Rodrigues type which can be cast into the following update formula for the cell axis $\boldsymbol{n}_0^{k+1}$:

$$\boldsymbol{n}_0^{k+1} = \cos(\Delta t \, \omega) \boldsymbol{n}_0^k + \sin(\Delta t \, \boldsymbol{\omega}) \times \boldsymbol{n}_0^k + [1 - \cos(\Delta t \, \omega)](\boldsymbol{n}^\omega \cdot \boldsymbol{n}_0^k) \boldsymbol{n}^\omega.$$

In the algorithmic formulation, we shall tacitly assume that $\boldsymbol{n}_0^{k+1} \times \boldsymbol{n}^\omega \approx \boldsymbol{n}_0^k \times \boldsymbol{n}^\omega$, so that the rotation vector $\boldsymbol{\omega}$ can be approximated as a function of the known orientation $\boldsymbol{n}_0^k$. Accordingly, $\boldsymbol{\omega}$ can then be updated explicitly as $\boldsymbol{\omega} = (\boldsymbol{n}_0^k \times \boldsymbol{n}_\lambda^{\max})/\tau_\omega$ which is a reasonable assumption in the context of the gradual alignment postulated herein.

4 Example – Energy Contour Plots

To gain a better understanding of the influence of anisotropy, we elaborate the contour plots of the free energy function introduced in (4) and (5)–(7); see Holzapfel et al. (2000), Holzapfel and Ogden (2003) and Holzapfel et al. (2004). Three representative fixed unit cell orientations are elaborated for the general transversely isotropic network model in Fig. 3, the isotropic model in Fig. 4, and the transversely isotropic model in Fig. 5. Therein the cell orientations are characterized through the angle between the unit cell axis $\boldsymbol{n}_0$ and the horizontal axis as $\angle\,\boldsymbol{n}_0 = 0°, 30°, 45°$. Moreover, in Fig. 6, we study the same test cases for the reorientation model. The model parameters are adopted from Garikipati et al. (2004) as $L = 25.277$, $A = 1.3775$, $c = 7 \times 10^{21}$, $k = 1.38 \times 10^{-23}$, $\theta = 310$, $a = 9.3$, $b = 6.2$, $\gamma = 1000$ and $\beta = 4.5$.

Figures 3–6 show the isolines of the free energy plotted in the principal stretch space. The principal stretches vary from $0.0 < \lambda_1, \lambda_2 < 8.0$, thus covering both the tensile and compressive regimes. The dotted lines $\lambda_1 = 1.0$ and $\lambda_2 = 1.0$ represent the axes of uniaxial deformation and their intersection characterizes the undeformed state. The first two columns of Figs. 3–6 correspond to the plane strain case whereas the last two columns

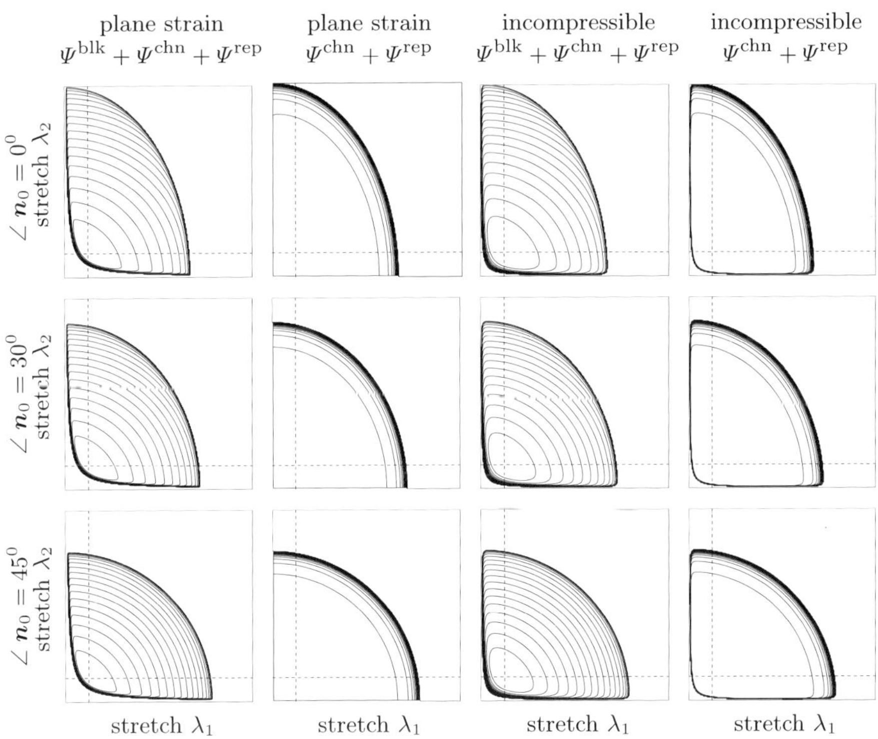

Fig. 3. Transversely isotropic chain network model – energy contour plots Ψ.

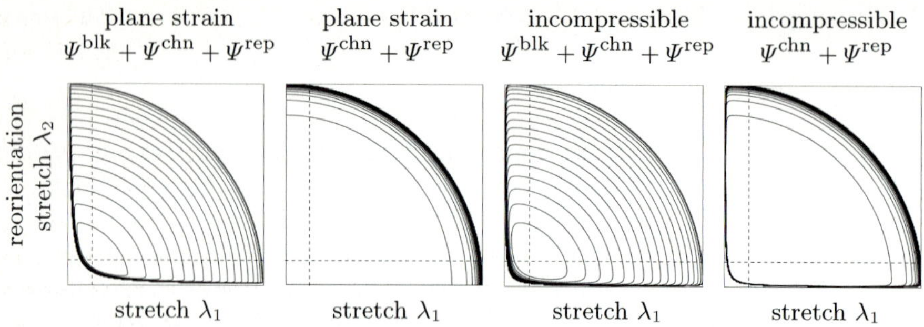

Fig. 4. Isotropic chain network model – energy contour plots Ψ.

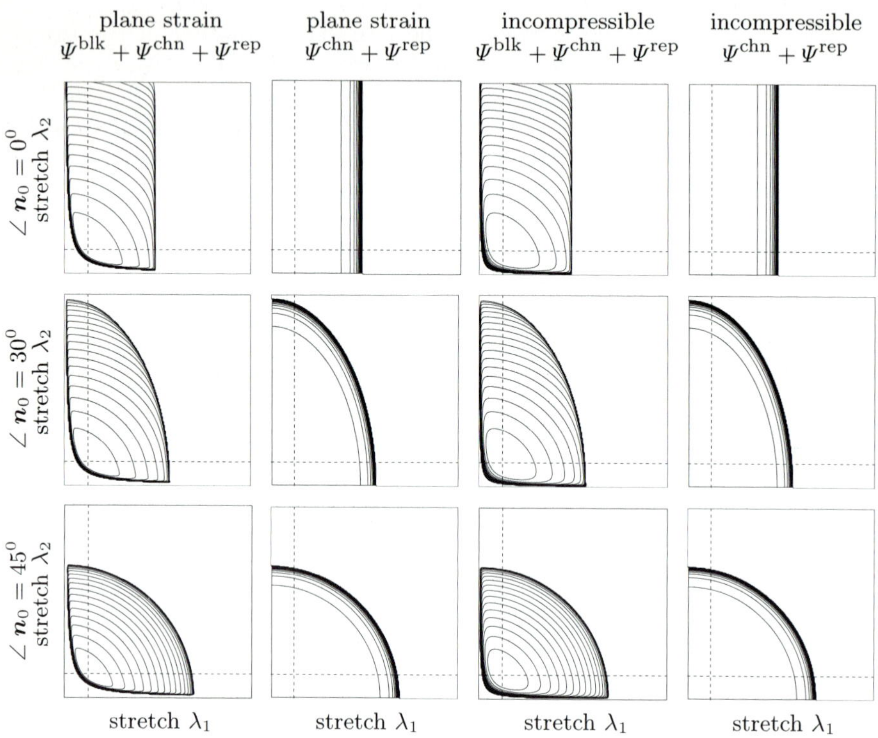

Fig. 5. Transversely isotropic chain model – energy contour plots Ψ.

correspond to the incompressible case. The first columns of each series, i.e. columns one and three, show the overall free energy $\Psi^{\text{blk}} + \Psi^{\text{chn}} + \Psi^{\text{rep}}$. In columns two and four we analyze the limiting case of a vanishing bulk term such that $\Psi^{\text{chn}} + \Psi^{\text{rep}}$ reflects the response of the chain and the repulsive contributions alone. All studies are bounded by the limit of validity $0 < r < L$ of the underlying single-chain model.

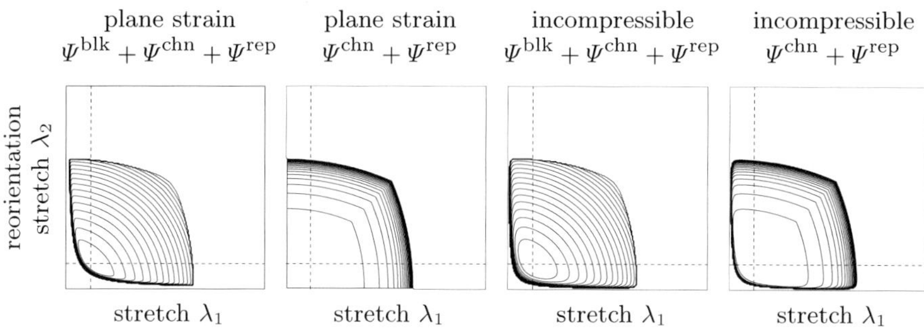

Fig. 6. Transversely isotropic chain network reorientation model – energy contour plots Ψ.

The first series of energy isolines given in Fig. 3 shows the response of the transversely isotropic chain network model with fixed unit cell axis. The effect of transverse isotropy becomes obvious since the isolines are not arranged radially symmetrically with respect to the minimum of the free energy. As expected, for angles $0° \geq \angle \boldsymbol{n}_0 > 45°$, the chain terms strengthen the material in the λ_1-direction. Thereby, directional strengthening is reflected through the densification of isolines corresponding to a lower locking stretch in the strong direction. The contour plots of the overall energies $\Psi^{\text{blk}} + \Psi^{\text{chn}} + \Psi^{\text{rep}}$ depicted in the first and third column show a smooth increase as the model reaches its limit of validity $r = L$ indicated by the outermost isolines. The isolines of the chain contributions $\Psi^{\text{chn}} + \Psi^{\text{rep}}$ depicted in the second and fourth column show a more pronounced densification close to the outermost isolines, which represent the limit of validity $r = L$ of the underlying wormlike-chain model. This isoline densification agrees nicely with the characteristic locking behavior close to the locking stretch $r = L$, as illustrated in Fig. 1, left.

Figure 4 depicts the corresponding isolines of the free energy Ψ of the isotropic chain network model with the particular choice of $a \equiv b$. The resulting chain network is no longer transversely isotropic since $\Psi^{\text{rep}}(I_1)$ and $\Psi^{\text{blk}}(I_1)$ no longer depend on the unit cell orientation $\boldsymbol{n}_0$. Isotropy is clearly demonstrated because the isolines are arranged radially symmetrically with respect to the minimum of the free energy.

Figure 5 illustrates the second special case of the general model by neglecting the network effects ($b \equiv 0$). Lacking the directional smoothing effect of the chain network, the transversely isotropic chain model of Fig. 5 with $\Psi^{\text{chn}}(I_4)$ and $\Psi^{\text{rep}}(I_4)$ shows a pronounced directional dependence. In particular, the chain contributions alone depicted in columns two and four are highly anisotropic. Non-symmetry of the isolines with respect to the minimum of the free energy is far more pronounced for the transversely isotropic chain model of Fig. 5 than for the corresponding network model of Fig. 3. Finally, we elaborate the case of remodeling. Accordingly, the unit cell normal $\boldsymbol{n}_0$ is aligned with the maximum principal strain direction $\boldsymbol{n}_\lambda^{\max}$. Figure 6 nicely reflects the symmetry with respect to the stretch space diagonal below which $\boldsymbol{n}_0$ is aligned with the principal direction related to λ_1, i.e. $\angle\boldsymbol{n}_0 = 0°$ and above which $\boldsymbol{n}_0$ is aligned with the λ_2 direction, i.e. $\angle\boldsymbol{n}_0 = 90°$. Remarkably, the reorientation model reveals the densest contour lines. Remodeling with respect to the principal strain axes, which was found to explain characteristic phenomena in biomechanics, obviously renders the strongest response of all test cases analyzed within the present study.

5 Example – Remodeling in Living Tendon

Finally, the suggested model will be applied to simulate a real biomechanical boundary value problem motivated by an experiment of engineered functional tendon constructs. Experimental observations of Calve et al. (2005) show that in the absence of loading, the *in vitro* grown tendon construct reveals the typical characteristics of an embryonic tendon demonstrated by the absence of a collagenous scaffold. It is only upon mechanical loading that collagen fibrils form within the tendon and orient themselves along the loading direction.

These observations motivated a preliminary finite element study in which we simulate the remodeling process in a cylindrical model tendon. The tendon material is described with the transversely isotropic wormlike-chain model for which an initially random fiber orientation is assumed to represent the neonatal state. The current fiber orientation $\boldsymbol{n}_0$ is thus introduced as an internal variable which is stored locally on the integration point level. The two wormlike-chain parameters, i.e. the contour length and the persistence length, take values $L = 2.50$ and $A = 1.82$, respectively. The bulk parameters take the values of $\gamma^{\text{blk}} = 100$ and $\beta = 4.5$, the chain density is chosen as $\gamma^{\text{chn}} = 7 \times 10^{21}$ and the absolute temperature as $\theta = 310$. The dimensions of the transversely isotropic unit cell are taken as $a = 2.43$ and $b = 1.85$. The relaxation time for the remodeling procedure is chosen as $\tau = 1.0$.

First, the model tendon is gradually stretched to about 200% of its initial length. The final load is then held constant to allow for fiber reorientation (see Fig. 7, left). The right-hand part of Fig. 7 shows the temporal evolution of the elongation of the tendon. During the loading phase, the elongation increases smoothly. Then, in the reorientation period, the tendon obviously stiffens considerably due to fiber reorientation. Accordingly, the overall elongation reduces until a final biological equilibrium state is reached.

Figure 8 depicts five representative stages of the remodeling history. The left figure shows the initial state at the beginning of the loading history. The arrows indicate the initial orientation of the unit cells $\boldsymbol{n}_0$ (in the biomechanical sense, the directions of pronounced collagen fiber orientation) that have been assigned randomly to each individual integration point. The contour levels ranging from white to black depict the fiber load angle α

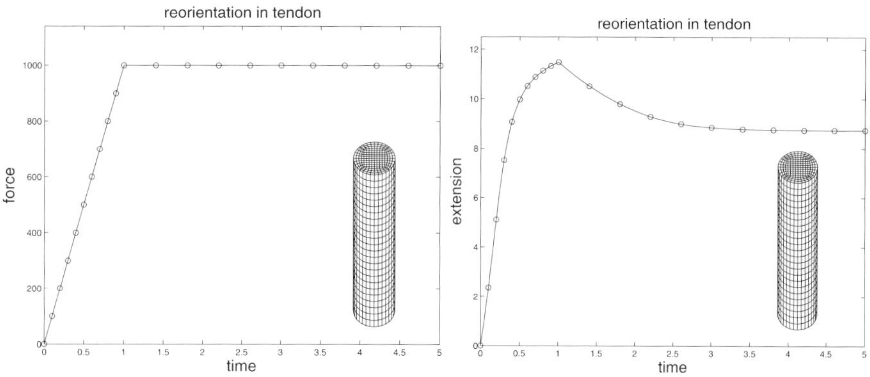

Fig. 7. Remodeling of soft biological tissue – prescribed loading and extension.

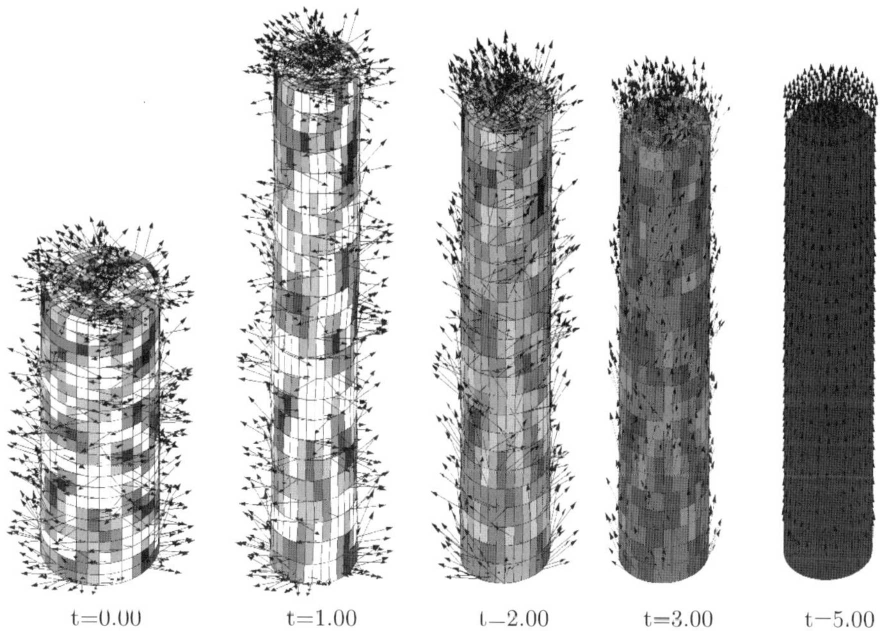

Fig. 8. Remodeling of soft biological tissue – strain-based fiber reorientation of initially randomly oriented collagen fibers in a cylindrical tendon.

varying from $\alpha = 90°$, i.e. the strong cell axis being orthogonal to the loading axis, to $\alpha = 0°$, i.e. a full alignment of the cell axis with the axis of loading.

Figure 8 from left to right nicely documents the history of remodeling. While the tendon is loaded uniaxially, a clear reorientation of the collagen fibers with respect to the loading axis can be observed. In this sense, the fifth figure on the right represents the final biological equilibrium state. No further remodeling takes place since all fibers are aligned with the maximum principal strain direction.

6 Discussion

A transversely isotropic chain network model has been discussed which is particularly suited for simulating remodeling of collagen fibers in soft biological tissue. A wormlike-chain model was applied to represent the individual collagen chains. The overall chain network is characterized by a transversely isotropic eight-chain unit cell. The present model can be understood as a generalization of the classical eight-chain model that captures the original isotropic eight-chain model and a unidirectional fiber reinforced model as special cases.

To incorporate biomechanical remodeling, the characteristic axis of the transversely isotropic unit cell was allowed to rotate with respect to the maximum principal strain direction. A theoretical framework of remodeling and its algorithmic treatment have been introduced. The characteristic unit cell axis has been introduced as an internal variable on the integration point level of a finite element realization. Contour plots of the free energy function in the principal stretch space have shown that, indeed, the reorientation approach yielded the stiffest material response. Finally, the approach presented has been compared qualitatively with the experimental findings of *in vitro* engineered tendon constructs. Being essentially based on micromechanical considerations, the model derived is governed by a limited number of physically-motivated material parameters. As such, it is believed to be ideally suited to simulate not only the passive behavior of soft biological tissues but also their active response to changes in the mechanical loading environment.

References

Arruda, E. M., and Boyce, M. C. (1993). A three-dimensional constitutive model for the large stretch behavior of rubber elastic materials. *J. Mech. Phys. Solids* 41:389–412.

Bischoff, J. E., Arruda, E. M., and Grosh, K. (2002). A microstructurally based orthotropic hyperelastic constitutive law. *J. Appl. Mech.* 69:570–579.

Boyce, M. C., and Arruda, E. M. (2000). Constitutive models of rubber elasticity: A review. *Rubber Chem. Technol.* 73:504–523.

Bustamante, C., Smith, S., Marko, J. F., and Siggia, E. D. (1994). Entropic elasticity of lambda-phage DNA. *Science* 265:1599–1600.

Calve, S., Dennis, R. G., Kosnik, P. E., Baar, K., Grosh, K., and Arruda, E. M. (2005). Engineering of functional tendon. *Tissue Eng.* in press.

Cowin, S. C. (1995). Optimization of the strain energy density in linear anisotropic elasticity. *J. Elasticity* 34:45–68.

Driessen, N. J. B., Peters, G. W. M., Huyghe, J. M., Bouten, C. V. C., and Baaijens, F. P. T. (2003). Remodelling of continuously distributed collagen fibres in soft connective tissue. *J. Biomech.* 36:1151–1158.

Flory, P. J. (1969). *Statistical Mechanics of Chain Molecules.* New York: Wiley – Interscience.

Garikipati, K., Arruda, E. M., Grosh, K., Narayanan, H., and Calve, S. (2004). A continuum treatment of growth in biological tissue: The coupling of mass transport and mechanics. *J. Mech. Phys. Solids* 52:1595–1625.

Holzapfel, G. A., and Ogden, R. W., eds. (2003). *Biomechanics of Soft Tissue in Cardiovascular Systems.* Wien – New York: Springer-Verlag.

Holzapfel, G. A., Gasser, T. C., and Ogden, R. W. (2000). A new constitutive framework for arterial wall mechanics and a comparative study of material models. *J. Elasticity* 61:1–48.

Holzapfel, G. A., Gasser, T. C., and Ogden, R. W. (2004). Comparison of a multi-layer structural model for arterial walls with a Fung-type model, and issues of material stability. *J. Biomech. Eng.* 126:264–275.

Holzapfel, G. A. (2000). *Nonlinear Solid Mechanics. A Continuum Approach for Engineering.* Chichester: John Wiley & Sons.

Humphrey, J. D. (2002). *Cardiovascular Solid Mechanics. Cells, Tissues, and Organs.* New York: Springer-Verlag.

Kratky, O., and Porod, G. (1949). Röntgenuntersushung gelöster Fadenmoleküle. *Recl. Trav. Chim. Pays-Bas.* 68:1106Ű–1123.

Kuhl, E., Garikipati, K., Arruda, E. M., and Grosh, K. (2005a). Remodeling of biological tissue: Mechanically induced reorientation of a transversely isotropic chain network. *J. Mech. Phys. Solids.* in press.

Kuhl, E., Menzel, A., and Garikipati, K. (2005b). On the convexity of transversely isotropic chain network models. *Phil. Mag. Lett.* in press.

Marko, J. F., and Siggia, E. D. (1995). Stretching DNA. *Macromolecules* 28:8759–8770.

Menzel, A. (2005). Modelling of anisotropic growth in biological tissues. A new approach and computational aspects. *Biomech. Model. Mechanobio.* 3:147–171.

Miehe, C., Göktepe, S., and Lulei, F. (2004). A micro-macro approach to rubber-like materials – Part I: The non-affine micro-sphere model of rubber elasticity. *J. Mech. Phys. Solids* 52:2617–2660.

Sgarra, C., and Vianello, M. (1997). Rotations which make strain and stress coaxial. *J. Elasticity* 47:217–224.

Treloar, L. R. G. (1975). *The Physics of Rubber Elasticity.* Oxford: Oxford University Press, 3rd edition.

Vianello, M. (1996). Coaxiality of strain and stress in anisotropic linear elasticity. *J. Elasticity* 42:283–289.

Anisotropic Remodelling of Biological Tissues

A. Menzel

University of Kaiserslautern, Chair of Applied Mechanics, Germany
amenzel@rhrk.uni-kl.de

Two different modelling concepts for anisotropic remodelling of biological tissues are discussed. Both approaches are based on the introduction of an additional field into the free energy, often described as a structural tensor, which enables us to model anisotropic elastic response. On the one hand, we directly set up evolution equations for the referential density and the structural tensor so that biological adaptation is captured. On the other hand, an incompatible growth map is incorporated that also evolves with the structural tensor field. The proposed remodelling framework of the overall adaptation process also addresses an alignment of the principal directions of the structural tensor with respect to the principal stretch directions.

1 Introduction

The modelling of hard and soft biological tissues nowadays constitutes a challenging task in continuum and computational mechanics. The reader is referred to the monographs by Cowin and Humphrey (2001), Humphrey (2002) and Holzapfel and Ogden (2003) for an overview on this research area. The incorporation of growth and/or remodelling effects and adaptation is of cardinal importance. Following the classification given by Rachev in (Holzapfel and Ogden (2003), pp. 221–271) we address a change in mass, structure and mechanical properties in this contribution but neglect any change in geometry of the body of interest. The mass change, associated with the evolution of the referential density, is therefore assumed to be driven solely by a source term; additional flux terms are neglected. Classical rate-dependent effects, such as viscoelasticity, are not discussed in this work since adaptation takes place on a larger time scale than typical creep or relaxation phenomena. Two different phenomenological modelling concepts are compared in what follows and we restrict attention to transversely isotropic bodies throughout. Taking the engineering point of view, this representative symmetry admits an interpretation in terms of a fibre-reinforced composite which nicely serves for the modelling of hard tissues, such as bones, and collagenous soft tissues. Even though special emphasis is placed on phenomenological settings in this contribution, similar concepts can also be applied to micro-mechanically motivated formulations such as chain models; see Kuhl et al. (2005a) and the contribution by Kuhl et al. (2005b) in this monograph or the review article by Beatty (2004).

On one hand, an incompatible growth map is defined which is additively decomposed into a spherical and a rank one contribution. The latter term is also introduced into the Helmholtz free energy density and consequently affects the elastic properties of the material. Mass change is directly related to the determinant of the growth map for the particular model of interest while the change in structure is captured via the correlated evolution of an appropriate deformation and structural tensor. The approach dates back to the contribution by Rodriguez et al. (1994) and was further elaborated in, e.g., Klisch et al. (2001) and Lubarda and Hoger (2002). This particular formulation is classified as 'Framework I' in what follows.

On the other hand, the theory of open systems is adopted. Since mass is a non-conserved quantity a source term for the referential density is incorporated. Practically speaking, the stress field that enters the balance of linear momentum is weighted by a relative referential density. Density evolution of some body of interest without deformation consequently results in vanishing stress fields. This is a fundamental difference compared to the previous approach where density evolution affects the underlying kinematics. For detailed overviews of the theory of open system mechanics the reader is referred to the pioneering work by Cowin and Hegedus (1976) and the contributions by in Epstein and Maugin (2000), Kuhl and Steinmann (2003a,b) and Kuhl et al. (2003). In this work, however, we adopt the approach that has recently been developed in Menzel (2005) in which a structural tensor is introduced into the Helmholtz free energy density, and its evolution obviously affects the elastic properties of the material. This particular formulation is classified as 'Framework II' in what follows.

2 Essential Kinematics

To set the stage, we briefly reiterate some essential kinematics of nonlinear continuum mechanics. The reader is referred to the monographs by, e.g., Ogden (1997) or Holzapfel (2000) for a detailed outline.

Let $\boldsymbol{X} \in \mathcal{B}_0 \subset \mathbb{E}^3$ denote placements of particles of the body B in the reference configuration and $\boldsymbol{x} = \boldsymbol{\varphi}(\boldsymbol{X}, t) \in \mathcal{B}_t \subset \mathbb{E}^3$ the corresponding non-linear motion that determines the positions of particles in the spatial configuration. The underlying motion gradient is assumed to be multiplicatively decomposed according to

$$\partial_{\boldsymbol{X}} \boldsymbol{\varphi} = \boldsymbol{F} = \boldsymbol{F}_{\mathrm{e}} \cdot \boldsymbol{F}_{\mathrm{g}} \,, \quad \boldsymbol{F}^{-1} = \boldsymbol{f} = \boldsymbol{f}_{\mathrm{g}} \cdot \boldsymbol{f}_{\mathrm{e}} \,, \quad J, J_{\mathrm{e}}, J_{\mathrm{g}}, j, j_{\mathrm{g}}, j_{\mathrm{e}} > 0, \qquad (1)$$

wherein $\boldsymbol{F}_{\mathrm{g}}$ denotes the incompatible linear growth map, which transforms infinitesimal line elements in the tangent space $T\mathcal{B}_0$ to the tangent space $T\mathcal{B}_{\mathrm{g}}$, and the notations $J, \ldots$ stand for the determinants $\det \boldsymbol{F}, \ldots$, respectively. It is now straightforward to introduce Finger-type deformation fields, namely

$$\begin{aligned} \boldsymbol{b} &= \boldsymbol{F} \cdot \boldsymbol{G}^{-1} \cdot \boldsymbol{F}^{\mathrm{t}} = \lambda_i^b \quad \boldsymbol{m}_i \otimes \boldsymbol{m}_i, \qquad \mathrm{L}_t \boldsymbol{b} = \boldsymbol{0}, \\ \boldsymbol{b}_{\mathrm{e}} &= \boldsymbol{F}_{\mathrm{e}} \cdot \boldsymbol{G}_{\mathrm{g}}^{-1} \cdot \boldsymbol{F}_{\mathrm{e}}^{\mathrm{t}} = \lambda_i^{b_{\mathrm{e}}} \boldsymbol{m}_{\mathrm{e}\,i} \otimes \boldsymbol{m}_{\mathrm{e}\,i}, \ \mathrm{L}_t \boldsymbol{b}_{\mathrm{e}} = -2[\boldsymbol{l}_{\mathrm{g}} \cdot \boldsymbol{b}_{\mathrm{e}}]^{\mathrm{sym}}, \end{aligned} \tag{2}$$

with $\boldsymbol{m}_i \cdot \boldsymbol{g} \cdot \boldsymbol{m}_j = \boldsymbol{m}_{\mathrm{e}\,i} \cdot \boldsymbol{g} \cdot \boldsymbol{m}_{\mathrm{e}\,j} = \delta_{ij}$ and $\boldsymbol{G}$, $\boldsymbol{G}_{\mathrm{g}}$, $\boldsymbol{g}$ constituting (covariant) metric tensors with respect to the material, intermediate and spatial configuration. The Lie derivative is represented by $\mathrm{L}_t \bullet = \boldsymbol{\varphi}_* \mathrm{D}_t(\boldsymbol{\varphi}^* \bullet)$, where D_t characterizes the material time derivative, φ_* and φ^* are, respectively the pushforward and pullback operators between $\mathcal{B}_0$ and $\mathcal{B}_t$, and the spatial growth velocity gradient reads $\boldsymbol{l}_{\mathrm{g}} = \boldsymbol{F}_{\mathrm{e}} \cdot \mathrm{D}_t \boldsymbol{F}_{\mathrm{g}} \cdot \boldsymbol{f} = -\boldsymbol{F} \cdot \mathrm{D}_t \boldsymbol{f}_{\mathrm{g}} \cdot \boldsymbol{f}_{\mathrm{e}}$. We place special emphasis on spatial arguments in this contribution; for a detailed outline of the multiplicative decomposition with respect to the reference and intermediate configurations, see Menzel and Steinmann (2003a).

Finally, recall that different mass densities can be introduced, the underlying connection being defined by the correlated transformation rules for infinitesimal volume elements

$$\mathrm{d}m = \rho_0 \,\mathrm{d}V = \rho_{\mathrm{g}} \,\mathrm{d}V_{\mathrm{g}} = \rho_t \,\mathrm{d}v \,, \qquad \rho_0 = J_{\mathrm{g}} \,\rho_{\mathrm{g}} = J \,\rho_t \,, \quad \rho_{\mathrm{g}} = J_{\mathrm{e}} \,\rho_t . \tag{3}$$

Note that standard balance equations for biological materials have to be modified since the referential density ρ_0 is in general not a conserved quantity. However, we do not reiterate these relations in this contribution but refer the reader to Kuhl and Steinmann (2003a), Menzel (2005) and references cited therein.

3 Constitutive Framework

Let the (covariant) Helmholtz free energy density ψ_0 be defined in terms of an appropriate deformation tensor, such as $\boldsymbol{b}$ or $\boldsymbol{b}_{\mathrm{e}}$. A symmetric second-order field, say $\boldsymbol{A} = \boldsymbol{A}^{\mathrm{t}} \in \mathcal{B}_0$ or $\boldsymbol{A}_{\mathrm{g}} = \boldsymbol{F}_{\mathrm{g}} \cdot \boldsymbol{A} \cdot \boldsymbol{F}_{\mathrm{g}}^{\mathrm{t}}$ or $\boldsymbol{a} = \boldsymbol{F} \cdot \boldsymbol{A} \cdot \boldsymbol{F}^{\mathrm{t}} \in \mathcal{B}_t$, which is specified later on, is additionally introduced and will appear, for instance, as a second argument defining ψ_0 and also connects $\boldsymbol{b}$ and $\boldsymbol{b}_{\mathrm{e}}$. Neglecting an explicit dependence of ψ_0 on material placements $\boldsymbol{X}$ and the temperature field we adopt the representation

$$\psi_0(\rho_{\mathrm{g}}, \boldsymbol{b}_{\mathrm{e}}, \boldsymbol{a}) = \psi_0^{\mathrm{iso}}(\rho_{\mathrm{g}}, \boldsymbol{b}_{\mathrm{e}}) + \psi_0^{\mathrm{ani}}(\rho_{\mathrm{g}}, \boldsymbol{b}_{\mathrm{e}}, \boldsymbol{a}) = \psi_0^{\mathrm{iso}}(\rho_0, I_{1,2,3}) + \psi_0^{\mathrm{ani}}(\rho_0, I_{1,\ldots,10}), \tag{4}$$

with the particular case $\boldsymbol{b}_{\mathrm{e}} \to \boldsymbol{b}$ and $\rho_{\mathrm{g}} \to \rho_0$ for $\boldsymbol{F}_{\mathrm{g}} \to \boldsymbol{I}$ being obvious, where $\boldsymbol{I}$ denotes the second order material identity. The framework of isotropic tensor functions gives us the invariants

$$\begin{aligned} I_i &= \boldsymbol{i} : [\boldsymbol{g} \cdot \boldsymbol{b}_{\mathrm{e}}]^i \,, & I_{3+i} &= \boldsymbol{i} : [\boldsymbol{b}_{\mathrm{e}}^{-1} \cdot \boldsymbol{a}]^i \,, & I_7 &= \boldsymbol{i} : [\boldsymbol{g} \cdot \boldsymbol{a}] \,, \\ I_8 &= \boldsymbol{i} : [\boldsymbol{g} \cdot \boldsymbol{a} \cdot \boldsymbol{g} \cdot \boldsymbol{b}_{\mathrm{e}}] \,, & I_9 &= \boldsymbol{i} : [\boldsymbol{b}_{\mathrm{e}}^{-1} \cdot \boldsymbol{a} \cdot \boldsymbol{g} \cdot \boldsymbol{a}] \,, & I_{10} &= \boldsymbol{i} : [\boldsymbol{g} \cdot \boldsymbol{a} \cdot \boldsymbol{g} \cdot \boldsymbol{a}] , \end{aligned} \tag{5}$$

where $i = 1, 2, 3$ and $\boldsymbol{i}$ characterizing the second-order spatial identity. Purely isotropic response is recovered for a vanishing or spherical tensor field $\boldsymbol{A}$ or simply by neglecting ψ_0^{ani}. The hyperelastic format of the Kirchhoff stress tensor $\boldsymbol{\tau}$ consequently reads

$$\boldsymbol{\tau} = 2\,\partial_{\boldsymbol{g}}\psi_0 = 2\,\boldsymbol{g}^{-1} \cdot [\,\partial_{\boldsymbol{b}_{\text{e}}}\psi_0 \cdot \boldsymbol{b}_{\text{e}} + \partial_{\boldsymbol{a}}\psi_0 \cdot \boldsymbol{a}\,], \tag{6}$$

which admits an interpretation as an extension of the celebrated (isotropic) Murnaghan-Truesdell formula to anisotropic response; see Menzel and Steinmann (2003a,b) for further details.

3.1 Framework I: Incompatible Growth Map

In this section we briefly review a transversely isotropic model, similar to the formulation recently advocated by Lubarda and Hoger (2002), which is essentially based on an incompatible growth map. The evolution of $\boldsymbol{F}_{\text{g}}$ is stress driven and $\boldsymbol{A} = \boldsymbol{n}_0 \otimes \boldsymbol{n}_0$ with $\boldsymbol{n}_0 = \text{constant}$ and $\|\boldsymbol{n}_0\| = \boldsymbol{n}_0 \cdot \boldsymbol{G} \cdot \boldsymbol{n}_0 = 1$ is assumed in what follows.

The fundamental Ansatz of this framework relies on the assumption that the growth map is additively decomposed into a spherical and a rank-one contribution. Moreover, the direction of the referential vector of the rank-one term is constrained to remain constant under the action of $\boldsymbol{F}_{\text{g}}$. Thus,

$$\boldsymbol{F}_{\text{g}} = \vartheta\,\boldsymbol{I} + [\,1 - \zeta^{-1}\,\vartheta\,]\,\boldsymbol{n}_{\text{g}} \otimes \boldsymbol{n}_0 \cdot \boldsymbol{G} \quad \text{with} \quad \boldsymbol{F}_{\text{g}} \cdot \boldsymbol{n}_0 = \boldsymbol{n}_{\text{g}} = \zeta\,\boldsymbol{n}_0 \quad \text{and} \quad \|\boldsymbol{n}_0\| = 1. \tag{7}$$

The inverse growth map then reads $\boldsymbol{f}_{\text{g}} \doteq \vartheta^{-1}\,\boldsymbol{I} + [\,\zeta^{-2} - \zeta^{-1}\,\vartheta^{-1}\,]\,\boldsymbol{n}_0 \otimes \boldsymbol{n}_{\text{g}} \cdot \boldsymbol{G}_{\text{g}}$, and by application of eqs (1) and (2) we obtain the elastic tangent map and the elastic Finger tensor as

$$\boldsymbol{F}_{\text{e}} = \vartheta^{-1}\,\boldsymbol{F} + [\,\zeta^{-2} - \zeta^{-1}\,\vartheta^{-1}\,]\,\boldsymbol{n}_t \otimes \boldsymbol{n}_{\text{g}} \cdot \boldsymbol{G}_{\text{g}} \quad \text{and} \quad \boldsymbol{b}_{\text{e}} = \vartheta^{-2}\,\boldsymbol{b} + [\,\zeta^{-2} - \vartheta^{-2}\,]\,\boldsymbol{a}, \tag{8}$$

where $\boldsymbol{a} = \boldsymbol{F} \cdot [\,\boldsymbol{n}_0 \otimes \boldsymbol{n}_0\,] \cdot \boldsymbol{F}^{\text{t}} = \boldsymbol{n}_t \otimes \boldsymbol{n}_t$. Based on these relations, the spatial growth velocity gradient takes the representation

$$\boldsymbol{l}_{\text{g}} = \vartheta^{-1}\,\text{D}_t\vartheta\,\boldsymbol{i} + [\,\zeta^{-1}\,\text{D}_t\zeta - \vartheta^{-1}\,\text{D}_t\vartheta\,]\,\boldsymbol{n}_t \otimes \boldsymbol{n}_0 \cdot \boldsymbol{G} \cdot \boldsymbol{f} \quad \text{so that} \quad \boldsymbol{l}_{\text{g}} \cdot \boldsymbol{b}_{\text{e}} = \boldsymbol{b}_{\text{e}} \cdot \boldsymbol{l}_{\text{g}}^{\text{t}}. \tag{9}$$

As an interesting side aspect, one additionally observes that the computation of the Lie derivative of the elastic Finger tensor results in

$$\text{L}_t\boldsymbol{b}_{\text{e}} = -\,2\,\vartheta^{-3}\,\text{D}_t\vartheta\,\boldsymbol{b} - 2\,[\,\zeta^{-3}\,\text{D}_t\zeta - \vartheta^{-3}\,\text{D}_t\vartheta\,]\,\boldsymbol{a} = -\,2\,\boldsymbol{l}_{\text{g}} \cdot \boldsymbol{b}_{\text{e}}. \tag{10}$$

In order to motivate the driving forces for the evolution of ϑ and ζ, we first place emphasis on the pointwise isothermal dissipation inequality

$$\mathcal{D}_0 = [\,\boldsymbol{g}\cdot\boldsymbol{\tau}\,] : \boldsymbol{l}_{\rm g} = [\,\boldsymbol{g}\cdot\boldsymbol{\tau}\cdot\boldsymbol{b}_{\rm e}^{-1}\,] : [\,\boldsymbol{l}_{\rm g}\cdot\boldsymbol{b}_{\rm e}\,] = -\tfrac{1}{2}\,[\,\boldsymbol{g}\cdot\boldsymbol{\tau}\,] : [\,{\rm L}_t\boldsymbol{b}_{\rm e}\cdot\boldsymbol{b}_{\rm e}^{-1}\,] \quad (11)$$

and, second, introduce two (sub)sets of invariants, say

$$H_i = \boldsymbol{i} : [\,\boldsymbol{g}\cdot\boldsymbol{\tau}\,]^i\,, \qquad H_{3+i} = \boldsymbol{i} : [\,\boldsymbol{g}\cdot\boldsymbol{\tau}\cdot\boldsymbol{g}\cdot\boldsymbol{a}\,]^i, \qquad i = 1,2,3, \quad (12)$$

by analogy with the definition of a plasticity-type yield function as an isotropic tensor function in terms of $\boldsymbol{\tau}$ and $\boldsymbol{a}$. With these relations in hand, we finally adopt the saturation-type evolution equations

$$\begin{aligned} {\rm D}_t\vartheta &= k_\vartheta\;[\,[\,1+\nu_\vartheta\,]\,H_1 - \nu_\vartheta\,H_4\,]\,, \\ k_\vartheta &= \begin{cases} k_\vartheta^+\left[\dfrac{\vartheta^+ - \vartheta}{\vartheta^+ - 1}\right]^{m_\vartheta^+} & \text{for } H_4 > \dfrac{\nu_\vartheta\,H_1}{1+\nu_\vartheta} \\[2ex] k_\vartheta^-\left[\dfrac{\vartheta - \vartheta^-}{1 - \vartheta^-}\right]^{m_\vartheta^-} & \text{for } H_4 < \dfrac{\nu_\vartheta\,H_1}{1+\nu_\vartheta}\,, \end{cases} \end{aligned} \quad (13)$$

$$\begin{aligned} {\rm D}_t\zeta &= k_\zeta\;[\,[\,1+\nu_\zeta\,]\,H_1 - \nu_\zeta\,H_4\,]\,, \\ k_\zeta &= \begin{cases} k_\zeta^+\left[\dfrac{\zeta^+ - \zeta}{\zeta^+ - 1}\right]^{m_\zeta^+} & \text{for } H_4 > \dfrac{\nu_\zeta\,H_1}{1+\nu_\zeta} \\[2ex] k_\zeta^-\left[\dfrac{\zeta - \zeta^-}{1 - \zeta^-}\right]^{m_\zeta^-} & \text{for } H_4 < \dfrac{\nu_\zeta\,H_1}{1+\nu_\zeta}\,. \end{cases} \end{aligned} \quad (14)$$

The density field can be computed via 'some kind of post-processing': following Lubarda and Hoger (2002), the referential density source takes the representation

$$\begin{aligned} {\rm D}_t\rho_0 &= R_0 = \rho_0\,J_{\rm g}\,{\rm D}_tJ_{\rm g} \\ &= \rho_0\,\boldsymbol{f}_{\rm g}^{\rm t} : {\rm D}_t\boldsymbol{F}_{\rm g} = \rho_0\,\boldsymbol{i}^{\rm t} : \boldsymbol{l}_{\rm g} = \rho_0\,[\,\zeta^{-1}\,{\rm D}_t\zeta + 2\,\vartheta^{-1}\,{\rm D}_t\vartheta\,] \end{aligned} \quad (15)$$

together with the reasonable initial conditions $\rho_0|_{t_0} = \rho_0^* > 0$ and $\vartheta|_{t_0} = \zeta|_{t_0} = 1$. The solution of the integral of ${\rm D}_t\rho_0 = R_0$ over the time interval of interest yields the relation

$$\rho_0 = \zeta\,\vartheta^2\,\rho_0^* = J_{\rm g}\,\rho_0^* \quad \text{so that} \quad \rho_{\rm g} = j_{\rm g}\,\rho_0 = \rho_0^*, \quad (16)$$

and it is straightforward to verify $J_{\rm g} = \zeta\,\vartheta^2$ via the Sherman-Morrison-Woodbury theorem; see, e.g., Householder (1975). Apparently, $\rho_{\rm g}$ remains constant during the deformation process, which motivates the classification of this model as 'density preserving growth' (with respect to the intermediate configuration). The determinant of the growth map then has the interpretation of a relative referential density $\xi = \rho_0/\rho_0^* = J_{\rm g}$.

3.2 Framework II: Structural Tensor

In this section we briefly reiterate a transversely isotropic model which is essentially based on the introduction of a structural tensor, $\boldsymbol{A} = \lambda^A\,\boldsymbol{n}_0\otimes\boldsymbol{n}_0$, and has recently been elaborated in Menzel (2005). The evolution of the referential density ρ_0 as well as the norm of the structural tensor λ^A is energy driven and $\boldsymbol{n}_0$ = constant, where $\|\boldsymbol{n}_0\| = \boldsymbol{n}_0 \cdot \boldsymbol{G} \cdot \boldsymbol{n}_0 = 1$ is assumed in what follows.

Since the growth map is enforced to remain constant, i.e. $\boldsymbol{F}_{\rm g} = \boldsymbol{I}$, the Helmholtz free energy density depends on ρ_0, $\boldsymbol{b}$ and $\boldsymbol{a}$. To be specific, we choose the density weighted format

$$\psi_0(\rho_0, \boldsymbol{b}, \boldsymbol{a}) = \xi^n \left[\psi_0^{\rm iso}(\rho_0, \boldsymbol{b}) + \psi_0^{\rm ani}(\rho_0, \boldsymbol{b}, \boldsymbol{a}) \right] \quad \text{with} \quad \xi = \rho_0 / \rho_0^*, \tag{17}$$

where $\rho_0^* > 0$ denotes some fixed initial value for the density field. The associated hyperelastic stress tensors turn out to be density weighted as well. For the referential density evolution, we adopt the formulation advocated by Harrigan and Hamilton (1994), namely

$$\mathrm{D}_t \rho_0 = R_0 = \xi^{-m}\, \psi_0 - {}^{\rho_0}\psi_0^*, \tag{18}$$

which has been proven to meet existence and uniqueness of a global energy minimum if $m > n$ (at least within a small strain and isotropic setting). The material parameter ${}^{\rho_0}\psi_0^* > 0$ is often designated as the density stimulus. The structural tensor $\boldsymbol{a}$ is a function of λ^A, $\boldsymbol{n}_0$ and $\boldsymbol{F}$. In what follows we assume $\boldsymbol{n}_0$ = constant and choose an energy driven evolution for the fibre diameter λ^A (equivalent to a measure of fibre strength) by analogy with the evolution of the referential density ρ_0. However, it is assumed that only the energy correlated to the stretch (to the power of two) along the fibre direction drives λ^A. The modified driving quantity is consequently introduced as $\psi_0^a = \psi_0(\rho_0, \boldsymbol{b}^a, \boldsymbol{a})$ with $\boldsymbol{b}^a = \boldsymbol{g}^{-1} + \left[1 - [\boldsymbol{n}_t \cdot \boldsymbol{g} \cdot \boldsymbol{n}_t]^{-1} \right] \boldsymbol{n}_t \otimes \boldsymbol{n}_t$, which enables us to define

$$\mathrm{D}_t \lambda^A = [\, \Lambda^a(\rho_0, \boldsymbol{b}^a, \boldsymbol{a}) \,]^{-l}\, \psi_0^a - {}^{\lambda^A}\psi_0^*, \tag{19}$$

with $l > 1$. The function $[\, \Lambda^a \,]^{-l}$ is chosen by analogy with ξ^{-m} in eq. (17) in order to obtain saturation type behaviour, i.e. $[\, \Lambda^a \,]^{-l} \searrow 0$ for $\lambda^A \nearrow \infty$. For further details, including remarks on the dissipation inequality, see Menzel (2005).

3.3 Reorientation

Until now, the fibre direction $\boldsymbol{n}_0$ has been assumed to remain constant during the deformation process. In this section we reiterate a formulation, fully outlined in Menzel (2005), that aligns the fibre direction with the predominant principal stretch direction in a viscous manner. In the case that the alignment is completed, the correlated stress and deformation tensors commute. The proposed framework can be applied to 'Framework I', based on the growth map, as well as to 'Framework II' - the structural tensor approach.

A general representation for the fibre direction evolution, as represented by the Lie derivative $\mathrm{L}_t \boldsymbol{n}_t = \boldsymbol{F} \cdot \mathrm{D}_t \boldsymbol{n}_0$, is provided by the Ansatz

$$\mathrm{L}_t \boldsymbol{n}_t = [\, \boldsymbol{b} \cdot \boldsymbol{w} - [\, \boldsymbol{n}_t \cdot \boldsymbol{w} \cdot \boldsymbol{n}_t \,] \, \boldsymbol{i} \,] \cdot \boldsymbol{n}_t \quad \text{so that} \quad \mathrm{L}_t \boldsymbol{n}_t \cdot \boldsymbol{b}^{-1} \cdot \boldsymbol{n}_t = 0, \tag{20}$$

where the properties $\boldsymbol{n}_t \cdot \boldsymbol{b}^{-1} \cdot \boldsymbol{n}_t = 1$ and $\|\boldsymbol{n}_0\| = 1$ have been used; see, e.g., Imatani and Maugin (2002). The second-order tensor $\boldsymbol{w}$ should possess real eigenvalues, so that $\boldsymbol{b} \cdot \boldsymbol{w}$ also possesses real eigenvalues; see, e.g., Ericksen (1960). Since eq. (20) formally coincides with the structure of an eigenvalue problem, it is obvious that $\boldsymbol{n}_t$ aligns according to one of the principal directions of $\boldsymbol{w}$. The choice of $\boldsymbol{w}$ affects the reorientation of $\boldsymbol{n}_t$. Different second-order tensors can be chosen for $\boldsymbol{w}$; e.g., the Mandel type stress field $\boldsymbol{g} \cdot \boldsymbol{\tau} \cdot \boldsymbol{b}^{-1}$. In this work, however, we apply an alignment with respect to appropriate stretch tensors such that correlated stress fields and deformation tensors finally commute. To be specific, $\boldsymbol{n}_t$ is assumed to follow the predominant stretch direction, i.e. $\boldsymbol{w} = \boldsymbol{b}^{-1} \cdot \boldsymbol{m}_3 \otimes \boldsymbol{m}_3 \cdot \boldsymbol{g}$ for $\lambda_3^b > \lambda_2^b \geq \lambda_1^b$ and $\mathrm{L}_t \boldsymbol{n}_t = \boldsymbol{0}$ otherwise. This motivates the particular Ansatz

$$\mathrm{L}_t \boldsymbol{n}_t = \boldsymbol{\omega} \times [\, \boldsymbol{b}^{-1} \cdot \boldsymbol{n}_t \,] = [\, -\boldsymbol{\varepsilon}_t \cdot \boldsymbol{\omega} \,] \cdot [\, \boldsymbol{b}^{-1} \cdot \boldsymbol{n}_t \,] \quad \text{with} \quad \boldsymbol{\omega} = \frac{\pi}{2 t^* \lambda_3^b} \, \boldsymbol{n}_t \times \boldsymbol{m}_3, \tag{21}$$

where $\boldsymbol{\omega}$ and $\boldsymbol{\varepsilon}_t$ denote the underlying angular velocity and a (mixed-variant) permutation tensor of third order. The parameter t^* acts similarly as a relaxation time. Moreover, recalling eq. (8), we note that if $\boldsymbol{a}$ and $\boldsymbol{b}$ commute then $\boldsymbol{a}$ and $\boldsymbol{b}_\mathrm{e}$ commute too. Hence, we do not distinguish between $\boldsymbol{b}$ and $\boldsymbol{b}_\mathrm{e}$ in eqs (20) and (21) since an alignment of $\boldsymbol{a}$ with $\boldsymbol{b}$ automatically renders an alignment of $\boldsymbol{a}$ with $\boldsymbol{b}_\mathrm{e}$. By analogy with eq. (20), it is straightforward to verify the orthogonality $\mathrm{L}_t \boldsymbol{n}_t \cdot \boldsymbol{b}^{-1} \cdot \boldsymbol{n}_t = 0$ in eq. (21) via the '$\varepsilon\,\delta$-rule'; to be specific,

$$\mathrm{L}_t \boldsymbol{n}_t = \frac{\pi}{2\, t^* \, \lambda_3^b} \left[\, [\, \boldsymbol{n}_t \cdot \boldsymbol{b}^{-1} \cdot \boldsymbol{n}_t \,] \, \boldsymbol{m}_3 - [\, \boldsymbol{n}_t \cdot \boldsymbol{b}^{-1} \cdot \boldsymbol{m}_3 \,] \, \boldsymbol{n}_t \,\right]. \tag{22}$$

For further details, including remarks on the dissipation inequality, see Menzel (2005).

4 Numerical Examples

The subsequent numerical examples are based on the Ansatz

$$\rho_0 \, \psi^{\mathrm{iso}} = \rho_0 \, \psi^{\mathrm{nh}} - \mu \ln(J) + \frac{\lambda}{2} \ln^2(J), \qquad \rho_0 \, \psi^{\mathrm{ani}} = \frac{\alpha}{2\beta} \left[\, \exp(\beta \, [\, I_7 - I_4 \,]^2) - 1 \,\right] \tag{23}$$

for the Helmholtz free energy density with $\rho_0 \, \psi^{\mathrm{nh}} = \mu \, [\, I_1 - 3 \,]/2$ and $\lambda, \mu, \alpha, \beta, J > 0$ as well as $J = \det(\boldsymbol{F}) = [\, 2 I_3 - 3 I_1 I_2 + I_1^3 \,]^{\frac{1}{2}} / \sqrt{6}$. The anisotropic contribution has been adapted from Holzapfel et al. in

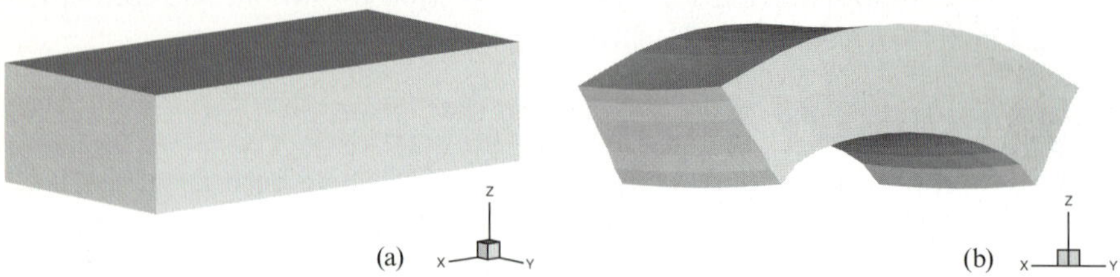

Fig. 1. Elastic response: undeformed (a) and deformed (b) specimen at time step 2.

(Cowin and Humphrey (2001), pp. 1–48). Throughout, fully implicit techniques are chosen for the integration of the underlying evolution equations. To be specific, for ϑ, ζ, ρ_0 and λ^A an Euler backward scheme is applied. Moreover, an exponential technique, such as that represented by the Euler-Rodrigues formula, is adopted for the integration of $\mathrm{D}_t\boldsymbol{n}_0$, which determines $\mathrm{L}_t\boldsymbol{n}_t = \boldsymbol{F}\cdot\mathrm{D}_t\boldsymbol{n}_0$; see Menzel (2005) for further details. In order to visualize non-commutativity of two second-order tensors, such as correlated stretch and stress fields for general anisotropic response, a scalar-valued quantity is introduced, namely

$$\delta(\boldsymbol{b},\boldsymbol{\tau}) = \|\,\boldsymbol{b}\cdot\boldsymbol{g}\cdot\boldsymbol{\tau} - \boldsymbol{\tau}\cdot\boldsymbol{g}\cdot\boldsymbol{b}\,\| / [\,\|\,\boldsymbol{b}\,\|\;\|\,\boldsymbol{\tau}\,\|\,]\,. \tag{24}$$

4.1 Bending of an Isotropic Specimen

To set the stage, we first consider an isotropic specimen of dimensions $4\times2\times2$. Bending type loading enables us to address different characteristics under tension and compression for frameworks I and II. Dirichlet boundary conditions are applied within 2 time steps so that the longitudinal (prescribed) displacements at the top and bottom edges at the (y, z) ends of the body correspond to $\pm1/3$; see Fig. 1, where the deformation associated with purely elastic response is highlighted. Then, these displacement constraints are fixed for the subsequent 13 times steps. Due to the underlying symmetry, one quarter of the specimen is modelled with $12\times6\times6$ finite elements (Q1). The chosen material parameters are $\rho_0^* = 1$, $\mu = 3.103$, $\lambda = 27.793$, $\alpha = 0$, $\Delta t = 0.1$ and $\vartheta|_{t_0} = 1$, $k_\vartheta^+ = 1$, $\vartheta^+ = 1.5$, $m_\vartheta^+ = 1.5$, $k_\vartheta^- = 1$, $\vartheta^- = 0.5$, $m_\vartheta^- = 1.5$, $\nu_\vartheta = 0$, $\zeta = 1$ (for Framework I) as well as $n = 2.5$, $m = 3$, ${}^{\rho_0}\psi_0^* = 0.2$ (Framework II).

Figures 2 (Framework I) and 3 (Framework II) illustrate the contribution of the relative referential density. Besides the differences in the results for this field, we additionally observe a pronounced increase in deformation for the last 13 time steps within Framework I.

4.2 Tension of an Anisotropic Specimen

We now consider a specimen under tension and additionally activate the anisotropic fibre response and reorientation. The discretization of the specimen

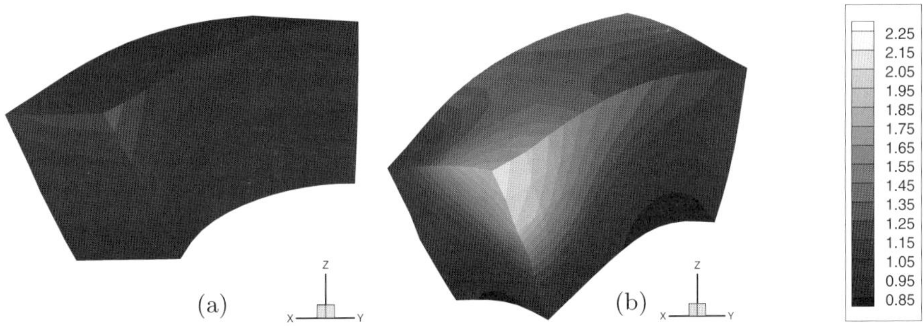

Fig. 2. Framework I: relative referential density $\xi = \vartheta^2$ at time steps 2 (a) and 15 (b).

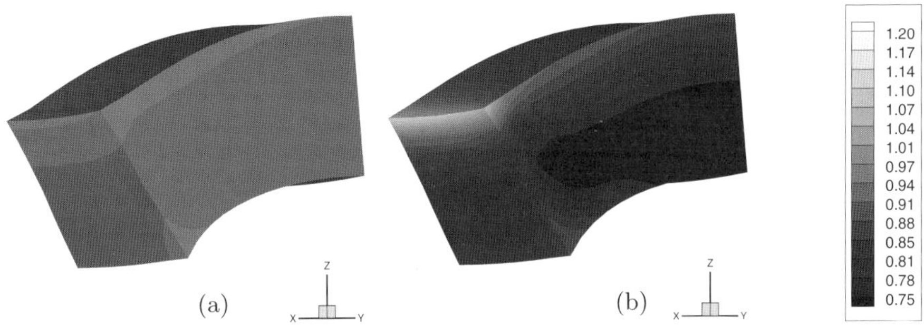

Fig. 3. Framework II: relative referential density $\xi = \rho_0 / \rho_0^*$ at time steps 2 (a) and 15 (b).

of dimensions $2 \times 1 \times 1$ is performed via $12 \times 6 \times 6$ finite elements (Q1E9). One end of the body is totally clamped while the displacements of the finite element nodes at the other end are enforced to be identical. The deformation process is force driven, so that the body might elongate or contract under a prescribed load – whether the material softens or hardens. To be specific, we increase the applied longitudinal force within the first two time steps up to an amount of 2 and fix this force for the subsequent 13 time steps; see Fig. 4, where the deformation for purely elastic response is highlighted. The chosen material parameters are $\rho_0^* = 1$, $\mu = 3.103$, $\lambda = 27.793$, $\alpha = 2$, $\beta = 1$, $\boldsymbol{n}_0|_{t_0} = [\boldsymbol{e}_1 + \boldsymbol{e}_2]/\sqrt{2}$, $\Delta t = 0.1$, $t^* = 10$ and $\vartheta|_{t_0} = 1$, $k_\vartheta^+ = 0.05$, $\vartheta^+ = 1.5$, $m_\vartheta^+ = 1.5$, $k_\vartheta^- = 0.05$, $\vartheta^- = 0.5$, $m_\vartheta^- = 1.5$, $\nu_\theta = 0.5$, $k_\zeta^+ = 0.05$, $\zeta^+ = 1.5$, $m_\zeta^+ = 1.5$, $k_\zeta^- = 0.05$, $\zeta^- = 0.5$, $m_\zeta^- = 1.5$, $\nu_\zeta = 0.5$ (for Framework I) as well as $n = 2.5$, $m = 3$, $l = 1.1$, ${}^{\rho_0}\psi_0^* = 0.2$, ${}^{\lambda^A}\psi_0^* = 0.1$ (Framework II).

Figures 5 (Framework I) and 6 (Framework II) illustrate the contribution of the relative referential density, the fibre parameter and two anisotropy measures. The non-coaxiality of stretch and stress, stemming from the anisotropic elastic response, is represented by $\delta(\boldsymbol{b}, \boldsymbol{\tau})$, while the fibre reorientation results in non-vanishing contributions $\delta(\boldsymbol{A}_0, \boldsymbol{A}_0|_{t_0})$, i.e. the (referential) initial and (referential) actual structural tensors do not commute. Besides the differences

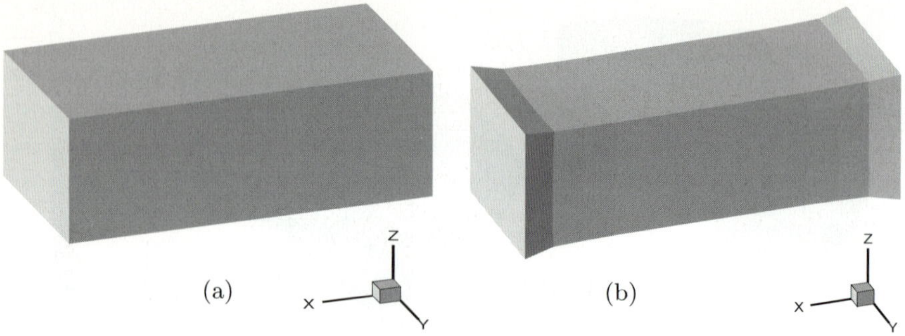

Fig. 4. Elastic response: undeformed (a) and deformed (b) specimen at time step 2.

in the results for these fields, we additionally observe a pronounced increase in deformation for the last 13 time steps within Framework I and a similar displacement field for elastic behaviour in Framework II (the specimen slightly contracts in Framework II).

5 Discussion

The multiplicative decomposition of the deformation gradient, as it has been applied within Framework I, is well-established in, e.g., crystal plasticity. In the present context, however, $\boldsymbol{F}_{\mathrm{g}}$ is assumed to be symmetric such that no growth spin occurs. Moreover, the correlated growth velocity gradient is not trace-less, which enabled us to incorporate volumetric growth. Inherently, growth, or remodelling and/or adaptation, means that for the considered formulation the deformation field entering the constitutive law is modified. Volumetric growth and isochoric contributions can therefore be addressed separately. Moreover, Framework I allows modelling of a density increase at constant volume, a volume increase at constant density and combinations thereof. In this work, a constant density with respect to the intermediate configuration has been taken into account.

The theory of open systems, as it has been applied within Framework II, gave us the density weighted stress fields at hand. Furthermore, the fibre strength evolution affects the structural tensor and hence also the stress tensors. Deformation measures, however, have not been directly modified by either the density field or the fibre strength.

Both frameworks were based on the same reorientation model, namely that the fibre aligns according to the predominant stretch direction in a viscous manner. Besides the kinematic fields entering the constitutive functions, one main difference between the two formulations is in the forces driving the adaptation process – Framework I is essentially stress driven while Framework II is energy driven. For a more realistic comparison of these approaches, both formulations should probably be based on similar driving forces, and

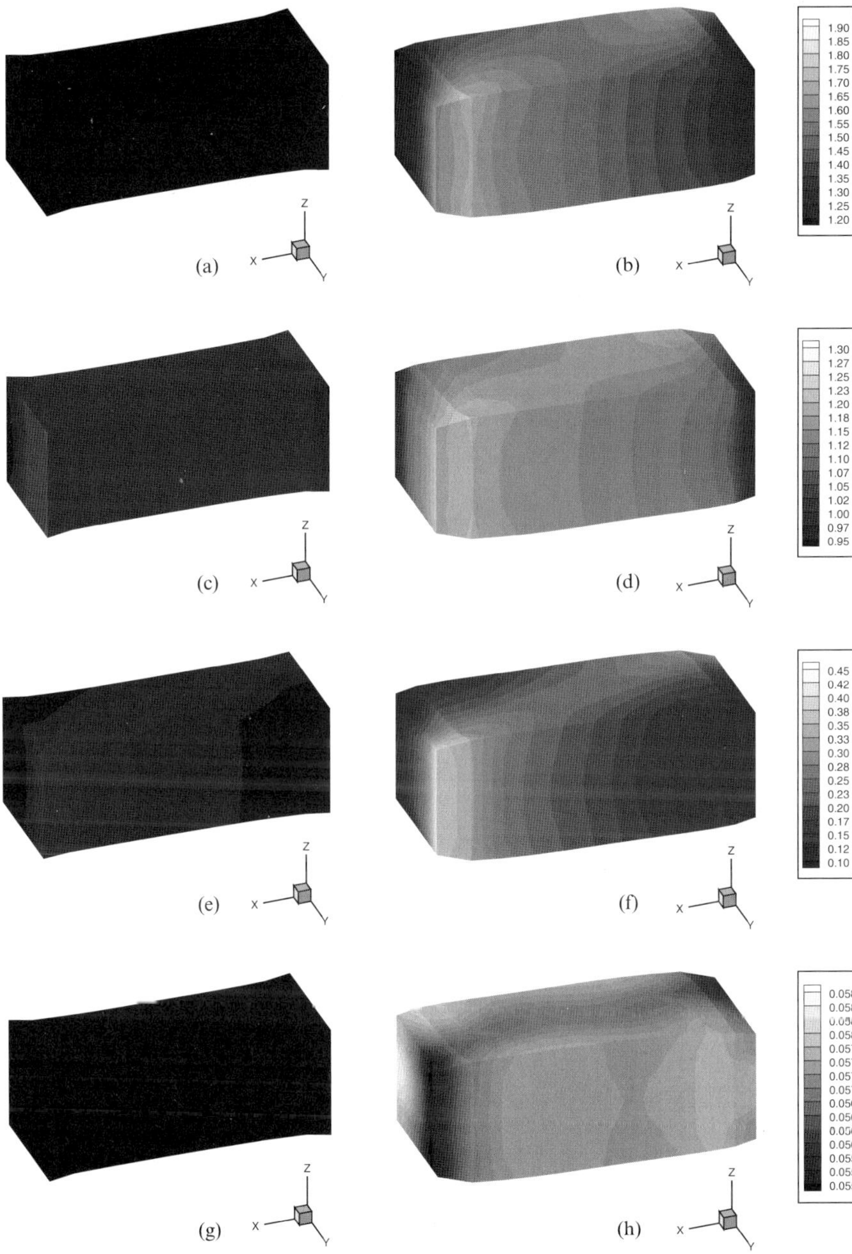

Fig. 5. Framework I: relative referential density $\xi = \zeta\,\vartheta^2$ at time steps 2 (a) and 15 (b), fibre parameter ζ at time steps 2 (c) and 15 (d), anisotropy measure $\delta(\boldsymbol{b}, \boldsymbol{\tau})$ at time steps 2 (e) and 15 (f), anisotropy measure $\delta(\boldsymbol{A}_0, \boldsymbol{A}_0|_{t_0})$ at time steps 2 (g) and 15 (h).

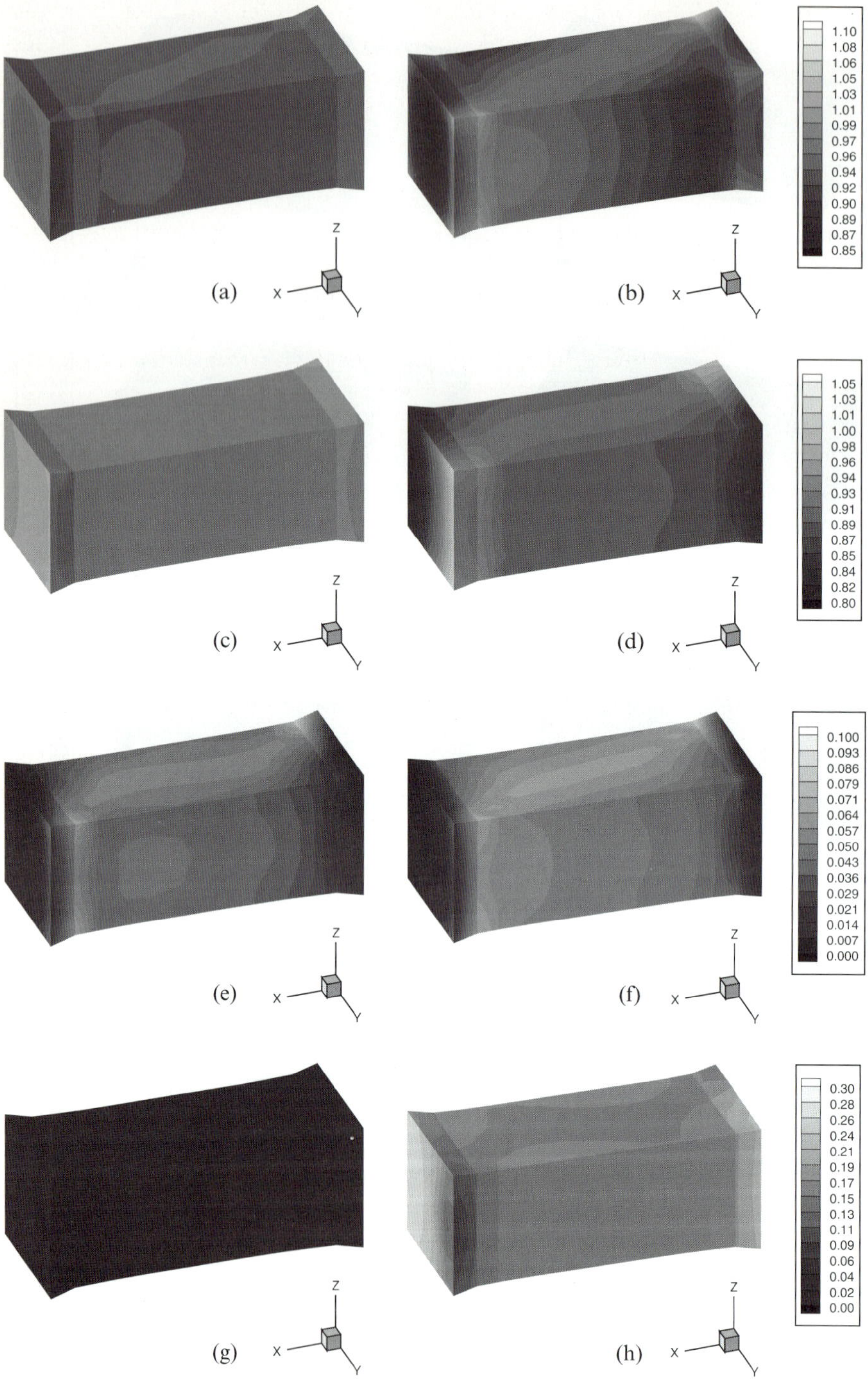

Fig. 6. Framework II: relative referential density $\xi = \rho_0/\rho_0^*$ at time steps 2 (a) and 15 (b), fibre parameter λ^A at time steps 2 (c) and 15 (d), anisotropy measure $\delta(\boldsymbol{b}, \boldsymbol{\tau})$ at time steps 2 (e) and 15 (f), anisotropy measure $\delta(\boldsymbol{A}_0, \boldsymbol{A}_0|_{t_0})$ at time steps 2 (g) and 15 (h).

Framework II should also be enabled to admit different responses in tension and compression. Finally, the calibration of the underlying material parameters constitutes future research.

References

Beatty, M. F. (2004). An average–stretch full–network model for rubber elasticity. *J. Elasticity* 70:65–86.

Cowin, S. C., and Hegedus, D. H. (1976). Bone remodeling I: Theory of adaptive elasticity. *J. Elasticity* 6:313–326.

Cowin, S. C., and Humphrey, J. D., eds. (2001). *Cardiovascular Soft Tissue Mechanics.* Dordrecht: Kluwer Academic Publishers.

Epstein, M., and Maugin, G. A. (2000). Thermomechanics of volumetric growth in uniform bodies. *Int. J. Plasticity* 16:951–978.

Ericksen, J. L. (1960). Tensor fields. In Flügge, S., ed., *Encyclopedia of Physics*, volume III/1. Springer. 794–858.

Harrigan, T. P., and Hamilton, J. J. (1994). Necessary and sufficient conditions for global stability and uniqueness in finite element simulations of adaptive bone remodelling. *Int. J. Solids Structures* 31:97–107.

Holzapfel, G. A., and Ogden, R. W., eds. (2003). *Biomechanics of Soft Tissue in Cardiovascular Systems.* Wien – New York: Springer-Verlag.

Holzapfel, G. A. (2000). *Nonlinear Solid Mechanics. A Continuum Approach for Engineering.* Chichester: John Wiley & Sons.

Householder, A. S., ed. (1975). *The Theory of Matrices in Numerical Analysis.*

Humphrey, J. D. (2002). *Cardiovascular Solid Mechanics. Cells, Tissues, and Organs.* New York: Springer-Verlag.

Imatani, S., and Maugin, G. (2002). A constitutive model for material growth and its application to three–dimensional finite element analysis. *Mech. Res. Commun.* 29:477–483.

Klisch, S. M., Van Dyke, T. J., and Hoger, A. (2001). A theory of volumetric growth for compressible elastic biological materials. *Math. Mech. Solids* 6:551–575.

Kuhl, E., and Steinmann, P. (2003a). Mass– and volume specific views on thermodynamics for open systems. *Proc. R. Soc. Lond. A* 459:2547–2568.

Kuhl, E., and Steinmann, P. (2003b). Theory and numerics of geometrically nonlinear open system mechanics. *Int. J. Numer. Meth. Eng.* 58:1593–1615.

Kuhl, E., Menzel, A., and Steinmann, P. (2003). Computational modeling of growth – A critical review, a classification of concepts and two new consistent approaches. *Comput. Mech.* 32:71–88.

Kuhl, E., Menzel, A., and Garikipati, K. (2005a). On the convexity of transversely isotropic chain network models. *Phil. Mag. Lett.* in press.

Kuhl, E., Menzel, A., Garikipati, K., Arruda, E. M., and Grosh, K. (2005b). Modeling and simulation of remodeling in soft biological tissues. In Holzapfel, G. A., and Ogden, R. W., eds., *Mechanics of Biological Tissue*, 77–89. Heidelberg: Springer-Verlag.

Lubarda, V. A., and Hoger, A. (2002). On the mechanics of solids with a growing mass. *Int. J. Solids Structures* 39:4627–4664.

Menzel, A., and Steinmann, P. (2003a). On the spatial formulation of anisotropic multiplicative elasto–plasticity. *Comput. Meth. Appl. Mech. Eng.* 192:3431–3470.

Menzel, A., and Steinmann, P. (2003b). A view on anisotropic finite hyper–elasticity. *Eur. J. Mech. A/Solids* 22:71–87.

Menzel, A. (2005). Modelling of anisotropic growth in biological tissues. A new approach and computational aspects. *Biomech. Model. Mechanobio.* 3:147–171.

Ogden, R. W. (1997). *Non-linear Elastic Deformations.* New York: Dover.

Rodriguez, E. K., Hoger, A., and McCulloch, A. D. (1994). Stress-dependent finite growth in soft elastic tissues. *J. Biomech.* 27:455–467.

A Mechanobiological Formulation of Bone Healing

M. Doblaré, J.M. García-Aznar, M.J. Gómez-Benito

University of Zaragoza, Aragón Institute of Engineering Research (I3A), Spain
mdoblare@unizar.es

The main purpose of this work is to present a theoretical continuum formulation for modelling growth, differentiation and damage of living tissues, valid for both hard and soft tissues. The governing equations have been treated following classical continuum mechanics, considering the influence of the extracellular matrix concentrations and cellular populations. Finally, the present formulation has been particularized to the modelling of bone fracture healing.

1 Introduction

Theoretical analysis of the development of living tissues has been an important research topic for many years. Early works included a theoretical model of the mass evolution in a living bone by Cowin and Hegedus (1976) that accounted for the fluid phase via irreversible sources and fluxes of momentum, energy and entropy. The kinematics of growth was first described analytically by Skalak et al. (1982). Growth is usually described in the biomechanics literature for two different applications: hard and soft tissues. Hard tissues, such as bone, undergo only small deformations and growth is developed by surface mass apposition. On the other hand, soft tissues such as blood vessels and tendons experience large deformations and suffer volumetric growth, as proposed by Skalak et al. (1982), that leads to internal residual stresses. An extensive review of the most significant work in this field can be found in Humphrey (1995), Taber (1995) and Cowin (2004). Nevertheless, it is interesting to remark that several works with strong theoretical fundamentals have recently been developed in this field. Oster et al. (1983) proposed a model for formation in which mechanical interactions between cells and extracellular matrix determine the pattern growth. Lubarda and Hoger (2002) proposed a model of the mechanics of solids with a growing mass using the framework of finite deformation continuum thermodynamics. Kuhl and Steinmann presented a theoretical and numerical model for open system thermodynamics with the application to hard tissues (Kuhl and Steinmann (2003)) and healing (Kuhl and Steinmann (2004)). Garikipati et al. (2004) proposed a complete framework for mass transport and mechanics for the modelling of living tissue growth.

All these works have focused mainly on the modelling of tissue growth as a consequence of the mechanical forces that the tissue supports, but these forces may also act within tissues, injuring them or regulating different biological processes such as tissue differentiation and remodelling. So, other models have focused on the study of the mechanoregulation rules that control tissue differentiation (Prendergast et al. (1997), Carter et al. (1998), García-Aznar et al. (2005)) and the effect of the tissue damage in all these processes (García-Aznar et al. (2005)).

In this work, we propose a general formulation for growth, differentiation and damage of living tissues controlled by the mechanical environment.

Biological tissues consist of solid and fluid phases that are usually analyzed in the context of a multiphasic theory. The model proposed by Garikipati et al. (2004), for instance, considers this fact in a general way. We have followed their approach, although incorporating the biological processes associated with the specialized cell populations using a classical continuum formulation. In fact, a tissue can be described from a macroscopic point of view as a continuum mixture of cells and extracellular matrices (ECMs) that perform specialized functions (Cowin (2004)).

The present paper is organized as follows. Section 2 treats the fundamental balance equations for mass and cells, and the kinematics of growth. In Section 3 we study the particularization of the formulation to the modelling of the healing course of a bone fracture. A numerical example is provided in Section 4, and, finally, a discussion and some concluding remarks are included in Section 5.

2 Global Formulation Framework

We consider the tissue to be composed of N species (different tissue components and fluid) and M types of cells (each specie in general containing a subset of such cells). We assume one of the species as the reference solid, while the rest of the phases move relatively with respect to it.

We define concentrations of the species $\rho_0^i = \overline{\rho}_0^i(V_m^i - h_0^i)$ as masses per unit volume in Ω_0, where $\overline{\rho}_0^i$ is the real tissue density, V_m^i is the volume fraction of specie i in the initial configuration and h_0^i is a scalar variable that quantifies the volume fraction of microcracks for each specie i. At the same time we define the cellular concentration of each cell type j as c_0^j.

We can also express the different variables in the current configuration Ω_t. However, in order not to make the paper excessively lengthy, we have only presented the Lagrangian form of the equations.

Summing over all the species, it is clear that

$$\sum_i v_m^i = \sum_i V_m^i = 1, \quad h_0 = \sum_i h_0^i, \tag{1}$$

where h_0 is the total volume of microcracks in the initial configuration.

2.1 Balance of Mass

If we work in the reference configuration, the balance of mass is written as

$$\frac{\partial \rho_0^i}{\partial t} = \frac{\partial \overline{\rho}_0^i}{\partial t}(V_m^i - h_0^i) + \overline{\rho}_0^i \frac{\partial V_m^i}{\partial t} - \overline{\rho}_0^i \frac{\partial h_0^i}{\partial t} = \Pi^i - \boldsymbol{\nabla} \cdot \boldsymbol{M}^i, \qquad i = 1, ..., N, \tag{2}$$

where Π^i is the net production of mass for the specie i per unit volume in the reference configuration, $\boldsymbol{M}^i$ is the flux of transported mass per unit time and unit surface in the reference configuration.

Moreover, the fluid is assumed to be incompressible and with no microcracks, which implies $\partial \bar{\rho}^f / \partial t = 0$, $h_0^f = 0$, and, therefore

$$\bar{\rho}_0^f \frac{\partial V_m^f}{\partial t} = \Pi^f - \boldsymbol{\nabla} \cdot \boldsymbol{M}^f. \tag{3}$$

From eq. (2) it becomes clear that the rate of variation of the volume fraction $\partial V_m^i / \partial t$ is determined from the net mass production Π^i, the flux of mass $\boldsymbol{M}^i$, the real tissue density evolution $\partial \overline{\rho}_0^i / \partial t$ and the rate of variation of the microcrack volume $\partial h_0^i / \partial t$. Next, we specify the evolution of each of these variables.

Production of Mass. We assume that the production of mass, without taking into account the species density variation, is due to the activity of the cells associated with each specie. Thus, we propose the expression

$$\Pi^i - \frac{\partial \overline{\rho}_0^i}{\partial t}(V_m^i - h_0^i) = c_0^i B_{pr}^i \overline{\rho}_0^i - \sum_{j=1, j \neq i}^{N} c_0^j B_j^i \overline{\rho}_0^i, \tag{4}$$

where B_{pr}^i is the rate of volume production of specie i per unit cell, with initial density $\bar{\rho}_0^i$ (we have used the same superscript for associated cells and species, that is, for the cell type that produces such specie), B_j^i is the volume of removed specie i by cells of j type per unit cell and unit time. It is interesting to remark that dead cells neither produce nor remove any volume, and therefore $B_{pr}^d = 0$ and $B_d^j = 0$.

Real Tissue Density Evolution. Normally, most growth models assume that the real tissue density is kept constant during growth and remodelling (Garikipati et al. (2004)). However, there are tissues in which the real tissue density changes with time. One typical example is bone tissue, where the extracellular matrix is mineralized in a process that can last 6 months or more. This variation $\partial\overline{\rho}^i/\partial t$ therefore has to be characterized in order to complete the model.

Damage Growth and Removal. There is increasing evidence that fatigue microdamage contributes to the degradation of the mechanical properties of living tissues. Damage intensity is normally evaluated by two alternative measures: the macroscopic mechanical degradation d, or a microscopic measure h of the density of internal microcracks. Continuum damage in elastic materials, for example, is experimentally related to the loss of stiffness by means of $d = 1 - E/E_0$, where E_0 corresponds to a reference value of the elastic modulus associated with an ideal undamaged material. This variable d corresponds to the continuum damage variable usually used in the isotropic theory of continuum damage mechanics, and is thereby restricted to the interval $[0, 1]$.

The two variables are physically related and but the relationship is not yet fully understood and depends on many factors, such as load state, porosity, etc. Only particular correlations have been determined in bone tissue; for example, that proposed by Burr et al. (1998), for which there is a linear relationship between stiffness loss and crack effective area for intact canine femurs subjected to bending. Then, using this result as a first approach, we assume a linear relationship between h and d through the expression $h = kd$, with k a constant that has to be determined experimentally.

In contrast to inert materials, damage in living tissues may undergo repair, so the rate of damage variation can be expressed as a balance of damage growth and repair for each specie i:

$$\frac{\partial h_0^i}{\partial t} = \left.\frac{\partial h_0^i}{\partial t}\right|_G - \left.\frac{\partial h_0^i}{\partial t}\right|_R . \tag{5}$$

Damage increase $\partial h_0^i/\partial t|_G$ is due to external mechanical loads that we consider independent of the biological processes involved. Therefore, the definition of the corresponding evolution requires that it is known from experiment how damage accumulates with load. Damage repair $\partial h_0^i/\partial t|_R$ is induced by remodelling or growth and is clearly dependent on the biological process developed within each specific specie through the evolution of $\partial V_m^i/\partial t$. As a first approach, we assume that damage h_0^i is uniformly distributed inside a representative macroscopic volume of the extracellular matrix of the tissue V_m^i and that damage repair is produced in a random way. Thus, when matrix is removed, the cracked volume proportion with respect to total volume is reduced, while when new undamaged matrix is deposited, damage is kept constant:

$$\left.\frac{\partial h_0^i}{\partial t}\right|_R = \begin{cases} 0 & \text{if} \quad \dfrac{\partial V_m^i}{\partial t} > 0, \\[2ex] \dfrac{\partial V_m^i}{\partial t}\dfrac{h_0^i}{V_m^i} & \text{if} \quad \dfrac{\partial V_m^i}{\partial t} < 0. \end{cases} \tag{6}$$

2.2 Balance of Cells

In a similar way, we define the balance of cells in the reference configuration Ω_0 as

$$\frac{\partial c_0^i}{\partial t} = \Pi_c^i - \nabla \cdot \boldsymbol{M}_c^i, \qquad i = 1, ..., M, \tag{7}$$

where c_0^i is the cell concentration of cell type i (number of cells per unit volume) in the current configuration, Π_c^i is the net production of cells per unit volume and time, and $\boldsymbol{M}_c^i$ is the cell flux of type i per unit time and per unit surface in the reference configuration. The net production of cells of type i can have different origins: proliferation, differentiation from different cells to this kind of cell or differentiation of this type of cell to other types (including dead cells). Thus, we can express the production of cells as

$$\Pi_c^i = \Pi_{c_{pr}}^i + \sum_{j=1, j\neq i}^{M} c_0^j \alpha_j^i - \sum_{j=1, j\neq i}^{M} c_0^i \alpha_i^j, \tag{8}$$

$$\Pi_c^i = \Pi_{c_{pr}}^i + \sum_{j=1, j\neq i}^{M} \left(c_0^j \alpha_j^i - c_0^i \alpha_i^j \right) = c_0^i \alpha_{pr}^i + \sum_{j=1, j\neq i}^{M} \left(c_0^j \alpha_j^i - c_0^i \alpha_i^j \right), \tag{9}$$

where $\Pi_{c_{pr}}^i$ is the production of cells of type i due to mitosis (proliferation) per unit time and volume in the initial configuration, α_{pr}^i is the same variable but per unit cell instead of per unit volume, and α_j^i is the generation of cells of type i due to differentiation of type j per unit cell and unit time. All these factors $\alpha_{pr}^i, \alpha_j^i$ are dependent on the mechanical and biochemical environment, being functions that have to fulfil the second law of thermodynamics. It is clear that

$$\sum_{i=1}^{M} \Pi_c^i = \sum_{i=1}^{M} c_0^i \alpha_{pr}^i, \tag{10}$$

and

$$\sum_{i=1}^{M} \sum_{j=1, j\neq i}^{M} (c_0^j \alpha_j^i - c_0^i \alpha_i^j) = 0. \tag{11}$$

Balance of Linear Momentum. We shall distinguish between the balance of linear momentum for species and for cells. Following the approach presented by Garikipati et al. (2004) we can express the local balance of linear momentum in the reference configuration for a single specie i and a single type of cell j as

$$\rho_0^i \frac{\partial}{\partial t}(\boldsymbol{V} + \boldsymbol{V}^i) = \rho_0^i(\boldsymbol{g} + \boldsymbol{L}^i) + \nabla \cdot \boldsymbol{P}^i - [\nabla(\boldsymbol{V} + \boldsymbol{V}^i)] \cdot \boldsymbol{M}^i, \tag{12}$$

$$\boldsymbol{0} = c_0^j \boldsymbol{L}_c^j + \nabla \cdot \boldsymbol{P}_c^j, \tag{13}$$

where $\boldsymbol{L}^i$ is the force per unit mass exerted upon i by all the other species and cells present in the tissue, $\boldsymbol{L}_c^j$ is the force per unit cell exerted upon the cells j by the other species and cells present in the tissue, $\boldsymbol{g}$ is the body force per unit mass, $\boldsymbol{P}^i$ is the first Piola-Kirchhoff stress tensor corresponding to specie i, $\boldsymbol{P}_c^j$ is the same tensor but corresponding to cells j, $\boldsymbol{V}$ is the velocity of the reference solid and $\boldsymbol{V}+\boldsymbol{V}^i$ is the total velocity for each of the remaining species.

Balance of Angular Momentum. This balance for the species and cells is similar to that of Garikipati et al. (2004):

$$\boldsymbol{P}^i\boldsymbol{F}^T = \boldsymbol{F}\boldsymbol{P}^{iT}, \qquad \boldsymbol{P}_c^i\boldsymbol{F}^T = \boldsymbol{F}\boldsymbol{P}_c^{iT}. \tag{14}$$

Balance of Energy. In addition to the terms previously introduced, some more have now to be included: e^i, the internal energy per unit mass of specie i; e_c^j, the internal energy per unit cell of type j; r^i, the heat supply to specie i per unit mass of that specie (we have not considered the heat supply per unit cell); the partial heat flux vector of i, $\boldsymbol{Q}^i$. These are all defined in Ω_0. Interaction energetic terms appear between species and cells: the energy per unit mass transferred to the specie i by all other species and cells is $\tilde{e}^i$ and the energy per unit cell transferred to the cell type j by all other species and cells is $\tilde{e}_c^j$. Summing over all tissues i and cells j, the local rate of change of energy becomes

$$\begin{aligned}
&\sum_j \left(c_0^j \frac{\partial e_c^j}{\partial t} \right) + \sum_i \left(\rho_0^i \frac{\partial e^i}{\partial t} \right) \\
&= \sum_j \left(\boldsymbol{P}_c^j : \dot{\boldsymbol{F}} + \boldsymbol{P}_c^j : \nabla \boldsymbol{V}_c^j - c_0^j \boldsymbol{L}_c^j \cdot \boldsymbol{V}_c^j - \Pi_c^j e_c^j - \nabla e_c^j \cdot \boldsymbol{M}_c^j \right) \\
&+ \sum_i \Big(\boldsymbol{P}^i : \dot{\boldsymbol{F}} + \boldsymbol{P}^i : \nabla \boldsymbol{V}^i - \nabla \cdot \boldsymbol{Q}^i + \rho_0^i r^i - \rho_0^i \boldsymbol{L}^i \cdot \boldsymbol{V}^i \\
&- \Pi^i (e^i + \tfrac{1}{2} \| \boldsymbol{V}^i \|^2) - \nabla e^i \cdot \boldsymbol{M}^i \Big).
\end{aligned} \tag{15}$$

The Entropy Inequality: Clausius-Duhem Form. Let η^i be the entropy per unit mass of species i, η_c^j the entropy associated with the cells per unit cell and θ the absolute temperature. In this case, the local entropy production inequality, also known as the Clausius-Duhem inequality, is

$$\begin{aligned}
&\sum_i \rho_0^i \left(\frac{\partial e^i}{\partial t} - \theta \frac{\partial \eta^i}{\partial t} \right) + \sum_j c_0^j \left(\frac{\partial e_c^j}{\partial t} - \theta \frac{\partial \eta_c^j}{\partial t} \right) + \sum_i \Pi^i (e^i + \tfrac{1}{2} \| \boldsymbol{V}^i \|^2) \\
&+ \sum_i \frac{\nabla \theta \cdot \boldsymbol{Q}^i}{\theta} + \sum_i \left(\rho_0^i \frac{\partial}{\partial t} (\boldsymbol{V} + \boldsymbol{V}^i) - \rho_0^i \boldsymbol{g} - \nabla \cdot \boldsymbol{P}^i + \nabla (\boldsymbol{V} + \boldsymbol{V}^i) \cdot \boldsymbol{M}^i \right) \cdot \boldsymbol{V}^i \\
&- \sum_j (\nabla \cdot \boldsymbol{P}_c^j) \cdot \boldsymbol{V}_c^j - \sum_i \left(\boldsymbol{P}^i : \dot{\boldsymbol{F}} + \boldsymbol{P}^i : \nabla \boldsymbol{V}^i - (\nabla e^i - \theta \nabla \eta^i) \cdot \boldsymbol{M}^i \right) \\
&- \sum_j \left(\boldsymbol{P}_c^j : \dot{\boldsymbol{F}} + \boldsymbol{P}_c^j : \nabla \boldsymbol{V}_c^j - \Pi_c^j e_c^j - (\nabla e_c^j - \theta \nabla \eta_c^j) \cdot \boldsymbol{M}_c^j \right) \leq 0.
\end{aligned} \tag{16}$$

Kinematics of Growth. Following the standard literature on biological growth (Skalak et al. (1982), Taber (1995)) we assume that each specie i can grow in a different way, so the global deformation gradient can be defined by a multiplicative decomposition for each specie as

$$\boldsymbol{F} = \boldsymbol{F}^{e^i} \boldsymbol{F}^{g^i}, \tag{17}$$

where $\boldsymbol{F}^{g^i}$ is the growth tensor and $\boldsymbol{F}^{e^i}$ is the actual deformation tensor due to external loads.

Constitutive Relations from the Clausius-Duhem Inequality. As usual in field theories of continuum physics, we use the Clausius-Duhem inequality(16) to obtain constitutive laws and thermodynamic constraints on the evolution and on the relation between the different variables involved. It is usual to express the Clausius-Duhem inequality in terms of the specific Helmholtz free energy. The specific internal energy can be expressed as $e^i = \psi^i + \theta\eta^i$, and, for the cells, as $e^i_c = \psi^i_c + \theta\eta^i_c$, where we assume in a general way that the mass-specific Helmholtz free energy is $\psi^i = \hat{\psi}^i(\boldsymbol{F}^{e^i}, \theta, \rho^i_0, d^i)$ and the cellular-specific Helmholtz free energy is $\psi^i_c = \hat{\psi}^i_c(\boldsymbol{F}^{e^i}, \theta, c^i_0)$. Applying now the chain rule, regrouping some terms and considering a uniform temperature distribution, (16) becomes

$$\begin{aligned}
&\sum_i \left(\rho^i_0 \frac{\partial \psi^i}{\partial \boldsymbol{F}^{e^i}} - \boldsymbol{P}^i \boldsymbol{F}^{g^{iT}}\right) : \dot{\boldsymbol{F}}^{e^i} + \sum_j \left(c^j_0 \frac{\partial \psi^j_c}{\partial \boldsymbol{F}^{e^j}} - \boldsymbol{P}^j_c \boldsymbol{F}^{g^{jT}}\right) : \dot{\boldsymbol{F}}^{e^j} \\
&+ \sum_i \rho^i_0 \left(\frac{\partial \psi^i}{\partial \theta} + \eta^i\right) \frac{\partial \theta}{\partial t} + \sum_j c^j_0 \left(\frac{\partial \psi^j_c}{\partial \theta} + \eta^j_c\right) \frac{\partial \theta}{\partial t} \\
&+ \sum_i \Pi^i (\psi^i + \theta\eta^i + \tfrac{1}{2} \| \boldsymbol{V}^i \|^2) + \sum_i \rho^i_0 \frac{\partial \psi^i}{\partial d^i} \dot{d}^i \\
&+ \sum_i \left(\rho^i_0 \frac{\partial}{\partial t} (\boldsymbol{V} + \boldsymbol{V}^i) - \rho^i_0 \boldsymbol{g} - \nabla \cdot \boldsymbol{P}^i + \nabla (\boldsymbol{V} + \boldsymbol{V}^i) \cdot \boldsymbol{M}^i\right) \cdot \boldsymbol{V}^i \\
&- \sum_j (\nabla \cdot \boldsymbol{P}^j_c) \cdot \boldsymbol{V}^j_c - \sum_i \left(\boldsymbol{P}^i : (\boldsymbol{F}^{e^i} \dot{\boldsymbol{F}}^{g^i} + \nabla \boldsymbol{V}^i) - (\nabla \psi^i) \cdot \boldsymbol{M}^i\right) \\
&+ \sum_i \left(\rho^i_0 \frac{\partial \psi^i}{\partial \rho^i_0} \dot{\rho}^i_0\right) + \sum_j \left(c^j_0 \frac{\partial \psi^j_c}{\partial c^j_0} \dot{c}^j_0\right) - \sum_j \left(\boldsymbol{P}^j_c : (\boldsymbol{F}^{e^j} \dot{\boldsymbol{F}}^{g^j} + \nabla \boldsymbol{V}^j_c)\right. \\
&\left.- (\nabla \psi^j_c) \cdot \boldsymbol{M}^j_c\right) + \sum_j \Pi^j_c (\psi^j_c + \theta \eta^j_c) \leq 0.
\end{aligned} \tag{18}$$

Using standard arguments in continuum mechanics (see Garikipati et al. (2004)) we obtain constitutive relations for the first Piola-Kirchhoff stress tensor, the entropy and the species fluxes, respectively

$$\boldsymbol{P}^i \boldsymbol{F}^{g^{iT}} = \rho^i_0 \partial \psi^i / \partial \boldsymbol{F}^{e^i}, \qquad \eta^i = -\partial \psi^i / \partial \theta,$$

$$\rho^i_0 \boldsymbol{V}^i = -\frac{\tilde{\boldsymbol{D}}^i}{\rho^i_0} \left(\rho^i_0 \frac{\partial \boldsymbol{V}}{\partial t} - \rho^i_0 \boldsymbol{g} - \nabla \cdot \boldsymbol{P}^i + (\nabla \boldsymbol{V}) \boldsymbol{M}^i + \rho^i_0 \boldsymbol{F}^{-T} \nabla \psi^i\right).$$

In a similar way, we can obtain the constitutive relations for the cell behaviour:

$$\boldsymbol{P}_c^j \boldsymbol{F}^{g^{jT}} = c_0^j \frac{\partial \psi_c^j}{\partial \boldsymbol{F}^{e^j}}, \tag{19a}$$

$$\eta_c^j = -\frac{\partial \psi_c^j}{\partial \theta}, \tag{19b}$$

$$c_0^i \boldsymbol{V}_c^i = -\frac{\boldsymbol{H}^i}{c_0^i}(c_0^i \boldsymbol{F}^{-T} \nabla \psi_c^i - \nabla \cdot \boldsymbol{P}_c^i), \qquad \boldsymbol{w} \cdot \boldsymbol{H}^i \boldsymbol{w} \geq 0, \forall \boldsymbol{w} \in \mathbb{R}^3. \tag{19c}$$

Equation (19a) specifies a constitutive relation for the cellular stress $\boldsymbol{P}_c^j$ that defines the interaction between cell and ECM (Oster et al. (1983)) from the cellular-specific Helmholtz free energy ψ_c^j. In a similar way, the entropy η_c^j associated with the cells per unit of cell is determined in (19b). Equation (19c) is newly proposed in the present work. The cellular flux is defined as the product of a positive semi-definite mobility tensor $\boldsymbol{H}^i$ for each cell type and the corresponding thermodynamic driving force. The first term of this driving force specifies that cells tend to move in the direction opposite to their energy gradient, whereas the second term indicates the existence of a force that controls cell movement associated with the divergence of $\boldsymbol{P}_c^i$.

In summary, the complete definition of this model requires that the following functions be characterized for all the species: the production of mass Π^i by means of B_j^i, B_{pr}^i (see eq. (4)), the real tissue density evolution $\partial \bar{\rho}_0^i/\partial t$, the mechanical damage growth $\partial h_0^i/\partial t|_G$ and the mass-specific Helmholtz free energy ψ^i; for all the cells, the production of cells Π_c^i by means of $\alpha_i^j, \alpha_{pr}^i$; and, finally, the cellular-specific Helmholtz free energy ψ_c^i.

3 Particularization: Modelling Bone Fracture Healing

The bone fracture healing process is a good example that can serve to study the factors that control tissue growth and differentiation from progenitor cells. Recently, we have proposed a model that allows the first stage of this complex event to be simulated, including the effect of the different cellular populations and tissue growth (Gómez-Benito et al. (2005), García-Aznar et al. (2005)). In fact, this model is a particularization of the general formulation presented above.

We assume that the fracture site is invaded by proliferating and migrating stem cells in the first stage of bone healing. These cells may differentiate into cartilage cells, bone cells or fibroblasts, or may die, depending on mechanical stimulus. After this, these specialized cells begin to synthesize the extracellular organic matrices of the different tissues. During healing, callus size and shape are constantly modified under the influence of several factors. Different interfragmentary movements have been observed to produce different callus sizes, suggesting that mechanical stimulation is a particularly determinant factor. The model assumes two callus growth mechanisms: mesenchymal stem cell proliferation and chondrogenesis. The callus stiffness is characterized through the volume of extracellular matrix corresponding to each tissue type and through its geometry.

Therefore, considering all these characteristics of bone fracture healing, the general formulation previously presented has been used with the following general assumptions.

(i) We assume that all processes that are involved in bone fracture healing (tissue growth, differentiation and damage) are guided by a single mechanical stimulus: the second invariant of the deviatoric strain tensor ($\varphi \equiv J_2$).

(ii) We distinguish between different species: bone (b), cartilage (c), fibrous tissue (f), granulation tissue (g) and debris tissue (d). Hence, we consider the cells associated with these respective tissues: bone cells (c_0^b), cartilage cells (c_0^c), fibroblasts (c_0^f), mesenchymal stem cells (c_0^s) and dead cells (c_0^d).

(iii) For simplicity, all the tissues are considered to be composed of a fixed proportion of the following substances: collagen (mainly type I, II and III), ground substance, mineral and water, and when new tissue is created, it maintains this composition. Therefore, we assume for all the tissues that the time evolution of the real tissue density is null ($\partial \overline{\rho}_0^i / \partial t = 0$).

(iv) As a first approach, we assume that the only tissues that can be mechanically damaged are the granulation tissue and the debris tissue ($h_0^i|_G = 0.95$, $i = s, d$; if $\varphi > \varphi_{\text{damage}}$). Obviously, other tissues are also susceptible to damage, although under normal physiological loads the corresponding damage level is usually negligible. Therefore, we consider that $h_0^i = 0$, $i = b, c, f$.

(v) The mass production is determined from B_j^i, B_{pr}^i, constants that were specified for each cell type in a previous work (Gómez-Benito et al. (2005)).

(vi) The dynamics of cells, proliferation and differentiation, are determined by means of functions $\alpha_i^j, \alpha_{pr}^i$ that depend on the mechanical stimulus φ and are indicated in Table 1.

(vii) We have simplified the growth law by working with small strains. The deformation gradient tensor $\boldsymbol{F}$ can therefore be approximated by the additive rule

$$\boldsymbol{F} = \boldsymbol{F}^g \boldsymbol{F}^e \simeq \mathbf{1} + \nabla \boldsymbol{U}^g + \nabla \boldsymbol{U}^e. \tag{20}$$

We assume that callus growth is isotropic and due to two different processes: mesenchymal cell proliferation and endochondral ossification. Therefore, we propose the expression

$$\nabla \boldsymbol{U}^g = [g^v_{\text{endochondral}}(\varphi, t) + f^v_{\text{proliferation}}(c_0^s, \varphi) - 1]\mathbf{1}, \tag{21}$$

where $g^v_{\text{endochondral}}(\varphi, t)$ controls the rate of callus growth due to endochondral ossification and $f^v_{\text{proliferation}}(c_0^s, \varphi)$ defines the rate of callus growth due to proliferation (see Table 2).

(viii) All tissues are assumed to be biphasic and isotropic. The material properties of the different tissues, except for mature bone, are determined using a mixture rule that is prescribed as a function of the apparent density (García-Aznar et al. (2005)). Permeability is also different for each type of

Table 1. Functions that describe the evolution of the cellular populations (Gómez-Benito et al. (2005)); t_m^i is the time that stem cells need to differentiate into specialized cells (i); φ_i is the stimulus level which indicates when cell i can appear; p_{mi} is the mineralization degree; p_{mi}^m is the minimal mineralization degree needed for endochondral ossification to take place; c^{bm} is the concentration of bone cells that indicates when blood supply is completed at the ossification front and osteoblasts can differentiate directly from stem cells; β_{pr} defines stem cell proliferation.

	α_i^s	α_i^c if $\varphi_b < \varphi < \varphi_c$	α_i^b if $\varphi_{lim} < \varphi < \varphi_b$	α_i^f if $\varphi_c < \varphi < \varphi_f$	α_i^d	α_{pr}^i
α_s^j	–	$\begin{cases} 1 & \text{if } t > t_m^c \\ 0 & \text{otherwise} \end{cases}$	$\begin{cases} \frac{D_b}{c_0^s}\nabla^2 c_0^b & \text{if } c_0^b < c^{bm} \ \& \ t > t_m^b \\ 1 & \text{if } c_0^b > c^{bm} \ \& \ t > t_m^b \\ 0 & \text{otherwise} \end{cases}$	$\begin{cases} 1 & \text{if } t > t_m^f \\ 0 & \text{otherwise} \end{cases}$	$\begin{cases} 1 & \text{if } \varphi > \varphi_d \\ 0 & \text{otherwise} \end{cases}$	$\frac{\beta_{pr} \cdot \varphi}{\varphi + \varphi_{pr}}$
α_c^j	0	–	$\begin{cases} \frac{D(\varphi)}{c_0^c}\nabla^2 c_0^b & \text{if } p_{mi} > p_{mi}^m \ \& \ c_0^b < c^{bm} \\ 1 & \text{if } p_{mi} > p_{mi}^m \ \& \ c_0^b > c^{bm} \\ 0 & \text{otherwise} \end{cases}$	0	0	0
α_b^j	0	0	–	0	0	0
α_f^j	0	0	0	–	0	0

Table 2. Functions that describe the growth of the callus: c^{smx} corresponds to a saturation value of the stem cell concentration; c_0^{cm} is the equilibrium value of cartilage cell concentration; φ_{cal} is the stimulus level below which cartilage begins to calcify; k_{hyper} controls the volume increase during cartilage calcification (Gómez-Benito et al. (2005)).

$g^v_{\text{endochondral}}(\varphi, t)$, if $(\varphi < \varphi_{\text{cal}})$	$f^v_{\text{proliferation}}(c_s, \varphi)$
$\begin{cases} -\frac{1}{c_0^c} \cdot \frac{\partial c_0^c}{\partial t} = -\frac{1}{c_0^c}\left(\frac{\varphi - \varphi_{\text{ncal}}}{\varphi_b - \varphi_{\text{cal}}} \cdot k_{\text{hyper}}\right) & \text{if } (c_0^c < c_0^{cm}) \\ 0 & \text{otherwise} \end{cases}$	$\begin{cases} 0 & \text{if } (c_0^s < c^{smx}) \\ \alpha_{pr}^s \frac{c_0^s}{c^{smx}} & \text{if } (c_0^s = c^{smx}) \end{cases}$

tissue. In soft tissues, the permeability was computed according to its composition and porosity, following the model proposed by Levick (1987). On the other hand, for hard tissues, we assumed a constant permeability function that distinguishes between cortical and trabecular bone, and calcified cartilage (see García-Aznar et al. (2005)).

(ix) All species (except the fluid) are assumed to be non-diffusive ($\boldsymbol{M}^i = \boldsymbol{0}$, if $i \neq j$).

(x) The Helmholtz free energy for the extracellular solid matrix for all tissues is defined as a quadratic $\psi = \boldsymbol{\varepsilon}^e \cdot \boldsymbol{C}\boldsymbol{\varepsilon}^e/(2\rho_0)$, with $\boldsymbol{\varepsilon}^e$ the infinitesimal strain tensor of the solid phase and $\boldsymbol{C}$ the tensor of elastic moduli. In granulation and debris tissues, we also considered the possibility of damage accumulation by using the modified Helmholtz free energy function $\psi = (1-d)\boldsymbol{\varepsilon}^e \cdot \boldsymbol{C}^0\boldsymbol{\varepsilon}^e/(2\rho_0)$, with $\boldsymbol{C}^0$ the tensor of elastic moduli for the undamaged tissue.

(xi) The cellular-specific Helmholtz free energy is zero for all cells $\psi_c^j = 0$, except for stem cells $\psi_c^j = \mu c_0^j$, with μ normally known as the *cellular electro-chemical potential.* Therefore, considering that $\boldsymbol{V}_c^j = \boldsymbol{F}\boldsymbol{M}_c^j/c_0^j$, the constitutive relation (19c) is then particularized to

$$\boldsymbol{M}_c^j = -\boldsymbol{H}^j \nabla \psi_c^j = -\boldsymbol{H}^j \mu \nabla c_0^j = -\boldsymbol{D}^j \nabla c_0^j, \tag{22}$$

which identifies the cell diffusion tensor $\boldsymbol{D}^j = \boldsymbol{H}^j \mu$.

4 A Numerical Example: 3D Bone Fracture Healing

The above formulation has been implemented in a Finite Element commercial code in combination with an automatic mesh generation program (Barber et al. (1996)). Continuous update of the mesh is required because fracture callus is progressively growing during healing, modifying its shape and geometry. At the same time, all cellular events are simulated by means of a combination of several different analyses (see Fig. 1).

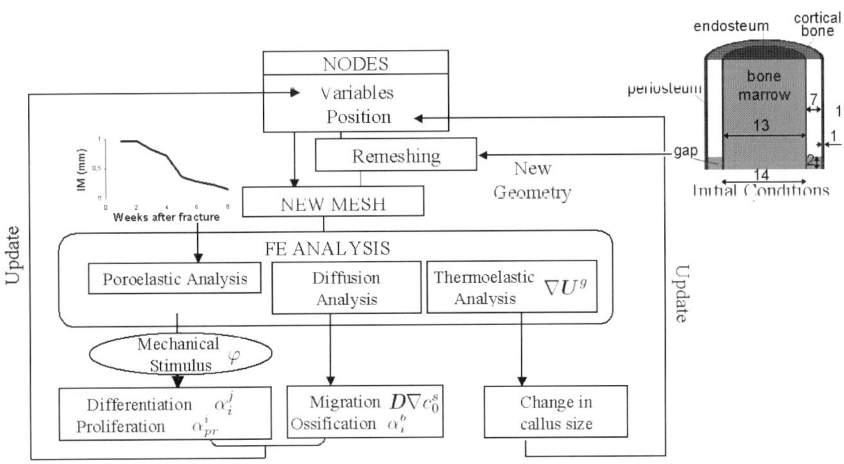

Fig. 1. Scheme of the numerical implementation.

In order to study the potential of this approach, we have analyzed the evolution of the callus of a simplified human tibia fracture (see Fig. 2). The healing fracture has been studied for an axial interfragmentary movement determined experimentally (Claes et al. (1997)); see Fig. 2.

The model simulates the different cellular events that are normally described in fracture healing (see Fig. 2) and the progressive evolution of the callus (see Fig. 3).

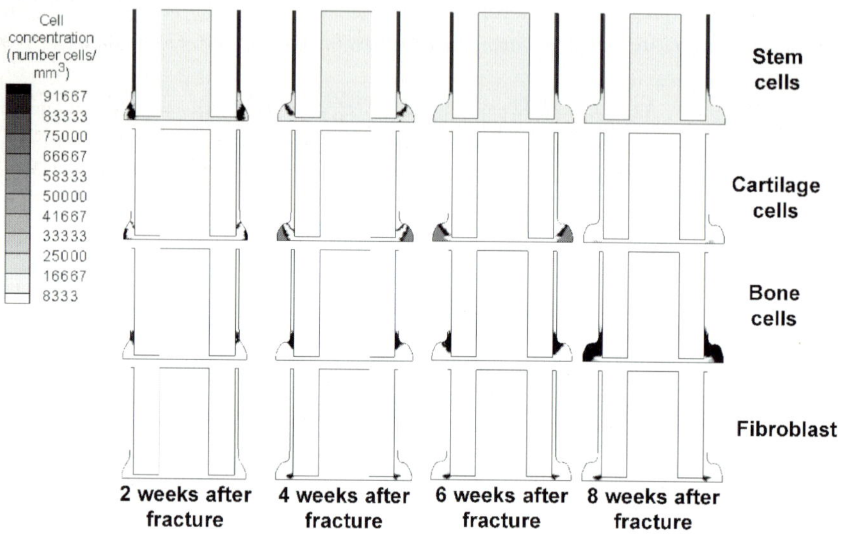

Fig. 2. Cellular distribution at different times of the healing.

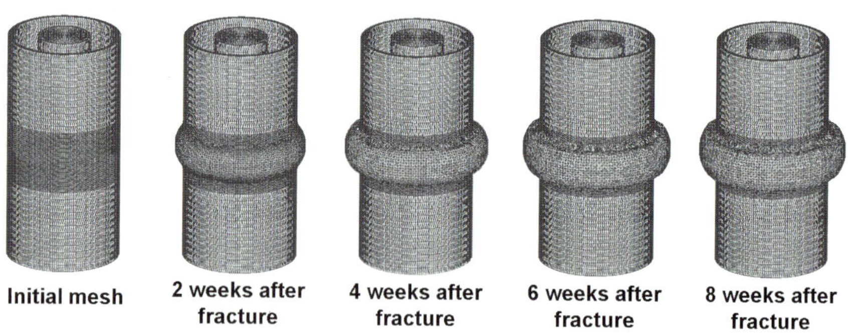

Fig. 3. Meshes that show the evolution of the callus.

5 Discussion and Conclusions

We have presented in this paper a general constitutive theory for growth, differentiation and damage of living tissues. This model is assumed to be controlled by mechanical influences, although other effects could also be included, such as chemical reactions, growth factors, and so on. The governing equations have been derived by considering the extracellular matrix concentration and the cellular population.

Following the framework of this formulation, one theoretical particularization has been developed to show its potential: the modelling of bone fracture healing. It is a complex process where growth, differentiation and damage act on the tissue.

Although the particularization shown here presents some simplifications (small deformations, no residual stresses, null cell-matrix interaction and so on) the global formulation proposed is sufficiently general to be used in other biomechanical applications, such as limb lengthening, tendons and vessel growth and remodelling.

References

Barber, C. B., Dobkin, D. P., and Huhdanpaa, H. (1996). The quickhull algorithm for convex hulls. *ACM Trans. Math. Software* 22:469–483.

Burr, D. B., Turner, C. H., Naick, P., Forwood, M. R., Ambrosius, W., Hasan, M. S., and Pidaparti, R. (1998). Does microdamage accumulation affect the mechanical properties of bone? *J. Biomech.* 31:337–345.

Carter, D. R., Beaupre, G. S., Giori, N. J., and Helms, J. A. (1998). Mechanobiology of skeletal regeneration. *Clin. Orthop.* S355:S41–S55.

Claes, L., Augat, P., Suger, G., and Wilke, H. J. (1997). Influence of size and stability of the osteotomy gap on the success of fracture healing. *J. Orthop. Res.* 15:577–584.

Cowin, S. C., and Hegedus, D. H. (1976). Bone remodelling I: Theory of adaptive elasticity. *J. Elasticity* 6:313–326.

Cowin, S. C. (2004). Tissue growth and remodeling. *Ann. Rev. Biomed. Eng.* 6:77–107.

García-Aznar, J. M., Kuiper, J. H., Gómez-Benito, M. J., Doblaré, M., and Richardson, J. B. (2005). Computational simulation of fracture healing. *J. Biomech.* in press.

Garikipati, K., Arruda, E. M., Grosh, K., Narayanan, H., and Calve, S. (2004). A continuum treatment of growth in biological tissue: the coupling of mass transport and mechanics. *J. Mech. Phys. Solids* 52:1595–1626.

Gómez-Benito, M. J., García-Aznar, J. M., Kuiper, J. H., and Doblaré, M. (2005). Influence of fracture gap size on the pattern of long bone healing: A computational study. *J. Theoret. Biol.* in press.

Humphrey, J. D. (1995). Mechanics of the arterial wall: Review and directions. *Crit. Rev. Biomed. Eng.* 23:1–162.

Kuhl, E., and Steinmann, P. (2003). Theory and numerics of geometrically non-linear open system mechanics. *Int. J. Numer. Meth. Eng.* 58: 1593–1615.

Kuhl, E., and Steinmann, P. (2004). Computational modeling of healing: An application of the material force method. *Biomech. Model. Mechanobio.* 2:187–203.

Levick, J. R. (1987). Flow through interstitium and other fibrous matrices. *Q. J. Exp. Physiol.* 72:409–438.

Lubarda, V. A., and Hoger, A. (2002). On the mechanics of solids with a growing mass. *Int. J. Solids Structures* 39:4627–4664.

Oster, G. F., Murray, J. D., and Harris, A. K. (1983). Mechanical aspects of mesenchymal morphogenesis. *J. Embryol. Exp. Morph.* 78:83–125.

Prendergast, P. J., Huiskes, R., and Soballe, K. (1997). Biophysical stimuli on cells during tissue differentiation at implant interfaces. *J. Biomech.* 6: 539–548.

Skalak, R., Dasgupta, G., Moss, M., Otten, E., Dullemeijer, P., and Vilmann, H. (1982). Analytical description of growth. *J. Theoret. Biol.* 94:555–577.

Taber, L. A. (1995). Biomechanics of growth, remodelling, and morphognesis. *Appl. Mech. Rev.* 48:487–545.

Continuum Models of Growth with Emphasis on Articular Cartilage

S.M. Klisch

Cal Poly State University, Department of Mechanical Engineering, USA
sklisch@calpoly.edu

Continuum theories of growth that have led to cartilage growth mixture models are summarized. Several theoretical features that contrast with those presented by other authors are discussed. An attempt is made to resolve two outstanding issues: the incremental boundary-value problem and its implementation, and invariance requirements on the growth law.

1 Introduction

Articular cartilage is a multiphasic tissue composed of molecules that serve distinct biomechanical roles. The fluid-saturated extracellular solid matrix includes proteoglycans (PGs) that primarily resist compressive loading and a crosslinked collagen (COL) network that primarily resists tensile and shear loading. A key feature of cartilage growth is that cell metabolism can be regulated by mechanical stimuli. In recent years, my colleagues and I have developed cartilage growth mixture models that allow for the specification of multiple constituents that may grow differentially, i.e. at different rates. Since the apparent density of the molecular constituents can change, the growth theory for incompressible elastic solids (Rodriguez et al. (1994)) was first extended to compressible elastic solids (Klisch et al. (2001)), which led to growth theories for thermoelastic solids and mixtures (Klisch et al. (2000), Klisch and Hoger (2003)). We have proposed several cartilage growth mixture models and solved boundary-value problems to explore their capabilities and to learn how to implement them (Klisch et al. (2000, 2003, 2005)). Continuum theories of growth have been presented recently by other authors for thermoelastic materials (Epstein and Maugin (2000), Lubarda and Hoger (2002), Menzel (2005)) and mixtures (Humphrey and Rajagopal (2002), Garikipati et al. (2004)).

Here, the theoretical development of our continuum growth theory is summarized and extended. Although the main elements of the theory first presented in Klisch et al. (2000) are adhered to, there are several issues that have not been resolved. Here, an attempt is made to resolve the structure of the incremental problem to be used in numerical approaches and the invariance conditions on the growth law. Also, several issues are more thoroughly discussed, including the experimental prescription for the elastic and growth tensors, the balance of mass, the work-energy principle in the mechanical theory, and objections to the Coleman-Noll approach in the thermodynamical theory.

2 Growing Elastic Materials

In this section, the purely mechanical theory of a growing compressible elastic material presented in Klisch et al. (2001) is discussed. The structure of that theory was motivated by how it may be applied in practice. More specifically, tissue explants may be harvested at different stages of the growth process and the tissue's compositional, geometric, and material properties can be experimentally characterized relative to a pre-determined reference configuration. Thus, a fixed reference configuration is used that can be identified with an experimental configuration. Furthermore, since a multiplicative decomposition of the deformation gradient tensor into elastic and growth tensors is employed, specific constitutive assumptions are made that lead to an experimental prescription for the elastic and growth tensors so that the theory may be rendered 'testable'.

2.1 Kinematics

Consider a growing body $\mathcal{B}$ with a fixed reference configuration $\kappa_0(\mathcal{B})$ and a present configuration $\kappa(\mathcal{B})$ at time t, with the body unloaded in $\kappa_0(\mathcal{B})$[1]. A particle of $\mathcal{B}$ occupies positions $\mathbf{X} \in \kappa_0(\mathcal{B})$ and $\mathbf{x} \in \kappa(\mathcal{B})$. The motion for $\mathcal{B}$ is defined as

$$\mathbf{x} = \chi(\mathbf{X}, t). \tag{1}$$

Upon assuming that the set of material points in $\kappa_0(\mathcal{B})$ is dense in the set of all points in $\kappa(\mathcal{B})$ and noting that the set of material points in a continuum is not countable, invertibility of (1) may be assumed. Consequently, any material point that is added during the growth process is associated with a unique material point in $\kappa_0(\mathcal{B})$ through the inverse mapping. The deformation gradient tensor is

$$\mathbf{F} = \frac{\partial \chi}{\partial \mathbf{X}}(\mathbf{X}, t) \tag{2}$$

and is assumed to obey $J = \det \mathbf{F} > 0$ and the decomposition

$$\mathbf{F} = \mathbf{F}_{\mathrm{e}}\mathbf{F}_{\mathrm{g}}, \tag{3}$$

[1]See Klisch et al. (2001) for details on accounting for a residual stress field in $\kappa_0(\mathcal{B})$.

where $\mathbf{F}_e$ is the elastic tensor and $\mathbf{F}_g$ is the growth tensor. The tensor $\mathbf{F}_g$ quantifies the amount and orientation of mass deposition while $\mathbf{F}_e$ ensures continuity of the growing body and may include a superposed elastic deformation. In order to identify experimental prescriptions for $\mathbf{F}_e$ and $\mathbf{F}_g$, it is assumed that the mass density, internal energy, and stress functions in $\kappa(\mathcal{B})$ are independent of $\mathbf{F}_g$[2]. In particular, removal of the external loads acting on $\kappa(\mathcal{B})$ quantifies the superposed elastic deformations. Then, $\mathbf{F}_e\mathbf{F}_g$ can be calculated by measuring the total deformation gradient tensor relative to $\kappa_0(\mathcal{B})$. Then, $\mathbf{F}_e$ can be measured by performing a series of destructive experiments that are designed to relieve any residual stress. Finally, $\mathbf{F}_g$ can be computed from (3) as $\mathbf{F}_e^{-1}\mathbf{F}$. Taking a material time-derivative of (3) and defining the elastic and growth velocity gradient tensors

$$\mathbf{L}_e = \dot{\mathbf{F}}_e\mathbf{F}_e^{-1}, \quad \mathbf{L}_g = \dot{\mathbf{F}}_g\mathbf{F}_g^{-1}, \tag{4}$$

with symmetric parts $(\mathbf{D}_e, \mathbf{D}_g)$, the velocity gradient tensor associated with $\mathbf{F}$ is

$$\mathbf{L} = \dot{\mathbf{F}}\mathbf{F}^{-1} = \mathbf{L}_e + \mathbf{F}_e\mathbf{L}_g\mathbf{F}_e^{-1}, \tag{5}$$

with symmetric part $\mathbf{D}$. Also, the following equations can be derived using the above results (where $\mathbf{I}$ is the identity tensor):

$$\dot{J}/J = \mathbf{D}\cdot\mathbf{I}, \quad \dot{\overline{\det \mathbf{F}_g}}/\det \mathbf{F}_g = \mathbf{D}_g\cdot\mathbf{I}. \tag{6}$$

2.2 Balance Equations

To derive balance equations, the material deposited during growth is assumed to have the same mechanical properties as the original material. Consequently, the mass density, linear momentum, angular momentum, internal energy, and kinetic energy of the deposited material are the same as that of the original material. Furthermore, the mechanical response functions of the deposited material are the same as those of the original material. Then, the balance equations for mass, linear momentum, angular momentum, and work-energy on $\kappa(\mathcal{B})$ are[3]

$$\dot{\rho} + \rho\,\mathrm{div}\,\mathbf{v} = \rho c, \tag{7}$$

$$\mathrm{div}\,\mathbf{T} + \rho\mathbf{b} = \rho\dot{\mathbf{v}}, \tag{8}$$

$$\mathbf{T} = \mathbf{T}^T, \tag{9}$$

$$\rho\dot{\varepsilon} = \mathbf{T}\cdot\mathbf{D} + \rho\beta, \tag{10}$$

where ρ is the mass density, $\mathbf{v}$ is the velocity, c is the rate of mass deposition per unit current mass (called the mass growth function), $\mathbf{T}$ is the Cauchy stress tensor, $\mathbf{b}$ is the body force density, ε is the internal energy density, and β is

[2]See Chen and Hoger (2000) for a mathematical justification of this assumption.
[3]See Klisch et al. (2001) for these equations stated in integral form.

the rate of growth energy per unit current mass. Due to the introduction of c, an additional equation is needed. Recalling that ρ is assumed independent of $\mathbf{F}_{\mathrm{g}}$, the balance of mass (7) is decomposed into an elastic continuity equation

$$\rho \det \mathbf{F}_{\mathrm{e}} = \rho_0, \tag{11}$$

where ρ_0 is the constant density in $\kappa_0(\mathcal{B})$, and a growth continuity equation

$$\det \mathbf{F}_{\mathrm{g}} = \exp\left\{\int_{\tau=t_0}^{t} c\,\mathrm{d}\tau\right\} \quad \text{or} \quad c = \mathbf{D}_{\mathrm{g}} \cdot \mathbf{I}. \tag{12}$$

There exist several fundamental differences between the theory proposed in this section and those presented elsewhere. Several theories have modeled growth using an evolving reference configuration (Epstein and Maugin (2000), Menzel (2005)) in which ρ_0 is time-dependent. Also, in Epstein and Maugin (2000) and Garikipati et al. (2004) a growth mass flux through the boundary of a control element was introduced. In deriving (7), we only considered a volumetric mass source due to the biological aspects of cartilage growth. For example, COL synthesis involves mRNA processing, assembly of procollagen molecules inside chondrocytes, secretion through the cell membrane and into the extracellular matrix, and aggregation into ordered fibrillar structures stabilized by crosslinks. Thus, mobile COL fibrils that enter into a control volume should not be considered as the same material as the COLs that are fixed to the extracellular matrix until they become crosslinked to the COL network. Consequently, the role of mobile COL fibrils, and other precursor molecules, can be made explicit by identifying them as another constituent in a growing mixture. Then, their diffusion into the control volume can be modeled by a continuity equation (7) and their conversion into the immobile COL network can be modeled as a volumetric mass exchange term (i.e. ρc). Finally, we have assumed that the molecules added during growth have the same stress as the existing material; of course, the growth process may lead to a change in the local stress. The reader is referred to Humphrey and Rajagopal (2002) for an alternative approach in which the deposited material may have a 'deposition-stress' that differs from that of the neighboring material.

2.3 Constitutive Equations

A growth law that describes the time-rate of change of $\mathbf{F}_{\mathrm{g}}$ must be specified. Since the relationship between growth and mechanical stimuli is poorly understood, we represent the growth law as $\hat{\mathbf{G}}(M)$ and assume

$$\dot{\mathbf{F}}_{\mathrm{g}} = f\{\hat{\mathbf{G}}(M)\}, \tag{13}$$

where $\hat{\mathbf{G}}(M)$ is an experimentally determined function of mechanical stimuli M. The determination of the function f will be addressed in Section 3. Recalling that the internal energy and stress are assumed independent of $\mathbf{F}_{\mathrm{g}}$

and using standard invariance arguments, we consider constitutive equations $\varepsilon = \hat{\varepsilon}(\mathbf{C}_e)$ and $\mathbf{T} = \tilde{\mathbf{T}}(\mathbf{F}_e)$, where $\mathbf{C}_e = \mathbf{F}_e^T \mathbf{F}_e$. Consequently, (10) becomes

$$\left\{ 2\rho \mathbf{F}_e \frac{\partial \hat{\varepsilon}}{\partial \mathbf{C}_e} \mathbf{F}_e^T - \mathbf{T} \right\} \cdot \mathbf{L}_e - \left\{ \mathbf{F}_e^T \mathbf{T} \mathbf{F}_e^{-T} \right\} \cdot \mathbf{L}_g - \rho\beta = 0. \tag{14}$$

To obtain necessary and sufficient conditions for (14) to hold, one must also assume that $\mathbf{L}_e$ can be specified independently from $\mathbf{L}_g$ and β; consequently,[4]

$$\mathbf{T} = 2\rho \mathbf{F}_e \frac{\partial \hat{\varepsilon}}{\partial \mathbf{C}_e} \mathbf{F}_e^T, \quad \mathbf{T} \cdot \left\{ \mathbf{F}_e \mathbf{L}_g \mathbf{F}_e^{-1} \right\} + \rho\beta = 0. \tag{15}$$

These results state that the growing material is Green-elastic and that the stress power due to growth is balanced by the rate of change of the growth energy. It is emphasized that for Green-elasticity to be a necessary condition, $\mathbf{L}_e$ and $\mathbf{L}_g$ must be independent of each other. Recalling (4) and (13), this leads to the requirement that *the growth law is independent of the elastic strain rate*. This conclusion contradicts observations for biological tissues; consequently, in the context of the theory presented here, Green-elasticity may not be a necessary condition for the work-energy principle. This issue is discussed further in Section 4.3.

3 Incremental Equations and Invariance Requirements

The theory presented above employed a fixed reference configuration; however, the biological aspects of growth suggest that the growth law should be defined relative to the current configuration. In this section, the equations for an increment of growth are developed and used to derive the manner in which the stress constitutive equation evolves during growth and the relationship between $\dot{\mathbf{F}}_g$ and $\hat{\mathbf{G}}(M)$ proposed in (13). Also, the incremental equations serve as the foundation for numerical approaches that will be needed for solving growth boundary-value problems when there exists a residually stressed configuration or when growth occurs in a nonhomogeneous manner. Since numerical approaches may be based on the solution to the equilibrium growth boundary-value problem, as discussed in Klisch et al. (2005), attention is restricted to equilibrium configurations. The development here extends that presented in Klisch et al. (2001).

3.1 Definition of the Incremental Growth Problem

Consider a subset $\mathcal{P}$ of the body, i.e. $\mathcal{P} \subset \mathcal{B}$, that is unloaded and stress-free in a reference configuration $\kappa_0(\mathcal{P})$ at time t_0. Referring to Fig. 1, suppose

[4]In the thermodynamical theory, $(15)_2$ is replaced by an inequality; see Klisch and Hoger (2003).

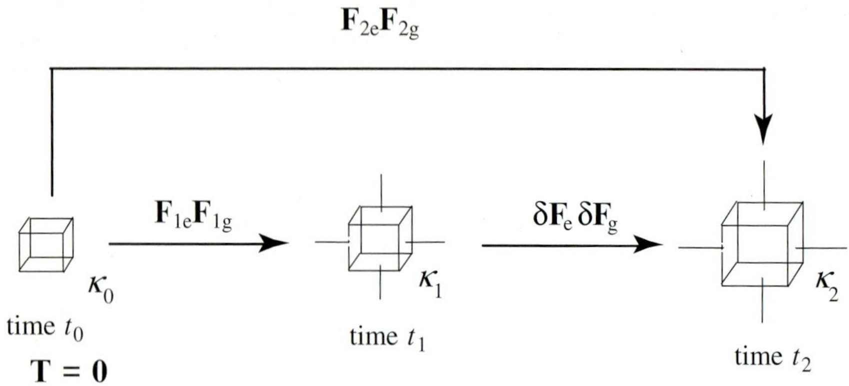

Fig. 1. The incremental growth boundary-value problem.

that the deformation gradient tensor $\mathbf{F}_1 = \mathbf{F}_{1e}\mathbf{F}_{1g}$ that maps $\kappa_0(\mathcal{P})$ to the current configuration $\kappa_1(\mathcal{P})$ is known at time t_1 and let the incremental deformation gradient tensor $\delta\mathbf{F}$ that maps $\kappa_1(\mathcal{P})$ to $\kappa_2(\mathcal{P})$ at time t_2 obey the decomposition

$$\delta\mathbf{F} = \delta\mathbf{F}_e\delta\mathbf{F}_g, \tag{16}$$

where $\delta\mathbf{F}_e$ and $\delta\mathbf{F}_g$ are called the incremental elastic and growth tensors. Note that both $\kappa_1(\mathcal{P})$ and $\kappa_2(\mathcal{P})$ may be loaded due to internal loads in $\mathcal{B}$ that arise as residual stresses develop during growth. In applications, the body $\mathcal{B}$ will be unloaded at times t_1 and t_2. The tensor $\delta\mathbf{F}_g$ quantifies the amount and orientation of mass deposition relative to $\kappa_1(\mathcal{P})$ and the tensor $\delta\mathbf{F}_e$ ensures compatibility of the body. Finally, suppose that the deformation gradient tensor $\mathbf{F}_2 = \mathbf{F}_{2e}\mathbf{F}_{2g}$ maps $\kappa_0(\mathcal{P})$ to $\kappa_2(\mathcal{P})$ at time t_2. Using the chain rule, we obtain

$$\mathbf{F}_2 = \delta\mathbf{F}\mathbf{F}_1 = \delta\mathbf{F}_e\delta\mathbf{F}_g\mathbf{F}_{1e}\mathbf{F}_{1g}. \tag{17}$$

If $\mathcal{B}$ is residually stressed at time t_0, then $\kappa_0(\mathcal{P})$ corresponds to a stress-free virtual configuration that may be found through a series of destructive experiments. Then, the first increment of growth may be defined relative to $\kappa_1(\mathcal{P})$, where $\mathbf{F}_1$ is the residual elastic deformation gradient tensor. In application, the subset $\mathcal{P}$ may correspond to a finite element. The solution to the incremental growth problem involves determining $\delta\mathbf{F}_g$ and $\delta\mathbf{F}_e$.

3.2 The Growth Law and the Incremental Growth Tensor

The biological aspects of growth suggest that a growth law should govern the tensorial description of mass deposition relative to $\kappa_1(\mathcal{P})$, which is considered as the reference configuration for the incremental growth problem. The growth law is denoted by $\mathbf{G} = \hat{\mathbf{G}}(\mathbf{X}, t)$, where $\mathbf{X}$ is the position vector in $\kappa_1(\mathcal{P})$, and corresponds to the time rate of change of the incremental growth tensor $\delta\mathbf{F}_{\mathrm{g}}$. Using a Taylor series expansion about t_1 for the increment $\Delta t = t_2 - t_1$, we obtain, to first order,

$$\delta\mathbf{F}_{\mathrm{g}}(\mathbf{X}, t_1) = \mathbf{I} + \hat{\mathbf{G}}(\mathbf{X}, t_1)\Delta t. \tag{18}$$

Using (18), the growth continuity equation (12) evaluated in $\kappa_1(\mathcal{P})$ may be written in terms of the incremental growth tensor as, to first order,

$$\det\delta\mathbf{F}_{\mathrm{g}}(\mathbf{X}, t_1) = 1 + c(\mathbf{X}, t_1)\Delta t \quad \text{or} \quad c(\mathbf{X}, t_1) = \overline{\det\delta\mathbf{F}_{\mathrm{g}}}^{\boldsymbol{\cdot}} = \hat{\mathbf{G}}(\mathbf{X}, t_1) \cdot \mathbf{I}. \tag{19}$$

3.3 The Incremental Elastic Tensor

The equilibrium equation $\operatorname{div}\mathbf{T} = \mathbf{0}$ and the stress constitutive equation are used to determine $\delta\mathbf{F}_{\mathrm{e}}$ for $\mathcal{P}$ while satisfying the boundary conditions on the body $\mathcal{B}$. Suppose that the stress constitutive functions for deformations out of $\kappa_0(\mathcal{P})$ and $\kappa_1(\mathcal{P})$ are of the general forms $\hat{\mathbf{T}}_0(\mathbf{F}_{2\mathrm{e}})$ and $\hat{\mathbf{T}}_1(\delta\mathbf{F}_{\mathrm{e}})$, respectively. Usually, the function $\hat{\mathbf{T}}_0(\mathbf{F}_{2\mathrm{e}})$ will be known; consequently, we need to determine $\hat{\mathbf{T}}_1(\delta\mathbf{F}_{\mathrm{e}})$. The derivation is not trivial because the chain rule cannot be used, as in the theory of small deformations superimposed on large deformations, due to the decomposition (16). The physical idea that must be satisfied is that the stress in $\kappa_2(\mathcal{P})$, $\mathbf{T}_2$, must be the same when evaluated using either of these stress constitutive functions:

$$\mathbf{T}_2 = \hat{\mathbf{T}}_0(\mathbf{F}_{2\mathrm{e}}) = \hat{\mathbf{T}}_1(\delta\mathbf{F}_{\mathrm{e}}). \tag{20}$$

For orthotropic, transversely isotropic, and isotropic materials, if $\mathbf{F}_{2\mathrm{e}}$ and $\hat{\mathbf{T}}_1(\delta\mathbf{F}_{\mathrm{e}})$ satisfy

$$\mathbf{F}_{2\mathrm{e}} = \delta\mathbf{F}_{\mathrm{e}}\mathbf{F}_{1\mathrm{e}}, \quad \hat{\mathbf{T}}_1(\delta\mathbf{F}_{\mathrm{e}}) = \hat{\mathbf{T}}_0(\delta\mathbf{F}_{\mathrm{e}}\mathbf{F}_{1\mathrm{e}}), \tag{21}$$

then these are sufficient conditions for (20) to hold[5]. Note that, using (17) and recalling $\mathbf{F}_2 = \mathbf{F}_{2\mathrm{e}}\mathbf{F}_{2\mathrm{g}}$, condition $(21)_1$ implies

$$\mathbf{F}_{2\mathrm{g}} = \mathbf{F}_{1\mathrm{e}}^{-1}\delta\mathbf{F}_{\mathrm{g}}\mathbf{F}_{1\mathrm{e}}\mathbf{F}_{1\mathrm{g}}. \tag{22}$$

The proof of this assertion begins by recalling the stress constitutive equation derived by Hoger (1997) for a residually stressed orthotropic material. In that work, the deformation gradient tensor, $\mathbf{F}^*$, relative to the virtual configuration is

$$\mathbf{F}^* = \mathbf{F}\mathbf{F}^0, \tag{23}$$

[5]This derivation was not presented in Klisch et al. (2001) and is based on an approach used to derive necessary and sufficient conditions for isotropic materials by van Dyke and Hoger (2000).

where $\mathbf{F}$ is the deformation gradient tensor relative to the residually stressed configuration and $\mathbf{F}^0$ maps the virtual configuration to the residually stressed configuration. For orthotropy, the set of basic polynomial invariants is

$$\mathbf{I}\cdot\mathbf{C},\ \mathbf{I}\cdot\mathbf{C}^2,\ \mathbf{I}\cdot\mathbf{C}^3,\ (\mathbf{k}_1\otimes\mathbf{k}_1)\cdot\mathbf{C},\ (\mathbf{k}_2\otimes\mathbf{k}_2)\cdot\mathbf{C},\ (\mathbf{k}_1\otimes\mathbf{k}_1)\cdot\mathbf{C}^2,\\ (\mathbf{k}_2\otimes\mathbf{k}_2)\cdot\mathbf{C}^2, \tag{24}$$

where $\mathbf{k}_1$ and $\mathbf{k}_2$ are unit normal vectors to two planes of reflectional symmetry. The stress constitutive equation derived in Hoger (1997) is

$$\begin{aligned}\mathbf{T} = \mathbf{R}\{&\alpha_0\mathbf{I} + \alpha_1\mathbf{U}\mathbf{B}^0\mathbf{U} + \alpha_2(\mathbf{U}\mathbf{B}^0\mathbf{U})^2\\ &+\sum_{i=1}^{2}\Omega_i^2\alpha_{3i}(\mathbf{U}\mathbf{B}^0\mathbf{U})^{-1/2}\mathbf{U}(\mathbf{k}_i^0\otimes\mathbf{k}_i^0)\mathbf{U}(\mathbf{U}\mathbf{B}^0\mathbf{U})^{-1/2}\\ &+\sum_{i=1}^{2}\Omega_i^2\alpha_{4i}[(\mathbf{U}\mathbf{B}^0\mathbf{U})^{-1/2}\mathbf{U}(\mathbf{k}_i^0\otimes\mathbf{k}_i^0)\mathbf{U}(\mathbf{U}\mathbf{B}^0\mathbf{U})^{1/2}\\ &\qquad+(\mathbf{U}\mathbf{B}^0\mathbf{U})^{1/2}\mathbf{U}(\mathbf{k}_i^0\otimes\mathbf{k}_i^0)\mathbf{U}(\mathbf{U}\mathbf{B}^0\mathbf{U})^{-1/2}]\\ &+\sum_{i=1}^{2}\Omega_i^2\alpha_{5i}(\mathbf{U}\mathbf{B}^0\mathbf{U})^{1/2}\mathbf{U}(\mathbf{k}_i^0\otimes\mathbf{k}_i^0)\mathbf{U}(\mathbf{U}\mathbf{B}^0\mathbf{U})^{1/2}\}\mathbf{R}^T,\end{aligned} \tag{25}$$

where $\mathbf{B}^0 = \mathbf{F}^0\mathbf{F}^{0T}$, $(\alpha_0, \alpha_1, \alpha_2, \alpha_{3i}, \alpha_{4i}, \alpha_{5i})$ are scalar functions of the invariants (24) evaluated at $\mathbf{C} = \mathbf{F}^{*T}\mathbf{F}^*$, and $\mathbf{k}_i^0$ is the direction of $\mathbf{k}_i$ in the residually stressed configuration that satisfies

$$\mathbf{k}_i^0 = \mathbf{F}^0\mathbf{k}_i/\Omega_i, \quad \Omega_i = 1/|\mathbf{F}^{0^{-1}}\mathbf{k}_i^0|. \tag{26}$$

Upon assuming the relations (21), the stress $\mathbf{T}_2$ in $\kappa_2(\mathcal{P})$ can be calculated using (25) in two manners. First, to use $\hat{\mathbf{T}}_1(\delta\mathbf{F}_e) = \hat{\mathbf{T}}_0(\delta\mathbf{F}_e\mathbf{F}_{1e})$ we set $\mathbf{F}^0 = \mathbf{F}_{1e}$ and $\mathbf{F} = \delta\mathbf{F}_e$ in (25). Second, to use $\hat{\mathbf{T}}_0(\mathbf{F}_{2e})$ we set $\mathbf{F}^0 = \mathbf{I}$, $\mathbf{F} = \mathbf{F}_{2e} = \delta\mathbf{F}_e\mathbf{F}_{1e}$, $\mathbf{k}_i^0 = \mathbf{k}_i$, and $\Omega_i = 1$. After a straightforward calculation, it follows that the stress in $\kappa_2(\mathcal{P})$ calculated in each of these manners is the same. Thus, the assertion for orthotropic materials holds, and the proof for transversely isotropic and isotropic materials follows by restricting the invariants in (24) accordingly.

3.4 Invariance Requirements on the Growth Law

In order to construct physically realistic growth laws, the growth law should satisfy invariance requirements. Epstein and Maugin (2000) discussed invariance requirements in their thermomechanical theory of growth, which differs from that presented here, when both the present and the reference configurations are subject to invariance. Here, the concept that the growth law should be invariant of the reference configuration will be explicitly employed by analyzing invariance requirements for the growth law $\mathbf{G}$ that is used to calculate the incremental growth tensor in (18). Invariance of the intermediate configuration defined by $\mathbf{F}_{\mathrm{g}}$, as used in the plasticity literature, is not invoked because a satisfactory manner in doing so has not been accomplished.

First, the relationship (13) between $\dot{\mathbf{F}}_{\mathrm{g}}$, defined relative to $\kappa_0(\mathcal{P})$, and the growth law $\mathbf{G}$, defined relative to $\kappa_1(\mathcal{P})$, is derived[6]. Recall the definition

$$\dot{\mathbf{F}}_{\mathrm{g}} = \lim_{\Delta t \to 0} \frac{1}{\Delta t}[\mathbf{F}_{\mathrm{g}}(t+\Delta t) - \mathbf{F}_{\mathrm{g}}(t)], \tag{27}$$

and note that $\mathbf{F}_{\mathrm{g}}(t+\Delta t) = \mathbf{F}_{2\mathrm{g}}$ and $\mathbf{F}_{\mathrm{g}}(t) = \mathbf{F}_{1\mathrm{g}}$. Using (18) and (22), we obtain

$$\dot{\mathbf{F}}_{\mathrm{g}} = \mathbf{F}_{1\mathrm{e}}^{-1}\mathbf{G}\mathbf{F}_{1\mathrm{e}}\mathbf{F}_{1\mathrm{g}}, \tag{28}$$

which makes explicit the function f presented in (13). Consider a growth law

$$\mathbf{G} = \hat{\mathbf{G}}(\mathbf{F}_{\mathrm{g}}, \mathbf{F}_{\mathrm{e}}, \mathbf{L}_{\mathrm{e}}, \mathbf{T}, \mathbf{a}), \tag{29}$$

where $\mathbf{a}$ corresponds to any of the diffusive velocities defined for a growing mixture in Section 4. Due to a rigid body motion superposed on $\kappa_1(\mathcal{P})$, standard results lead to the transformation rules

$$\begin{aligned} &\mathbf{F}_{\mathrm{g}}^{+} = \mathbf{F}_{\mathrm{g}}, \quad \mathbf{F}_{\mathrm{e}}^{+} = \mathbf{Q}\mathbf{F}_{\mathrm{e}}, \quad \mathbf{L}_{\mathrm{e}}^{+} = \mathbf{Q}\mathbf{L}_{\mathrm{e}}\mathbf{Q}^{T} + \dot{\mathbf{Q}}\mathbf{Q}^{T}, \quad \mathbf{T}^{+} = \mathbf{Q}\mathbf{T}\mathbf{Q}^{T}, \\ &\quad \mathbf{a}^{+} = \mathbf{Q}\mathbf{a}, \end{aligned} \tag{30}$$

where $\mathbf{Q}$ is an arbitrary proper orthogonal tensor. Furthermore, using $(4)_2$, (28), and $(30)_1$, a straightforward calculations leads to

$$\mathbf{G}^{+} = \mathbf{Q}\mathbf{G}\mathbf{Q}^{T}. \tag{31}$$

Then, (29)–(31) lead to the invariance requirement

$$\mathbf{Q}\hat{\mathbf{G}}\mathbf{Q}^{T} = \hat{\mathbf{G}}(\mathbf{F}_{\mathrm{g}}, \mathbf{Q}\mathbf{F}_{\mathrm{e}}, \mathbf{Q}\mathbf{L}_{\mathrm{e}}\mathbf{Q}^{T} + \dot{\mathbf{Q}}\mathbf{Q}^{T}, \mathbf{Q}\mathbf{T}\mathbf{Q}^{T}, \mathbf{Q}\mathbf{a}). \tag{32}$$

Choosing $\mathbf{Q} = \mathbf{I}$ and $\dot{\mathbf{Q}} = -\mathbf{W}_{\mathrm{e}}$ (i.e. the skew part of $\mathbf{L}_{\mathrm{e}}$) one finds that $\mathbf{L}_{\mathrm{e}}$ must be replaced by $\mathbf{D}_{\mathrm{e}}$ in (29). Consequently, the invariance requirement becomes

$$\mathbf{Q}\hat{\mathbf{G}}\mathbf{Q}^{T} = \hat{\mathbf{G}}(\mathbf{F}_{\mathrm{g}}, \mathbf{Q}\mathbf{F}_{\mathrm{e}}, \mathbf{Q}\mathbf{D}_{\mathrm{e}}\mathbf{Q}^{T}, \mathbf{Q}\mathbf{T}\mathbf{Q}^{T}, \mathbf{Q}\mathbf{a}). \tag{33}$$

4 Growing Mixtures

Here, we discuss the theory of Klisch and Lotz (2000) and Klisch and Hoger (2003), which extended the growth theory to thermoelastic materials and mixtures. Since it is not clear which of the many components of biological tissues are important to include as distinct constituents, we allowed for an arbitrary number of growing elastic materials. These constituents may grow via mass exchange with the other growing constituents or volumetric mass growth from unspecified sources.

---

[6]This analysis was not presented in Klisch et al. (2001).

4.1 Kinematics

Consider a growing mixture $\mathcal{B}$ of ν constituents at a common temperature, $\nu-1$ growing elastic materials C^α and an inviscid fluid C^ν, with a fixed reference configuration $\kappa_0(\mathcal{B})$ and a present configuration $\kappa(\mathcal{B})$ at time t. At each point $\mathbf{x} \in \kappa(\mathcal{B})$, there exists one material particle of each constituent. The motion of C^α and the common temperature of $\mathcal{B}$ are defined by smooth mappings

$$\mathbf{x}^\alpha = \chi^\alpha(\mathbf{X}^\alpha, t), \quad \theta = \Theta(\mathbf{X}^\alpha, t), \tag{34}$$

where $\theta > 0$ is the absolute temperature of the mixture. The velocities $\mathbf{v}^\alpha$ and relative (i.e. diffusive) velocities $\mathbf{a}^\alpha$ are defined as

$$\mathbf{v}^\alpha = \dot{\chi}(\mathbf{X}^\alpha, t), \quad \mathbf{a}^\alpha = \mathbf{v}^\alpha - \mathbf{v}^\nu. \tag{35}$$

Deformation gradient tensors are defined as in (2) for each C^α and are assumed to obey the decomposition

$$\mathbf{F}^\alpha = \mathbf{F}^\alpha_{\mathrm{e}} \mathbf{F}^\alpha_{\mathrm{g}}, \tag{36}$$

where $\mathbf{F}^\alpha_{\mathrm{e}}$ and $\mathbf{F}^\alpha_{\mathrm{g}}$ are the elastic and growth tensors with the same interpretations as for the growing elastic material. If two or more of the growing constituents are internally constrained (see Section 5), then the tensors $\mathbf{F}^\alpha_{\mathrm{e}}$ may include an elastic component that maintains the constraint. The mass density, free energy, and stress functions for each C^α are assumed to be independent of $\mathbf{F}^\alpha_{\mathrm{g}}$. In theory, this assumption leads to an experimental prescription for $\mathbf{F}^\alpha_{\mathrm{e}}$ and $\mathbf{F}^\alpha_{\mathrm{g}}$. For each C^α, the kinematical rate equations (4)–(6) hold.

4.2 Balance Equations

The classical mixture balance laws are modified to include partial mass growth functions c^α and growth energy terms β^α for each C^α. The balance equations corresponding to (7)–(10) are

$$\dot{\rho}^\alpha + \rho^\alpha \operatorname{div} \mathbf{v}^\alpha = \rho^\alpha c^\alpha, \quad \dot{\rho}^\nu + \rho^\nu \operatorname{div} \mathbf{v}^\nu = 0, \tag{37}$$

$$\operatorname{div} \mathbf{T}^\alpha + \boldsymbol{\pi}^\alpha + \rho^\alpha \mathbf{b}^\alpha = \rho^\alpha \dot{\mathbf{v}}^\alpha, \tag{38}$$

$$\mathbf{T}^\alpha - \mathbf{T}^{\alpha T} = \boldsymbol{\Lambda}^\alpha, \tag{39}$$

$$\rho^\alpha \dot{\varepsilon}^\alpha = \rho^\alpha r^\alpha - \operatorname{div} \mathbf{q}^\alpha + \gamma^\alpha + \mathbf{T}^\alpha \cdot \mathbf{D}^\alpha + \rho^\alpha \beta^\alpha, \tag{40}$$

$$\rho^\nu \dot{\varepsilon}^\nu = \rho^\nu r^\nu - \operatorname{div} \mathbf{q}^\nu + \gamma^\nu + \mathbf{T}^\nu \cdot \mathbf{D}^\nu, \tag{41}$$

where ρ^α, $\mathbf{v}^\alpha$, $\mathbf{T}^\alpha$, $\mathbf{b}^\alpha$, ε^α, r^α, $\mathbf{q}^\alpha$, $\mathbf{D}^\alpha$ are partial densities, velocities, Cauchy stresses, body forces, internal energies, external heat supplies, heat flux vectors, and rate of deformation tensors, respectively, while $\boldsymbol{\pi}^\alpha$, $\boldsymbol{\Lambda}^\alpha$, γ^α are internal supplies of linear momentum (i.e. diffusive force), angular momentum, and internal energy[7]. Recalling that ρ^α are assumed independent of $\mathbf{F}^\alpha_{\mathrm{g}}$, the elastic and growth continuity equations are obtained:

$$\rho^\alpha \det \mathbf{F}^\alpha_{\mathrm{e}} = \rho^\alpha_0, \quad \det \mathbf{F}^\alpha_{\mathrm{g}} = \exp\left\{ \int_{\tau=t_o}^{t} c^\alpha \mathrm{d}\tau \right\}. \tag{42}$$

The balance of mass and momentum equations for the mixture require that

$$\sum_{\alpha=1}^{\nu-1} \rho^\alpha c^\alpha = \rho c, \quad \sum_{\alpha=1}^{\nu} \boldsymbol{\pi}^\alpha = \mathbf{0}, \quad \sum_{\alpha=1}^{\nu} \boldsymbol{\Lambda}^\alpha = \mathbf{0}, \tag{43}$$

where c is the mass growth function for the mixture. The balance of energy equation for the mixture becomes

$$\sum_{\alpha=1}^{\nu} \left[\rho^\alpha \dot{\varepsilon}^\alpha - \rho^\alpha r^\alpha + \operatorname{div} \mathbf{q}^\alpha + \boldsymbol{\pi}^\alpha \cdot \mathbf{v}^\alpha - \mathbf{T}^\alpha \cdot \mathbf{L}^\alpha \right] - \sum_{\alpha=1}^{\nu-1} \rho^\alpha \beta^\alpha = 0. \tag{44}$$

4.3 Constitutive Equations: Objections to the Coleman-Noll Approach and an Alternative Treatment

Coleman and Noll (1963) formulated a continuum approach to thermodynamics. They assumed the existence of an entropy function and derived necessary and sufficient conditions (i.e. constitutive restrictions) for the Clausius-Duhem inequality to hold. For a thermoelastic material, these necessary and sufficient conditions state that the Cauchy stress and temperature are derived from a scalar free energy function. The derivation of necessary conditions was not trivial; in particular, it was assumed that the stress, temperature, and free energy were independent of the rates of strain and entropy change. The validity of introducing entropy as a primitive variable was first questioned by Rivlin (1973). Rivlin's approach was summarized for thermoelastic solids (Casey and Krishnaswamy (1998)) and extended to continuum mixture theory (Krishnaswamy and Batra (1997), Klisch (2002)). In those papers, a process was identified that was assumed to be path-independent (or reversible); consequently, the balance of energy equation leads to a prescription for the entropy functions. Then, necessary and sufficient conditions are obtained for the balance of energy to hold for reversible processes, and subsequent application of the Clausius-Duhem inequality leads to a residual inequality. Using that approach, one obtains the identification of a series of experiments that allows the measurement of the entropy function. However, in Klisch and Hoger (2003) we established that there may exist materials that do not experience reversible growth processes.

[7] The material time derivatives are defined by following material particles of each C^α.

Another objection to the Coleman-Noll approach arises for growing thermoelastic materials and mixtures. Recall from Section 2.3 that the growth law cannot depend on the rate of elastic strain for Green-elasticity to be a necessary condition for the work-energy principle to hold. This observation extends to growing thermoelastic materials and mixtures. In fact, in the pioneering paper of adaptive elasticity by Cowin and Hegedus (1976), it was stated that the remodeling rate equation had to be independent of elastic strain rate in order to derive necessary constitutive restrictions using the Coleman-Noll approach. Furthermore, if the Coleman-Noll approach is used in the theory of growing mixtures presented here, then the growth laws must be independent of the diffusive velocities in order to establish necessary conditions. These restrictions on the growth law contradict experimental results; for example, both strain rate (Cowin (2004)) and fluid diffusive velocity (Buschmann et al. (1999)) have been proposed as mechanical stimuli that drive the growth process.

Due to these objections, we formulated an alternative thermodynamical treatment of growth. Our approach was motivated by the idea that, at any time during growth, the growth process may be stopped and experiments may be performed to determine the thermomechanical properties. When the growth process is stopped, the material behaves as a thermoelastic material or mixture, for which path-independent and reversible processes exist. This thermoelastic material or mixture was called the *generating material.* Thus, the constitutive restrictions for the growing material were obtained from a single generating material. The reader is referred to Klisch and Hoger (2003) for full details. Also, in the theory of Epstein and Maugin (2000), a *thermoelastic point* was defined which is conceptually similar to the generating material first proposed in Klisch et al. (2000).

5 A Cartilage Growth Mixture Model

Here, we outline a cartilage growth mixture (CGM) model with the following simplifying assumptions.

(i) The mixture is composed of an inviscid fluid, representing the water constituent (W), and a solid matrix (S) composed of three constituents: a growing PG elastic material, a growing COL elastic network, and 'others' representing the non-collageneous proteins (OTH). The 'other' may grow through mass increase via cellular metabolism or matrix degradation and through mass decrease via conversion into functional PG or COL molecules.

(ii) The PG, COL, and OTH constituents are bound to the extracellular matrix and, consequently, are constrained to experience the same overall motion. This is a limitation as some of the PGs are soluble and mobile in the tissue matrix; however, the general theory allows for the specification of mobile constituents.

(iii) The mixture is constrained to be intrinsically incompressible, as is commonly employed in the cartilage biomechanics literature.

(iv) The PG stress depends on the PG, COL, and W densities. This assumption is based on the two-compartmental model of Basser et al. (1998) in which the PG stress is calculated from an effective fixed charge density.

(v) The COL stress depends only on the elastic deformation of the COL network. This assumption is made as there is insufficient data to quantify how other matrix molecules affect the mechanical properties of the COL network.

(vi) The determinate stresses in the W and OTH constituents are both zero. Assuming the determinate fluid stress to be zero is equivalent to assuming a constant fluid free energy function (Klisch and Lotz (2000)). We assume that the OTH constituent does not contribute directly to the tissue's mechanical properties.

The governing equations for this CGM are obtained from those presented above; the reader is referred to Klisch et al. (2005) for full details. Due to the introduction of the two constraints, there exist indeterminate constraint response functions. In the continuum mixture theory approach, the SM Cauchy stress tensor $\mathbf{T}^{\mathrm{S}}$ and diffusive force $\boldsymbol{\pi}^{\mathrm{S}}$ are assumed to obey the decompositions

$$\mathbf{T}^{\mathrm{S}} = \mathbf{T}^{\mathrm{PG}} + \mathbf{T}^{\mathrm{COL}} + \mathbf{T}^{\mathrm{OTH}}, \quad \boldsymbol{\pi}^{\mathrm{S}} = \boldsymbol{\pi}^{\mathrm{PG}} + \boldsymbol{\pi}^{\mathrm{COL}} + \boldsymbol{\pi}^{\mathrm{OTH}}. \tag{45}$$

The immobility constraint produces constraint responses in $\mathbf{T}^{\alpha}$ and $\boldsymbol{\pi}^{\alpha}$ which cancel upon addition when forming $\mathbf{T}^{\mathrm{S}}$ and $\boldsymbol{\pi}^{\mathrm{S}}$. Consequently, the equilibrium equations that must be satisfied reduce to

$$\operatorname{div} \mathbf{T}^{\mathrm{S}} = \mathbf{0}, \quad \operatorname{div} \mathbf{T}^{\mathrm{W}} = \mathbf{0}, \tag{46}$$

while the immobility constraint responses ensure that equilibrium is satisfied for all constituents. Also, there exists an arbitrary Lagrange multiplier term (i.e. the fluid pore pressure) in each $\mathbf{T}^{\alpha}$ that arises due to the intrinsic incompressibility constraint. To date, we have solved several boundary-value problems that have illustrated the main feature of the model: the CGM is capable of predicting the development of residual stresses and nonhomogeneous tissue composition (Klisch et al. (2001, 2005)) and the evolution of the tissue's mechanical properties during growth (Klisch et al. (2000, 2003)).

One of the biggest challenges lies in identifying an accurate stress constitutive equation for cartilage; such an equation is needed before the CGM can be properly validated. For even simple types of applied mechanical loading, the highly nonhomogeneous nature of cartilage leads to a complex strain field: thus, an accurate stress constitutive equation must be validated with multiple loading protocols. Since such an accurate equation for cartilage does not exist, we have recently developed a 6-parameter bimodular orthotropic second-order stress constitutive equation that can describe the asymmetric tensile-compressive response as well as the anisotropic tensile response (Klisch et al. (2004)). Another challenge lies in formulating efficient numerical implementations of the theory; we have developed a cartilage growth FEM using the incremental approach outlined here. We have used that FEM with a growth law that depends on the magnitude of the diffusive fluid velocity to quantify the nonhomogeneous growth that may occur in an in vitro dynamic confined compression protocol (Bingham et al. (2005)). However, the CGM model has not been rigorously validated; for example, the 'balance of stress' hypothesis $(45)_1$ has not been experimentally verified. The validation of these, and other, continuum growth models with carefully designed experiments appears to be a major challenge in the near future.

Acknowledgements. This work was supported by NSF and the Office of Naval Research. I would like to acknowledge several helpful discussions with Professors van Dyke and Hoger related to the unpublished material presented in Section 3.

References

Basser, P. J., Schneiderman, R., Bank, R. A., Wachtel, E., and Maroudas, A. (1998). Mechanical properties of the collagen network in human articular cartilage as measured by osmotic stress technique. *Arch. Biochem. Biophys.* 351:207–219.

Bingham, M., Davol, A., Sah, R. L., and Klisch, S. M. (2005). A nonlinear finite element model of cartilage growth under in vitro dynamic compression. In *ASME Summer Bioengineering Conference.* in review.

Buschmann, M. D., Kim, Y. J., Wong, M., Frank, E., Hunziker, E. B., and Grodzinsky, A. J. (1999). Stimulation of aggrecan synthesis in cartilage explants by cyclic loading is localized to regions of high interstitial fluid flow. *Arch. Biochem. Biophys.* 366:1–7.

Casey, J., and Krishnaswamy, S. (1998). A characterization of internally constrained thermoelastic materials. *Math. Mech. Solids* 3:71–89.

Chen, Y. C., and Hoger, A. (2000). Constitutive functions for elastic materials in finite growth and deformation. *J. Elasticity* 59:175–193.

Coleman, B. D., and Noll, W. (1963). The thermodynamics of elastic materials with heat conduction and viscosity. *Arch. Rat. Mech. Anal.* 13:167–178.

Cowin, S. C., and Hegedus, D. H. (1976). Bone remodeling I: A theory of adaptive elasticity. *J. Elasticity* 6:313–326.

Cowin, S. C. (2004). Tissue growth and remodeling. *Ann. Rev. Biomed. Eng.* 6:77–107.

Epstein, M., and Maugin, G. A. (2000). Thermomechanics of volumetric growth in uniform bodies. *Int. J. Plasticity* 16:951–978.

Garikipati, K., Arruda, E. M., Grosh, K., Narayanan, H., and Calve, S. (2004). A continuum treatment of growth in biological tissue: The coupling of mass transport and growth. *J. Mech. Phys. Solids* 52:1595–1625.

Hoger, A. (1997). Virtual configurations and constitutive equations for residually stressed bodies with material symmetry. *J. Elasticity* 48:125–144.

Humphrey, J. D., and Rajagopal, K. R. (2002). A constrained mixture model for growth and remodeling of soft tissues. *Math. Model. Meth. Appl. Sci.* 12:407–430.

Klisch, S. M., and Hoger, A. (2003). Volumetric growth of thermoelastic materials and mixtures. *Math. Mech. Solids* 8:377–402.

Klisch, S. M., and Lotz, J. C. (2000). A special theory of biphasic mixtures and experimental results for human annulus fibrosus tested in confined compression. *J. Biomech. Eng.* 122:180–188.

Klisch, S. M., Sah, R. L., and Hoger, A. (2000). A growth mixture theory for cartilage. In Casey, J., and Bao, G., eds., *Mechanics in Biology*. New York: The American Society of Mechanical Engineers (ASME). AMD-Vol. 242/BED-Vol. 46, pp. 229–242.

Klisch, S. M., van Dyke, T., and Hoger, A. (2001). A theory of volumetric growth for compressible elastic materials. *Math. Mech. Solids* 6:551–575.

Klisch, S. M., Chen, S. S., Sah, R. L., and Hoger, A. (2003). A growth mixture theory for cartilage with applications to growth-related experiments on cartilage explants. *J. Biomech. Eng.* 125:169–179.

Klisch, S. M., Holtrichter, S. E., Sah, R. L., and Davol, A. (2004). A bimodular second order orthotropic stress constitutive equation for cartilage. In *Transactions of the ASME: Advances in Bioengineering.*

Klisch, S. M., Sah, R. L., and Hoger, A. (2005). A cartilage growth mixture model for infinitesimal strains: Solutions of boundary-value problems related to in vitro growth experiments. *Biomech. Model. Mechanobio.* in press.

Klisch, S. M. (2002). A mixture of elastic materials with different constituent temperatures and internal constraints. *Int. J. Eng. Sci.* 40:805–828.

Krishnaswamy, S., and Batra, R. (1997). A thermomechanical theory of solid-fluid mixtures. *Math. Mech. Solids* 2:143–151.

Lubarda, V. A., and Hoger, A. (2002). On the mechanics of solids with a growing mass. *Int. J. Solids Structures* 39:4627–4664.

van Dyke, T., and Hoger, A. (2000). Personal communication.

Menzel, A. (2005). Modelling of anisotropic growth in biological tissues. A new approach and computational aspects. *Biomech. Model. Mechanobio.* 3:147–171.

Rivlin, R. S. (1973). Comments on some recent researches in thermomechanics. *Recent Advances in Engineering Science* 8:1–23.

Rodriguez, E. K., Hoger, A., and McCulloch, A. D. (1994). Stress-dependent finite growth in soft elastic tissues. *J. Biomech.* 27:455–467.

Humphrey, J. D., and Rajagopal, K. R. (2002). A constrained mixture model for growth and remodeling of soft tissues. *Math. Model. Meth. Appl. Sci.* 12:407–430.

Klisch, S. M., and Hoger, A. (2003). Volumetric growth of thermoelastic materials and mixtures. *Math. Mech. Solids* 8:377–402.

Klisch, S. M., and Lotz, J. C. (2000). A special theory of biphasic mixtures and experimental results for human annulus fibrosus tested in confined compression. *J. Biomech. Eng.* 122:180–188.

Klisch, S. M., Sah, R. L., and Hoger, A. (2000). A growth mixture theory for cartilage. In Casey, J., and Bao, G., eds., *Mechanics in Biology.* New York: The American Society of Mechanical Engineers (ASME). AMD-Vol. 242/BED-Vol. 46, pp. 229–242.

Klisch, S. M., van Dyke, T., and Hoger, A. (2001). A theory of volumetric growth for compressible elastic materials. *Math. Mech. Solids* 6:551–575.

Klisch, S. M., Chen, S. S., Sah, R. L., and Hoger, A. (2003). A growth mixture theory for cartilage with applications to growth-related experiments on cartilage explants. *J. Biomech. Eng.* 125:169–179.

Klisch, S. M., Holtrichter, S. E., Sah, R. L., and Davol, A. (2004). A bimodular second order orthotropic stress constitutive equation for cartilage. In *Transactions of the ASME: Advances in Bioengineering.*

Klisch, S. M., Sah, R. L., and Hoger, A. (2005). A cartilage growth mixture model for infinitesimal strains: Solutions of boundary-value problems related to in vitro growth experiments. *Biomech. Model. Mechanobio.* in press.

Klisch, S. M. (2002). A mixture of elastic materials with different constituent temperatures and internal constraints. *Int. J. Eng. Sci.* 40:805–828.

Krishnaswamy, S., and Batra, R. (1997). A thermomechanical theory of solid-fluid mixtures. *Math. Mech. Solids* 2:143–151.

Lubarda, V. A., and Hoger, A. (2002). On the mechanics of solids with a growing mass. *Int. J. Solids Structures* 39:4627–4664.

van Dyke, T., and Hoger, A. (2000). Personal communication.

Menzel, A. (2005). Modelling of anisotropic growth in biological tissues. A new approach and computational aspects. *Biomech. Model. Mechanobio.* 3:147–171.

Rivlin, R. S. (1973). Comments on some recent researches in thermomechanics. *Recent Advances in Engineering Science* 8:1–23.

Rodriguez, E. K., Hoger, A., and McCulloch, A. D. (1994). Stress-dependent finite growth in soft elastic tissues. *J. Biomech.* 27:455–467.

Part II

Micromechanics, Cells and Matrix

Tensile Properties and Local Stiffness of Cells

K. Hayashi

Osaka University, Department of Mechanical Science and Bioengineering, Japan
hayashi@me.es.osaka-u.ac.jp

Cytoskeletal structure is closely related to cell function, and it is reflected in the mechanical properties of cells. Therefore, it is very important to determine the properties for the study of the mechanisms of tissue and organ physiology and diseases, and also for better understanding of cell biomechanics and physiology. For these reasons, we have studied tensile properties and local stiffness of cells (fibroblasts, vascular endothelial cells, and smooth muscle cells) using a specially designed tensile test system and atomic force microscopes, respectively. These studies showed that (i) there are large differences in tensile properties and local stiffness among cells, (ii) the distribution of stiffness in a cell is not uniform, and (iii) the local stiffness of cells is affected by diseases like atherosclerosis.

1 Introduction

It is now well recognized that cells change their shape, structure, and mechanical properties in response to mechanical stress (Frangos (1993)). Cytoskeletal structure is closely related to cell function, and is reflected in the mechanical properties of cells. Therefore, the determination of the properties should contribute much to the study of the mechanisms of tissue and organ physiology, diseases, and other events that occur in the body. Moreover, it is basically very important to know the mechanical properties of cells to understand cell physiology and cell mechanics.

Various methods and techniques have been applied to the determination of the mechanical properties of cells, which are roughly divided into two categories: (i) measurement of the properties of a single, whole cell, and (ii) measurement of the local properties in a cell. The mechanical properties of a whole cell have been studied mainly on blood cells and muscle cells using the methods of micropipette aspiration of a whole cell (Hochmuth et al. (1993)), compression and stretch of a whole cell with a pair of micro-plates

(Thoumine and Ott (1997)), and stretch of a whole cell with a pair of micropipettes (Glerum et al. (1990), Palmer et al. (1996), Miyazaki et al. (2000), Matsumoto et al. (2000)). Local mechanical properties have been studied on a variety of cells using the techniques of, for example, cell poking (Zahalak et al. (1990)), micropipette aspiration of local cell surface layer (Sato et al. (1987)), twisting of embedded (Valberg and Feldman (1987)) or surface-attached (Wang and Ingber (1994)) magnetic particles with external magnetic field, bending of an extended portion of an adherent cell with micro needles (Albrecht-Buehler (1987)), scanning acoustic microscopy (Bereiter-Hahn et al. (1995)), and atomic force microscopy (Hoh and Schoenenberger (1994), Miyazaki and Hayashi (1999)).

We have been using a tensile test method and atomic force microscopy (AFM) for studying tensile properties and local stiffness of cells, respectively (Hayashi (2003)). This article first deals with our newly developed tensile test system for cells, and shows several results obtained from fibroblasts, vascular endothelial cells, and smooth muscle cells using this test system. Then, our recent AFM studies on the local stiffness of cultured smooth muscle cells, and also of endothelial cells on intact normal and atherosclerotic arteries are briefly described.

2 Tensile Properties of Cells

Simple tensile tests are useful for determining the basic mechanical properties of cells. However, only a few studies have been done on the tensile properties of cells. For example, Thoumine and Ott (1997) and Palmer et al. (1996) studied the viscoelastic and contractile properties of fibroblasts and cardiac myocytes, respectively, and also described their tensile properties. However, they did not determine the tensile force-elongation relations or the strength of these cells. Glerum et al. (1990) determined the tensile properties of smooth muscle cells obtained from the pig urinary bladder and human uterus. They knotted the ends of a single cell around the tips of a pair of micropipettes, and applied force to the cell by moving one of the micropipettes. Their gripping technique (knotting), however, can be used only for relatively long and stiff cells. More recently, we have developed a novel tensile test system applicable to a variety of cells (Miyazaki et al. (2000)), and used it to determine the tensile properties of fibroblasts, smooth muscle cells, and vascular endothelial cells. Although Matsumoto et al. (2000) also designed a tensile tester similar to our system, they applied it only to aortic smooth muscle cells obtained from the rat and bovine.

2.1 Tensile Test System

The test system is composed of a thermostatic test chamber, an inverted microscope attached with a CCD camera, two micromanipulators, a direct drive

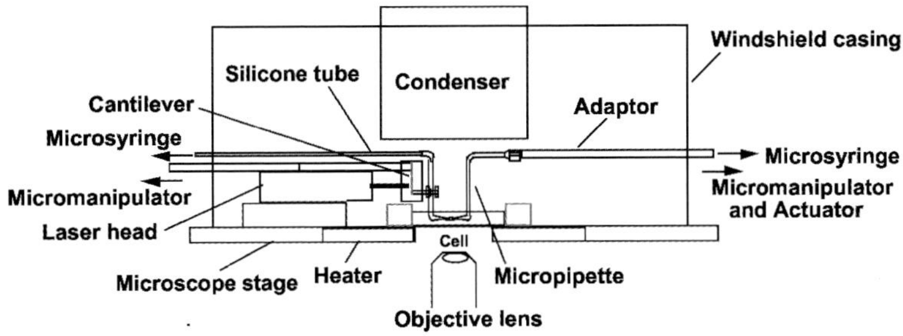

Fig. 1. Major part of the tensile test system for cells (Miyazaki et al. (2002)).

linear actuator, a cantilever-type load cell, and a video dimension analyzer (VDA) (Miyazaki et al. (2000)). The test chamber has a window of glass cover slip at the bottom, and is mounted on the microscope (Fig. 1). The microscope is equipped with a confocal laser scanner to observe the internal structure of cells. A cell is attached to the fine tips of the two micropipettes using cell adhesives. One of the micropipettes is connected to a micromanipulator, which is moved by the linear actuator to apply force to the cell. The other micropipette is attached to the cantilever-type load cell; load applied to the cell is determined from the deflection of the cantilever measured by means of a laser displacement meter. The elongation of the cell is obtained from the distance between the tips of the two micropipettes, which is measured from their CCD images using the VDA.

The maximum stroke is 10 mm, and the rate of displacement can be changed from 1 μm to 10 mm per second. The data demonstrated in this article were obtained at the rate of 6 μm/sec. The accuracy of force measurement is $+0.05\,\mu$N, and the resolution of displacement measurement is 0.24 μm.

2.2 Tensile Properties

Tensile tests were performed on fibroblasts (FBs), vascular endothelial cells (ECs), and vascular smooth muscle cells (VSMCs).

FBs were isolated from the rabbit patellar tendon by an enzymatic digestion method, and sub-cultured in Dulbecco's Modified Eagle Medium (DMEM) supplemented with 10% fetal calf serum (FCS) till the passage number 9 (Miyazaki et al. (2000)). They were spherical under non-loaded condition; their non-loaded diameter measured immediately before tensile tests were $20.6 \pm 3.7\,\mu$m (mean $\pm$ standard deviation for 6 cells from 6 animals). Shape of the cells was irregular during tensile testing, which indicates that their internal structure is inhomogeneous. The averaged load-elongation relation of those 6 cells was almost linear until fracture (Fig. 2). The maximum strength of fibroblasts was approximately 1 μN, and the elongation to fracture was 85 μm.

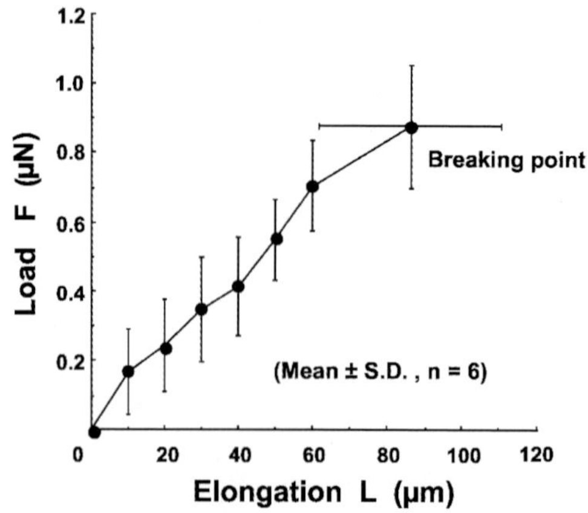

Fig. 2. Load-elongation relationship of fibroblasts (Miyazaki et al. (2000)).

Table 1. Initial dimensions and tensile test results of vascular endothelial cells (ECs): (mean±SD, $n = 5$; Nabeshima (2002)).

Cell	Length (μm)	Width (μm)
Arterial EC	20.1±6.2	3.3±1.7
Venous EC	18.9±3.1	2.8±0.7

Cell	Maximum load (μN)	Elongation (μm)	Stiffness (N/m)*
Arterial EC	0.22±0.09	18.3±8.8	0.04±0.04
Venous EC	0.19±0.07	14.3±3.2	0.03±0.02

(* at 0-3μm elongation)

Vascular ECs were obtained from the common carotid artery and the external jugular vein in rats by means of a mechanical method, and then stored in Hanks' balanced salt solution (HBSS). They had elongated shape before tensile tests, having the length of approximately 20 μm and width of 3 μm (Table 1); see Nabeshima (2002). Force was applied in their longitudinal direction during tensile tests; the tests were completed within 6 hours after the collection of cells. There were almost no differences in size and tensile properties between arterial and venous ECs. In comparison to FBs, vascular ECs had much lower strength and elongation at fracture, but higher stiffness in small elongation range.

VSMCs have two phenotypes: synthetic and contractile phenotypes. We determined the tensile properties of these two phenotypes of smooth muscle cells obtained from the rabbit thoracic aorta (Miyazaki et al. (2002)).

Synthetic phenotype of cells were isolated with an explant (scratching) method, and then sub-cultured in DMEM supplemented with 10% FCS up to twice for 2 weeks. On the other hand, contractile phenotype of cells were isolated by enzymatic digestion (collagenase), and cultured in the above-mentioned solution only for 3 days. To confirm phenotypes, we observed smooth muscle myosin heavy chain (SM-MHC), a contractile phenotype marker, using a confocal laser scanning microscope. Abundant SM-MHCs were observed in the cells that were isolated with the enzymatic digestion method. On the other hand, almost no myosin heavy chains were observed in the cells obtained from explants. Moreover, we studied the response of these cells to norepinephrine (10^{-5} M), and observed that the cells isolated by enzymatic digestion contracted and formed remarkable membranous evagination on cell surfaces. However, explanted cells did not exhibit these phenomena. These results indicated that the cells isolated with the enzymatic digestion method were of contractile phenotype, and the cells obtained from the explant method were of synthetic phenotype.

Non-loaded VSMCs of both phenotypes were spherical. Their diameters were approximately 30 μm; there was almost no difference between the two phenotypes (Fig. 3). Almost the same load-elongation curves were observed at smaller elongation than 15 μm. At larger elongation, however, the slope of the curve became higher in contractile than in synthetic phenotypes of cells, and the difference in the stiffness between 10 and 25 μm elongation was statistically

Phenotype	Initial diameter (μm)
Contractile	31.0 ± 3.9
Synthetic	32.3 ± 7.4

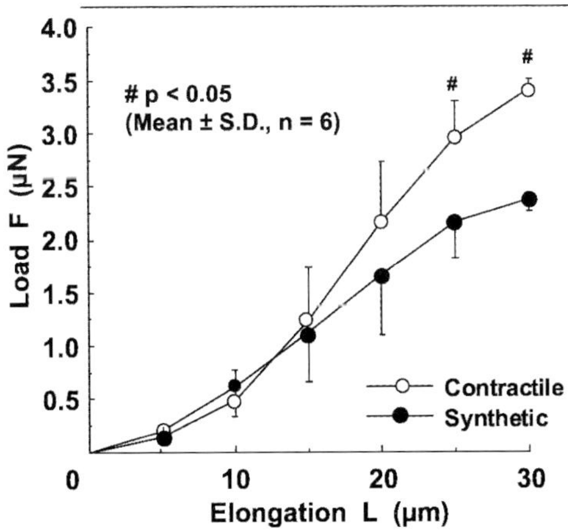

Fig. 3. Load-elongation relationships of vascular smooth muscle cells of contractile and synthetic phenotypes (Miyazaki et al. (2002)).

significant. These cells did not break even at the force of 2 μN, which indicates that VSMCs are much stronger than FBs and vascular ECs. The stiffness of these VSMCs were much higher than that observed by Matsumoto et al. (2000) from bovine VSMCs. This difference may be attributable to the difference in strain rate between the two experiments, as they stated in their report.

To see the reason for such higher stiffness in contractile phenotype of cells than in synthetic phenotype of cells, we treated both cells with FITC-phalloidin for the fluorescence observation of F-actin with confocal laser microscopy. Actin bundles were much thicker in contractile cells than in synthetic ones, which suggests that there are differences in cytoskeletal structure and contractile apparatus between the two phenotypes of cells. These differences may be one of the reasons for the difference in tensile properties.

Then, we studied the effect of contraction on the tensile properties of VSMCs (Kajino (2003), Kajino et al. (2003)). The contraction induced by the administration of norepinephrine (10^{-5} M) to HBSS significantly decreased the diameter (15.1 $\pm$ 1.5 mm) and increased the stiffness (Fig. 4), which is essentially similar to the phenomenon observed in contracted arterial wall. The active, contracted cells did not break even at the tensile force of 6.5 μN. Tensile curves obtained under a physiologically normal condition (in HBSS) were almost the same as those under a passive, relaxed condition (administration of 10^{-4} M papaverine to HBSS) (Kajino (2003); data not shown).

Table 2 shows a summary of the tensile properties of the three cells. There were large differences in strength among these cells. As stated above, FBs were broken at the force of approximately 1 μN with large elongation to failure. However, VSMCs did not break even at 2 μN, and they had much larger stiffness compared with FBs. Although FBs and synthetic phenotype of VSMCs

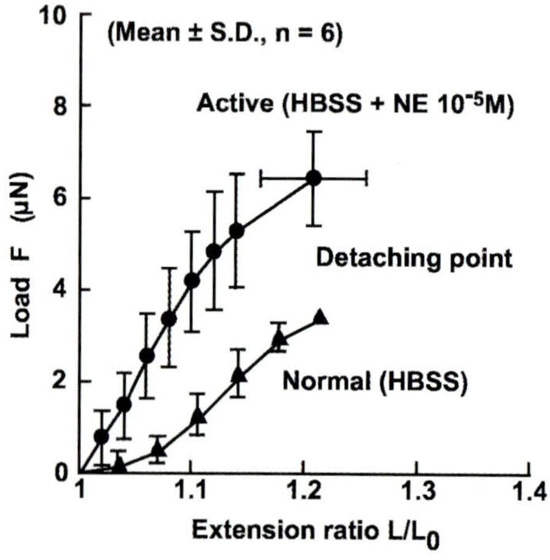

Fig. 4. Effect of activation on tensile characteristics of vascular smooth muscle cells (Kajino et al. (2003)).

Table 2. Summary of tensile test results of fibroblasts (FBs), vascular endothelial cells (ECs), and vascular smooth muscle cells (VSMCs): (mean±SD).

Cell	Maximum load (μN)	Elongation (μm)	Stiffness (N/m)
FB	0.9±0.1	85.7±15.3	0.02±0.01*
EC (Arterial)	0.2±0.1	18.3±8.8	0.04±0.04**
EC (Venous)	0.2±0.1	14.3±3.2	0.03±0.02**
VSMC (S)	> 2.4	> 30	0.09±0.03#
VSMC (C, Normal)	> 3.4	> 30	0.17±0.03#
VSMC (C, Active)	> 6.5	> 12	0.88±0.45##

(S=Synthetic; C=Contractile; Stiffness at 30-50 (*), 0-3 (**), 10-25 (#), and 2-5 (##) μm elongation)

have the same function of synthesizing extracellular matrix components, there was a great difference in tensile properties. VSMCs had the highest strength, and vascular ECs were weakest. Stiffness was significantly larger in VSMCs than in the other two cell types. The stiffness of VSMCs is larger in contractile phenotype than in synthetic phenotype, and was greatly increased by contraction.

Such large differences in tensile properties among cells might be attributable to differences in cytosekeletal structures like the distribution and density of stress fibers and contractile apparatus. For example, actin bundles seemed to be thicker and denser in VSMCs than in FBs (Miyazaki et al. (2002)). Moreover, the network of stress fibers in vascular ECs seemed to be less organized compared with the other two cell types. However, the details are not known yet, and are issues for future studies.

3 Local Stiffness in Cells

As stated above, there are several methods and techniques for the measurement of local properties in a cell. Of those methods, atomic force microscopy (AFM) can easily be used for the study of the local stiffness of cells.

AFM has been developed for micro- to nano-scale topography, and is very useful for the high resolution imaging of such biological specimens as cells, proteins, and DNA with no specific treatment of samples (Lal and John (1994)).

Measurements of local mechanical properties of cells are also possible with AFM nano-indentation techniques (Weisenhorn et al. (1993)). This technique has been applied to several kinds of cultured cells; for example, MDCK cells (Hoh and Schoenenberger (1994)), myocytes (Shroff et al. (1995)), carcinoma cells (Goldmann and Ezzell (1996)), 3T6 cells (Ricci et al. (1997)). We have applied this technique to cultured vascular smooth muscle cells (VSMCs) of synthetic and contractile phenotypes. Moreover, it is also powerful for the

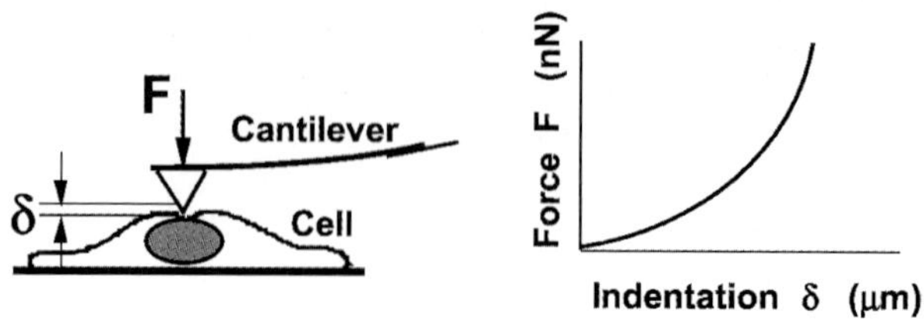

Fig. 5. Atomic force microscopic measurement of local force-indentation relationship in a cell.

measurement of the local stiffness of intact cells. To our knowledge, however, there has been no report on the AFM measurement of the mechanical properties of living cells in situ. Thus, we have determined the local stiffness of intact endothelial cells (ECs) in fresh vascular segments obtained from aortic bifurcations in healthy and atherosclerotic rabbits.

3.1 AFM Stiffness Measurement

If we indent a very tiny pyramidal tip of an AFM indenter (V-shaped silicon nitride cantilever) into a cell, we can obtain force-indentation data and then local stiffness (Fig. 5). Force is automatically calculated from the cantilever deflection measured with a laser beam and a photodiode, while indentation is obtained from the difference between the displacement of the AFM sample stage and the movement of the cantilever tip caused by cantilever deflection.

Force (F)-indentation (δ) relations are expressed by Miyazaki and Hayashi (1999) as

$$F = a\left(\exp(b\delta) - 1\right), \tag{1}$$

where a and b are constants. The slope of the force-indentation curve is written as:

$$\mathrm{d}F/\mathrm{d}\delta = bF + ab = bF + c. \tag{2}$$

These equations are empirical, and provide a measure of the structural stiffness of cells, but not the material stiffness. The parameter a is an index associated with the shape of force-indentation curves; this parameter is related to the spatial change of inhomogeneous structure inside a cell. The parameter c (= ab) is the initial modulus, and can be used to represent the stiffness of a cell. The parameter b corresponds to the rate of modulus change induced by stress, which is related to structural inhomogeneity.

3.2 Local Stiffness of Vascular Smooth Muscle Cells

In this section, the stiffness of VSMCs of two phenotypes, synthetic and contractile phenotypes, are dealt with (Ohara (2000)). Cells were obtained with the same methods as those applied to the cells used for tensile tests (see above). First, the surface topograph of each cell was obtained by the scanning of an AFM indenter tip (Fig. 6), from which the highest point of the cell was determined. Usually this point is just over the cell nucleus. At this point, a force-indentation curve was obtained.

An example of the results is shown in Fig. 7. The parameter c representing cell stiffness was significantly higher in the cells of contractile phenotype than

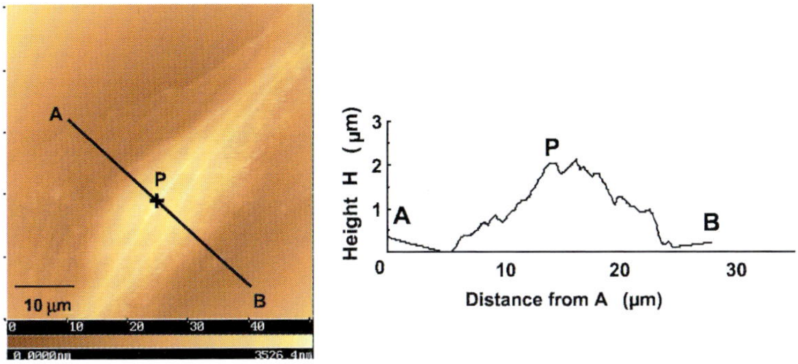

Fig. 6. Atomic force microscopic image of a vascular smooth muscle cell (left) and its profile along a line A-B (right) (Ohara (2000)).

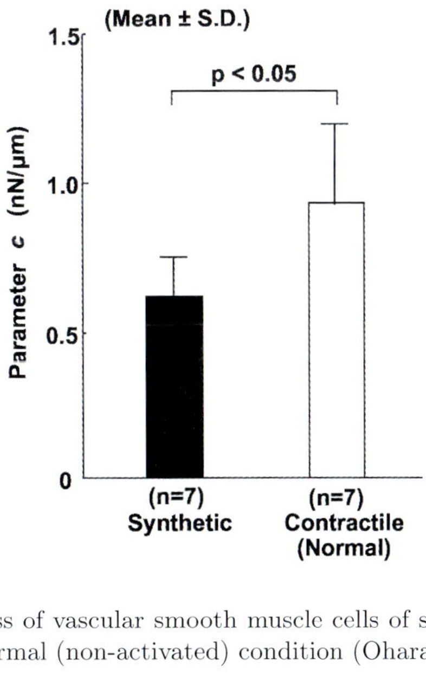

Fig. 7. Local stiffness of vascular smooth muscle cells of synthetic and contractile phenotypes under normal (non-activated) condition (Ohara (2000)).

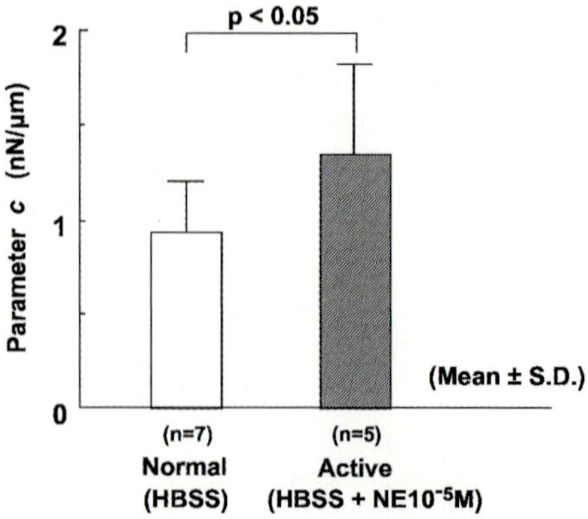

Fig. 8. Effect of activation on stiffness of vascular smooth muscle cells (HBSS = Hanks' balanced salt solution; NE = norepinephrine) (Ohara (2000)).

in those of synthetic phenotype. The higher stiffness in contractile phenotype of VSMCs than in synthetic ones is essentially similar to the results obtained from tensile tests of single whole cells that were explained above (see Fig. 3). However, the AFM stiffness of these cells calculated from parameters b and c (approximately 5 N/m at 2.5 μN force) was one order in magnitude larger than the stiffness determined from the tensile tests (0.1 to 0.2 N/m at 0.5 to 3 μN force; see Table 2). This difference is attributable to the fact that AFM stiffness is primarily affected by stress fibers concentrated in cell surface layers as well as by stiff cell nuclei, while tensile stiffness reflects the whole structure of a cell including stress fibers, nucleus, and intracellular liquid.

The activation of contractile VSMCs by the administration of norepinephrine to the concentration of 10^{-5} M significantly increased AFM local stiffness (Fig. 8). This phenomenon was also similar to that observed in the tensile tests of single whole cells (Fig. 4).

Figure 9 shows an example of stiffness distribution in a VSMC. The photo on the left-hand side is a topograph of the cell (Kawamura et al. (2002)). The variation of cell height along the line A–E is shown in the figure on the right-hand side. If we compare the trace of cell height with the AFM image, the height of the cell increases at the locations B, C and D (right) where there seems to be actin bundles or stress fibers (left). It is interesting to see from the right figure that stiffness is high at these locations, which implies that local stiffness measured by AFM sensitively reflects underlying cytoskeletal structures.

3.3 Local Stiffness in Endothelial Cells in Normal and Atherosclerotic Wall

As far as the author knows, there has been no report on the AFM measurement of local stiffness of living cells in situ. Therefore, we applied the AFM technique

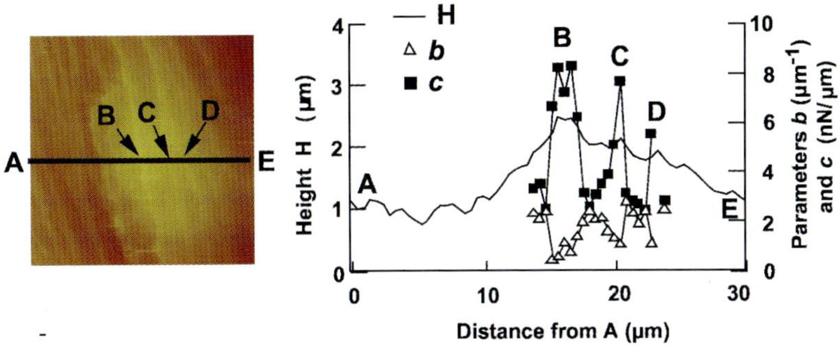

Fig. 9. Distribution of stiffness in a vascular smooth muscle cell (Kawamura et al. (2002)).

to the measurement of local stiffness in intact ECs in fresh aortic bifurcations obtained from normal rabbits and cholesterol-fed ones. The details of the technique have been reported elsewhere (Miyazaki and Hayashi (2003)).

The first part of this section deals with the stiffness of ECs in healthy, normal aortic bifurcations (Miyazaki and Hayashi (1999)). Small arterial segments (3 mm × 3 mm) were cut out from 3 locations: the abdominal aorta, and the medial and lateral sites of the common iliac artery. The lateral site of the common iliac artery is considered to be a predilection site for atherosclerotic lesions because of low wall shear stress (Hayashi et al. (1996)). On the other hand, high wall shear stress is applied to the medial site, which is considered to be a location of relatively low incidence of atherosclerosis.

Our preliminary experiments demonstrated that force-indentation curves obtained from different locations in a cell were very variable, depending upon location and possibly being affected by inhomogeneous cell structure (Miyazaki and Hayashi (1999)). However, we obtained force-indentation curves from the highest areas in cells, which were estimated with AFM topography, and their scatter was much less than that observed at different locations in a cell; slopes of the curves obtained from central, high areas, probably over cell nuclei, were greater compared to peripheral areas. Based on these results, we obtained data from the highest points in cells for comparison among arterial sites.

Figure 10 shows averaged force-indentation curves at the three sites in the aortic bifurcation. The slope of the curve, in particular the initial slope, was larger in the medial wall than in the abdominal aorta and the lateral wall, which indicates that ECs on the medial wall having minimal incidence of atherosclerosis are relatively stiffer than those at the other sites. This difference might be attributable to the development of cytoskeletal structures accompanied by abundant stress fibers at the medial site.

The end part of this article is devoted to the local stiffness of ECs in aortic bifurcations obtained from atherosclerotic animals (Higaki (2003)). Matured, male Japanese white rabbits were fed 1% cholesterol diet for 16 to 32 weeks.

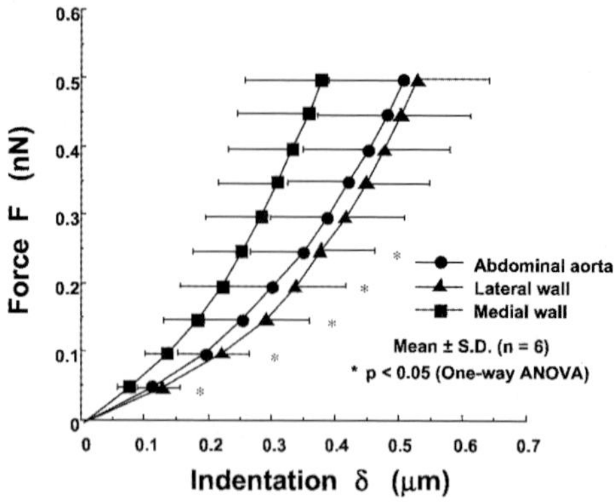

Fig. 10. Force-indentation curves of endothelial cells at three different sites in the aortic bifurcation of normal rabbits (Miyazaki and Hayashi (1999)).

Age-matched, regular chow-fed rabbits were used to obtain control data. The micrographic observation of luminal surfaces stained with Sudan IV indicated that atherosclerotic plaques were more or less developed at all the three sites in the aortic bifurcation. Moreover, AFM surface topography demonstrated that ECs in the abdominal aorta and at the medial site of the iliac artery were elongated in the axial direction of each vessel. However, cells at the lateral site were more round or oval than those at the other two sites. This difference in cell morphology may be attributable to differences in the magnitude and direction of wall shear stress. There were almost no differences in cell morphology between atherosclerotic and normal arteries and also between locations on atherosclerotic plaques and off plaques.

Force-indentation curves of ECs in the normal, control vessels (Fig. 11, left) were almost the same as those shown in Fig. 10, which implies high reproducibility of the experiments. As already mentioned above, the curves obtained from the medial site of the iliac artery were significantly shifted towards the left compared with those obtained from the abdominal aorta and at the lateral site of the iliac artery. Even in cholesterol-fed animals, the results obtained from locations apart from atherosclerotic plaques were very similar to control data (Fig. 11, center). However, the locations on plaques at the medial site gave very different results, that is, force-indentation curves from these locations were very similar to those from the other two sites (Fig. 11, right). These results indicate that only the stiffness of endothelial cells on plaques at the medial site is affected by atherosclerosis.

Again, eq. (2) was applied to these force-indentation curves for the calculation of the values of parameters c and b, which represent initial stiffness and stiffening rate, respectively. Then, those values were plotted against the area fraction of sudanophilic lesions that represents the degree of atherosclerosis (Hayashi et al. (1994)). Parameter b at all the three sites and parameter

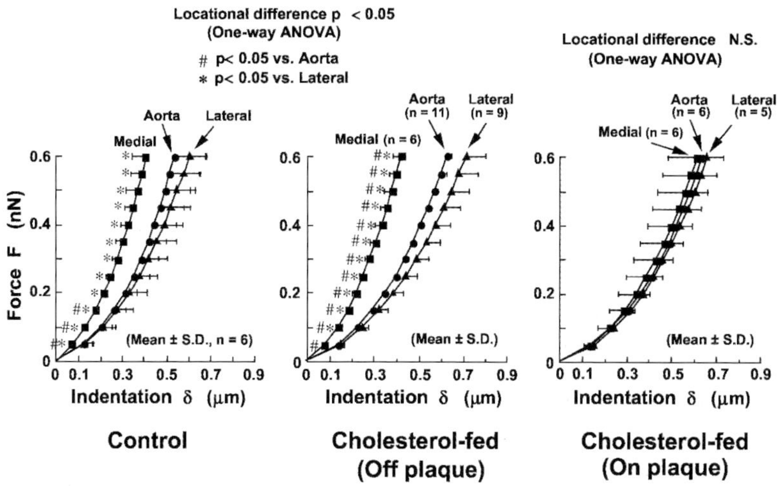

Fig. 11. Force-indentation curves of endothelial cells at three different sites of the aortic bifurcation in atherosclerotic rabbits (Higaki (2003)).

c in the aorta and at the lateral site of the iliac artery were not affected by atherosclerosis (data not shown). However, only parameter c at the medial site decreased with increase in the area fraction of lesions. That is, the stiffness of ECs at this site significantly decreases with the progression of atherosclerosis, which might be due to lipid uptake and/or some disorganization of the cytoskeletal structure, such as stress fibers. At this moment, however, mechanisms for this phenomenon are not known.

4 Conclusions

From these results, we can say:

(i) there are large differences in tensile properties and local stiffness among cells, depending upon cell kind, location, phenotype, passive or active state, and so on. Fibroblasts have the tensile strength of approximately 1 μN, while vascular smooth muscle cells do not break even at 2 μN. Vascular endothelial cells are much weaker than these two cells. The stiffness of vascular smooth muscle cells, which is significantly larger than those of the other two cells, is larger in contractile phenotype than in synthetic phenotype, and is greatly increased by contraction.

(ii) The distribution of stiffness in a cell is not uniform, possibly due to inhomogeneous cytoskeletal structure. The stiffness of cells is higher in the central areas over cell nuclei than in peripheral areas, and greater at the locations where there exist actin bundles or stress fibers underneath.

(iii) The stiffness of vascular endothelial cells at the medial site of the iliac artery significantly decreases with the progression of atherosclerosis, probably due to lipid uptake and/or some disorganization of cytoskeletal structure, such as stress fibers.

Acknowledgements. The author appreciates Hiroshi Miyazaki, Yoshitaka Hasegawa, Yuji Ohara, Michitaka Higaki, Shohei Kajino, Akihide Kawamura, and Yuki Nabeshima for their great contribution to the experiments. This work was financially supported in part by Grant-in-Aids for Scientific Research (A)(2) (nos. 09358020, 12308047, 15200036) and (B)(2) (no.13558111) from the Ministry of Education, Culture, Sports, Science and Technology, Japan.

References

Albrecht-Buehler, G. (1987). Role of cortical tension in fibroblast shape and movement. *Cell Motil. Cytoskel.* 7:54–67.

Bereiter-Hahn, J., Karl, I., Luers, H., and Voth, M. (1995). Mechanical basis of cell shape: Investigations with the scanning acoustic microscope. *Biochem. Cell Biol.* 73:337–348.

Frangos, J. A., ed. (1993). *Physical Forces and the Mammalian Cell.* New York: Elsevier Science & Technology Books.

Glerum, J. J., Van Mastrigt, R., and Van Koeveringe, A. J. (1990). Mechanical properties of mammalian single smooth muscle cells. III. Passive properties of pig detrusor and human a terme uterus cells. *J. Muscle Res. Cell Motil.* 11:453–462.

Goldmann, W. H., and Ezzell, R. M. (1996). Viscoelasticity in wild-type and vinculin-deficient (5.51) mouse F9 embryonic carcinoma cells examined by atomic force microscopy and rheology. *Exp. Cell Res.* 226:234–237.

Hayashi, K., Ide, K., and Matsumoto, T. (1994). Aortic walls in atherosclerotic rabbits – Mechanical study. *J. Biomech. Eng.* 116:284–293.

Hayashi, K., Yanai, Y., and Naiki, T. (1996). A 3d-lda study of the relation between wall shear stress and intimal thickness in a human aortic bifurcation. *J. Biomech. Eng.* 118:273–279.

Hayashi, K. (2003). Mechanical properties of soft tissues and arterial walls. In Holzapfel, G. A., and Ogden, R. W., eds., *Biomechanics of Soft Tissue in Cardiovascular Systems*, 15–64. Wien: Springer-Verlag. CISM Courses and Lectures No. 441, International Centre for Mechanical Sciences.

Higaki, M. (2003). *Atomic force microscopic measurement of local stiffness of vascular endothelial cells in atherosclerotic rabbits.* MS dissertation, Department of Systems and Human Science, Graduate School of Engineering Science, Osaka University. Japanese text with English abstract, figures, and tables.

Hochmuth, R. M., Ting-Beall, H. P., Beaty, B. B., Needham, D., and Son Tay, R. T. (1993). Viscosity of passive human neutrophils undergoing small deformations. *Biophys. J.* 64:1596–1601.

Hoh, J. H., and Schoenenberger, C. A. (1994). Surface morphology and mechanical properties of MDCK monolayers by atomic force microscopy. *J. Cell Sci.* 107:1105–1114.

Kajino, S., Miyazaki, H., and Hayashi, K. (2003). Tensile properties of contracted or relaxed vascular smooth muscle cells. In *Proc. 79th Annual Conf. JSME Kansai Branch*, I–37–I–38. JSME. Japanese text with English abstract and figures.

Kajino, S. (2003). *Tensile properties of contracted or relaxed vascular smooth muscle cells.* MS dissertation, Department of Systems and Human Science, Graduate School of Engineering Science, Osaka University. Japanese text with English abstract, figures, and tables.

Kawamura, A., Miyazaki, H., and Hayashi, K. (2002). Effects of the phenotype modulation on the local stiffness of vascular smooth muscle cells. In *Proc. Mech. Eng. Cong., 2002, Japan*, volume 1, 103–104. JSME. Japanese text with English abstract, figures, and tables.

Lal, R., and John, S. A. (1994). Biological applications of atomic force microscopy. *Am. J. Physiol.* 266:C1–C21.

Matsumoto, T., Sato, J., Yamamoto, M., and Sato, M. (2000). Smooth muscle cells freshly isolated from rat thoracic aortas are much stifer than cultured bovine cells: Possible effect of phenotype. *JSME Intern. J., Ser. C* 43: 867–874.

Miyazaki, H., and Hayashi, H. (1999). Atomic force microscopic measurement of the mechanical properties of intact endothelial cells in fresh arteries. *Med. Biol. Eng. Comput.* 37:530–536.

Miyazaki, H., and Hayashi, K. (2003). Measurement of mechanical properties of intact endothelial cells in fresh arteries. In Braga, P. C., and Ricci, D., eds., *Atomic Force Microscopy: Biomedical Methods and Applications.* Totowa, New Jersey: Human Press. 307–313.

Miyazaki, H., Hasegawa, Y., and Hayashi, K. (2000). A newly designed tensile tester for cells and its application to fibroblast. *J. Biomech.* 33:97–104.

Miyazaki, H., Hasegawa, Y., and Hayashi, K. (2002). Tensile properties of contractile and synthetic vascular smooth muscle cells. *JSME Intern. J., Ser. C* 45:870–879.

Nabeshima, Y. (2002). *Tensile properties of vascular endothelial cells.* BA thesis, Department of Systems, Graduate School of Engineering Science, Osaka University. Japanese text with English figures and tables.

Ohara, Y. (2000). *Atomic force microscopic measurement of mechanical properties of vascular smooth muscle cells.* MS dissertation, Department of Mechanical Science, Graduate School of Engineering Science, Osaka University. Japanese text with English figures and tables.

Palmer, R. E., Brady, A. J., and Roos, K. P. (1996). Mechanical measurements from isolated cardiac myocytes using a pipette attachment system. *Am. J. Physiol.* 270:C697–C704.

Ricci, D., Tedesco, M., and Grattarola, M. (1997). Mechanical and morphological properties of living 3T6 cells probed via scanning force microscopy. *Microsc. Res. Tech.* 36:165–171.

Sato, M., Levesque, M. J., and Nerem, R. M. (1987). An application of the micropipette technique to the measurement of the mechanical properties of cultured bovine aortic endothelial cells. *J. Biomech. Eng.* 109:27–34.

Shroff, S. G., Saner, D. R., and Lal, R. (1995). Dynamic micromechanical properties of cultured rat atrial myocytes measured by atomic force microscopy. *Am. J. Physiol.* 269:C286–C292.

Thoumine, O., and Ott, A. (1997). Time scale dependent viscoelastic and contractile regimes in fibroblasts probed by microplate manipulation. *J. Cell Sci.* 110:2109–2116.

Valberg, P. A., and Feldman, H. A. (1987). Magnetic particle motions within living cells. measurement of cytoplasmic viscosity and motile activity. *Biophys. J.* 52:551–561.

Wang, N., and Ingber, D. E. (1994). Control of cytoskeletal mechanics by extracellular matrix, cell shape, and mechanical tension. *Biophys. J.* 66: 2181–2189.

Weisenhorn, A. L., Khorsandi, M., Kasas, S., Gotzos, V., and Butt, H. J. (1993). Deformation and height anomaly of soft surfaces studied with an AFM. *Nanotechnology* 4:106–113.

Zahalak, G. I., McConnaughey, W. B., and Elson, E. L. (1990). Determination of cellular mechanical properties by cell poking, with an application to leukocytes. *J. Biomech. Eng.* 112:283–294.

Microfluid-dynamics in Three-dimensional Engineered Cell Systems

F. Boschetti,[1] M.T. Raimondi,[2] F. Migliavacca,[2] M. Cioffi,[2] R. Pietrabissa[2]

[1] Politecnico di Milano, Department of Structural Engineering, Italy
federica.boschetti@polimi.it
[2] Politecnico di Milano, Department of Bioengineering, Italy
manuela.raimondi@polimi.it, francesco.migliavacca@polimi.it
margherita.cioffi@polimi.it, riccardo.pietrabissa@polimi.it

Bioreactors allowing direct perfusion of culture medium through tissue-engineered constructs may overcome diffusion limitations associated with static culturing, and may provide flow-mediated mechanical stimuli. The hydrodynamic stress imposed on cells in these systems will depend not only on the culture medium flow rate but also on the scaffold three-dimensional (3D) micro-architecture. We performed computational fluid dynamics (CFD) simulations of the flow of culture medium through 3D porous scaffolds, cultured in a direct perfusion bioreactor, with the aim of predicting the shear stress acting on cells adhering on the scaffold walls as a function of various parameters that can be set in a tissue-engineering experiment. We developed three CFD models: Model 1 was built from histological sections of a fibre scaffold, Model 2 was built from micro-computed tomography reconstruction of a porous foam, and Model 3 was based on an idealized geometry of the actual porous foam. The simulations predicted different distributions of the shear stresses acting on the scaffold walls for each scaffold geometry modelled. In contrast, the simulations predicted the same value of median shear stress in all of the three models. Our results provide a basis for the completion of more exhaustive quantitative studies to further assess the relationship between perfusion, under known micro-fluid dynamic conditions, and tissue growth *in vitro*.

1 Introduction

The generation of autologous cartilaginous grafts requires the development of procedures not only to expand human chondrocytes quickly but also to promote their re-differentiation in a reproducible way and to maintain their chondrogenic potential in 3D culture systems (Jakob et al. (2001)). The conditions which favour *in vitro* chondrogenesis include high cell density and maintenance of spherical cell shape by reduced cell-matrix interaction. This condition may be achieved if the cells are carried by a 3D scaffold, depending on the scaffold material, geometry and pore size (Hutmacher (2000)). Other conditions which are believed to favour *in vitro* chondrogenesis include efficient nutrient delivery and mechanical stimulation (Ingber (2000), Martin et al. (2000)). Direct-perfusion systems may provide an effective strategy for overcoming current diffusion limitations associated with culturing 3D tissues *in vitro* by enhancing nutrient delivery and catabolites removal within the constructs and by providing flow-mediated mechanical stimuli (Dunkelman et al. (1995), Pazzano et al. (2000), Mizuno et al. (2001), Davisson et al. (2002)). However, in most of the direct-perfusion systems currently available the hydrodynamic environment imposed on cells is neither characterized nor controlled. This condition is required to quantify the response to flow-mediated mechanical stimuli of these 3D cell systems.

We have developed a direct-perfusion bioreactor in which the culture medium flows through porous cellular constructs (Raimondi et al. (2002, 2004)). In this paper we present CFD models of the hydrodynamic environment for different scaffolds. CFD modelling provides a method for relating the flow of culture medium applied to cells on a macroscopic level to the shear stress imposed on the cells at a microscopic level.

2 Description of the Computational Models

For all the models, the culture medium was modelled as an incompressible, homogeneous, Newtonian fluid with density, ρ, equal to $1000\,\mathrm{kg\,m^{-3}}$. The viscosity, μ, measured at 37°C by means of a Series 100 capillary viscometer (Cannon-Fenske, State College, PA, USA), was set at $8.2 \times 10^{-4}\,\mathrm{kg\,m^{-1}s^{-1}}$. Cells were not included in the models. Boundary conditions were set in order to simulate a tissue engineering experiment described in Raimondi et al. (2002, 2004). No-slip boundary conditions were applied to the scaffold surfaces, under the hypothesis of rigid, impermeable walls. A flat velocity profile was applied at the inlet, corresponding to an average flow rate, Q, of $0.5\,\mathrm{cm^3\,min^{-1}}$ through a scaffold of 15 mm diameter, while null total stress was applied at the outlet. The fluid-induced shear stresses acting at the wall surface of the scaffolds were calculated and assumed as an estimate of the shear stresses acting on the membranes of the cells. The Reynolds number was very low for all the models, ranging from 3×10^{-4} to 6×10^{-2}.

Typical dimensions for the modelled scaffolds are listed in Table 1.

2.1 Model 1 – Fibre Scaffold

This scaffold consisted of fibres, $15\,\mu\mathrm{m}$ in diameter, randomly distributed in a 3D space traversed by a constant flow of culture medium (Fig. 1). The scaffold porosity was approximately 88%. The simplified geometry adopted to

Table 1. Dimensions of the model scaffolds.

Model	Scaffold	Geometry	Typical Dimension
1	Fibre	Reconstructed	Fibre diameter = 15 μm
2	Foam	Reconstructed	Pore size = 100 μm
3	Foam	Idealized	Pore size = 50–150 μm

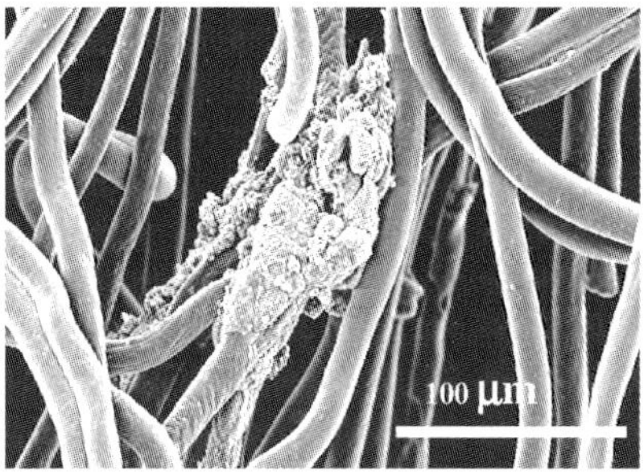

Fig. 1. Scanning electron microscopy image of the fibre scaffold seeded with human articular chondrocytes.

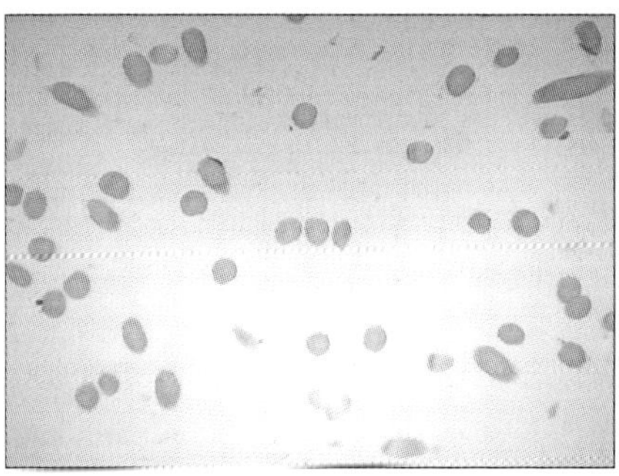

Fig. 2. A representative light-microscopy image of a cross section of the fibre scaffold at original magnification 40×.

set up the CFD model consisted of a two-dimensional domain representing a $400 \times 300\,\mu\text{m}^2$ cross section of a histological sample of a construct, where a group of fibres were arranged in a cross-sectional-like pattern.

Fourteen light microscopy images of the histological sections were analyzed; a representative image is shown in Fig. 2. Details for this model are described elsewhere (Raimondi et al. (2002)). The inlet velocity for this model was 44 μm/s.

Post-processing of the CFD Results. The shear stresses acting on the fibre contours were extracted from the CFD results. At different flow rate values the shear stress, τ, acting on each edge element of a fibre contour was extracted from the simulation results. The statistical distributions and the median values from all the images were determined.

2.2 Model 2 – Porous Foam

The second scaffold modelled is a biodegradable polyestherurethane foam (Degrapol®) with a porosity of roughly 77% and a nominal pore diameter of 100 μm. As shown in Fig. 3, cells preferably adhere on the inner surfaces of the scaffold pores; cell dimensions were assumed to be negligible compared with the pore size of the foam.

The scaffold micro-geometry was reconstructed from micro-computed tomography (μCT) images of the scaffold, acquired using a Skyscan 1072 desktop X-ray high resolution micro-tomograph (Skyscan, Aartselaar, Belgium). A 1 mm thick sample was acquired in 250 micro-tomograph slices; each slice had a section of 17 mm^2 and a thickness of 4 μm. Tomograph images were stored in BMP files.

The 3D model was reconstructed and meshed with AMIRA 3.1 software (TGS, San Diego, CA, USA). For each μCT image three sub-areas of the originally measured data were considered, creating three sub-volumes, A, B, C, in the central part of the specimen (Fig. 4) to eliminate preparation artifacts on the surface. The three cubes, 400 μm in side, were used separately as geometric models for CFD simulations. The foam was separated from the background with a thresholding procedure, which maintained the scaffold topology. After a smoothing procedure, the fluid volume was visualized by a triangulation of the surface that divided the foam from the void, which would be occupied by the fluid in the CFD simulation. For each cube, the fluid domain was created by meshing the fluid volume in roughly 9,000 tetrahedrons. The commercial finite-volume code Fluent (FLUENT, Lebanon, NH, USA) was used to set up

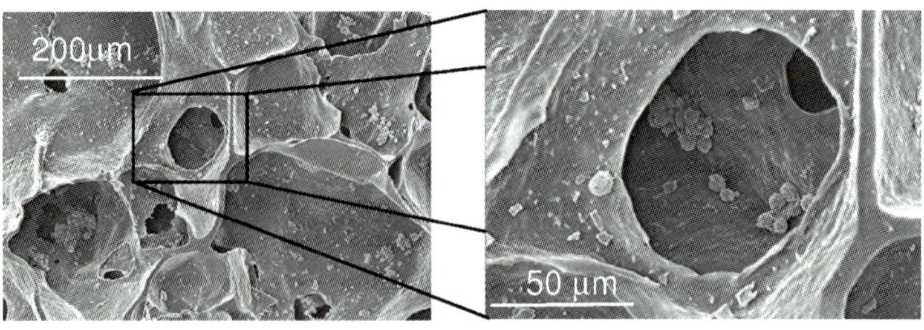

Fig. 3. Scanning electron microscopy image of a porous foam seeded with human articular chondrocytes.

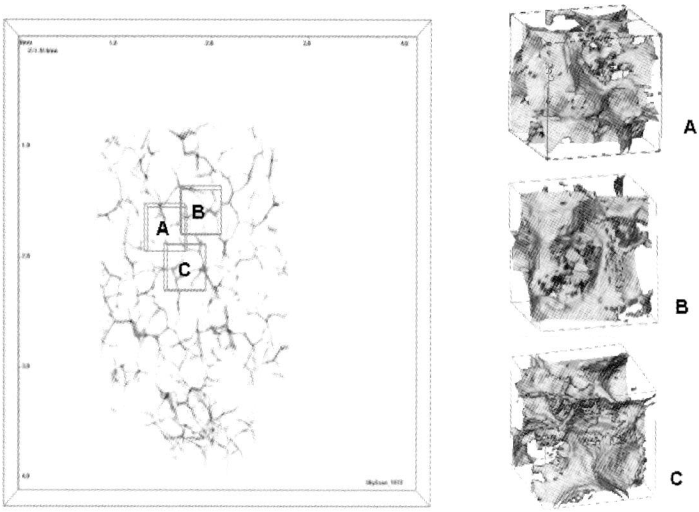

Fig. 4. Method to set up Model 2: for each micro-CT image, three areas of the originally measured data were considered, to create three volumes of interest, 400 μm in side (cubes A, B, and C), in the central part of the scaffold.

and solve the problem and to analyze the results. The mesh of each cube was refined up to reach about 600,000 tetrahedrons for each model. The surface belonging to the central region of each cube (roughly 4,000 faces) was considered as the reference surface for the evaluation of the shear stresses acting on the scaffold walls. The inlet velocity for this model was 53 μm/s.

Post-processing of the CFD Results. The shear stresses were analyzed on the surface belonging to the central region of each cube, to minimize boundary effects. The shear stress statistical distribution and the median value from all of the cubes were determined.

2.3 Model 3 – Idealized Porous Foam

This CFD model was based on a simplification of the geometry of the polymeric scaffold shown in Fig. 3. A micro-domain of the scaffold can be idealized as made of 27 sub-units, arranged in a honeycomb-like pattern. Each subunit was obtained by subtracting a solid sphere from a concentric solid cube (Fig. 5).

The characteristic lengths of the solid sphere and of the solid cube were varied in order to obtain different pore sizes and porosities, respectively. We built three groups of models corresponding to three diameters of the solid sphere, 50, 100 and 150 μm. Each group was made of four models corresponding to 59, 65, 77, and 89% porosity. The porosity was defined as the ratio of

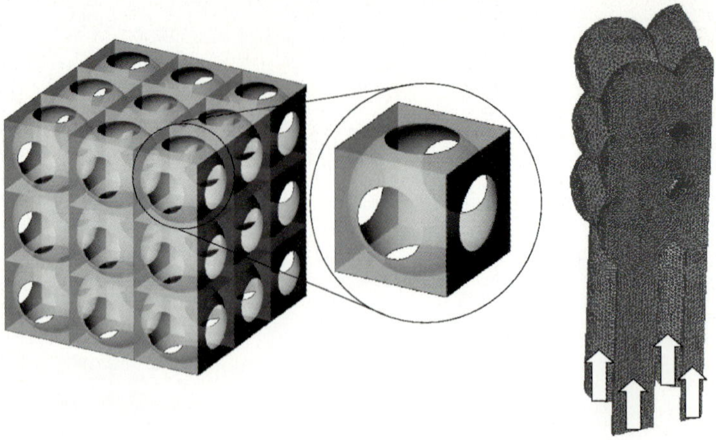

a) b)

Fig. 5. Method used to set up Model 3: (a) honeycomb structure representing a neighbourhood of a single pore; (b) mesh of the complete domain. The white arrows show the flow direction.

void volume to total volume. The reference model had a porosity of 77%, a sphere diameter of 100 μm and a cube dimension of 86.2 μm.

The ACIS-based, solid modeller Gambit (Fluent, Lebanon, NH, USA) was used to set up the model and build the mesh. Due to the existing symmetries, only one quarter of the model was studied. A 4-node tetrahedral mesh was used, with a higher density in correspondence with the central sub-unit. The meshes presented a number of elements up to 270,000 depending on porosity and pore size, with a corresponding number of nodes up to 50,000. The commercial finite-element code FIDAP (Fluent, Lebanon, NH, USA) was used to set up and solve the problem and to analyze the results.

The inlet velocity ranged between 105 and 689 μm/s depending on the porosity. The steady state Navier-Stokes equations were solved by FIDAP.

Post-processing of the CFD Results. The results of the computations were analyzed in the central subunit of the honeycomb structure. The distribution of the shear stress τ was determined for the internal walls of the central subunit. Shear stress data were processed to obtain the value of τ and its frequency distribution. The mode value of τ is defined as $\tilde{\tau}$ and represents the value of shear stress to which most of the cells adhered to the scaffold walls would be subjected at the beginning of culture.

Since the Reynolds number is very low, the convective terms are negligible and the shear stress is linearly related to the imposed inlet flow rate. Therefore, the calculated shear stresses were used to extrapolate $\tilde{\tau}$ for different flow rates and scaffold sizes.

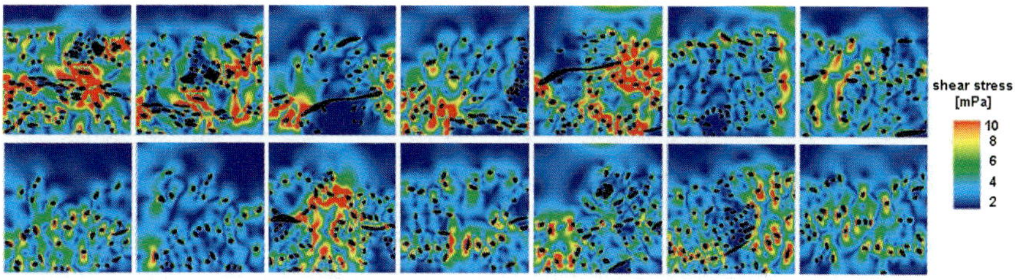

Fig. 6. Results of the CFD simulations for the 14 images analyzed (Model 1): the shear stress field is mapped for each fluid domain obtained from the histological images.

3 Results of the CFD Simulations

Because of the extremely low Reynolds number, recirculation and separation zones are not present in the domains, and the median shear stress values range linearly for recirculation flow rates in the range 0 to $2.25\,\mathrm{cm}^3\times\ \mathrm{min}^{-1}$.

3.1 Model 1 – Fibre Scaffold

The calculated field of shear stresses is presented in Fig. 6 for all the 14 images analyzed.

The streamline contours resulting from the simulations clearly show the existence of low flow regions where the fibres are massed and, conversely, the existence of preferential flow paths. Since the velocity is null on the fibre contours, the higher values of shear stress τ are localized at the fibre contour segments that face regions of high flow rate. Due to the unequal distribution of the fibres within the scaffold structure, the values of τ vary considerably, not only on each fibre contour but also among fibres. The distribution of contour shear stresses shows a large spread on low values; however, the tail extends up to maximal values on the order of 80 mPa (0.8 dyne/cm^2). The median τ calculated at the fibre contour, at a flow rate of $0.5\,\mathrm{cm}^3\times\mathrm{min}^{-1}$, corresponding to a $44.2\,\mu\mathrm{m}\times\mathrm{s}^{-1}$ inlet velocity, is 3 mPa.

Fig. 7 shows the calculated shear stress values imposed on cells in the direct-perfusion bioreactor as functions of the culture medium flow rate and of the inlet velocity.

3.2 Model 2 – Porous Foam

Figure 8 shows the wall shear stresses mapped on the inner surface of each cubic sub-model.

Here, the shear stress varies between 0 and 40 mPa among 95% of the surface in the domain. The mode and the median value range from 2 to 4 mPa for the three models. The global distribution of the sum of the three cubic sub-models shows a median value of 3 mPa.

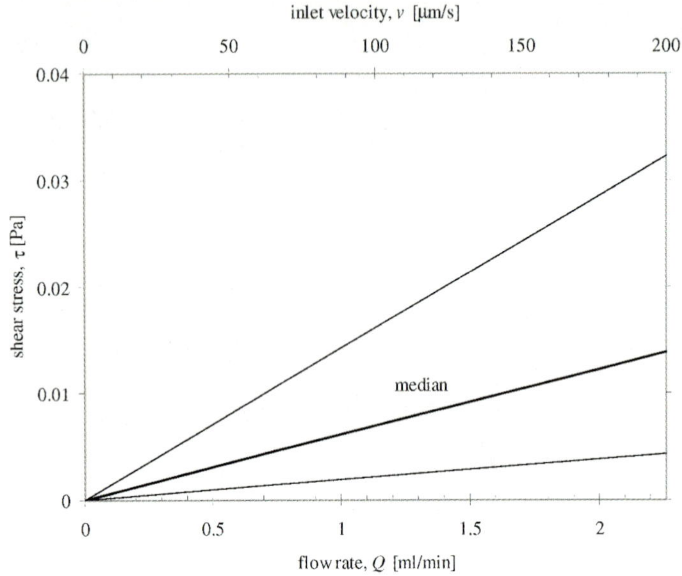

Fig. 7. Extrapolation of results from CFD simulations (Model 1): calculated shear stress level imposed on cells in the direct-perfusion bioreactor as a function of the culture medium flow rate and of the inlet velocity. The light lines represent the 25th and the 75th percentiles of the calculated shear stress distribution.

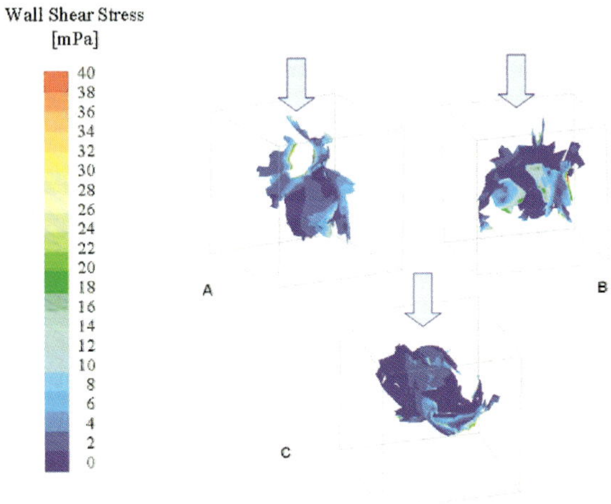

Fig. 8. Results of the CFD simulations (Model 2): the wall shear stress field is mapped on the inner surface for three cubic sub-volumes obtained from the μCTs.

3.3 Model 3 – Idealized Porous Foam

The shear stress distributions on the walls of the central pore are mapped in Fig. 9 for all the porosities and pore sizes modelled.

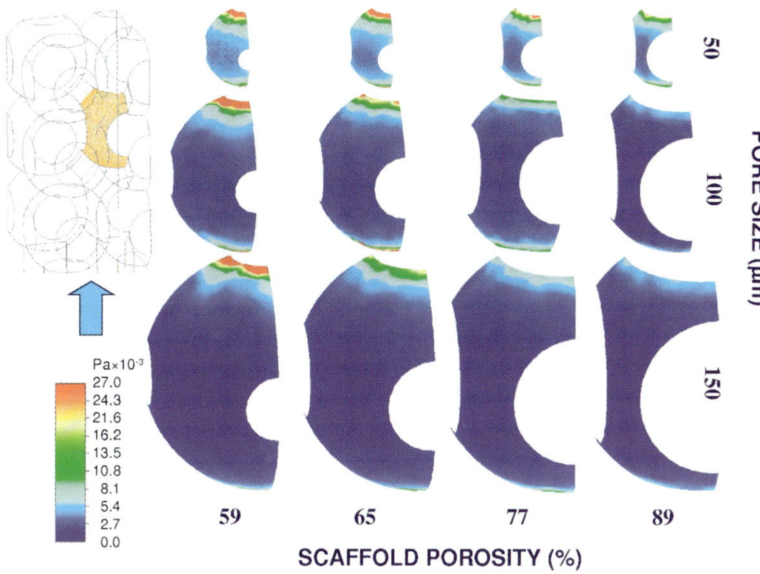

Fig. 9. Results of the CFD simulations (Model 3): the wall shear stress field is mapped on the inner surface of the central pore for increasing porosity and pore dimension.

Although the shear stress τ reaches values up to 70 mPa at the pore inlet and outlet, the shear stress acting on most of the pore wall is much lower, generally around 3 mPa. The distribution of the wall shear stresses shows that, for increasing porosities, the mode τ value, $\tilde{\tau}$, is roughly constant, whereas the stress distribution spreads, consistently with the presence of smaller wall areas subject to uniform stresses. In contrast, for increasing pore size, the values of $\tilde{\tau}$ decrease, consistently with the presence of lower velocity gradients, while the stress distribution narrows, consistently with the presence of larger wall areas subject to uniform stresses. All the above results were obtained for a circular scaffold, 15 mm in diameter, perfused by a flow rate of $0.5\,\text{cm}^3 \times \text{min}^{-1}$. The values of τ extrapolated for other scaffold diameters and flow rates are plotted in Fig. 10. Here, the ratio τ/Q is given as a function of the diameter of the perfused section, D, for different values of pore diameter. The porosity is not accounted for since it does not affect $\tilde{\tau}$, as stated above.

4 Discussion and Conclusion

The aim of the studies here described was to characterize the hydrodynamic field imposed on cells in cellular constructs cultured in a direct-perfusion bioreactor. The CFD simulations numerically predicted the shear stresses induced at the internal walls in relation to three scaffold geometries.

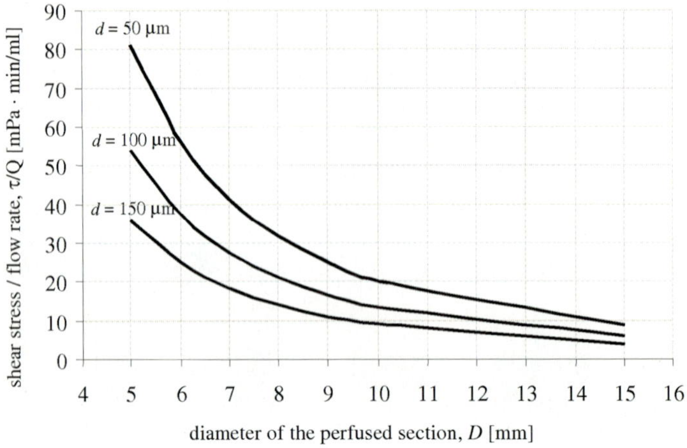

Fig. 10. Shear stress level imposed on cells in the direct-perfusion bioreactor, extrapolated from the results of Model 3, normalized by the flow rate, as a function of the diameter of the perfused section, D, for different values of pore diameter, *d*.

In fibre scaffolds, configured as the ones represented in Fig. 1, cells adhere preferentially at the fibre intersections. Since the fibres are randomly massed, cells are often located in regions of lower flow, where the shear stresses predicted by the simulations showed a spread on low values. Results from Models 2 and 3 indicated that an open-pore structure, rather than a fibre structure, is likely to optimize hydrodynamic stimulation and mass transport to the cells.

Although the three models represented different geometries, they all predicted the same level of median shear stress. Since the Reynolds numbers are very low, the actual micro-geometry of the scaffold did not significantly affect the level of shear stress acting on the inner scaffold walls. Thus, simplified models may correctly predict the level of shear stress imposed on cells adhering to the scaffold walls. In contrast to this finding, the simulations showed a great variability of results between the three scaffolds modelled in terms of the distributions of the shear stresses. This variability can be explained by the great differences in micro-geometry among the three modelled scaffolds. In the case of an irregular geometry (Models 1 and 2) the stress distribution spreads, consistently with the presence of smaller wall areas subject to uniform stresses. In contrast, for a microstructure showing a regular geometry (Model 3), in specific conditions the simulations showed a very narrow stress distribution. Thus, a scaffold with a homogeneous distribution of pores, with low porosity and high pore size, would probably allow a precise control of the induced wall shear, consistently with the presence of larger wall areas subject to uniform stresses.

The most severe simplification for Model 1 was the two-dimensional flow modelling. In such a configuration, the fibre axes extend infinitely in the direction normal to the plane of the model, which is rather unrealistic. For Models 2 and 3, the microstructure was modelled in 3D, hopefully giving more realistic predictions of the flow dynamics inside the perfused scaffold. Substantial simplifications were, however, introduced in order to reduce the computational resources needed to model the actual scaffold microstructure.

One of the major criticisms of the present models can be easily identified in the lack of experimental validation of the estimated shear stress distribution. Although any experimental data would add value to this modelling study, in fact it is nearly impossible to measure shear stress distributions inside such microstructures in an experimental set-up. In addition, calls for numerical methodology to help in this issue have recently been made by leading scientists in tissue engineering (see, for instance, Ivan Martin, 'The optimal flow conditions of a bioreactor should not be determined through a trial-and-error approach, but rather should be supported by simulation methods', Martin et al. (2004)).

Neglect of cells in the models also raise major concern about the reliability of the results, especially for Model 1, which shows a fibre diameter comparable to the size of a round cell. Nevertheless, after scaffold seeding few sparse cells are visible on the scaffold walls, and cell spreading (flattening) is observed as an outcome of cell adhesion. Cell rounding-up is prevented with the help of cytoskeletal elements following cell adhesion. We hypothesized that rare flattened structures of a few tenths of microns in thickness spread on the external surface of solid fibres 15 microns in diameter would generate only a small disturbance in the flow pattern generated by the fibres themselves. This hypothesis was assumed also for the porous foams, although in this case even round cells would be negligible if compared to the pore size ($\sim 100\,\mu$m).

In conclusion, computational modelling can be used to quantify, at a microscopic level, the shear stress artificially applied to cells in three-dimensional engineered cell systems in bioreactors. This method, coupled with experimental analysis, allows quantification of the effects of fluid-dynamic shear on the growth modulation of tissue-engineered cartilage constructs, to potentially enhance tissue growth *in vitro*. Furthermore, these results can be used to derive new scaffold design criteria, such as indications of microstructures that would optimize the hydrodynamic field imposed on cells.

Acknowledgements. This research was supported in part by the Italian Ministry of University and Research (DD.MM.13 and 14, dated 26 and 27 January 2001) and by Politecnico di Milano (Young Researchers Award 2000). The porous foams were supplied without charge by ETH (Zurich, CH). The μCT equipment was made available courtesy of the Istituto Ortopedico Rizzoli (Bologna, Italy) without charge by the company Assing (Monterotondo, Roma, Italy); the equipment was operated by Dr. Egon Perilli.

References

Davisson, T., Sah, R. L., and Ratcliffe, A. (2002). Perfusion increases cell content and matrix synthesis in chondrocyte three-dimensional cultures. *Tissue Eng.* 8:807–816.

Dunkelman, N. S., Zimber, M. P., LeBaron, R. G., Pavelec, R., Kwan, M., and Purchio, A. F. (1995). Cartilage production by rabbit articular chondrocytes on polyglycolic acid scaffolds in a closed bioreactor system. *Biotechnol. Bioeng.* 46:299–305.

Hutmacher, D. W. (2000). Scaffolds in tissue engineering bone and cartilage. *Biomaterials* 21:2529–2543.

Ingber, D. E. (2000). Mechanical and chemical determinants of tissue development. In Lanza, R. P., Langer, R., and Vacanti, J., eds., *Principles of Tissue Engineering*. San Diego: Academic Press, 2nd edition.

Jakob, M., Demarteau, O., Schafer, D., Hintermann, B., Dick, W., Herberer, M., and Martin, I. (2001). Specific growth factors during the expansion and redifferentiation of adult human articular chondrocytes enhance chondrogenesis and cartilaginous tissue formation in vitro. *J. Cell Biochem.* 81:368–377.

Martin, I., Obradovic, B., Treppo, S., Grodzinsky, A. J., Langer, R., Freed, L. E., and Vunjak-Novakovic, G. (2000). Modulation of the mechanical properties of tissue engineered cartilage. *Biorheology* 37:141–147.

Martin, I., Wendt, D., and M.Herberer. (2004). The role of bioreactors in tissue engineering. trends in biotechnology. *Trends in Biotechnology* 22: 80–86.

Mizuno, S., Allemann, F., and Glowacki, J. (2001). Effects of medium perfusion on matrix production by bovine chondrocytes in three-dimensional collagen sponges. *J. Biomed. Mater. Res. A* 56:368–375.

Pazzano, D., Mercier, K. A., Moran, J. M., Fong, S. S., DiBiasio, D. D., Rulfs, J. X., Kohles, S. S., and Bonassar, L. J. (2000). Comparison of chondrogensis in static and perfused bioreactor culture. *Biotechnol. Prog.* 16:893–896.

Raimondi, M. T., Boschetti, F., Falcone, L., Fiore, G. B., Remuzzi, A., Marazzia, M., Marinoni, E., and Pietrabissa, R. (2002). Mechanobiology of engineered cartilage cultured under a quantified fluid dynamic environment. *Biomech. Model. Mechanobio.* 1:69–82.

Raimondi, M. T., Boschetti, F., Falcone, L., Migliavacca, F., Remuzzi, A., and Dubini, G. (2004). The effect of media perfusion on three-dimensional cultures of human chondrocytes: integration of experimental and computational approaches. *Biorheology* 41:401–410.

Nonlinear Constitutive Models for Cochlear Outer Hair Cells

K. Grosh, N. Deo

University of Michigan, Department of Mechanical Engineering, USA
grosh@umich.edu

Cochlear outer hair cells (OHC) are unique in nature as they serve as both actuators and sensors. The importance of OHCs for hearing rests in their ability to actively modulate the sensory epithelium of the cochlea at acoustic frequencies. In our experiments, we have demonstrated *in vivo* electromotility of these cells up to 100 kHz. *In vitro* results on isolated OHCs (typical dimensions of which are 10mm diameter and a length of 40-70mm) also show electromotility to ultrasonic frequencies. Furthermore, these controlled experiments have demonstrated several important features that we seek to replicate in our model: (i) nonlinear capacitance, (ii) nonlinear stiffness, and (iii) electromotility, each of which has been shown to vary as a function of voltage and internal cell turgor pressure. Our approach is to build the model from a two-state model of the active protein in the cell wall membrane. The state transition is dictated by a Boltzmann function of the difference in the energy in the two states in terms of both elastic and electrical energy terms. This approach directly couples the domains in a thermodynamically consistent fashion. When this material law is used as part of the OHC cylindrical shell model, our results replicate the nonlinear stiffness, saturating nonlinear electromotility and capacitance seen in the experimental data. At the time of this writing, a set of parameters that accurately replicates the dependence on cell turgor pressure have not been found. The modeling approach is general and can be applied to other active materials including non-biological (e.g., ferroelectrics and piezoelectrics) and biological systems (e.g., muscle). This connection will be discussed.

1 Introduction

The sensitivity and frequency selectivity of the mammalian cochlea is believed to be a consequence of the electromotile nature of outer hair cells (OHCs) inside the organ. Located in the organ of Corti between the basilar membrane and the reticular lamina, OHCs are in an ideal position to influence basilar

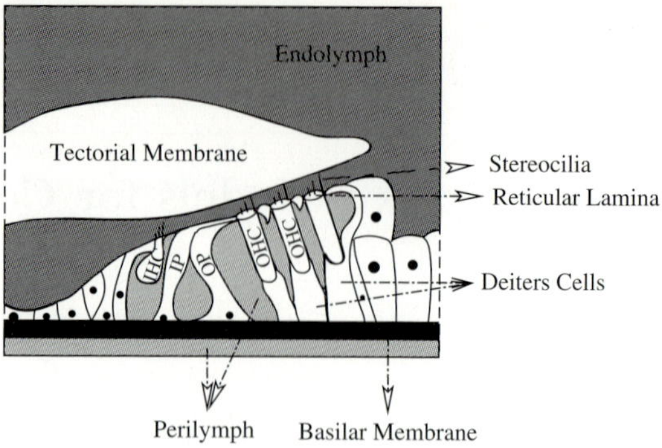

Fig. 1. Cross-section of the cochlea showing the organ of Corti. Labels are as follows: OHC – Outer Hair Cell, IHC – Inner Hair Cell, IP – Inner Pillar cell, OP – Outer Pillar cell. This picture repeats itself about 3000 times in a typical cochlea. The Reticular Lamina separates the endolymph (potential approximately +80mV) from the perilymph (potential approximately 0mV). Motion of the basilar membrane causes relative shear between the tectorial membrane (a gel-like structure) and the reticular lamina, causing the stiff stereocilia atop the OHCs to bend. This allows current to flow through tension-gated channels in the stereocilia into the OHCs. The resulting transmembrane voltage change causes change in OHC stiffness and length change which feed back to the basilar membrane motion.

membrane motion (see Fig. 1). Arranged in 3 or 4 rows, a typical mammalian cochlea has around 30,000 OHCs. *In vitro* experiments on OHCs have revealed nonlinear dependence of OHC transmembrane capacitance, axial stiffness, and axial strain on the transmembrane voltage and cell turgor pressure (see Deo and Grosh (2004) for a review). A change in voltage across the lateral membrane of the OHC causes a change in its length, while application of axial force on the OHC causes charge to develop across its lateral membrane. This piezoelectric like behavior has been seen *in vitro* at frequencies up to 70 kHz (Frank et al. (1999)) and over 100 kHz *in vivo* (Grosh et al. (2004)). The nonlinear stiffness, capacitance and motility exhibited by OHCs is reflected in the various nonlinear effects seen in experiments on the cochlea, such as two-tone distortion emissions or masking.

The source of electromotility of the OHCs is, at least in part, believed to be a motor protein called prestin (Zheng et al. (2000)). It is hypothesized that prestin, or some complex containing prestin, undergoes conformational change when subjected to forcing or transmembrane voltage change. A previous model (the area motor model) based on this hypothesis has been successful in replicating capacitance and length changes with respect to transmembrane voltage change (Iwasa (1994, 2001)). We extend the area motor model to include state dependent compliance of the protein. Stiffness, capacitance, and motility predictions of the new model and its response to simultaneous excitation by mechanical and electrical inputs compare favorably with *in vitro* results. A simpler version of the model obtained by expanding its equations in Taylor series and retaining only the first-order nonlinear terms, manages to capture all the essential features seen in the complete model for small voltage and deflections seen under normal operating conditions. Such a simplified model will be easier to construct from experiments and also easier to implement in a predictive model of the cochlea. This paper represents a summary of our work on OHCs published in the Biophysical Journal (Deo and Grosh (2004)) and the talk given at the IUTAM Symposium on Mechanics of Biological Tissue.

2 Model Assumptions and Derivation

The model problem consists of a simplified model of the patch clamp or micropipet experiment, as pictured in Fig. 2. The OHC is assumed to be a right circular cylinder of length L, radius r, and thickness h. The applied loads are considered axisymmetric and it is assumed that the principal axes of the material properties coincide with the cylindrical coordinates. All bending stresses on the OHC are neglected. Since radius to thickness ratio of the OHC is very high, all radial strains are ignored. The motor proteins are assumed to be spread uniformly on the cylindrical OHC surface. It is assumed that the motor protein flips between two states (extended and contracted configuration). The flipping between two configurations is accompanied by some charge transfer 'q' across the protein wall as well as a change in the resting length

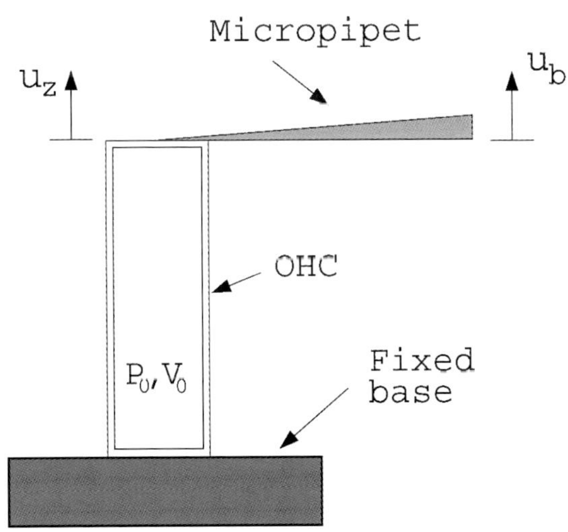

Fig. 2. OHC experiment: initially pressurized, stimulated by applied voltage about the resting potential. Mechanical loading is through a micropipet (modeled as a linear spring) or an externally applied constant force. Resting internal fluid pressure, P_0; transmembrane voltage, V_0.

and compliance of the material. The probability that a motor will flip state is governed by a Boltzmann function P_e, which is the probability of being in the extended state. The OHC is subjected to fluid pressure p from the inside, external axial tension T_{ext}, and transmembrane voltage V. The axial and circumferential strain for the OHC can then be written as a sum of strains of motors in the short state plus strains of motors in the long state. Thus,

$$\varepsilon_z = (s_z^s T_z + s_{zc}^s T_c)(1 - P_e) + (s_z^l T_z + s_{zc}^l T_c) P_e + n a_z P_e\,, \tag{1}$$

$$\varepsilon_c = (s_{zc}^s T_z + s_c^s T_c)(1 - P_e) + (s_{zc}^l T_z + s_c^l T_c) P_e + n a_c P_e\,, \tag{2}$$

where a_z and a_c are the anisotropic area changes from the shortened to the extended state (Iwasa (1993), Iwasa and Adachi (1997)), (s_z^s, s_{zc}^s, s_c^s) and (s_z^l, s_{zc}^l, s_c^l) are the compliance coefficients in the short and long state, respectively, $T_z = \sigma_{zz} h$, $T_c = \sigma_{\theta\theta} h$, σ_{zz} is the stress in the axial direction, $\sigma_{\theta\theta}$ is the stress in the circumferential direction and n is the motor density of proteins per unit area. Similarly, the charge Q is decomposed as

$$Q = C_{\text{lin}} V + n q P_e, \tag{3}$$

where Q is the total charge across the membrane per unit surface area and C_{lin} is the linear capacitance per unit surface area. The expression for the Boltzmann probability is

$$P_e = 1 \Big/ \left[1 + \exp\left(\frac{\Delta F}{k_B \theta}\right)\right]. \tag{4}$$

The argument of the Boltzmann function ΔF is the jump in the Gibbs free energy considering the material to be completely in one state or the other (Achenbach et al. (1986)), k_B is the Boltzmann constant and θ is the temperature. It is assumed that the density of states is reached instantaneously (i.e. there is no rate dependence in the switching of states). This simplifies the analysis (since the evolution of the states need not be computed) and is justified by the high frequency response of the OHCs (Frank et al. (1999)).

The Gibbs energy barrier which governs the probability of switching state is given by

$$\Delta F = \Delta\psi - T_z \Delta\varepsilon_z - T_c \Delta\varepsilon_c - V \Delta Q, \tag{5}$$

where ψ is the Helmholtz free energy. All Δ quantities represent the change from short to long state. The compliance coefficients for one motor protein can be written as the values of coefficients for a unit area divided by the number of proteins in a unit area. We can then write,

$$\Delta\psi = \tfrac{1}{2}\frac{\Delta s_z}{n} T_z^{\,2} + \frac{\Delta s_{zc}}{n} T_z T_c + \tfrac{1}{2}\frac{\Delta s_c}{n} T_c^{\,2} + F_0, \tag{6}$$

$$\Delta\varepsilon_z = a_z + \frac{\Delta s_z}{n} T_z + \frac{\Delta s_{zc}}{n} T_c, \tag{7}$$

$$\Delta\varepsilon_c = a_c + \frac{\Delta s_c}{n} T_c + \frac{\Delta s_{zc}}{n} T_z, \tag{8}$$

$$\Delta Q = q, \tag{9}$$

$$\Delta F = F_0 - qV - T_z a_z - T_c a_c - \tfrac{1}{2}\frac{\Delta s_z}{n} {T_z}^2 - \frac{\Delta s_{zc}}{n} T_z T_c - \tfrac{1}{2}\frac{\Delta s_c}{n} {T_c}^2. \tag{10}$$

Here, F_0 is some constant representing any non-mechanical energy transfer during change of state.

A Gibbs free energy function can be derived for this model. Let G be the free energy function such that

$$\frac{\partial G}{\partial T_z} = -\varepsilon_z, \quad \frac{\partial G}{\partial T_c} = -\varepsilon_c, \quad \frac{\partial G}{\partial V} = -Q. \tag{11}$$

Using eqs (1)–(10) we can integrate expressions given in eq. (11) to get

$$\begin{aligned} G = & -\tfrac{1}{2} s_z T_z^2 - \tfrac{1}{2} s_c T_c^2 - T_z T_c s_{zc} - \tfrac{1}{2} C_{\text{lin}} V^2 - n T_z a_z \\ & - n T_c a_c - nqV + n k_B \theta \log P_e + G_0, \end{aligned} \tag{12}$$

where G_0 is a constant of integration.

After the initial turgor pressure is applied to the cell, the deformation is assumed to be incompressible and the volume constraint $\varepsilon_z + 2\varepsilon_c = \epsilon_{\text{vol}}$ is applied. Here ϵ_{vol} is a constant computed from the quasi-static problem for a given resting potential and turgor pressure. Using this constraint we can solve eqs (1), (2) and (3) to get expressions for the axial stiffness and capacitance (expressions not listed here).

3 Comparison with Experiments

3.1 Stiffness and Dynamics Simulations

The original area motor model succeeded in capturing capacitance and stiffness changes with respect to voltage, but failed to capture stiffness variations. The extended model shows similar capacitance and motility predictions (see Deo and Grosh (2004)) and manages to capture stiffness variations as well. Figure 3 shows stiffness versus voltage prediction of the extended model. In Fig. 4 this result is compared with data from He and Dallos (1999).

The chosen parameters give a good match to the data. More importantly, the model shows the right qualitative behavior. It exhibits asymptotic maximum stiffness at extreme hyperpolarization and an asymptotic minimum at extreme depolarization, which is clearly evident in experiments of He and Dallos (2000). A resting voltage of -55 mV was selected for the model to match the resting voltage observed by He and Dallos (1999) in their stiffness experiments. The maximum and minimum stiffness values were taken from data published by He and Dallos (2000) (see Fig. 5c in He and Dallos (2000)) and

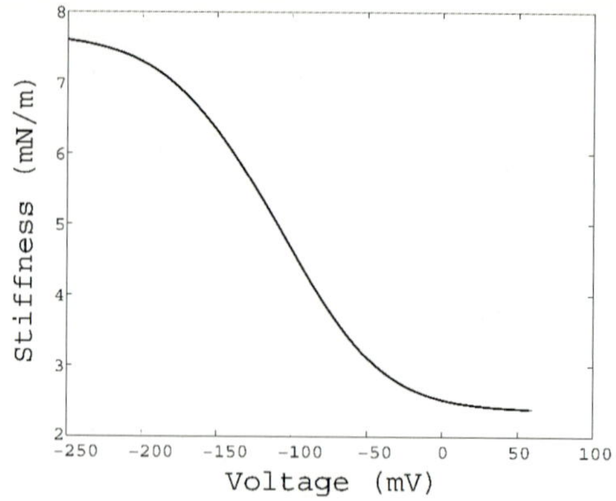

Fig. 3. Stiffness versus voltage for modified area motor model (see Appendices A and B for parameter values). Results are for free extension of the cell.

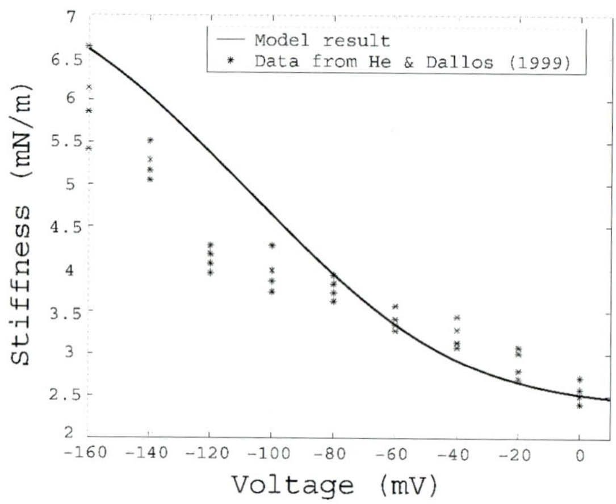

Fig. 4. Stiffness prediction of extended area motor model compared with data from He and Dallos (1999) (see Appendices A and B for parameter values). The simulation and experimental data use a resting voltage $\approx -55\,\text{mV}$.

the values for elastic coefficients published by Iwasa (2001) were used as a guide in selecting the elastic coefficients in the short and long state (see Appendices A and B for parameter values). Stiffness data from He and Dallos (2000) were not compared since their experiments were done using the microchamber setup in which the exact transmembrane voltage was not known.

Dynamic simulations were performed for conditions similar to the experiments of He and Dallos (1999); see Fig. 2. The boundary conditions for the model were as follows. One end of the OHC was held fixed while the other was attached to a spring, which simulated contact with a micropipet. The base of the micropipet was oscillated at a sinusoidal frequency f_B with amplitude

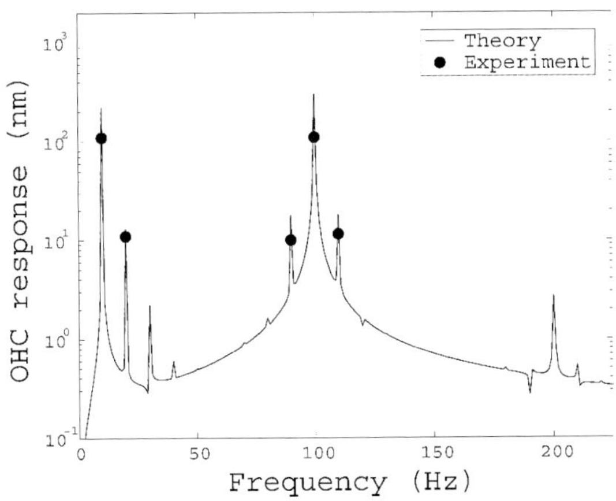

Fig. 5. FFT of ε_z versus time as predicted by the modified area motor model using $f_V = 10\,\mathrm{Hz}$, $f_B = 100\,\mathrm{Hz}$. OHC preload was $1\,\mu\mathrm{m}$. Voltage signal was about $-70\,\mathrm{mV}$ with $V_{\mathrm{amp}} = 40\,\mathrm{mV}$, $B_{\mathrm{amp}} = 1\,\mu\mathrm{m}$. Probe (spring) stiffness was $4.73\,\mathrm{mN/m}$. Other parameter values are the same as those used for Fig. 4. Results are compared with the experimental results of He and Dallos (2000).

B_{amp}. The transmembrane voltage was simultaneously excited at a different frequency, f_V, with amplitude V_{amp}. The micropipet (spring) was preloaded to simulate the preloading in the experiment, by an amount B_0. In Fig. 5, results for the frequency response predicted using the extended model are shown along with an experimental result from He and Dallos (2000). The response predicted by the extended model at sum and difference tones around f_B is pronounced, closely matching experimental results (Fig. 3F from He and Dallos (2000)) for the same set of mechanical and electrical excitation frequencies. A characteristic noticed in experiments is the absence of harmonics of the probe frequency f_B. In the present mathematical model the stiffness depends on P_e, so it also depends on the strains. This implies that harmonics of the fundamental will be generated by mechanical excitation. In Fig. 5 we can see a peak at 200 Hz which is the first harmonic of the mechanical excitation frequency. However, note that the amplitude is very low, below the noise floor of measurements, estimated to be around 5 nm for this experimental data set. Further investigations on the model response to pure mechanical excitation alone also showed a very low amplitude mechanical harmonic. These results agree with the experimental findings that the OHC stiffness is weakly nonlinear with respect to mechanical strain but more strongly nonlinear with respect to voltage.

3.2 Turgor and Temperature Effects

The effect of external force and temperature on OHC behavior hasn't been studied extensively. There are some studies on the effect of turgor pressure on OHCs in the literature. Data from Gale and Ashmore (1994) and Santos-Sacchi and Navarrete (2002) indicate capacitance versus transmembrane voltage curves shifting along the voltage axes and sharpening or flattening with change in maximum capacitance when turgor pressure is changed. The flattening effect of the capacitance curve at higher turgor pressures is also seen in the model. This is due to the changing turgor pressure in the OHC as the voltage is varied and the quadratic terms in the energy barrier. However, the extended area motor model does not show significant changes in the maximum capacitance as the turgor pressure is varied, and the shifting in curves obtained is very sensitive to the parameters chosen. This could be due to an area effect which is not modeled correctly due to the cylindrical assumption made for the OHC shape. At higher pressures, the OHC will tend to assume a more spherical shape and reduce its overall surface area, which will lead to lower capacitance. There could be other issues too; for instance, uneven distribution of motor proteins in the cell surface. We have assumed an even distribution.

Simulations done for varying temperatures do not show much change in the capacitance curves. The only temperature dependent term in the equations is P_e which is quite insensitive to temperature changes. It will be possible to model both temperature and turgor effects with this model if we allow 'q', the charge transferred, to vary with the stress and temperature. However, at present, there is no known or suspected physical basis for such a variation to occur.

4 A Simplified Nonlinear Model

In vivo the voltage changes and strain undergone by OHCs vary only over a small range. The maximum fluctuations of the transmembrane potential are around 10 mV (see Fig. 6 in Dallos (1985)), while strain variations can be as high as 0.6% for a 70 μm long OHC at high sound level inputs (see Khanna and Hao (1999)). A simpler model can be constructed which possesses sufficient richness to describe *in vivo* nonlinearities and response. Experimentally, it is easier to measure strain rather than tension so an energy function with voltage and strains as the independent variables will be more useful. Using the Legendre transformation, we obtain

$$W(\varepsilon_z, \varepsilon_c, V) = G(T_z, T_c, V) + T_z\varepsilon_z + T_c\varepsilon_c. \tag{13}$$

Since ε_z and ε_c are related through the volume constraint, we can replace ε_c in our expressions for energy, stress and charge. After the substitution we can write the charge and external load as,

$$\frac{\partial W(\varepsilon_z, V)}{\partial V} = -Q, \quad \frac{\partial W(\varepsilon_z, V)}{\partial \varepsilon_z} = T_z - \frac{T_c}{2} = T_{\mathrm{ext}}. \tag{14}$$

Let V_0, Q_0, $T_{\text{ext}0}$, and ε_{z0} denote voltage, charge, tension, and strain respectively, at the resting state. Let $q = Q - Q_0$, $t = T_{\text{ext}} - T_{\text{ext}0}$, $v = V - V_0$, and $\epsilon = \varepsilon_z - \varepsilon_{z0}$ denote the shift from the values at resting state. Equations (14) can be expanded in Taylor series about their resting state to yield

$$q = Cv + e\epsilon + \frac{1}{2}C_2 v^2 + C_1 v\epsilon - \frac{1}{2}k_1\epsilon^2, \tag{15}$$

$$t = k\epsilon - ev + \frac{1}{2}k_2\epsilon^2 + k_1\epsilon v - \frac{1}{2}C_1 v^2. \tag{16}$$

$C, C_1, C_2, k, k_1, k_2, e$ are corresponding derivatives of W with respect to ε_z and V, evaluated at the resting state (ε_{z0}, V_0); for example, $e = -\partial^2 W/\partial V\partial\varepsilon|_{\varepsilon_0,V_0}$. All second-order terms are retained and higher-order terms are neglected. Simulations with the simple model show that it does a reasonably good job of matching the full model around the resting state. For instance, over a 15 mV range of voltage away from the resting voltage of −55 mV and strain variation of 0.5%, the simplified model matches the stiffness nonlinearity with respect to voltage with a relative error of less than 1.5%, while the capacitance matches with a relative error of less than 3%. The simplified model has first-order nonlinearities which yield stiffness and capacitance that are linearly dependent on the strain and voltage. As Fig. 6 shows, the second-order expansion possesses sufficient richness to capture experimentally observed dynamic behavior. Contributions arising from higher-order terms will be much smaller than contributions from the second-order terms included in this model and can be ignored.

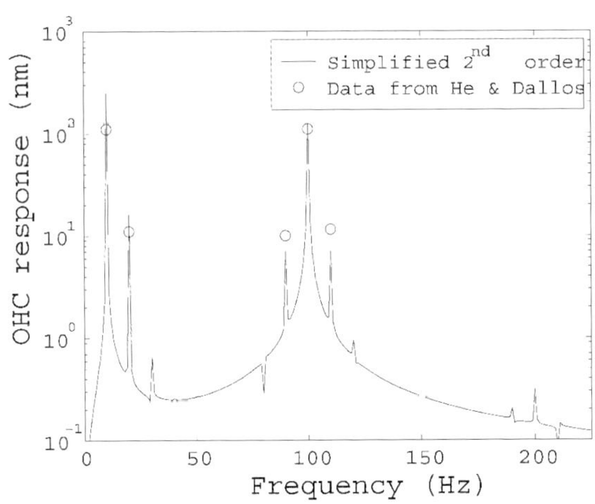

Fig. 6. Frequency spectrum of OHC response. Simultaneous excitation by mechanical and electrical input. Voltage stimulus is at 10 Hz with 40 mV amplitude while probe signal is at 100 Hz with 400 nm amplitude. Circles represent data from Fig. 3F in He and Dallos (2000). Note the match at sum and difference frequencies (90 Hz and 100 Hz). Probe (spring) stiffness used in the model is 4.73 mN/m.

5 Conclusions

OHCs play a key role in amplifying and sharpening the frequency selectivity of the basilar membrane motion. Damage to OHCs or their attached stereocilia reduces the sensitivity and frequency selectivity present in normal hearing (see Oxenham and Bacon (2003) for a review). In the present paper, we develop a micromechanically motivated constitutive model for the overall nonlinear electromechanical behavior of an isolated OHC. This model is appropriate for simulating the large voltage excursions typically utilized in *in vitro* experiments on isolated OHCs. Our model is able to replicate much of the response seen in those experiments, with the notable exception of turgor pressure and temperature (no model yet replicates all of these effects). The model's assumptions of axisymmetric loads and no bending effects are valid for comparison with *in vitro* data, although they may be inadequate for *in vivo* conditions since OHC orientation in the cochlea changes along the length of the basilar membrane (see Spector et al. (2002)). In defense of the present state of modeling, more experimental evidence is needed to aid further developments in the model. Of particular interest is obtaining capacitance, stiffness, and motility data on the same OHC under the same conditions where only very limited data has been obtained so far.

The ultimate technical goal of cochlear mechanics research is to provide a complete understanding of the operation of the cochlea based on individual models of the cochlear components. This understanding hopefully will enable better protection of hearing, non-invasive testing and the development of better prostheses. This necessarily entails incorporation of the isolated model of the OHC into a global model of the cochlea (see Fig. 1). As typical fluctuations of the voltage and strain are fairly small, a small signal model of the OHC is all that is typically needed for accurate (and also more efficient) implementation in a global prediction code. We show how to approximate such a small signal, although still nonlinear, model from the more computationally intensive complete model. Nonlinearities are measured in the basilar membrane response even with low level acoustic input – so they represent real phenomena in hearing. Using such a simplified nonlinear OHC model in a predictive cochlear model, along with a coupled nonlinear stereocilia model, one can look for answers to some of the important research questions in cochlear mechanics. For instance, which effect could be the most important for amplifying the motion of the basilar membrane – stereocilia force generation (as seen in reptiles, but not yet in mammals) or OHC gain? What effect does efferent stimulation have on the response of the cochlea? Does such stimulation, which affects resting voltage in an OHC, alter the operating point of the OHC, thereby reducing the efficiency of the gain or simply reduce the overall transducer currents? By having a predictive, parametric model of the OHC, we introduce the effect of the OHC not as an unknown parameter to be fitted in the global model, but rather as a quantitatively determined factor in the analysis. Complete models can then selectively turn on and turn off

Table 1. Parameter values used for the model.

Parameter	Value	Unit
L	70	μm
R	5	μm
q	-1.602×10^{-19}	C
n	9000	$(\mu\text{m})^{-2}$
a_c	-0.75	$(\text{nm})^2$
a_z	4.5	$(\text{nm})^2$
C_{lin}	21.9	pF

various hypothesized mechanisms (like stereocilia force generation at frequencies greater than 8 kHz). Hence, with these predictive models in hand one can develop a more trustworthy predictive model for the cochlea.

Appendix A. Parameters

The elastic coefficients used for the modified model are $d_z^s = 0.0143\,\text{Nm}^{-1}$, $d_c^s = 0.0211\,\text{Nm}^{-1}$, $c^s = 0.0143\,\text{Nm}^{-1}$, $d_z^l = 0.0464\,\text{Nm}^{-1}$, $d_c^l = 0.0685\,\text{Nm}^{-1}$, $c^l = 0.0464\,\text{Nm}^{-1}$. The rest of the parameter values used are given in Table 1. C_{lin} was obtained from Santos-Sacchi and Navarrete (2002). Iwasa (2001) was used as a guide for choosing the rest of the parameters. Other standard constants used are Boltzmann constant $k_B = 1.38 \times 10^{-23}\,\text{JK}^{-1}$ and temperature (for simulations with constant temperature) $\theta = 300\,\text{K}$. F_0 was adjusted as needed to have peak capacitance voltage around the desired voltage.

Appendix B. Compliance Coefficients

The compliance coefficients are obtained by inverting the matrix of elastic coefficients:

$$\begin{bmatrix} s_z & s_{zc} \\ s_{zc} & s_c \end{bmatrix} = \begin{bmatrix} d_z & c \\ c & d_c \end{bmatrix}^{-1} = \frac{1}{d_z d_c - c^2} \begin{bmatrix} d_c & -c \\ -c & d_z \end{bmatrix}.$$

The short and long coefficients are obtained by substituting the corresponding long and short elastic coefficients.

Acknowledgements. This work was supported by National Institutes of Health grant NIDCD R01–04084.

References

Achenbach, M., Atanackovic, T., and Muller, I. (1986). A model for memory alloys in plane-strain. *Int. J. Solids Structures* 22:171–193.

Dallos, P. (1985). Response characteristics of mammalian cochlear hair cells. *J. Neurosci.* 5:1591–1608.

Deo, N., and Grosh, K. (2004). Two state model for outer hair cell stiffness and motility. *Biophys. J.* 86:3519–3528.

Frank, G., Hemmert, W., and Gummer, A. (1999). Limiting dynamics of high-frequency electromechanical transduction of outer hair cells. *Proc. Natl. Acad. Sci. USA* 96:4420–4425.

Gale, J. E., and Ashmore, J. F. (1994). Charge displacement induced by rapid stretch in the basolateral membrane of the guinea-pig outer hair cell. *Proc. R. Soc. Lond. B* 255:233–249.

Grosh, K., Zheng, J. F., Zou, Y., de Boer, E., and Nuttall, A. L. (2004). High-frequency electromotile responses in the cochlea. *J. Acoust. Soc. Am.* 115:2178–2184.

He, D. Z., and Dallos, P. (1999). Somatic stiffness of cochlear outer hair cells is voltage-dependent. *Proc. Natl. Acad. Sci. USA* 96:8223–8228.

He, D. Z., and Dallos, P. (2000). Properties of voltage-dependent somatic stiffness of cochlear outer hair cells. *J. Assoc. Res. Otolaryngol.* 1:64–81.

Iwasa, K. H., and Adachi, M. (1997). Force generation in the outer hair cell of the cochlea. *Biophys. J.* 73:546–555.

Iwasa, K. H. (1993). Effect of stress on the membrane capacitance of the auditory outer hair cell. *Biophys. J.* 65:492–498.

Iwasa, K. H. (1994). A membrane motor model for the fast motility of the outer hair cell. *J. Acoust. Soc. Am.* 96:2216–2224.

Iwasa, K. H. (2001). A two-state piezoelectric model for outer hair cell motility. *Biophys. J.* 81:2495–2506.

Khanna, S. M., and Hao, L. F. (1999). Reticular lamina vibrations in the apical turn of a living guinea pig cochlea. *Hear. Res.* 132:15–33.

Oxenham, A. J., and Bacon, S. P. (2003). Cochlear compression: Perceptual measures and implications for normal and impaired hearing. *Ear & Hearing* 24:352–366.

Santos-Sacchi, J., and Navarrete, E. (2002). Voltage-dependent changes in specific membrane capacitance caused by prestin, the outer hair cell lateral membrane motor. *Pflügers Arch. – Eur. J. Physiol.* 444:99–106.

Spector, A. A., Ameen, M., and Schmiedt, R. A. (2002). Modeling 3-d deformation of outer hair cells and their production of the active force in the cochlea. *Biomech. Model. Mechanobio.* 1:123–135.

Zheng, J., Shen, W., He, D. Z. Z., Long, K. B., Madison, L. D., and Dallos, P. (2000). Prestin is the motor protein of cochlear outer hair cells. *Nature* 405:149–155.

Prediction of Changes in Cell-substrate Contact under Cyclic Substrate Deformation Using Cohesive Zone Modelling

J.P. McGarry, B.P. Murphy, P.E. McHugh

National University of Ireland, National Centre for Biomedical Engineering Science and the Department of Mechanical and Biomedical Engineering, Ireland
patrick.mcgarry@nuigalway.ie, bruce.Murphy@nuigalway.ie
peter.mchugh@nuigalway.ie

The goal of this study entails the development of computational models of *in vitro* procedures in order to obtain an enhanced understanding of the mechanisms underlying the response of adherent cells to mechanical stimuli. The well established phenomenon of cell alignment under conditions of cyclic substrate stretching is considered. Cell-substrate adhesion is modelled using cohesive zone formulations. Two-dimensional cohesive zone modelling of cyclic substrate stretching reveals a mechanism of cell debonding associated with the cyclic accumulation of viscoelastic strain. Cell alignment is investigated for two distinct modes of cyclic substrate deformation using three-dimensional cohesive zone modelling. It is found that the alignment of the cell-substrate contact areas corresponds to directions of cell alignment observed experimentally for both modes (Wang et al. (2001)).

1 Introduction

The *in vitro* reorientation of isolated cells under conditions of cyclic substrate stretching is widely reported in the literature; see, for example, Wang et al. (1995), Takemasa et al. (1998) and Neidlinger-Wilke et al. (2001). Such behavior is observed for a variety of cell phenotypes including fibroblasts, human melanocytes and endothelial cells. Wang et al. (2001) report that under uniaxial substrate stretching, where substrate contraction is prohibited, cells were found to align at 90° to the direction of stretching. In cases where lateral substrate contraction was permitted cells were found to align at 65° to the direction of stretching. Both alignment directions correspond to the direction of minimum substrate strain. However, it is not clear from such *in vitro* studies why such cell reorientation occurs. In the current study computational models are developed in order to elucidate the mechanisms underlying such behavior. Two cohesive zone models are considered for modeling of cell-substrate adhesion (Beltz and Rice (1991), Xu and Needleman (1993)). Cells are modeled as a viscoelastic continuum. Such an approach to cell modeling has been shown by Leipzig and Athanasiou (2004) to provide better predictions for initial creep response than the alternative bi-phasic modeling approach. Viscoelastic properties are fitted to the experimental measurements of Sato et al. (1990) for porcine endothelial cells.

2 Constitutive Relations

For elasticity, a hyperelastic description is used, where the stress can be derived from a strain energy density potential. A neo-Hookean description is adopted. In this case the form of the strain energy density potential, U, is

$$U = C_{10}(\bar{I}_1 - 3) + \frac{1}{D_1}(J-1)^2, \quad \bar{I}_1 = \bar{\lambda}_1^2 + \bar{\lambda}_2^2 + \bar{\lambda}_3^2, \tag{1}$$

where C_{10} and D_1 are material constants and J is the Jacobian of the deformation, $J = \det \mathbf{F}$. In the above, the $\bar{\lambda}_i$ are the principal deviatoric stretches, given by $\bar{\lambda}_i = J^{-\frac{1}{3}}\lambda_i$, where λ_i are the principal stretches of the deformation, which can be determined directly from the deformation gradient $\mathbf{F}$ as the eigenvalues of $\mathbf{F}^T \cdot \mathbf{F}$. The neo-Hookean description of hyperelasticity is chosen since it is a relatively straightforward representation of finite strain elasticity that can be specified in terms of two material constants C_{10} and D_1, that can easily be related to the initial or linear isotropic elastic constants, the elastic modulus E and Poisson's ratio ν. In specific terms, we use the relationships

$$C_{10} = \frac{E}{4(1+\nu)}, \quad D_1 = \frac{6(1-2\nu)}{E}. \tag{2}$$

The finite strain viscoelasticity used in this work is a generalization of small strain theory, and the basic formulation is most easily outlined beginning with the small strain theory. The viscoelastic material model. relates a time-varying shear strain, $\gamma(t)$, to shear stress, $\tau(t)$, by

$$\tau(t) = G_0 \int_0^t g_R(t-s)\,\dot{\gamma}(s)\,\mathrm{d}s \tag{3}$$

where $\dot{\gamma}(t)$ is the shear strain rate and the shear relaxation modulus, $g_R(t)$, is defined as

$$g_R(t) = \frac{G_R(t)}{G_0}, \tag{4}$$

$G_0 = G_R(0)$ being the instantaneous shear modulus and $G_R(t)$ the shear modulus at time t. It is assumed that the shear relaxation modulus is defined by a Prony series expansion such that

$$g_R(t) = 1 - \sum_{i=1}^{N} \overline{g}_i^p \left(1 - e^{-t/\tau_i^G}\right), \tag{5}$$

where $\overline{g}_i^p$ and τ_i^G are material constants and N is the number of terms in the Prony series (ABAQUS/Standard Users Manual, 2002; Hibbitt et al. (2002)). Equation (4) can be transformed by using integration by parts to yield

$$\tau\left(t\right) = G_0 \left(\gamma - \int_0^t \dot{g}_R\left(s\right)\gamma\left(t-s\right)\mathrm{d}s \right) = \tau_0(t) - \int_0^t \dot{g}_R\left(s\right)\tau_0\left(t-s\right)\mathrm{d}s, \quad (6)$$

where $\tau_0(t) = G_0\gamma$ is the instantaneous shear stress at time t. Where volumetric deformation of the viscoelastic material is also being considered the following expression, analogous to eq. (6), is used to define the pressure

$$p\left(t\right) = -K_0 \left(e^v - \int_0^t \dot{k}_R\left(s\right)e^v\left(t-s\right)\mathrm{d}s \right) = p_0(t) + \int_0^t \dot{k}_R\left(s\right)p_0\left(t-s\right)\mathrm{d}s, \quad (7)$$

where e^v is the volumetric strain, k_R is the bulk relaxation modulus, K_0 is the instantaneous bulk modulus and $p_0(t)$ is the instantaneous pressure at time t. As in eq. (5), k_R is represented using a Prony series

$$k_R\left(t\right) = 1 - \sum_{i=1}^{N} \overline{k}_i^p \left(1 - e^{-t/\tau_i^K}\right), \quad (8)$$

where $\overline{k}_i^p$ and τ_i^K are material constants and N is the number of terms in the Prony series. In ABAQUS it is assumed that $\tau_i^K = \tau_i^G$.

From this a generalization to finite strains can be made where the small strain elastic constitutive laws $\tau_0(t) = G_0\gamma$ and $p_0(t) = -K_0e^v$, involving the infinitesimal stresses $\tau_0(t)$ and $p_0(t)$, can be replaced with an appropriate finite strain constitutive law and an appropriate finite strain stress measure. In ABAQUS a generalization based on the expression

$$\begin{aligned}\boldsymbol{\tau}\left(t\right) = \; & \boldsymbol{\tau}_0(t) - \int_0^t \dot{g}_R\left(s\right)\mathbf{F}_t^{-1}(t-s)\cdot\boldsymbol{\tau}_0^D\left(t-s\right)\cdot\mathbf{F}_t(t-s)\mathrm{d}s \\ & - \int_0^t \dot{k}_R\left(s\right)\mathbf{F}_t^{-1}(t-s)\cdot\boldsymbol{\tau}_0^H\left(t-s\right)\cdot\mathbf{F}_t(t-s)\mathrm{d}s \end{aligned} \quad (9)$$

with

$$\boldsymbol{\tau}(t) = J\boldsymbol{\sigma}(t), \quad \boldsymbol{\tau}_0(t) = J\boldsymbol{\sigma}_0(t) = \boldsymbol{\tau}_0^D\left(t\right) + \boldsymbol{\tau}_0^H\left(t\right), \quad (10)$$

is used, where $\boldsymbol{\tau}(t)$ is the Kirchhoff stress tensor and $\boldsymbol{\tau}_0(t)$ is the instantaneous or elastic Kirchhoff stress that is decomposed into deviatoric, $\boldsymbol{\tau}_0^D(t)$, and volumetric, $\boldsymbol{\tau}_0^H(t)$, parts. In eq. (9) the transformation to finite strain necessitates the introduction of a relative deformation gradient tensor $\mathbf{F}_t$ that represents the deformation from the configuration at time $t-s$ to the configuration at time t. In using this constitutive law, the Cauchy stress $\boldsymbol{\sigma}$ can be recovered by the determination of $\boldsymbol{\tau}$ through eq. (9) and the inversion of eq. $(10)_1$, and in a similar fashion the instantaneous or elastic Cauchy stress can be determined by $\boldsymbol{\sigma}_0(t) = \boldsymbol{\tau}_0(t)/J$ (Belytschko et al. (2000), Hibbitt et al. (2002)).

Finite strain viscoelasticity as described above is used in this study to represent the cell material behavior and, as is the case for material elasticity, much experimental work has been performed to determine values for cell viscoelastic material properties. The cells are therefore represented by both hyperelasticity and viscoelasticity and as such could be described as hyper-viscoelastic. The formal linkage between the two constitutive descriptions is achieved by using hyperelasticity to specify the hyperelastic Cauchy stress at any time in the deformation history (any time t). This can then be identified as $\boldsymbol{\sigma}_0(t)$ and inserted into eqs (9) and (10) to determine the final stress ($\boldsymbol{\tau}(t)$ and hence $\boldsymbol{\sigma}(t)$) at time t due to both hyperelastic and viscous modes of deformation.

The cell substrate interface is described using cohesive zone models where by the energy of separation of two surfaces is expressed as a potential function, $\phi(\boldsymbol{\Delta})$, where $\boldsymbol{\Delta}$ is a vector of displacement jumps between the surfaces. Interface traction-separation relationships can be derived from the potential whereby

$$\mathbf{T} = \frac{\partial \phi\left(\boldsymbol{\Delta}\right)}{\partial \boldsymbol{\Delta}}, \tag{11}$$

where $\mathbf{T}$ is a vector of interface tractions. Two potential functions are considered in this study, the Xu-Needleman potential (12) and the Beltz-Rice potential (13), respectively

$$\begin{aligned}\phi\left(\Delta_n, \Delta_t\right) = \phi_n + \phi_n \exp\left(-\frac{\Delta_n}{\delta_n}\right) \\ \times \left\{\left[1 + \frac{\Delta_n}{\delta_n}\right](q-1) - \left[q + q\frac{\Delta_n}{\delta_n}\right]\exp\left(-\frac{\Delta_t^2}{\delta_t^2}\right)\right\},\end{aligned} \tag{12}$$

$$\begin{aligned}\phi\left(\Delta_n, \Delta_t\right) = \phi_n + \phi_n \exp\left(-\frac{\Delta_n}{\delta_n}\right) \\ \times \left\{\left[q + q\frac{\Delta_n}{\delta_n}\right]\sin^2\left(\frac{\pi\Delta_t}{\delta_t}\right) - \left[1 + \frac{\Delta_n}{\delta_n}\right]\right\},\end{aligned} \tag{13}$$

with

$$q = \frac{\phi_t}{\phi_n}, \tag{14}$$

where ϕ_n is the work of normal separation and ϕ_t is the work of tangential separation, Δ_n and Δ_t are the normal and tangential displacement jumps, respectively, across the interface, and δ_n and δ_t are normal and tangential characteristic lengths for the interface. Both cohesive zone models exhibit an exponential form of the traction-separation relationship in the normal direction. This is consistent with the form of the force-separation curve measured for isolated ligand-receptor bonds (Leckband and Israelachvili (2001)). However, different forms are exhibited in the tangential direction, with the Xu-Needleman potential again yielding an exponential traction-separation relationship and the Beltz-Rice potential yields a sinusoidal relationship. While the former probably represents a more realistic representation of the localized breaking of ligand-receptor bonds in the tangential direction, the latter allows for cell re-adhesion in regions remote from the original bonding sites. Equations (13) and (14) can be generalized to three dimensions by the substitution:

$$\Delta_t = \left(\Delta_m^2 + \Delta_s^2\right)^{1/2}, \tag{15}$$

where Δ_s, Δ_m and Δ_n form a right-handed set.

3 Finite Element Model

A two-dimensional finite element model of a generic cell and substrate is shown in Fig. 1. By manipulation of symmetry only half the cell geometry is modelled. Displacement boundary conditions are applied to the right-hand side of the substrate such that the substrate strain varies cyclically from 0% strain to 5% strain at a frequency of 1 Hz. The cell material is assumed to be incompressible with an elastic modulus of 5 kPa, an elastic modulus of 2.5 MPa and a Poisson's ratio of 0.4 are used to model the silicone substrate. A coarser mesh is used for the substrate since it is several orders of magnitude stiffer than the cell and would consequently exhibit very low gradients of strain. A cell of height 15 μm and of contact radius 8.57 μm is shown in Fig. 1. However, several cell heights and contact radii are considered in this study.

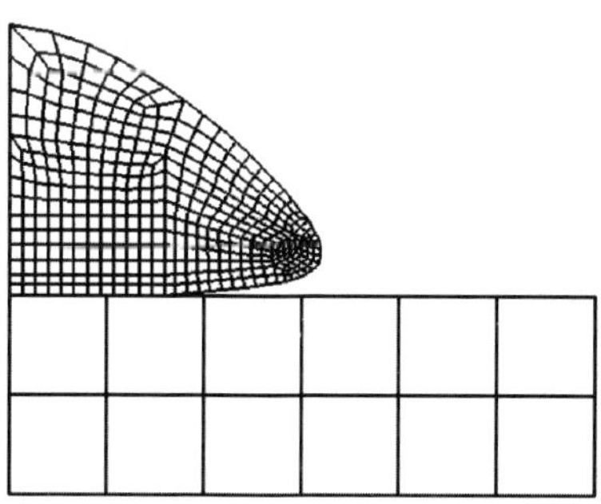

Fig. 1. Two-dimensional finite element cell-substrate model. The densely meshed section represents a cell attached to a coarsely meshed substrate. Cyclic strain is applied to the right-hand-side of the model in the horizontal direction.

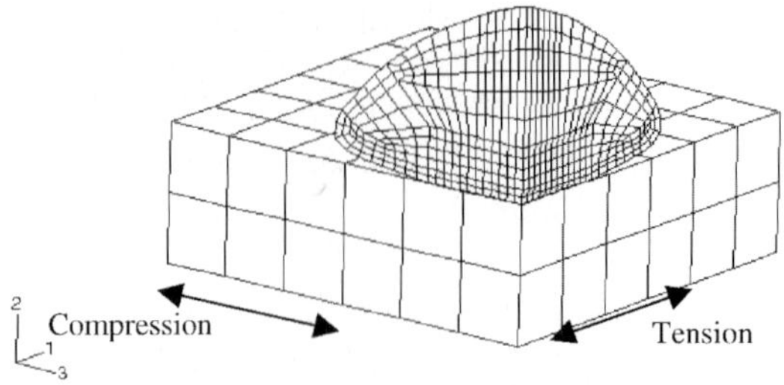

Fig. 2. Three-dimensional finite element cell-substrate model. The model is shown with the two planes of symmetry facing forwards.

By rotation of the two-dimensional cell geometry about its axis of symmetry a generic three-dimensional cell geometry is generated, as shown in Fig. 2. Again, by manipulation of symmetry only a quarter of the cell is modelled. Cyclic stretching of the substrate in the 1-direction is accompanied by cyclic compression of the substrate in the 3-direction due to the Poisson's effect. This is referred to as biaxial substrate deformation. Simulations are also performed in which lateral compression of the substrate is restrained. This is referred to as uniaxial substrate deformation. Both modes of deformation were used in the *in vitro* studies of cell reorientation by Wang et al. (2001).

4 Results

A brief summary of the main findings for two-dimensional and three-dimensional simulations is presented in this section. Interface strengths in the range of 500 Pa to 1000 Pa are used. This corresponds to a range of ligand-receptor bond strengths of 10 pN to 20 pN assuming a bond density of 50 sites/μm^2 (Thoumine et al. (1999)). Characteristic interface lengths of 25 nN are used in all simulations, based on experimentally measured ligand-receptor bond lengths.

Simulations are performed using a two-dimensional Xu-Needleman cohesive zone model. Following five cycles a tensile strain concentration develops above the cell-substrate contact opening as a result of cell viscoelasticity (see Fig. 3). This has the effect of lifting the cell from the substrate. Following 200 cycles the increasing magnitude of the tensile strain concentration results in debonding of a large section of the cell from the substrate. An equilibrium configuration is obtained following 200 cycles with no further debonding being computed. Quantitatively similar results are computed using the Beltz-Rice cohesive zone model.

As mentioned above, the cell geometry used is representative of a range of adherent cell types. Endothelial cells spread on a substrate can have heights

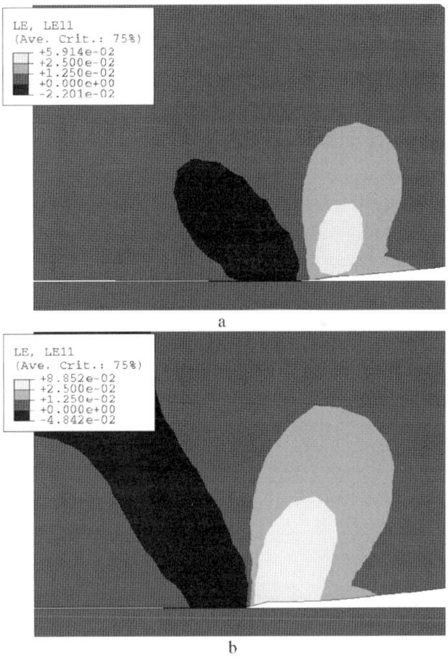

Fig. 3. Contours of strain component parallel to stretching direction computed using a two-dimensional Xu-Needleman model. Contour plots of the strain component parallel to the direction of substrate stretching are shown following five cycles and 200 cycles in Figs 3a and 3b respectively, Fig. 3b shows an increased zone of debonding.

as low as 5 μm (Davies et al. (1993)). In order to examine the influence of cell geometry on the mechanism of cell debonding models of spread cells of height 5 μm, with the entire cell base initially in contact with the substrate, are used. The debonded cell configuration is shown in Fig. 4a and Fig. 4b for cells with initial contact lengths of 8.5 μm and 12 μm, respectively. The mechanism of nodal debonding again entails the accumulation of a tensile strain concentration above the contact edge. While substantially more debonding occurs for the cell with a contact length of 12 μm the final contact length is very similar for both cases. However, it should be noted that for higher values of interface strength no debonding occurs for the cell with the lower contact length. In the case of the cell with a contact length of 12 μm a linear dependence of debonding on the interface strength is computed. In both cases the final contact length is very similar to that computed for the cell geometry shown in Fig. 1.

While two-dimensional models are used to investigate the mechanism of cell debonding, three-dimensional models must be used in order to investigate alignment of the cell substrate contact area. Strain contours for the case of

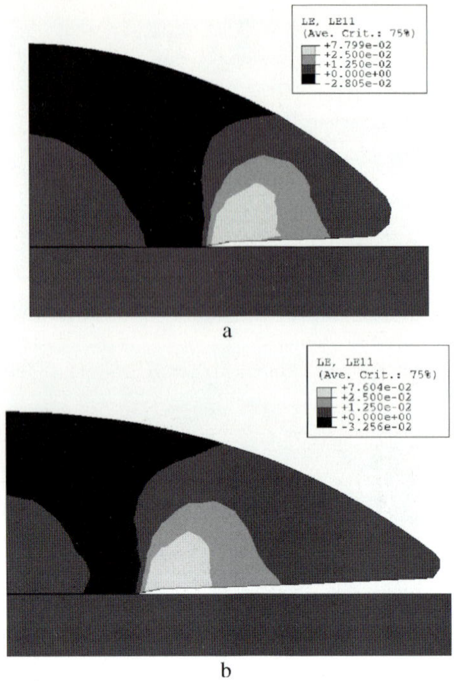

Fig. 4. Strain contours plotted on debonded cell geometry for cells of height 5 μm and radii of 8.5 μm (see Fig. 4a) and 12 μm (see Fig. 4b).

biaxial substrate deformation, where lateral contraction of the substrate due to the Poisson's effect occurs, are shown in Fig. 5. Contours of the strain component in the direction of substrate stretching are in Fig. 5a. As was revealed by two-dimensional models, the accumulation of tensile strain above the contact edge results in cell debonding in the direction of substrate stretching. Contours of the strain component in the direction of substrate contraction (Fig. 5b) reveal that a compressive strain accumulates above the contact edge, resulting in an upwards buckling of the cell. This leads to a small amount of cell debonding in the direction of substrate compression in the case of biaxial substrate deformation. In the case of uniaxial substrate deformation, where no lateral substrate deformation occurs, such an accumulation of compressive strain does not occur, with debonding being computed only in the direction of substrate stretching. The resultant contact areas for the case of biaxial and uniaxial substrate deformation are shown in Fig. 6a and Fig. 6b, respectively. Debonding in the compressive direction in the former case results in an alignment of the contact at 65° to the direction of stretching. In the case of uniaxial deformation, the contact area aligns perpendicular to the direction of substrate stretching. This complies with the directions of cell orientation observed *in vitro* by Wang et al. (2001) for such modes of substrate deformation.

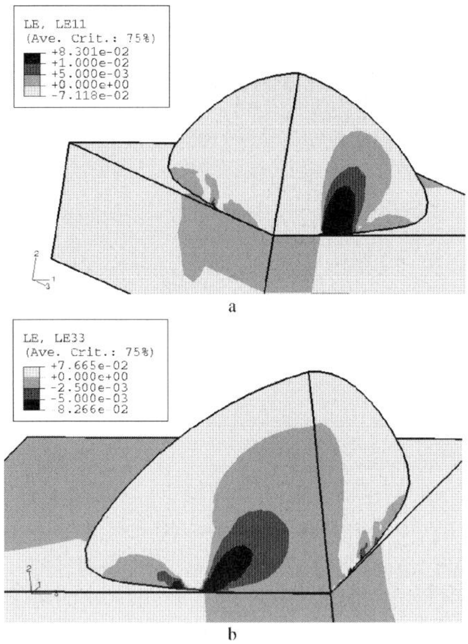

Fig. 5. Strain components in the direction of substrate stretching (see Fig. 5a), and substrate contraction (see Fig. 5b) for biaxial substrate deformation.

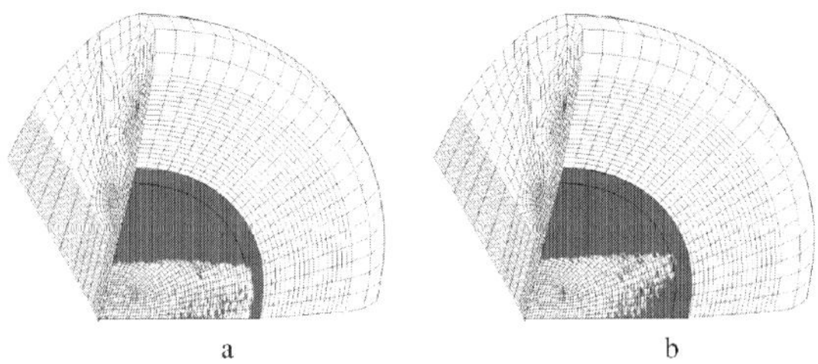

Fig. 6. Final cell-substrate contact areas: uniaxial substrate deformation (see Fig. 6a); biaxial substrate deformation (see Fig. 6b).

5 Discussion

A mechanism of cell debonding under cyclic substrate stretching where the accumulation of tensile strains due to cell viscoelasticity results in a lifting of the cell from the substrate is revealed by the computational models. The mechanism is shown to apply to a range of cell geometries. Development of compressive strain concentrations under biaxial substrate stretching results in a contact area alignment at 65° to the direction of stretching. The absence of such compressive strains in the case of uniaxial substrate stretching results in an alignment of the contact area perpendicular to the direction of stretching.

This complies with the directions of cell orientation observed *in vitro* by Wang et al. (2001) for such modes of substrate deformation. It should be noted that no significant alignment of the cell shape is computed by the computational models. This suggests that the computed breaking of cell-substrate adhesions acts as an initial step in the process of cell alignment. The associated unanchoring and subsequent rearrangement of the cell cytoskeleton has been shown to occur during cell alignment (Wang (2000)).

Acknowledgements. This research was supported by the Higher Education Authority of Ireland. Simulations were performed using the finite element code ABAQUS under and academic license from HKS Inc.

References

Beltz, G. E., and Rice, J. R. (1991). Dislocation nucleation versus cleavage decohesion at crack tips. In Lowe, T. C., Rollett, A. D., Follansbee, P. S., and Daehn, G. S., eds., *Modelling the Deformation of Crystalline Solids.* Warrendale, PA, USA: Proceedings TMS–AIME. 457–480.

Belytschko, T., Liu, W. K., and Moran, B. (2000). *Nonlinear Finite Elements for Continua and Structures.* New York: John Wiley and Sons.

Davies, P. F., Robotewskyj, A., and Griem, M. L. (1993). Endothelial cell adhesion in real time. *J. Clin. Invest.* 91:2640–2652.

Hibbitt, Karlsson, and Sorensen., eds. (2002). *ABAQUS/Standard User's Manual, Version 6.3.* Pawtucket, USA: Hibbitt, Karlsson and Sorensen, Inc. (HKS).

Leckband, D., and Israelachvili, J. (2001). Intermolecular forces in biology. *Quart. Rev. Biophys.* 34:105–267.

Leipzig, N. C., and Athanasiou, K. A. (2004). Unconfined creep compression of chondrocytes. *J. Biomech.* in press.

Neidlinger-Wilke, C., Grood, E. S., Wang, J. H.-C., Brand, R. A., and Claes, L. (2001). Cell alignment is induced by cyclic changes in cell length: studies of cells grown in cyclically stretched substrates. *J. Orthop. Res.* 19:286–293.

Sato, M., Theret, D. P., Wheeler, L. T., Ohshima, N., and Nerem, R. M. (1990). Application of the micropipette technique to the measurement of cultured porcine aortic endothelial cell viscoelastic properties. *J. Biomech. Eng.* 112:263–268.

Takemasa, T., Yamaguchi, T., Yamamoto, Y., Sugimoto, K., and Yamashita, K. (1998). Oblique alignment of stress fibers in cells reduces the mechanical stress in cyclically deforming fields. *Eur. J. Cell Biol.* 77:91–99.

Thoumine, O., Cardoso, O., and Meister, J.-J. (1999). Changes in the mechanical properties of fibroblasts during spreading: a micromanipulation study. *Eur. Biophys.* 28:222–234.

Wang, J. H.-C., Ip, W., Boissy, R., and Grood, E. S. (1995). Cell orientation response to cyclically deformed substrates: Experimental validation of a cell model. *J. Biomech.* 28:1543–1552.

Wang, J. H.-C., Goldschmidt-Clermont, P., Willie, J., and Yin, F. C.-P. (2001). Specificity of endothelial cell reorientation in response to cyclic mechanical stretching. *J. Biomech.* 34:1563–1572.

Wang, J. H.-C. (2000). Substrate deformation determined actin cytoskeleton reorganization: A mathematical modelling and experimental study. *J. Theor. Biol.* 202:33–41.

Xu, X.-P., and Needleman, A. (1993). Void nucleation by inclusion debonding in crystal matrix. *Model. Simul. Mater. Sci. Eng.* 1:111–132.

Micromechanics and Macromechanics of the Tensile Deformation of Nacre

H.J. Qi,[1] B.J.F. Bruet,[3] J.S. Palmer,[2] C. Ortiz,[3] M.C. Boyce[2]

[1] University of Colorado, Department of Mechanical Engineering, USA
qih@colorado.edu
[2] Massachusetts Institute of Technology, Department of Mechanical Engineering, USA
mcboyce@mit.edu
[3] Massachusetts Institute of Technology
Department of Material Science and Engineering, USA
cortiz@mit.edu

Many natural materials exhibit extraordinary combinations of mechanical properties which are achieved through highly tailored and organized hierarchical microstructures. In particular, materials which function as natural body armor, such as mollusk shells, possess a structure with important features and properties at a variety of length scales, from the various constituent building blocks to the overall integrated and synergistic mechanical behavior of their complex assemblies. In this study, the mechanical behavior of the inner 'brick-and-mortar' nacreous layer of mollusk shells was modeled by taking into account both the mechanical behavior of organic matrix and the geometrical arrangement of the mineral-rich tablets. The protein, *Lustrin A*, which is present in *Haliotis rufescens* (red abalone) nacre, has been shown to possess a modular structure consisting of ~ 10 domains linked in series. Axial force-extension experiments on the full organic matrix of this same species exhibit an irregular 'saw-tooth' type profile, whereupon numerous load drops are found to occur over the course of large axial extension (Smith et al. (1999)). This nanomechanical behavior has been attributed to the sequential unfolding of *Lustrin A* subunits and their corresponding rupture of sacrificial bonds. The micromechanical model developed here incorporates a new finite deformation constitutive law that assumes sequential force-induced unfolding of the individual protein domains in the organic matrix, as well as the complex spatial organization of the organic and inorganic components. Numerical simulations of tensile extension of representative volume elements of nacre show that progressive unfolding of the modules in the organic matrix provide a macroscopic 'softening' mechanism, thus mitigating load transfer to the aragonite tablets, as well as averting early failure of the adhesive layers. This softening mechanism also enables larger deformation of nacre without catastrophic failure, and therefore offers an effective avenue for energy dissipation.

1 Introduction

Over millions of years of evolution, mollusks have developed an exemplary biological body armor that exhibits a multiscale, hierarchical composite structure. Natural armor ideally must protect against penetration and crash impacts, in particular by absorbing and dissipating energy. Many mollusk shells employ a multi-layered system including a hard outer layer of prismatic calcite crystals as a first barrier to penetration impacts, and an inner layer of nacre which absorbs and dissipates energy through inelastic deformation and fracture processes (Fig. 1). Nacre is a biogenic composite consisting of alternating layers of mineral tablets separated by thin layers of a biomacromolecular matrix. Nacre is composed of ~95 wt. % aragonite (i.e. an orthorhombic phase of $CaCO_3$) in the form of polygonal tablets which have dimensions of ~5–20 μm (perpendicular to the aragonite unit cell *c*-axis) and ~0.5–1.0 μm in thickness (parallel to the aragonite unit cell *c*-axis). The remaining ~5 wt. % of nacre is composed of a protein-rich organic matrix ~30–50 nm in thickness which acts as adhesive layers connecting aragonite tablets, thus maintaining the structural integrity of the composite. Additionally, small (~30 nm in diameter) mineral bridges span the tablet layers. Pure aragonite is a very brittle material exhibiting linear elastic behavior prior to brittle failure and is rarely used as an engineering material. However, nacre, through its 'brick and mortar' composite structure of micron-scale tablets and sub-micron scale adhesive protein-rich layers, enables outstanding mechanical performance including an excellent combination of stiffness, strength, impact resistance, and toughness (Curry (1977), Weiner and Traub (1984), Jackson et al. (1988)). The ability of nacre to optimize mechanical performance by properly organizing materials at different length scales has attracted a large interest in developing biomimetic materials (e.g., He et al. (1997), Mayer and Sarikaya (2002)), which, in turn, necessitates a deep understanding of the biological mechanisms responsible for its superior properties.

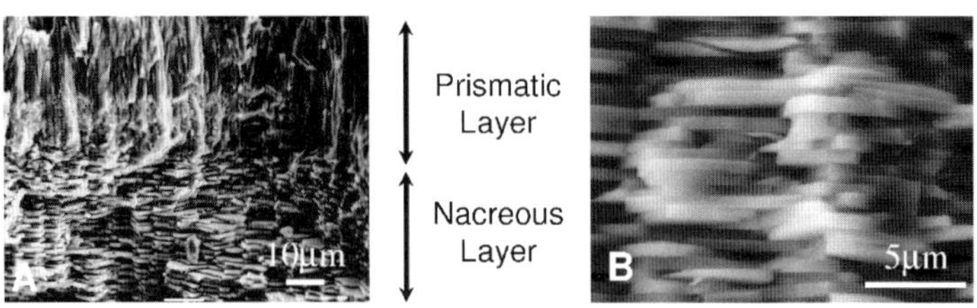

Fig. 1. SEM images of (A) the prismatic – nacreous boundary; (B) nacreous layer.

Hence, there has been much discussion recently of the deformation mechanisms of nacre in the literature. As emulated by modern advanced composites, the small length scale of the individual aragonite tablets is expected to reduce its defect content (i.e. the Griffith flaw argument), thus providing enhanced strength when compared to that of macroscale monolithic aragonite. This well-known and technologically exploited phenomenon was quantified for the case of the aragonite tablets in nacre using fracture mechanics analysis (Gao et al. (2003)). Jackson et al. (1988) previously noted that the observed tablet pull-out rather than tablet fracture under nacre tensile loading conditions suggested a high tensile strength of the micron-scale mineral tablets. The protein-rich adhesive layers facilitate the overall deformation processes by transferring load to the tablets. Evans et al. (2001) and Wang et al. (2001) studied inelastic deformation events in nacre which were attributed to the formation of dilatation bands and the interactions of nanoscale asperities during pull-out of the mineral tablets. In contrast, Katti et al. (2001, 2002) investigated the elastic and inelastic response of nacre by using a three-dimensional finite element model where the organic matrix was modeled as an elastoplastic material and suggested that the contact between neighboring mineral layers most likely has little effect on the overall elastic properties of nacre. Ji and Gao (2004) used the virtual internal bond (VIB) model for the protein matrix and found that the macromolecular layer could effectively enhance the toughness of nacre through crack shielding and impact protection, consistent with the experiments of Jackson et al. (1988).

Recently, nanoscale imaging and mechanical testing techniques have been used to study nacre, revealing new insights on the structure and properties of both the aragonite tablets and the organic matrix. Barthelat and Espinosa (2003) measured the nanoindentation properties of individual nacre tablets from *Haliotis rufescens* and through a comparison with finite element models suggested that the presence of intertablet organic material up to 3 layers below the tablet tested did not cause significant deviation on the elastic property of aragonite tablet measured. Bruet et al. (2005) conducted nanoindentation on individual aragonite tablets from the gastropod mollusk *Trochus Niloticus* and estimated a Young's modulus of ~ 90 GPa and a yield stress of ~ 10 GPa through a comparison with an elastic-perfectly plastic finite element model. AFM inspection of the indented region showed extensive pileup around the indentation zone and thus plastic deformation within the tablet; no micro/nanocracks were observed (Fig. 2). This suggested that the small length scale of the tablet provides a material with a reduced defect content and also suggested that occluded biomacromolecules may play a significant role in the deformation. Smith et al. (1999) used atomic force spectroscopy to test the mechanical axial force-extension behavior of organic matrix strands exposed on a fractured nacre surface. The axial force-extension behavior exhibited an irregular 'saw-tooth' character, so named because of the repeating pattern of a nonlinear force increase with extension followed by abrupt load drops. This type of nanomechanical profile is similar to that observed in numerous single

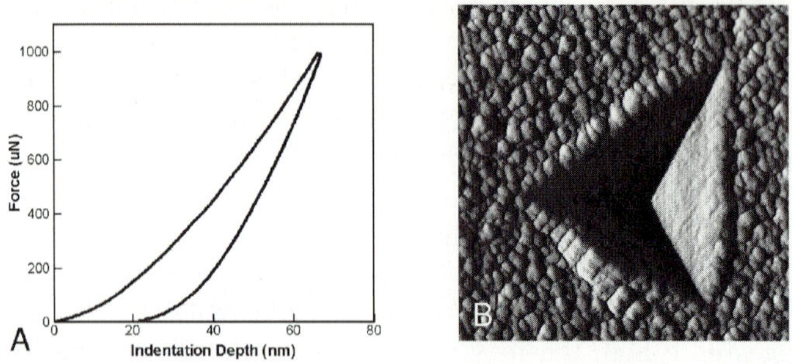

Fig. 2. Nanoindentation of an individual aragonite tablet: (A) averaged force vs indentation depth on individual freshly cleaved *T. Niloticus* nacre tablets; (B) tapping mode atomic force microscopy image of top down view of residual nanoindentation impression on individual freshly cleaved *T. Niloticus* nacre tablet; the horizontal dimension of B is 3.5 μm (Bruet et al. (2005)).

molecule tests on other protein molecules (including titin (Rief et al. (1997)), tenascin (Oberhauser et al. (1998)), and spectrin (Rief et al. (1999), Law et al. (2003)) and is attributed to unfolding of compacted domains along the biomacromolecular chain; this behavior is speculated to play a significant role in the mechanical behavior of nacre.

In this paper, the tensile deformation of nacre is simulated and studied utilizing micromechanical modeling which accounts for the basic features of the brick-and-mortar microstructure as described above. In order to understand the role of the organic matrix in contributing to the overall tensile deformation of nacre, a newly developed finite deformation constitutive model (Qi et al. (2004)) accounting for the influence of unfolding on the stress-strain behavior of the organic matrix is used to study the manner by which the adhesive layers mitigate load transfer to the aragonite tablets. The recent nanoindentation characterization of the tablets (Bruet et al. (2005)) is utilized to properly describe the mechanical properties of the tablets. The micromechanics and the macromechanics of the tensile deformation of nacre from the initial linear region into the nonlinear region of the macroscopic stress-strain behavior will thus be simulated.

2 Micromechanical Model Description

The micromechanical model has been formulated in terms of two distinct ingredients: (i) an idealized, but adequate, representation of the geometry of the microstructure together with appropriate periodic boundary conditions; and (ii) mathematical descriptions of the stress-strain behavior of the constituent materials.

2.1 Geometrical Representation of the Microstructure

The microstructure of nacre is one of a three-dimensional arrangement of parallel layers of somewhat randomly staggered polygonal aragonite tablets (planar dimension of $\sim 5\,\mu$m and thickness of $\sim 1\,\mu$m) connected by thin adhesive layers of protein rich matrix (thickness of ~ 40 nm, Figs 1B and 3), noting that small mineral bridges (diameter of ~ 30 nm) are also observed linking parallel layers. Here, in this early stage micromechanical representation, the structure is idealized as two-dimensional (accounting for key geometrical features through the thickness of the nacre). The representative volume element (RVE) considers a periodic staggered arrangement of tablets connected to one another by thin adhesive layers, as shown in Fig. 4. Periodic boundary conditions are utilized on all edges and thus only the region indicated by the dashed lines in Fig. 4A (enclosing the RVE) is numerically simulated. The model microstructure dimensions were taken from the SEM micrographs of Bruet et al. (2005) as: $L = 5.8\,\mu$m, $h_1 = 870$ nm, $t_1 = t_3 = 40$ nm. Other SEM images (e.g., Jackson et al. (1988)) reveal that the organic matrix forms ligands bridging two neighboring aragonite tablets during deformation. To account for this feature, small gaps of width of $h_2 = 1$ nm were introduced in the adhesive layer prior to the application of deformation. We thus are assuming that the

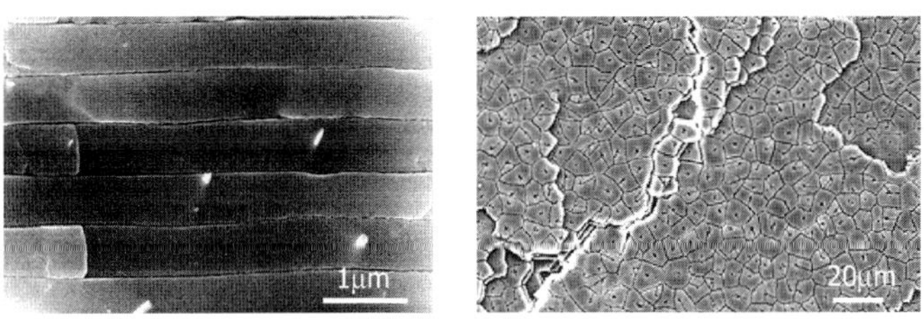

Fig. 3. Scanning electron microscopy (SEM) images of the nanoscale structure of nacre of gastropod mollusk *Trochus Niloticus*: side view (left) and top view (right) (Bruet et al. (2005)).

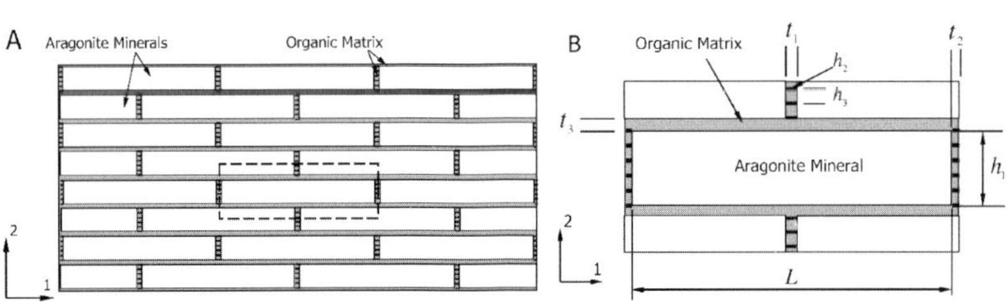

Fig. 4. A schematic of the micromechanical model for nacre: (A) the brick-and-mortar structure; (B) the representative volume element (RVE). Note that the RVE represents a repeating unit (enclosed by dashed line) in (A). The 1-axis is the tensile direction.

constraint imposed by the tablets during extension creates high hydrostatic tension in those adhesive layers normal to the tensile axis (the 1-direction) and some amount of cavitation and ligand formation occurs early on in the deformation. The distance between two neighboring gaps is h_3 and $h_3 \gg h_2$. The gap dimensions were deduced from the Jackson et al. (1988) SEM images of ligands.

Periodic boundary conditions are applied to ensure plane strain tension where the tensile loading axis is the 1-direction, the constraint direction is the 3-direction, and the free direction is the 2-direction. Note that the two-dimensional nature of this model means that, although a staggered arrangement of tablets is modeled in the 1-2 plane, the tablets are essentially columns extending in the 3-direction and subjected to a constraint of no deformation in the 3-direction. Details on appropriately applying the periodic boundary conditions and then obtaining the macroscopic tensile stress-strain behavior of the RVE are provided in, for example, Danielsson et al. (2002) and Sheng et al. (2004). We note that the two-dimensional and periodic nature of this model loses some of the deformation paths available in a fully three-dimensional model of the staggered tablet arrangement, but can still act to identify basic contributions of different features of the microstructure to the overall macroscopic behavior of the composite. Ongoing research addresses more detailed description of the three-dimensional and random multitablet nature of the tablet/mortar structure.

2.2 Constitutive Models of the Stress-Strain Behavior of the Constituent Materials

Aragonite Tablets. The aragonite tablets and their mineral bridges were modeled as an isotropic elastic-perfectly plastic material. Using the modulus and yield strength estimated from the nanoindentation tests of Bruet et al. (2005) on individual nacre tablets of *Trochus Niloticus*, the material is taken to have a modulus of 90 GPa and a yield stress of 10 GPa. We note that this seemingly high yield stress is consistent with the small size scale of the material volume under investigation. We further note that the size scale and the crystal structure of the aragonite tablets dictates an anisotropic behavior; however, this anisotropy is considered to be a higher order effect within the behavior under investigation in this paper and is being addressed in ongoing research.

Organic Adhesive Layers. The stress-strain behavior of the organic matrix is modeled using a newly developed constitutive model that accounts for the unfolding behavior of compacted domains along the macromolecular chains in the protein network (Qi et al. (2004)). This model is briefly described below.

Using statistical mechanics, the extensional force extension behavior of a macromolecular chain can be represented by a freely-jointed-chain (FJC) model (Kuhn and Grün (1942), Treloar (1958)), or a worm-like-chain (WLC) model (Kratky and Porod (1949), Fixman and Kovac (1973), Marko and Siggia (1995)). The FJC model is based on a statistical survey of possible configurations, thus entropy, of a macromolecular chain during stretching. The FJC model relates the axial force f to the chain end-to-end distance r of the macromolecular chain by

$$f = \frac{k_B T}{l}\beta, \tag{1}$$

where k_B is Boltzmann's constant, T is the absolute temperature, l is the Kuhn segment statistical length, $\beta = \mathcal{L}^{-1}(r/L)$ is the inverse Langevin function, where $\mathcal{L}(\beta) = \coth\beta - 1/\beta$ defines the Langevin function, and $L = Nl$ is the contour length of the macromolecular chain, where N is the number of 'rigid links' along the chain.

To predict unfolding of domains within an individual macromolecular chain during extension, a new criterion is employed whereby the evolution in orientation of rigid links relative to the chain stretch direction is monitored. The rigid links increasingly align towards the stretching direction during extension, thus continually varying their orientation (θ) distribution, where θ is defined as the azimuthal angle between each rigid link and the vector connecting the two chain ends. As the chain end-to-end displacement increases, the peak orientation angle, θ_p, decreases. Upon reaching a critical value of θ_p^{cr}, unfolding of a domain is triggered. The value of θ_p^{cr} is governed by the particular chemistry of bonding within the fold and is to be taken as a material parameter.

The single polymer chain behavior is then incorporated into a model for a three-dimensional (3D) macromolecular network solid. Here, a random 3D network is modeled utilizing an idealized 'eight-chain' network representation (Fig. 5) (Arruda and Boyce (1993)); this network idealization captures the essential features of a random network in an 'average' sense. The edges are taken to be aligned with the principal stretch directions. The strain energy density calculated based on this RVE subjected to stretch is

$$\bar{U} = vk_B TN\left[\frac{\lambda_c}{\sqrt{N}}\beta_c + \ln\left(\frac{\beta_c}{\sinh\beta_c}\right) - c\right], \tag{2}$$

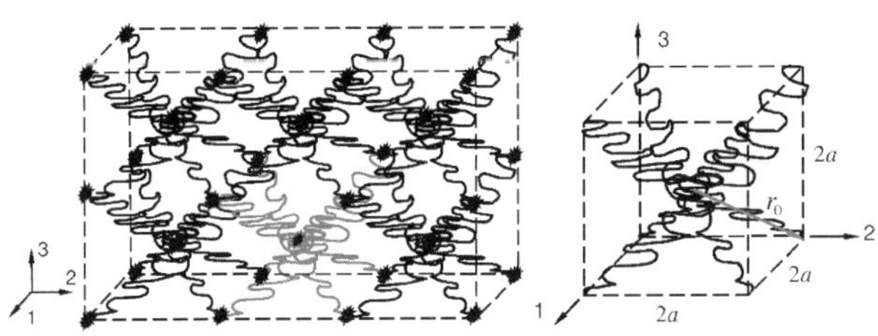

Fig. 5. Schematics of three-dimensional planar network: (A) an idealized macromolecule network with perfectly staggered crosslinks, before deformation; (B) an 8-chain network, which is the representative volume element (RVE) highlighted in gray in (A).

where v is the density of chains between crosslink sites, $\beta_c = \mathcal{L}^{-1}(r/L) = \mathcal{L}^{-1}(\lambda_c/\sqrt{N})$, $\lambda_c = [(\lambda_1^2 + \lambda_2^2 + \lambda_3^2)/3]^{1/2}$ is the chain stretch, and c is the contribution from the initial entropy of the macromolecule. Note that as molecular chain stretch evolves with the macroscopic stretch of the solid, the domains of the macromolecular chains will successively unfold as they reach the critical conditions. The unfolding of the domains acts to increase the contour length of the molecule (i.e. L evolves with stretch conditions) which in turn increases the entropy of the network creating a drop in the stress-strain curve. The network becomes less oriented as a result and the orientation evolution process begins again and the repeating 'saw-tooth' pattern is thus captured.

As an illustration, Fig. 6 shows the nominal stress versus stretch curve for simple tension on a three-dimensional solid composed of a network of modular macromolecules (i.e. molecules containing folded domains). The dashed line plot in Fig. 6 shows a simulation where unfolding does not occur during stretching. The dotted line plot in Fig. 6 shows a simulation for a homogeneous macromolecular network where all polymer chains between crosslinks have identical initial contour lengths and hence, all chains reach unfolding conditions at the same macroscopic stretch states. The solid line plot of Fig. 6 is a simulation which considers a nonuniform distribution in initial contour lengths where different polymer chains achieve unfolding conditions under slightly different stretch states. The material properties used for these simulations are listed in Table 1.

It is noted that in this case of simple tension, when the chain contour lengths are statistically distributed (solid line, non-uniform in Fig. 6), the stress-strain behavior undergoes a nonlinear increase, then transitions to a plateau region, and finally increases again at large stretches. The first

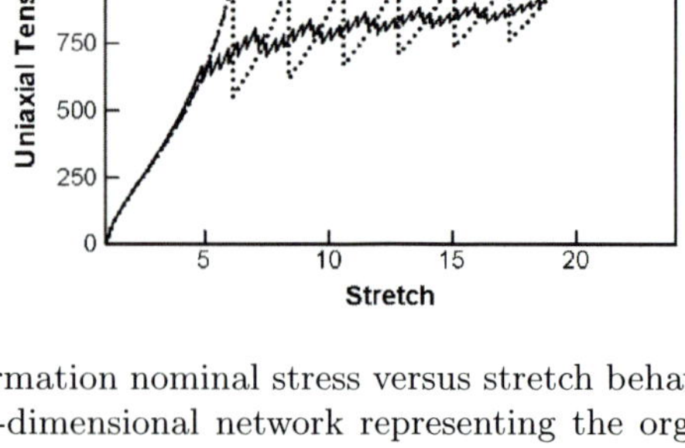

Fig. 6. Finite deformation nominal stress versus stretch behavior of simple tension of a modular three-dimensional network representing the organic matrix in nacre with and without consideration of domain unfolding.

Table 1. Material properties used for the simulations of the organic matrix.

Material Properties	
Persistence length, l (nm)	0.4
Initial contour length, L_i (nm)	10.0
Increase in contour length due to unfolding, ΔL (nm)	4.0
Critical peak angle, θ_p (degree)	30.0
Density of polymer chains between crosslinks, ν (m^{-3})	2.43×10^{28}

transition from a nonlinear increase to the plateau region corresponds to successive triggering of the unfolding of modules belonging to various polymer chains between crosslinks within the solid. The second transition from the plateau region to a second nonlinear increase corresponds to when all modules have unfolded and increased strain hardening ensues. These two transitions distinguish modular networks as having a unique 'softening' behavior, which has also been observed in rubbers and thermoplastic elastomers (Qi and Boyce (2004a,b)). The effect of this behavior on controlling load transfer to the aragonite tablets during nacre deformation will be illustrated and discussed in the results section.

3 Results and Discussion

Simulation results for the micromechanical model of nacre subjected to tensile loading are presented in terms of the macroscopic stress-strain behavior of the composite and contours of relevant field variables on deformed configurations of the simulated microstructure at different levels of macroscopic strain.

Figure 7 depicts the macroscopic tensile stress-strain behavior of the nacre micromechanical model. Two results are shown, one where the organic matrix is modeled neglecting the unfolding of the compacted domains and one where the effects of unfolding are included (the simple tensile stress-strain behaviors of these two matrix representations were shown in Fig. 6). The stress-strain results predict an initial modulus of 40 GPa, compared to the experimentally reported value of 60 GPa (Jackson et al. (1988)). Note that the Jackson et al. (1988) experiments were on *Pinctada* nacre as opposed to the *Trochus Niloticus* nacre from which we took the microstructural parameters; therefore, this comparison should not be perfect, but is given to demonstrate that the simulation results are quantitatively realistic. When unfolding is included in the organic matrix layers, the macroscopic stress-strain curve exhibits an apparent 'yield' stress of 135 MPa, corresponding to the Jackson et al. (1988) experimental tensile yield strength of 140 MPa which corresponded to tablet pullout as the tensile failure mechanism. The 'yield stress' and nonlinear transition corresponds to unfolding events and softening in the organic layers. These unfolding events, in turn, mitigate load transfer to the aragonite tablets, thus

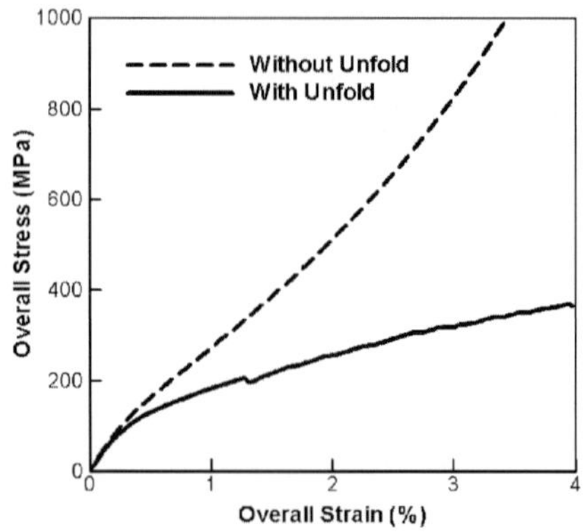

Fig. 7. Macroscopic tensile stress-strain curve for the micromechanical simulations of nacre considering an organic matrix layer with unfolding and one without unfolding.

preventing the stress levels in the aragonite from reaching fracture level conditions. The post-yield macroscopic strain hardening regime is due to the deformation response of the organic matrix layers parallel to the tensile direction, which shear and transfer load to the tablets via the well-known shear lag load transfer mechanism in fiber and platelet reinforced composites.

Figure 8 shows contours of the axial stress σ_{11} (in Pascals) in the aragonite tablets at the macroscopic axial strain of 0.2% (Fig. 8A) and 2.0% (Fig. 8C) for the case where unfolding of the organic matrix is taken into account. The corresponding stress contours for the case of a matrix without unfolding are shown in Figs 8B and 8D, respectively. At small tensile strain (0.2%) the simulated stress-strain curves for nacre with and without unfolding of the macromolecular component are approximately the same (Fig. 7). Hence, the stress contours in Figs 8A and 8B appear similar as well. For both the folding and unfolding cases, the tensile stress in the aragonite tablets increases along the tablet length from a value of 70 MPa at the tablet edges (edges normal to the loading direction) to a maximum tensile stress of $\sim$120 MPa. This stress buildup results from shearing of the organic adhesive layers between tablets acting to transfer additional load to the tablets, as shown in the tablet shear stress contours of Fig. 9.

At a larger overall tensile strain of $\sim$2.0%, which is in the 'post-yield' region of the stress-strain curve of Fig. 7 that includes unfolding, the aragonite axial stress builds from a value of about 140 MPa at the tablet edges to a very localized peak value of 870 MPa at the outer edge of the central region of the tablet with an average value of 600 MPa in that region. The buildup of stress from the edge to the center of the tablet is again through shear transfer from the matrix, as shown in the corresponding shear stress contours of Fig. 9. In contrast, the axial stress contours at 2.0% strain when matrix

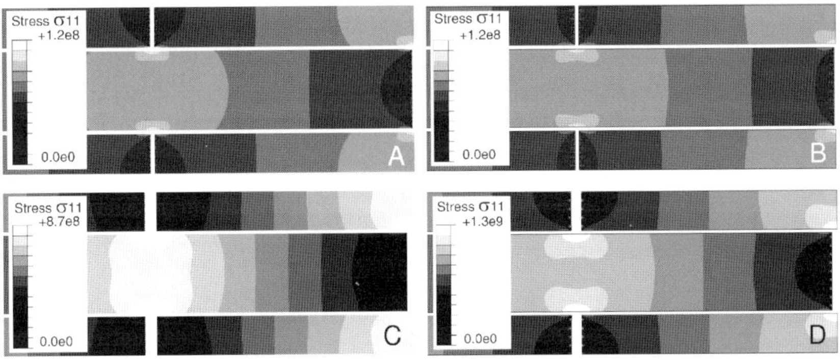

Fig. 8. Contours of the σ_{11} (in Pa) in the aragonite tablets for (A) the case of unfolding in organic matrix at 0.2% overall strain; (B) the case without unfolding at 0.2% overall strain; (C) the case of unfolding in organic matrix at 2.0% overall strain; (D) the case without unfolding at 2.0% overall strain. The 1-direction (the loading direction) is the horizontal direction. Note that, due to the dramatically different magnitudes in stress, different contour scales are needed and used in A, B, C, and D.

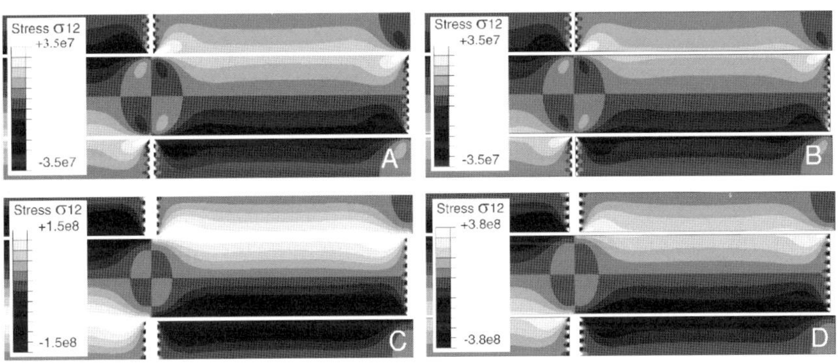

Fig. 9. Contours of σ_{12} (in Pa) in the aragonite tablets for (A) the case of unfolding in organic matrix at 0.2% overall strain; (B) the case without unfolding at 0.2% overall strain; (C) the case of unfolding in organic matrix at 2.0% overall strain; (D) the case without unfolding at 2.0% overall strain.

unfolding is not taken into account (Fig. 8D) show a build up from a tablet edge stress of 250 MPa to a central stress (over the entire central section) of over 1.1 GPa. This higher tablet stress and its occurrence over a larger volume of the aragonite create a scenario more favorable to tablet fracture. As macroscopic strain continues to increase, these values will continue to rise in a manner consistent with the steep slope of its macroscopic stress-strain curve of Fig. 7. In contrast, matrix unfolding substantially reduces the stress levels in

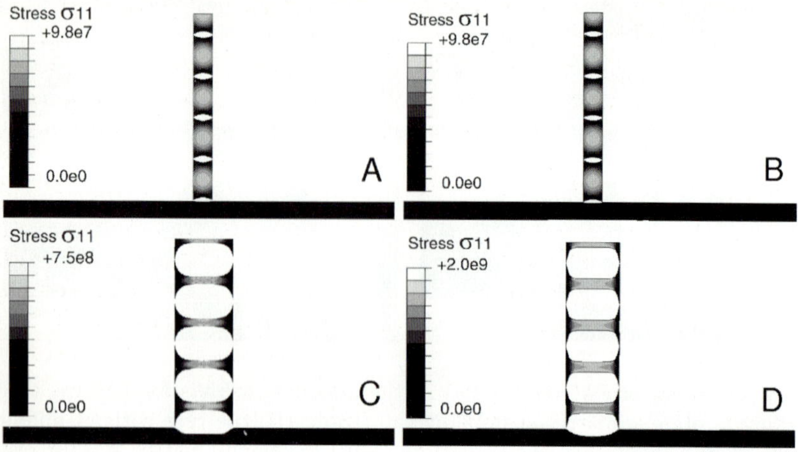

Fig. 10. Contours of stress component σ_{11} (in Pa) in the matrix ligaments for (A) the case of unfolding in organic matrix at 0.2% overall strain; (B) the case without unfolding at 0.2% overall strain; (C) the case of unfolding in organic matrix at 2.0% overall strain; (D) the case without unfolding at 2.0% overall strain.

the tablets and the manner in which the stress levels increase with increasing macroscopic strain, making tablet pullout the experimentally observed mode of tensile failure instead of tablet fracture.

To further illustrate the role of unfolding in controlling load transfer to the tablets during tensile loading conditions, contours of axial stress in the matrix layers normal to the tensile axis at 0.2% and 2% strain are shown in Fig. 10. Figures 10A and 10C show the axial stress in the matrix ligaments at 0.2% and 2.0% macroscopic strain when matrix unfolding is not considered; Figures 10B and 10D show the corresponding cases when unfolding is taken into account. The stress levels at 0.2 % strain are similar for both matrix cases ($\sim$100 MPa peak); however, at 2.0 % macroscopic strain the stress in the unfolding matrix are significantly lower than those in the non-unfolding case since unfolding essentially plateaus the stress level in the matrix ligaments. Thus, the stress levels transmitted to the tablets are mitigated by the unfolding events in the matrix ligaments.

4 Conclusion

This paper investigated the mechanics of nacre using a micromechanical representative volume element model, where the protein-rich organic matrix is modeled by a recently developed finite deformation constitutive model that accounts for the unfolding of protein molecules in the organic matrix. Numerical simulations of tensile deformation showed that in the presence of unfolding in the organic matrix, the macroscopic stress-strain behavior demonstrated non-linear behavior with an initial linear elastic region followed by an apparent 'yield' stress. This yield was due to triggering unfolding events in the organic layers normal to the applied tension and corresponded to a mitigation of load transfer to the aragonite tablets, thus preventing the stress levels in the aragonite from reaching fracture level conditions. As macroscopic strain continued to increase, the macroscopic post-yield behavior showed strain hardening. This hardening behavior was due to shear in the organic layers that transferred load to the tablets via the well-known shear lag load transfer mechanism in fiber and platelet reinforced composites. It is noted that this is a preliminary study where the simplified micromechanical model presented here captures the basic structural features of nacre and revealed new understanding of the role of the organic matrix softening mechanisms in controlling load transfer at the microscopic lengthscale. More detailed models which consider the three-dimensional and the random distribution of the tablet and mortar structure are currently under investigation.

Acknowledgements. This research was supported by the U. S. Army through the Institute for Soldier Nanotechnologies, under Contract DAAD-19-02-D0002 with the U.S. Army research office. The content does not necessarily reflect the position of the government and no official endorsement should be inferred. HJQ would also like to acknowledge the support from the start-up fund from the University of Colorado at Boulder.

References

Arruda, E. M., and Boyce, M. C. (1993). A three-dimensional constitutive model for the large stretch behavior of rubber elastic materials. *J. Mech. Phys. Solids* 41:389–412.

Barthelat, F., and Espinosa, H. D. (2003). Elastic properties of nacre aragonite tablets. In *Proceedings of the 2003 SEM Annual Conference and Exposition on Experimental and Applied Mechanics.* Paper 187.

Bruet, B. J. F., Panas, R., Tai, K., Frick, L., Ortiz, C., Qi, H. J., and Boyce, M. C. (2005) Nanoscale morphology and indentation of individual nacre tablets from the gastropod mollusk trochus niloticus. *J. Mat. Res.* in press.

Curry, J. D. (1977). Mechanical properties of mother of pearl in tension. *Proc. R. Soc. Lond. B* 196:443–463.

Danielsson, M., Parks, D. M., and Boyce, M. C. (2002). Three-dimensional micromechanical modeling of voided polymeric materials. *J. Mech. Phys. Solids* 50:351–379.

Evans, A. G., Suo, Z., Wang, R. Z., Aksay, I. A., He, M. Y., and Hutchinson, J. W. (2001). Model for the robust mechanical behavior of nacre. *J. Mater. Res.* 16:2475–2484.

Fixman, M., and Kovac, J. (1973). Polymer conformational statistics. III modified Gaussian models of stiff chains. *J. Chem. Phys.* 58:1564–1568.

Gao, H., Ji, B., Jager, I. L., Arzt, E., and Fratzl, P. (2003). Materials become insensitive to flaw at nanoscale: Lessons from nature. *Proc. Natl. Acad. Sci. USA* 100:5597–5600.

He, J. L., Wang, J., Li, W. Z., and Li, H. D. (1997). Simulation of nacre with TiN/Pt multilayers and a study of their mechanical properties. *Mat. Sci. Eng. B-Solid* B49:128–134.

Jackson, A. P., Vincent, J. F. V., and Turner, R. M. (1988). The mechanical design of nacre. *Proc. R. Soc. Lond. B* 234:415–440.

Ji, B., and Gao, H. (2004). A study of fracture mechanisms in biological nanocomposites via the virtual internal bond model. *Mat. Sci. Eng. A-Struct.* A366:96–103.

Katti, D. R., Katti, K. S., Sopp, J. M., and Sarikaya, M. (2001). 3D finite element modeling of mechanical response in nacre-based hybrid nanocomposites. *Comput. Theo. Polym. Sci.* 11:397–404.

Katti, D., Katti, K., Tang, J., and Sarikaya, M. (2002). Effect of nanostructure in nacre: a multiscale modeling approach. In *Proceedings of 15th ASCE Engineering Mechanics Conference.* Columbia Universtiy, New York, NY.

Kratky, O., and Porod, G. (1949). Röntgenuntersushung gelöster Fadenmoleküle. *Recl. Trav. Chim. Pays-Bas.* 68:1106Ű–1123.

Kuhn, W., and Grün, F. (1942). Beziehungen zwischen elastischen Konstanten und Dehnungsdoppelbrechung hochelastischer Stoffe. *Kolloid-Zeitschrift* 101:248–271.

Law, R., Carl, P., Harper, S., Dalhaimer, P., Speicher, D., and Discher, D. E. (2003). Cooperativity in forced unfolding of tandem spectrin repeats. *Biophys. J.* 84:533–544.

Marko, J. F., and Siggia, E. D. (1995). Stretching DNA. *Macromolecules* 28:8759–8770.

Mayer, G., and Sarikaya, M. (2002). Rigid biological composite materials: structural examples for biomimetic design. *Exp. Mech.* 42:395–403.

Oberhauser, A. F., Marszalek, P. E., Erickson, H. P., and Fernandez, J. M. (1998). The molecular elasticity of the extracellular matrix protein tenascin. *Nature* 393:181–185.

Qi, H. J., and Boyce, M. C. (2004a). Constitutive model for stretch-induced softening of the stress-Űstretch behavior of elastomeric materials. *J. Mech. Phys. Solids* 52:2187–2205.

Qi, H. J., and Boyce, M. C. (2004b). Stress-strain behavior of thermoplastic polyurethane. *Mech. Mat.* in press, web-release November, 2004.

Qi, H. J., Ortiz, C., and Boyce, M. C. (2004). A constitutive model for the stress-strain behavior of biomacromolecular networks containing folded domains. *Biophys. J.* submitted.

Rief, M., Gautel, M., Oesterhelt, F., Fernandez, J. M., and Gaub, H. E. (1997). Reversible unfolding of individual titin immunoglobulin domains by AFM. *Science* 276:1109–1112.

Rief, M., Pascual, J., Saraste, M., and Gaub, H. E. (1999). Single molecule force spectroscopy of spectrin repeats: low unfolding forces in helix bundles. *J. Mol. Biol.* 286:553–561.

Sheng, N., Boyce, M. C., Parks, D. M., Rutledge, G. C., Abes, J. I., and Cohen, R. E. (2004). Multiscale micromechanical modeling of polymer/clay nanocomposites and the effective clay particle. *Polymer* 45:487–506.

Smith, B. L., Schaffer, T. E., Viani, M., Frederick, J. B. T. N. A., Kindt, J., Belcher, A., Stucky, G. D., Morse, D. E., and Hansma, P. K. (1999). Molecular mechanistic origin of the toughness of natural adhesives, fibres and composites. *Nature* 399:761–763.

Treloar, L. R. G. (1958). *The Physics of Rubber Elasticity.* Oxford: Oxford at the Clarendon Pess.

Wang, R. Z., Suo, Z., Evans, A. G., Yao, N., and Aksay, I. A. (2001). Deformation mechanisms in nacre. *J. Mater. Res.* 16:2485–2493.

Weiner, S., and Traub, W. (1984). Macromolecules in mullusc shells and their functions in biomineralization. *Philos. Trans. R. Soc. Lond. B* 304:425–433.

Part III

Arteries in Health and Disease

Mechanical Properties of Atherosclerotic Tissues

P.D. Richardson

Brown University, Division of Engineering and Department of Molecular Pharmacology, Physiology and Biotechnology, USA
Peter_Richardson@brown.edu

The investigation of mechanical properties of atherosclerotic vessel walls was largely academic until 1980, when the mechanical properties began to be seen as more important for clinical issues. The reasons for the shift in perception are reviewed, and some characteristics of the mechanical properties, pertaining to elastic behavior and ultimate strength (and deformation) are outlined. Much research effort in the past dozen years has been devoted to investigating matrix metalloproteinases, as these are seen as contributing to weakening of plaque tissues, and some results of incubation with collagenase on mechanical properties are included. Imaging *in vivo* to permit local stress estimation has also garnered attention.

1 Introduction

Atherosclerosis is a disorder of blood vessels in humans, dominantly in arteries. It appears to begin early in life, perhaps even before birth. Its progression with age varies widely between individuals. It is usually widespread among arteries in those persons notably affected by it, though arteries in the arms are relatively spared. Its variability between individuals led to many anecdotal descriptions in the earlier pathological literature. Some features of its local arterial appearance are robust. Forms of atherosclerosis can be induced in some animal species which are used in experimental research but it is not indiginous in them to the extent it is to so many humans in the First World.

The composite structure of each of the principal layers of blood vessels (intima, media and adventitia), both as to structural proteins and cell contents, and organization, were well described for normal vessels by the 1950s, and the relative significance of elastin and collagen in determining the non-linear mechanical properties had been demonstrated, elastin being important at low stresses and elongations and collagen being dominant at higher stresses. Around this period there was also a vigorous development of analysis of the mechanical properties of rubber and other polymers. There is a popular conception that the arteries 'harden' with progression of age, implying a progressive shift in mechanical properties (it was also noted that the calcium content of arterial walls increases with age, the calcium sometimes appearing in small, plate-like segments that crunch under a scalpel, but commonly the calcium is more evenly and finely distributed). However, the definitive evidence for atherosclerosis is not based on a measure of mechanical properties of a blood vessel but upon features found in pathohistological examination of arterial tissue. Typical features include thickening (and sometimes thinning) of the principal layers of arteries, diffuse and focal inclusion of lipids including cholesterol, incorporation of fibrin from the blood, changes in population densities of cells, and attraction of monocytes from the blood to adhesion and intimal entry.

Clinical manifestations involving profound local atherosclerosis include claudication, some angina, myocardial infarctions and thrombotic strokes, and effects of regional ischemia. (Haemorrhagic strokes are associated with bleeding from blood vessels, typically within the subarachnoid space, not necessarily associated with atherosclerosis. Bow hunter's stroke is characterized by perfusion insufficiency associated with occlusion or stenosis of the vertebral artery associated with head movements. Neither of these types of stroke are, therefore, directly associated with atherosclerosis.) For many years the ischemia associated with atherosclerosis was read from pathology to be consequent upon thrombosis, often in association with arterial stenosis, and treatment focussed first upon diminishing the tendency for thrombi to form without turning patients into persons with very long bleeding times. A widely effective acute treatment of angina (still in use) involved administering vascular smooth-muscle relaxants, perceived as reducing stenosis as the arteries enlarge in diameter. A more chronically-effective treatment for reducing stenosis in coronary arteries is by balloon angioplasty, a procedure often accompanied by elective stenting. Stents are pre-fabricated lengths of cylindrical mesh structures inserted at a ballooned site for locally propping the vessel open.

2 Perceptions of Significant Aspects of Atherosclerosis to 1980

Atherosclerosis is widely regarded as a progressive disorder, although there may be episodes of quiescence and even recession.

For routine autopsy reports, there are standard grading templates based on examination of segments of vessels that have been slit open parallel to their central axes and pinned out (along the separated edges) on a wax tray so as to show their lumenal aspect. Such standards are very useful in encouraging a consistent level of description of gross pathology that presents itself in a great profusion of individual details. However, the procedure may have contributed to delayed recognition of positive remodeling.

For research investigations, histopathologic study of vessels is also the standard. The methods of preparation, however, have varied among investigators. The literature is rife with photographs of thin transverse sections of arteries, with standard tissue stains and, more recently, antibody-conjugated special stains. The typical pathologist has been most concerned to detect what can be found and differentiated chemically (the book by Woolf (1982) has photographs typical of the period up to about 1980). This priority, combined with use of thin transverse sections, did not support well those who seek to relate structure (tissue details) and function (load carrying) quantitatively, and in a three-dimensional context.

If one examines an aorta of a person who has died of natural causes at an age typical for that, the aorta has many atherosclerotic lesions visible at the lumen, starting from the aortic arch and continuing through the thoracic and abdominal sections. If anything, the proportion of intima covered increases as the distance downstream increases, and distal to the renal arteries the lesions are abundant. It is fairly common to find some plaques have ulcerated (an interesting illustration is provided by Texon (1980), Fig. 9, p. 17, with an aortic plaque cap rupture seen in a cross-section; see Fig. 1).

A three-dimensional perception regarding atherosclerosis arose among physiologists in a broad sense, in that it was observed that atherosclerotic

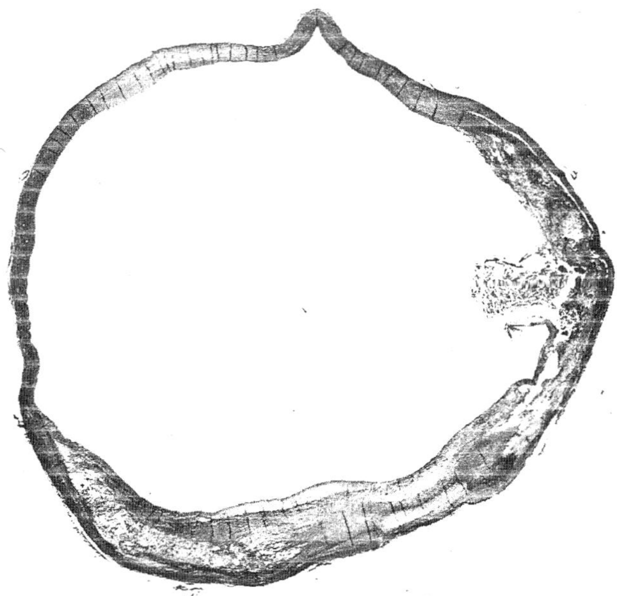

Fig. 1. Human thoracic aorta showing atherosclerotic plaque on dorsal wall (attachment lesion). Note rupture of superficial layers of plaque and embolization of atheromatous contents (adapted from Texon (1957)).

lesions were not randomly distributed, but are found with higher frequency in the vicinity of bends and of vessel junctions, particularly in arteries below about 6 mm diameter. This was soon attributed to aspects of the blood flow and two sets of investigations in this period stand out. One, published in the early 1960s, involved surgical creation of curves and serpentine axes for arteries that would otherwise be straight, in dogs, by transplantation of segments of carotid artery into the femoral artery, with observations of resultant plaque-like developments by Texon et al. (1962). The authors concluded that the study '*... further elaborates the concept that the laws of fluid mechanics are the primary factor in the etiology and localization of atherosclerosis*'. The other investigation, published in the late 1960s and early 1970s, focussed upon the possible role of the vessel-wall blood-flow shear rate, noting the latter was low in some regions in the vicinity of bifurcations where atheromatous lesions are more frequently found and where the flow is complicated, Caro et al. (1969, 1971). These three-dimensional approaches initiated experimental and computational studies of flows and transport in curved tubes and branchings which continue to the present time.

Another aspect of arterial behavior, well-known from the early times of quantitative physiological studies of the circulation, is that arteries can change their radius significantly as a consequence of activation of the smooth muscle cells in the media. In a review of static elastic properties of blood vessels, Canfield and Dobrin (1987) illustrate this by showing external radius as a function of transmural pressure for a dog carotid (see Fig. 2; their Fig. 16.4), and circumferential strain averaged for 160 dogs as a function of transmural pressure (see Fig. 3; their Fig. 16.6). Such vessels are roughly

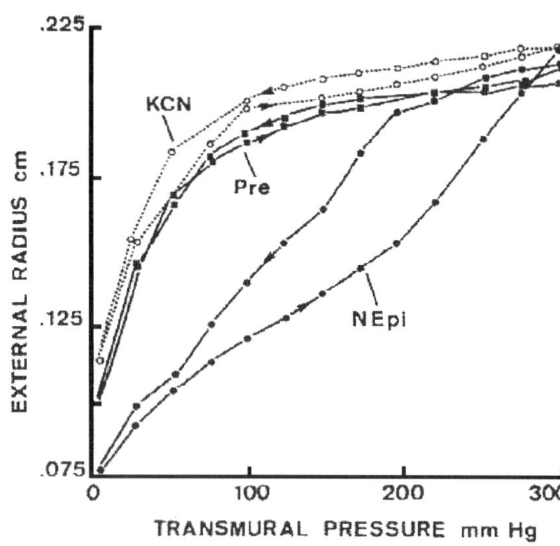

Fig. 2. Pressure-diameter curve for a dog carotid artery in relaxed state (Pre), after excitation of the muscle (NEpi), and after metabolic poisoning of the muscle (KCN) (adapted from Dobrin and Rovick (1969)).

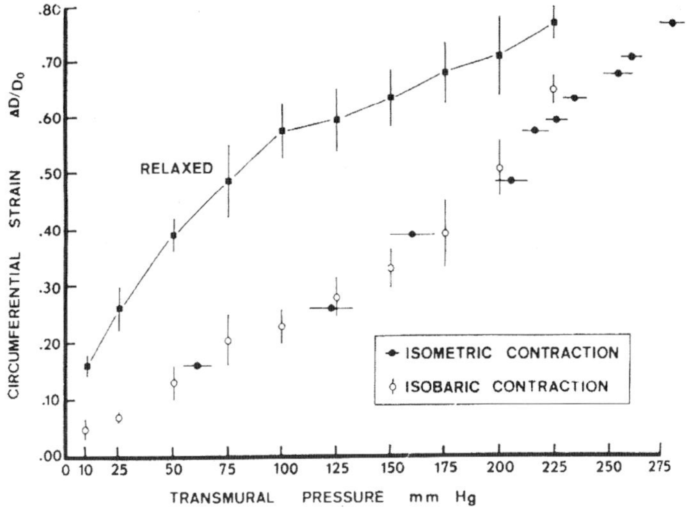

Fig. 3. Summary of isometric and isobaric contractions for 160 dog carotid arteries (adapted from Dobrin (1973)).

comparable in radius to adult human epicardial coronary arteries. The difference in circumferential strain between the relaxed state and maximal contraction with norepinephrine is about 35% at 100 mmHg transmural pressure, and diminishes slowly with physiological-range shifts of transmural pressure both above and below 100 mmHg. In a contracted state the total tensile circumferential force in a normal artery wall (required to counterbalance the internal arterial pressure) is diminished by the reduction in lumenal diameter, the thicknesses of the wall layers are increased (by conservation of wall tissue volume) and the tensile stresses in the intima and adventitia should be much reduced by the media taking over much (if not all) of the circumferential load. The frequent presence of arterial wall thickening due to atherosclerosis is likely to reduce the local change in radius between the relaxed and contracted states. However, even with stenosed coronary arteries there are indications that enough relaxation may be induced (for example, by sublingual nitroglycerine) to relieve ischemic pain through permitting greater blood flow.

3 Perceptions of Atherosclerosis Developed After 1980

Anecdotal information about fissures in plaque caps, dating from around 1966, being associated with local lumenal thrombi was extended by a few cardiac pathologists, Michael Davies of St George's Hospital Medical School in London in particular. Davies introduced fixation of occluded coronary arteries under a static perfusion pressure of 60 mmHg, seeking to make the tissue take a cross-sectional shape more typical of what it had in physiological conditions than obtained by fixing it with zero transmural pressure. In particular, he presented a longitudinal reconstruction, tediously obtained from serial transverse sections, at a meeting on 20 November 1980, with a title to the paper (co-authored by T. Thomas) when published (1981) (Davies and Thomas (1981)) which did not alert persons absent from the meeting that this was an important paper about plaque rupture. Breaks in the plaque cap were described as often large, from 4 mm to 15 mm long, based on a sample set of about a dozen coronary arteries. These lumenal breaks were of the order of the vessel diameter, and larger in size, but longitudinal with a helical component. The breaks could clearly provide an opening into the cavity in a lipid-filled plaque comparable in scale to the opening of a purse or wallet (an aspect given graphic emphasis by Davies (1998) in a later Atlas). The observation that thrombus could foot itself through that opening, and extend into the lumen from that anchorage (some extra space being provided by gross movement of lipid out into the lumen), while there are other examples of arterial obstruction by thrombus which could attach to a plaque without any opening in the cap being present, left the phenomenon of cap rupture perceived as a non-essential feature of acute symptoms.

Recognition of plaque fissuring or rupture as a possible event – and, more particularly, as a possibly frequent event – in the course of atherosclerotic arterial disease was not prompt. In his 1982 book on Atherosclerosis, the pathologist Neville Woolf (a close friend of Michael Davies) barely mentions it as such. Under the heading Plaque Haemorrhage, p. 69, he comments on some authors suggesting *'that the presence of iron within the plaque may be due to the entry of blood from the main arterial lumen into the substance of the plaque after splitting of the connective tissue 'cap'.'* Later, in the same chapter, for Figs. 4.32 and 4.33 he captions: *'A flap of cellular intimal tissue, somewhat resembling a hinged lid, has been raised from the plaque surface and there is an almost pure platelet thrombus occupying part of the lumen and extending into the plaque where platelet aggregates can be seen mixed with cholesterol clefts.'* In discussing plaque regression in the same book, Woolf (p. 291) comments: *'If these aims could be achieved then not only would stenosis of the arterial lumen be reduced but the chances of occlusive thrombosis occurring as a result of plaque splitting should be considerably diminished.'* Falk (1983) published a study describing plaque rupture with severe pre-existing stenosis. Davies and Thomas (1985) published a review paper with an unambiguous and more insistent title, *'Plaque fissuring – the cause of acute myocardial infarction, ischemic death and crescendo angina.'* After reviewing recent evidence, and referring to Peterson's work in the 1930s having given the perception that thrombus inside an arterial lesion could come from ruptures in the microcirculation which invades a lesion, Davies and Thomas remarked: *'It is difficult to escape the conclusion that the failure of pathologists to think in terms of dynamic processes and the belief what was seen at necropsy was an immutable reflection of events in life that occurred days before has seriously hindered the understanding by clinicians of how atheroma produces acute clinical symptoms.'*

Seymour Glagov, in his Abstract for the Vascular Disease Workshop, Aug 6-9, 1986, wrote: *'The lesion evolves not as a chronic, inexorable, proliferative response or a benign tumor, but as a defensive reaction preserving function as long as possible in the face of persisting extenuating circumstances'*, but with no mention of fissuring or rupture. At the same meeting Born and Richardson (1990) presented some measurements of mechanical properties for normal and atherosclerotic human coronary arteries. Keeny and Richardson (1987) published a paper comparing circumferential stress distributions in coronary artery walls with crescentic plaques for two cases, one where the plaque was lipid, and other where it was calcified. They found that while the stress distribution was affected by the extent of the plaque, by the nonlinear mechanical properties of the tissues, and by geometric factors associated with the mutual compatibility of local deformations, that in the presence of lipid plaque the maximum circumferential stress (due to the arterial blood pressure) was enhanced, and that lipid-filled plaques contra calcified plaques were at risk of tearing or fissuring. Richardson and Keeny (1989) reported measurements of anisotropy of human coronary artery intima.

Richardson et al. (1989) published a joint study which brought together analyses of stresses in computerized models of coronary arteries with plaques, predicting where the highest tensile stresses would occur in the caps of typical plaques due to arterial pressure, and making a comparison with a significant cohort of comparable clinical cases as to locations of plaque ruptures. This was taking a distinctly new direction by providing a possible physical explanation for formation of plaque cap fissures, with the obvious advantage that a well-known risk factor for heart attacks, hypertension, might be explained by it – higher blood pressures would give higher maximum tensile stresses in plaque caps, other things being equal, and with the increase in maximum stress rising faster than linearly with arterial pressure, just as the effect of the risk factor does. The authors mentioned that many fissures in atherosclerotic plaques in coronary arteries occurred where there were foam cells, likely to be of macrophage origin, and that this suggested the intimal layer may have been weakened by local factors including enzymatic degradation of the connective tissue matrix. (Jones et al. (2003) have recently surveyed literature relating to this.) This was followed up in a pair of papers from the same laboratories, Lendon et al. (1991) and Burleigh et al. (1992), the first comparing the mechanical behavior (extension and stress at failure) with macrophage content, and the second examining the structural components found to be affected, specifically collagen and sulphated glycosaminoglycans as the most significant, at different locations over human aortic plaque caps. Most citers of Lendon et al. (1991) have focussed on the presence of the macrophages and have sought matrix metalloproteinases, for several of which there are antigenic markers available. Far fewer investigators have sought to explore use of representations of the modified mechanical behavior.

One further paper from the group, Davies et al. (1993), addressed the relationship between plaque cap cellular content (macrophages and smooth muscle cells) and lipid pool proportion, showing that once the proportion of lipid is sufficiently high the macrophages significantly outnumber the smooth muscle cells, so the proteolytic potential can overwhelm the matrix-protein production potential. Davies (1998) subsequently prepared an excellent atlas on atherosclerotic plaques, with illustrations which demonstrated that once one knows what to look for it is possible to find clear examples of fissured plaque caps with various nuances on the basic concepts. Richardson (2002) has reviewed biomechanics of plaque rupture.

Taking aortic plaques only from adults over the age of forty, and excluding subjects with hypertension and diabetes, Lendon et al. (1993) observed the typical range of stress-strain relations (see Fig. 4). Proceeding with as large a group as practicable, Lendon et al. (1991) found from cohort of 40 plaques of which 18 had a tear in the cap (thereby classifying them as ulcerated) that the average extension required to reach mechanical failure was roughly double for the mechanical testing samples from the ulcerated plaques compared with the non-ulcerated plaques. Many citers of this seem to have been content to suppose that this is sufficient to demonstrate vulnerability being raised by the local influence of macrophages. However, in applying the tissue mechanical testing results to actual plaques, in the caps of which the mechanical properties are not necessarily uniform, it is important to know the location and relative extent of a portion of the plaque cap having modified mechanical properties in order to assess the vulnerability for it to lead to rupture.

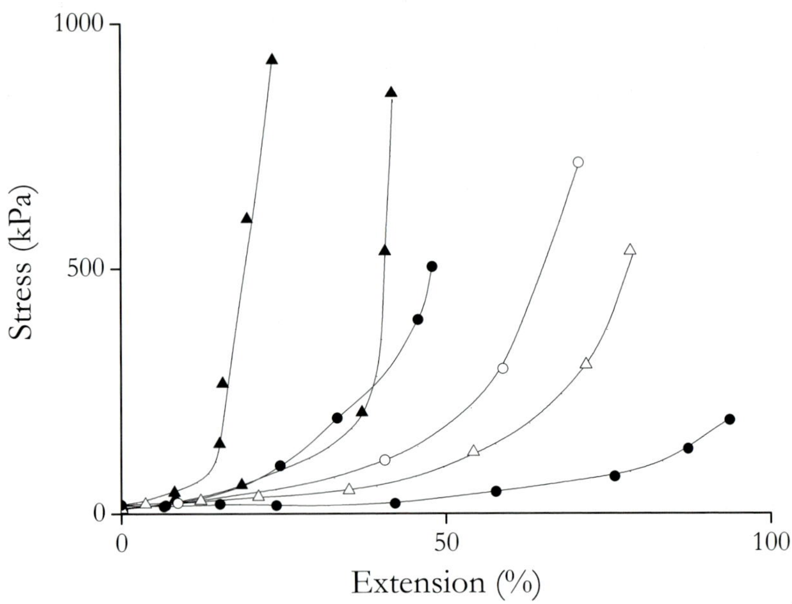

Fig. 4. Preliminary stress-strain relationships from non-ulcerated plaque caps (▲, $n = 2$), ulcerated plaque caps (● , $n = 2$), and areas of neighboring intima (△, ○ respectively) (adapted from Lendon et al. (1993)).

Thus, if there should be a small local patch (say 5% of the total area) in a plaque cap with the patch having corresponding mechanical properties associated with lytic damage – an extensibility beyond that required for rupture typical of non-ulcerated plaque – the 95% of plaque cap with stronger mechanical properties would serve to prevent the strain in the patch embedded in it from reaching the strain required for rupture under normal loading. It could be seen as a defensive reaction, in the sense suggested by Glagov, because it would allow load to be taken by the large proportion of stiffer plaque cap surrounding it and unaffected by the lytic effects, the small local patch being sheltered from a stress sufficient to rupture it. A region of a plaque cap characterized by a lowering of stress supported at modest extensions – a patch in the sense mentioned above – may deflect more readily in the direction perpendicular to its surface, while not showing large deflections circumferentially, under pulse pressure variations. Some methods are being applied to detect relative radial displacements (Schaar et al. (2004)) and one would expect that, other things being equal, relatively larger radial displacements would be easier to detect at lower mean arterial pressures for a given plaque of non-uniform mechanical properties. However, if the weaker patch increased in relative size the ability of neighboring plaque cap less affected by lytic effects to shelter it would be diminished. Burleigh et al. (1992) established that there are transverse variations in tissue constituents across an aortic plaque, as one goes from nearby intima, up over the shoulder to the center of a plaque cap, particularly in the collagen and GAGs content. Collagen content includes some freshly generated, as the smooth muscle cells can perform a repair process.

More broadly, it has come to be realized, largely in the post-1980 period, that remodeling of arterial walls in the presence of atherosclerotic lesions is itself an important aspect of atherosclerosis, and that positive remodeling (in which the lumen remains relatively constant in diameter but the wall thickens, thereby giving a locally larger perimeter to the adventitia) is just as liable to produce vulnerable lesions as is negative remodeling, and more dangerous in that it is not detected by angiography.

In reviewing the measurements of Burleigh et al. (1992), it needs to be recalled that ulcerated aortic plaques are likely to present tissue structures more mature – meaning further elaborated – than many coronary artery plaques. This is because they are not so harmful to their human hosts compared with how a typical ulcerated coronary artery plaque can be, and therefore can permit survival of their hosts and continue beyond plaque conditions that would likely be limiting in coronary arteries, in the sense that such plaques could cause severe cardiac events before progressing to the conditions found in some ulcerated aortic plaques. Plaques which ulcerate in the epicardial arteries are dangerous, because a thrombus can form and obstruct the lumen of the vessel. Plaques which ulcerate in the carotids are also hazardous, in providing emboli that can cause thrombotic strokes. Plaques which ulcerate in the aorta do not present the same clinical hazard to their hosts. If one assumes that some biochemical markers which appear in the bloodstream are largely independent of where the atherosclerotic sources are, the burden in the bloodstream from the aortic lesions – which are not the sources of acute clinical problems – may mask changes in mean values which arise from subtle changes in smaller arteries, such as the epicardial coronaries and the carotids.

4 Measurement of Mechanical Properties of Atherosclerotic Tissues

Atherosclerotic vessel walls differ from normal vessel walls because of the multiple effects of atherosclerosis on composition of the walls and on the local geometry of the vessels. Distinct effects on composition include: proteolytic damage; compensatory fibroblastic activity and extracellular assembly; net incorporation of substances (including lipids, calcium salts) not normally present to the same degree; alteration of cellular populations in the layers of the wall; micro-angiogenesis, for example, vaso vasorum, which can initiate other than from the lumen of a vessel. Effects on local geometry include: local thickening (hyperplasia) or thinning of vessel layers compared with their normal range; positive and negative remodeling; inclusion of concentrated pockets of lipids, or of fibro-fatty composites, or of hard calcium salts, often asymmetric with respect to the vessel axis; formation and re-composition of distinguishable components, such as plaque caps. Consequences of these atherosclerotic states on mechanical properties are clearly wide range. Furthermore, effects on the mechanical behavior of vascular structures in which these modified tissues lie are also wide range. The physiological (and sometimes traumatic) conditions to which these tissue structures are exposed are widely varied. In this context, a compact review must be selective.

Access to fresh human atherosclerotic artery tissue for measurement of mechanical properties is problematic. Tissue freshly removed inter-operatively or early post mortem is a goal for most freely-chosen tissue samples to provide the most reliable mechanical measurements. Endarterectomy specimens can give very fresh carotid artery tissue but its excision is determined by clinical considerations rather than optimization of testing specimen shape, and generally comes from persons in later decades of life. As untreated tissue ages in the early hours post mortem its mechanical properties may be modified as a consequence of autolysis of the cells it contains. In many circumstances post-mortem permission must be obtained to take human tissue for experimental purposes, and obtaining that permission may take so long that the freshness of the tissue for mechanical testing is compromised. Normal methods of tissue fixation, which introduce much cross-linking, alter mechanical properties – and dimensions – rather drastically. Abè et al. (1996) found relatively little data regarding arteries accessible to them for a data book on mechanical properties. However, recently there have been useful reviews of vessel constitutive models (Sacks and Sun (2003)) and the multiaxial mechanical behavior of biological materials, Greenleaf et al. (2003). More recently still, Holzapfel et al. (2004) have provided measurements for tissues of human iliac arteries.

In a Handbook of Bioengineering, Canfield and Dobrin (1987) provided the basics of small-strain and large-strain analyses, and also incremental elasticity in describing mechanical properties of blood vessels. They note Fung's long advocacy of exponential stress-strain relationships broadly for biological soft tissues. In the early 1980s the author, with colleagues, investigated such relationships for canine urinary bladders (Regnier et al. (1983)) with both whole-organ and uniaxial test-strip measurements, finding good and consistent strain-energy-function representation of exponential form. Some measurements were made also of the changes of mechanical properties of bladder tissue caused by timed exposure to collagenase (but not published at the time, and with C. Owens joining in the work); in these studies the bladders of male rats were removed suprapubically, emptied, rinsed with physiological solution and cut circumferentially to provide rings, two being taken from each bladder, then divided into 4 groups, each of 3 rings, and each incubated 20 hours at 18°C in Tyrode solution, group (i) as controls, (ii) with bacterial collagenase at 1 μgm/ml, (iii) same as (ii) but 10 times higher collagenase concentration, and group (iv) same as (ii) but with 100 times higher collagenase concentration; tissue of each ring was then tested as a uniaxial specimen while immersed in Tyrode solution at 37°C in a test chamber. The graphs of results showed group-average curves with very similar steep slopes over the upper 80% of the loading but offset at the 20% load level with elongation ratios of (i) 2.3, (ii) 2.5, (iii) 3.2, and (iv) 3.9.

In view of the effect of lytic enzymes (the series of matrix metalloproteinases (MMPs) in particular) on mechanical properties of plaque caps, and the fact that many investigators have examined the presence and distribution of MMPs in atherosclerotic tissue, there is the natural question of what these lytic enzymes would do to the mechanical properties of the stronger and more normal parts of plaque caps. A quick look in a reagents catalogue shows a problem: such individual MMPs as are available cost thousands-times more than collagenase, so unless the sizes of tissue specimens which would be bathed in them can be reduced, and still be tested for mechanical properties, the cost to explore this question would be very high. Further, to keep the period of exposure to the lytic agent down to a reasonable level *in vitro* the concentration has to be of the order of 1,000 times that believed typical *in vivo*, and the question of dose-response differences in that range is largely open.

Experiments to determine effects of a period of exposure to collagenase on mechanical tensile properties of human atherosclerotic plaque tissues were performed to a limited extent with samples of uninvolved intima in the same series of Lendon et al. (1991) and Burleigh et al. (1992). Multiple aortic intimal samples were taken from three subjects, aged 26, 69 and 75 years respectively, the former two subjects being males. Collagen incubation was for 12 hours at 37°C with collagenase from *Clostridium histolyticum* type VII (Sigma) in tris maleate buffer with a concentration of 160 units of enzyme per ml. After incubation, specimens were rinsed and placed in enzyme-free buffer containing ortho-phenanthroline. The stress-strain relations were measured for treated samples and for matched controls. Comparing results of the matched groups, the extension at failure was the mechanical measurement least affected by incubation with collagen. The stresses supported by the treated specimens rose with extension but rapidly diminished compared with the matched controls at the same extensions, except for the 26 year old, for whose tissues the stresses matched up to about 50% extension but then diverged to reach a maximum stress at failure for treated specimens only about 1/3 of the matched controls. For the specimens from the older subjects the ratio of maximum stresses reached was even smaller. The slopes of the stress-extension curves of the treated specimens in the upper reaches of their stress ranges were more mutually different than for the controls (looking at each of the latter beyond the last dip or plateau of the stress-elongation curves, the specimens not being 'pre-conditioned'). Recalling Lendon et al. (1993) (see Fig. 5), it can be noted that the cap tissue from near high concentrations of macrophages also had lower stresses at failure, but not as much reduced as found in the collagenase tests outlined above, and more particularly the extension at failure was roughly twice that found in the collagenase tests. This suggests that the macrophage-influenced weakening of plaque caps may not occur in a single lytic episode but over a period when some remodeling of the extra-cellular matrix can occur; and, if so, vulnerability may be found if caps in this period can be specifically detected.

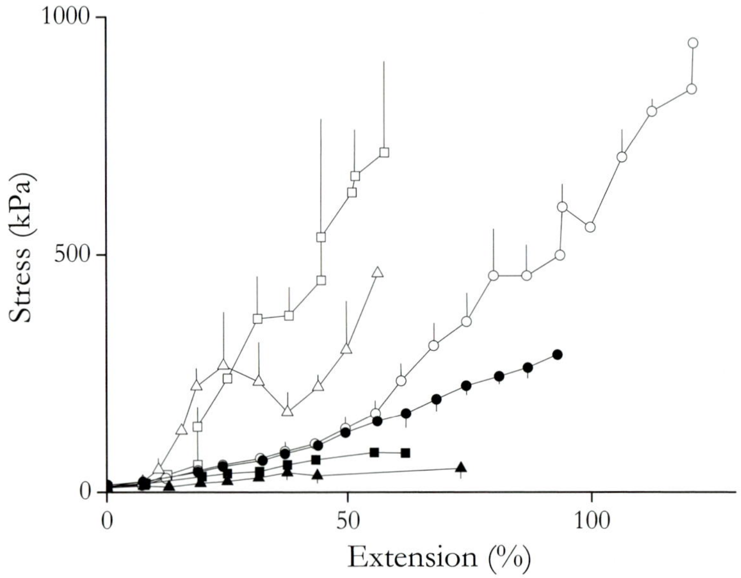

Fig. 5. The effect of collagenase on the stress-strain relationships of aortic intima of three subjects (adapted from Lendon et al. (1993)).

Because of the tissue-formation capabilities in vessel walls it is also useful to reflect on the matrix-formation processes as well. There are animal-model examples where considerable intimal thickening occurs in a matter of days following lumenal-surface injury, for example. However, *ex vivo* mechanical property measurements appear not to have been made for these.

Because fissuring of plaque caps is part of the tissue response to mechanical loading being examined, fracture mechanics – a distinct branch of mechanical property investigation – need to be considered. Problems in maintaining long life for machinery, and in designing successful armor for military vehicles and warships, provided an extra impetus to this to be investigated for application to metals. Over a few decades, starting from Griffith's 1920 hypothesis on cracks, and Irwin's 1948 use of work and energy methods in calculating crack propagation, with introduction of consideration of the work done in plastic deformation around a crack tip as well as the energy required to create new surface, a formal theory applicable to stiff elastic-plastic materials has developed, and dynamic effects have been included (Freund (1989)). Arterial tissue is more akin to polymer composites in its structure, and a fissure on a test specimen may develop with a form of craze (Williams (1980)). Theory matching practice for polymer composites is not as advanced as for metals. Additionally, live tissue with local mechanical deficits has the prospect of self-repair over time, a feature excluded in ordinary engineering fracture mechanics. Also, for metals there are non-destructive test methods for detecting advanced stages of crack propagation which, in limited circumstances, can provide warning of short functional life of components before failure. A corresponding method of non-destructive testing, applicable *in vivo*, which would indicate reduction in the ultimate strength of plaque tissue has not yet been developed.

The alternative to examining tissue as *ex vivo* samples is to examine it *in vivo*, and intact. This means, of course, there is not the same control over the loading and deformation conditions as there is *ex vivo*, nor is there the capability of separating the layers of a vessel wall, and therefore of measuring mechanical properties of the different layers separately. Imaging of arteries that are vulnerable, in the sense of potentially soon producing acute clinical morbidity or mortality, is primarily an issue of appropriate modalities to achieve a workable imaging resolution in space and time. Means for imaging elastic properties of biological tissues continue to be developed and evaluated (Schaar et al. (2004)). Ultrasound is currently a leading contender for critical arteries, although it has to be intravascular for *in vivo* use with coronary arteries. This is an approach which has considerable promise but faces growing competition.

5 Application of Mechanical Properties to Clinical Settings

The clinical settings of patients before an acute myocardial infarction are widely varied. However, the systolic blood pressure of all persons is affected by posture (lower when lying down than when sitting or standing, for example, in middle age they could be 85 mmHg versus 120 mmHg, for example). One of the few commonalities between persons is that they sleep lying down, and rise after waking to begin the activities of the day. Might rising from bed in the morning temporarily increase the risk of plaque rupture, because of the increase in systolic blood pressure which can raise the tensile stress in coronary plaque caps, and therefore the development of symptoms leading to identification of an myocardial infarction? Improved circumstantial evidence is given in a recent study by Tanaka et al. (2004) which extended previous observations of a roughly doubled incidence of myocardial infarctions in the three hours or so after rising by confirming that these involved plaque rupture. We do not know the speed of fissuring of plaque caps, which in any case may depend on specifics in each case, but recognizing thrombotic obstruction of a coronary artery could occur from initiation in a few minutes, fissuring may often take the order of an hour or so. Keeny and Richardson (1987) remark on the existence of intima-media shear stresses near the tips of crescentic lipid

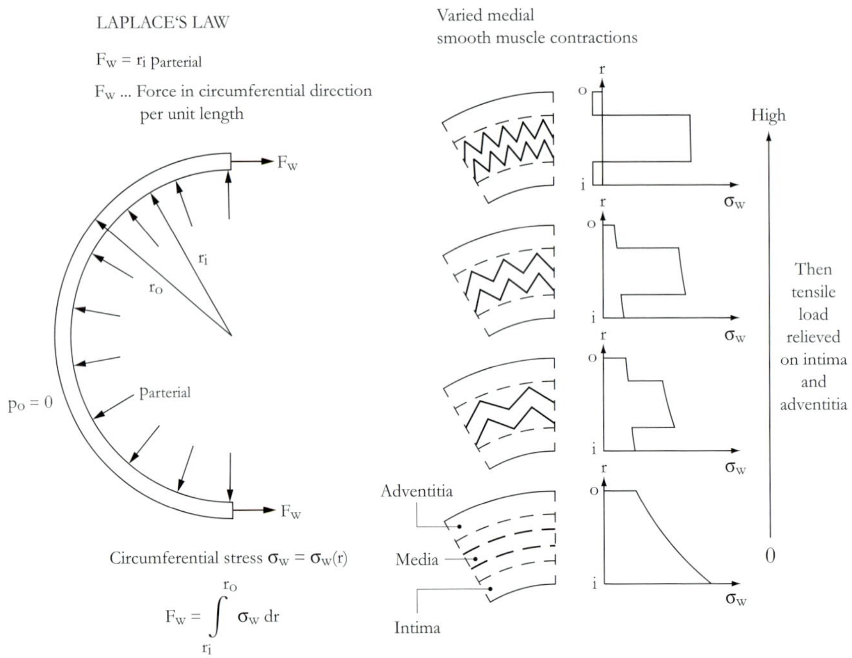

Fig. 6. Circumferential load and stress distribution in arterial walls. Left, upper and center: Laplace's Law and tensile load in wall. Left, lower: circumferential stress, as distributed radially across the thickness, must add up to be equal to the tensile load. Right: radial distribution of tensile stress in arterial wall depends on contractile action of medial smooth muscle, a range of variations is illustrated.

pools in plaques, and these may assist in enlargement of the circumferential arc of the plaque through (slow) delamination. The paradox that complete relaxation of the coronary artery media, while providing the most open lumen, also places higher tensile stresses on crescentic plaque caps for a given systemic blood pressure deserves further consideration (see Fig.6).

The prime clinical uses where some measure of mechanical tensile properties would likely improve outcome are in diagnostics (including follow-up of treatment), and in local decisions-to-treat (as for angioplasty and stenting). Measures of stiffening as well as thinning of caps of lipid-filled plaques would be useful information in diagnosis and decisions-to-treat. Follow-up, including checking on the success or failure of treatment beyond the question of decrease of any local stenoses, would likely be most demanding on fine resolution of changes in mechanical properties when tracking effects of pharmacological agents. This remains a challenge.

Acknowledgement. The preparation of this paper was supported in part by funds from the Jung Foundation, Hamburg, to whom many thanks.

References

Abè, H., Hayashi, K., and Sato, M., eds. (1996). *Data Book on Mechanical Properties of Living Cells, Tissues, and Organs.* New York: Springer-Verlag.

Born, G. V. R., and Richardson, P. D. (1990). Mechanical properties of human atherosclerotic lesions. In Glagov, S., Newman, W. P., and Schaffer, S., eds., *Pathology of the Human Atherosclerotic Plaque. Proceedings of Workshop on the Evolution of the Human Atherosclerotic Plaque*, 413–423. New York: Springer-Verlag.

Burleigh, M. C., Briggs, A. D., Lendon, C. L., Davies, M. J., Born, G. V. R., and Richardson, P. D. (1992). Collagen types I and III, collagen content, GAGs, and mechanical strength of human atherosclerotic plaque caps: spanwise variations. *Atherosclerosis* 96:71–81.

Canfield, T. R., and Dobrin, P. B. (1987). Static elastic properties of blood vessels. In Skalak, R., and Chien, S., eds., *Handbook of Bioengineering.* New York: McGraw-Hill.

Caro, C., Fitz-Gerald, J. M., and Schroter, R. C. (1969). Arterial wall shear and distribution of early atheroma in man. *Nature* 223:1159–1161.

Caro, C., Fitz-Gerald, J. M., and Schroter, R. C. (1971). Atheroma and wall shear. observations, correlation, and a proposal of a shear-dependent mass transfer mechanism for atherogenesis. *Philos. Trans. R. Soc. Lond. B* 177:109–159.

Davies, M. J., and Thomas, T. (1981). The pathological and microanatomy of occlusive thrombus formation in human coronary arteries. *Philos. Trans. R. Soc. Lond. B* 294:225–229.

Davies, M., and Thomas, A. (1985). Plaque fissuring – the cause of acute myocardial infarction, sudden death and crescendo angina. *Br. Heart J.* 53:363–373.

Davies, M. J., Richardson, P. D., Woolf, N., Katz, D. R., and Mann, J. (1993). Risk of thrombosis in human atherosclerotic plaques: role of extracellular lipid, macrophage, and smooth muscle cell content. *Br. Heart J.* 69:377–381.

Davies, M. J. (1998). *Atlas of Coronary Artery Disease.* Philadelphia: Lippincott-Raven.

Dobrin, P. B., and Rovick, A. A. (1969). Influence of vascular smooth muscle on contractile mechanics and elasticity of arteries. *Am. J. Physiol.* 217: 1644–1651.

Dobrin, P. B. (1973). Isometric and isobaric contraction of carotid arterial smooth muscle. *Am. J. Physiol.* 225:659–663.

Falk, E. (1983). Plaque rupture with severe pre-existing stenosis precipitating coronary thrombosis. Characteristics of coronary atherosclerotic plaque underlying fatal occlusive thrombi. *Br. Heart J.* 50:127–134.

Freund, L. B. (1989). *Dynamic fracture mechanics.* Cambridge, UK: Cambridge Univ. Press.

Greenleaf, J. F., Fatemi, M., and Insana, M. (2003). Selected methods for imaging elastic properties of biological tissues. *Ann. Rev. Biomed. Eng.* 5:57–78.

Holzapfel, G. A., Sommer, G., and Regitnig, P. (2004). Anisotropic mechanical properties of tissue components in human atherosclerotic plaques. *J. Biomech. Eng.* 126:657–665.

Jones, C. B., Sane, D. C., and Herrington, D. M. (2003). Matrix metalloproteinases: A review of their structure and role in acute coronary syndrome. *Cardiovasc. Res.* 59:812–823.

Keeny, S. M., and Richardson, P. D. (1987). Stress analysis of atherosclerotic arteries. Proc Ninth Annual Conf, Engineering in Medicine and Biology Soc. *IEEE* 3:1484–5.

Lendon, C. L., Davies, M. J., Born, G. V. R., and Richardson, P. D. (1991). Atherosclerotic plaque caps are locally weakened when macrophages density is increased. *Atherosclerosis* 87:87–90.

Lendon, C. L., Davies, M. J., Richardson, P. D., and Born, G. V. R. (1993). Testing of small connective tissue specimens for the determination of the mechanical behaviour of atherosclerotic plaques. *J. Biomed. Engr.* 15: 27–33.

Regnier, C. H., Kolsky, H., Richardson, P. D., Ghoniem, G. M., and Susset, J. G. (1983). The elastic behavior of the urinary bladder for large deformations. *J. Biomech.* 16:915–922.

Richardson, P. D., and Keeny, S. M. (1989). Anisotropy of human coronary artery intima. Proc 15th Ann NorthEast Bioengineering Conference. 205–206.

Richardson, P. D., Davies, M. J., and Born, G. V. R. (1989). Influence of plaque configuration and stress distribution on fissuring of coronary atherosclerotic plaques. *Lancet* 2(8669):941–944.

Richardson, P. D. (2002). Biomechanics of plaque rupture: progress, problems, and new frontiers. *Ann. Biomed. Eng.* 30:524–536.

Sacks, M. S., and Sun, W. (2003). Multiaxial mechanical behavior of biological materials. *Ann. Rev. Biomed. Eng.* 5:251–284.

Schaar, J. A., Regar, E., Mastik, F., McFadden, E. P., Saia, F., Disco, C., de Korte, C. L., de Feyter, P. J., van der Steen, A. F. W., and Serruys, P. W. (2004). Incidence of high-strain patterns in human coronary arteries - Assessment with three-dimensional intravascular palpography and correlation with clinical presentation. *Circulation* 109:2716–2719.

Tanaka, A., Kawarabayashi, T., Fukuda, D., Nishibori, Y., Sakamoto, T., Nishida, Y., Shimada, K., and Yoshikawa, J. (2004). Circadian variation of plaque rupture in acute myocardial infarction. *Am. J. Cardiol.* 93:1–5.

Texon, M., Imparato, A. M., Lord, J. W., and Helpern, M. (1962). Experimental production of arterial lesions. *AMA Arch. Intern. Med.* 110:50–52.

Texon, M. (1957). A hemodynamic concept of atherosclerosis, with particular reference to coronary occlusion. *AMA Arch. Intern. Med.* 99:418–427.

Texon, M. (1980). *Hemodynamic Basis of Atherosclerosis.* Washington DC: Hemisphere Publ. Corp.

Williams, J. G. (1980). *Stress analysis of polymers.* Chichester: Ellis Horwood, 2nd edition.

Woolf, N. (1982). *Pathology of Atherosclerosis.* London: Butterworth Scientific.

Towards a Computational Methodology for Optimizing Angioplasty Treatments with Stenting

G.A. Holzapfel[1,2], M. Stadler[2], Th.C. Gasser[1]

[1] Royal Institute of Technology, School of Engineering Sciences, 100 44 Stockholm, Sweden gh@hallf.kth.se, tg@hallf.kth.se

[2] Graz University of Technology, Institute for Structural Analysis – Computational Biomechanics, Austria
gh@biomech.tu-graz.ac.at

We propose a computational methodology that allows a set of stent parameters to be varied, with the aim of evaluating the difference in the mechanical environment within the wall before and after angioplasty with stenting. Proposed scalar quantities attempt to characterize the wall changes in the form of the contact pressure caused by the stent struts and the stresses within the individual components of the wall caused by the stent. These quantities are derived numerically and serve as indicators, which allow the determination of the correct size and type of the stent for each individual stenosis. In addition, the luminal change due to angioplasty may be computed as well. The methodology is demonstrated by using an image-based three-dimensional geometrical model of a postmortem specimen of a human iliac artery with a stenosis. To describe the material behavior of the artery, we considered mechanical data of eight different vascular tissues, which formed the stenosis. The constitutive models for the tissue components capture the typical mechanical characteristics under supra-physiological loading conditions. Three-dimensional stent models were parameterized in such a way as to enable new designs to be generated simply with regard to variations in their geometric structure.

1 Introduction

Balloon angioplasty is a well-established interventional procedure that aims to reduce the severity of atherosclerotic stenoses. Unfortunately, although most of the treatments are performed with stenting nowadays, the procedure still can fail due to resistance to dilation, thrombosis and stress-induced tissue growth such as neointimal hyperplasia. The latter biological process is triggered by inflammation, granulation and cellular proliferation, which are consequences of the changed stress situation in the arterial wall. Nevertheless, balloon angioplasty with stenting improves the short-term success rate and reduces the rate of restenosis in the longer term (see, for example, Betriu et al. (1999)). In recent years a number of approaches have been aimed at reducing or preventing restenosis; most notable is the use of drug-eluting stents, which demonstrate a remarkable efficacy and safety in preventing neointima formation and recurrent restenosis (see, for example, Degertekin et al. (2003)). It is important to note that the mechanical injury to the vessels is not eliminated with these recently developed procedures so that the present study remains meaningful for identifying the biomechanical nature of the stenting procedure.

Our present approach, which summarizes a recent study of Holzapfel et al. (2005), intends to predict the outcome of patient-specific angioplasty treatment on a computational basis. It allows insights into the development of mechanical stresses during inflation, and provides computational assessment of parametric stent designs. The angioplasty treatment can be performed with reference to several parameters, which can be varied on a computational basis to elicit the change of the mechanical environment during dilation and to find the optimal treatment for a particular stenosis type. As a starting point we alter (i) the types of stent cell, which are chosen to be similar to those in three commercial products, (ii) the geometry of stent struts, (iii) the radial mismatch between the smallest lumen diameter in the stenosis and the expanded stent diameter (nominal stent diameter), and analyze the relationship between these parameters. They were shown to be crucial for decreasing the risk of in-stent restenosis after stent placement. Note that two other contributions of this volume are also concerned with the important issue of stent design, i.e. the studies by Lally and Prendergast (2005) and Moreno et al. (2005).

The determination of an optimal treatment requires a quantification of the changes (here only mechanical) in the arterial wall during inflation and stent deployment. We propose the following strategy: first, evaluate mechanical field quantities, which are based on nonlinear finite element methods; second, evaluate indicators with the goal of replacing the field quantities by scalars, thereby enabling the establishment of a 'ranking' among the results. These indicators are derived numerically and allow a (better) characterization of

(i) the change of the intimal pressure caused by the struts of the stent,
(ii) the stress change within the arterial wall caused by the stent, and
(iii) the luminal change due to angioplasty.

The present approach allows us to analyze the 3D interaction between stent models and patient-specific wall models by considering image data and mechanical tests performed at body temperature, to quantify the change in the mechanical environment that occurs during stent placement, and to compare the effects of different parameters in order to identify optimal stent designs.

2 Methods

2.1 Indicators that Characterize the Outcome of Angioplasty with Stenting

Here we define three scalar indicators which should allow a judgement of the advantages and disadvantages of adapting one stent product or another.

Intimal Pressure Concentration Caused by the Stent Struts. The pressure, which occurs between the contact area of the stent struts and the intimal surface of the arterial wall is in part one determinant for restenosis. We propose that the (normalized) change of the intimal pressure caused by the stent struts, denoted by D_1, is expressed as

$$D_1 = \sum_{j=1}^{n_\mathrm{b}} \frac{\Delta p_j}{p_\mathrm{MAP}} \kappa_j, \quad \text{with} \quad \Delta p_j = p_{j,\mathrm{post}} - p_\mathrm{MAP}, \tag{1}$$

where j indicates the element number on the intimal surface, and n_b is the total number of elements forming the intimal surface. The factor κ_j is a measure for the surface of element j covered by the stent struts, while $p_{j,\mathrm{post}}$ is the contact pressure on the surface of element j after stenting due to the stent struts and the mean arterial pressure (for a more detailed description of $p_{j,\mathrm{post}}$ and κ_j see the Appendix in Holzapfel et al. (2005)). The mean arterial pressure before angioplasty is denoted by p_MAP. The pressure difference on the surface of element j after and before angioplasty is then given by Δp_j.

Stress Change within the Arterial Wall Caused by the Stent. To study the development of stresses within the arterial wall during angioplasty, we compute the Cauchy stress differences that arise in the finite elements. In particular, for an element with number i, the stress difference is

$$\Delta\sigma_i = (\sigma_\mathrm{post} - \sigma_\mathrm{pre})_i, \tag{2}$$

where $(\sigma_\mathrm{post})_i$ and $(\sigma_\mathrm{pre})_i$ denote the circumferential Cauchy stresses for element i after and before angioplasty, respectively. Note that $(\sigma_\mathrm{pre})_i$ is the Cauchy stress in element i at the mean arterial pressure.

The stress change within the arterial wall caused by the stent may then be defined by the measure (Holzapfel et al. (2002))

$$D_2 = \sum_{i=1}^{n} \Delta\sigma_i \Omega_i \Big/ \sum_{i=1}^{n} (\sigma_\mathrm{pre})_i \Omega_i, \tag{3}$$

where the stress difference is weighted by the volume Ω_i of the element i, in which it occurs. For the numerical studies we have assumed that only increases of wall stress during angioplasty are taken into account.

Luminal Change during Stenting. For the luminal change due to angioplasty, it is only meaningful to consider the smallest luminal area. Hence, we define a factor LG expressing the **L**uminal **G**ain, i.e.

$$LG = \frac{\Delta A}{A_{\text{pre}}}, \quad \text{with} \quad \Delta A = A_{\text{post}} - A_{\text{pre}}, \tag{4}$$

where A_{pre} is the smallest inner cross-section of the artery before angioplasty and stenting, and A_{post} is the cross-section at the same location as A_{pre} after angioplasty and stenting. Consequently, ΔA is the area difference and a measure of the area change during the angioplasty treatment.

2.2 Arterial Model

Material. An external iliac artery (female, 65.0 year old) was harvested during autopsy within 24 hours of death (detailed information about the related anamnesis is provided in Table 1 of Holzapfel et al. (2004), i.e. specimen I therein). The investigated artery has an atherosclerotic lesion of type V, according to Stary (2003), which contains mainly reparative smooth muscle cells and fibrous tissue and additionally two or more lipid pools of unequal size separated from each other by cells and fibrous tissue. Two cross-sectional macroscopic views of the stenotic iliac artery are provided in Fig. 1, sections A-A and C-C. The axial *in situ* pre-stretch, defined as the ratio of *in situ* length to *ex situ* length, was calculated to be 1.052. The use of autopsy material from human subjects was approved by the Ethics Committee, Medical University Graz, Austria.

In order to detect the 3D geometry for reconstruction purposes, we use hrMRI. The detailed procedure and the associated histological analyses required to identify the underlying tissue type are provided in Holzapfel et al. (2004). For this lesion eight different tissue types were considered: the nondiseased intima I-nos ('nos' stands for 'non-atherosclerotic'), fibrous cap I-fc (fibrotic part at the luminal border), fibrotic intima at the medial border I-fm, calcification I-c, lipid pool I-lp, nondiseased media M-nos, diseased fibrotic media M-f and adventitia A (Holzapfel et al. (2000b, 2002, 2004)). This classification has resulted in a separation of the diseased vessel wall (compare with Fig. 1) that is (solid) mechanically representative and that covers the gross histological composition of the stenosis. This separation is also physically feasible using surgical instruments.

Geometric Model. For each scanned cross-section the borders of the arterial components were traced automatically by a set of points. These points were then fitted by NURBS curves using a least-squares fitting procedure. Finally, the curves were combined along the arterial axis in order to identify the boundary surfaces of the different tissue components.

The geometry of the harvested stenotic iliac artery is shown in Fig. 1. In order to display the path of the lumen over the whole length of the specimen,

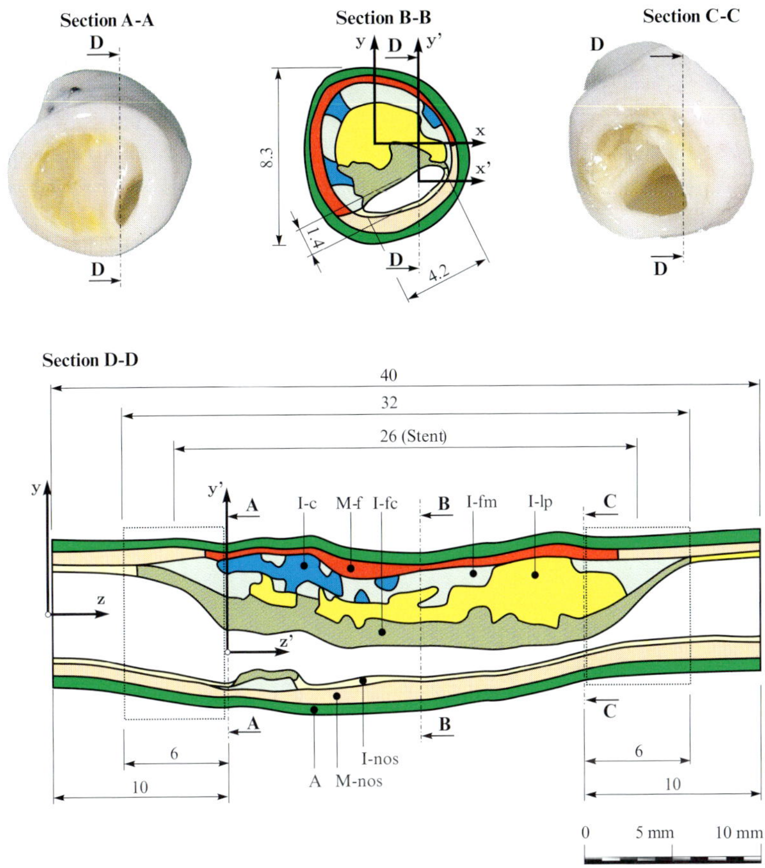

Fig. 1. Sections of the analyzed external iliac artery. Section B-B is the region with the smallest lumen diameter of 1.4 mm. The tissue components are: adventitia (A), nondiseased media (M-nos), nondiseased intima (I-nos), fibrous cap (I-fc), lipid pool (I-lp), calcification (I-c), fibrotic intima at the medial border (I-fm) and diseased media (M-f). The regions for the analysis of edge effects, as described in Section 3.2, are denoted by the areas with dotted frames (Section D-D).

which is tortuous, the cutting plane for Section D-D was chosen to be out of the center (for the location of D-D compare with the other sections).

Mechanical Tests, Physical and Numerical Models. *In vitro* mechanical tests were performed on a computer-controlled, screw-driven high-precision tensile testing machine. For the determination of the passive, quasi-static stress-stretch response of the individual tissue components, rectangular strip samples with axial and circumferential orientations were excised from the specimen and immersed in a calcium-free physiological 0.9% NaCl solution at 37°C ± 0.1°C. The lipid pool I-lp was not tested because of its 'butter-like' consistency. For the investigated artery the size of the tissue components I-nos and I-fm was not big enough to prepare appropriate strip samples for testing. A description of the mechanical tests and the stress-stretch responses obtained for 7 different tissue types from 9 human high-grade stenotic iliac arteries are documented in Holzapfel et al. (2004). All tissue samples were stretched far beyond the physiological loading domain up to failure so that the range of deformations induced by the stent was captured.

In order to model the material response each component of the stenotic artery is assumed to behave as a (nearly) incompressible composite reinforced with collagen fibres (for an overview of the biomechanics and mechanobiology of arterial walls see the books by, for example, Humphrey (2002) and Holzapfel and Ogden (2003)). The statistical distributions of fibre directions are characterized by two symmetrically arranged families of fibres whose mean directions are represented by the unit vectors $\mathbf{M}$ and $\mathbf{M}'$, and it is these directions that render the material properties anisotropic. As in Holzapfel et al. (2000a) we consider a strain-energy function relative to the reference configuration in the decoupled form

$$\Psi = U(J) + \overline{\Psi}(\bar{I}_1, \bar{I}_4, \bar{I}_6), \tag{5}$$

where the (mathematically motivated) function U is a purely volumetric contribution and $\overline{\Psi}$ is a purely isochoric contribution to the strain energy Ψ. In (5) $J = (\det \mathbf{C})^{1/2}$ is the local volume ratio, while $\mathbf{C} = \mathbf{F}^{\mathrm{T}}\mathbf{F}$ denotes the right Cauchy-Green tensor and $\mathbf{F}$ is the deformation gradient. In addition, $\bar{I}_1 = \operatorname{tr} \overline{\mathbf{C}}$ is the first invariant of the modified right Cauchy-Green tensor $\overline{\mathbf{C}} = J^{-2/3}\mathbf{C}$, and $\bar{I}_4 = \overline{\mathbf{C}} : \mathbf{M} \otimes \mathbf{M}$, $\bar{I}_6 = \overline{\mathbf{C}} : \mathbf{M}' \otimes \mathbf{M}'$.

For the function $\overline{\Psi}$ we use the particularization

$$\overline{\Psi}(\bar{I}_1, \bar{I}_4, \bar{I}_6) = \mu(\bar{I}_1 - 3) + \frac{k_1}{2k_2} \sum_{\alpha=4,6} \{\exp\{k_2[(1-\rho)(\bar{I}_1 - 3)^2 + \rho(\bar{I}_\alpha - 1)^2]\} - 1\}, \tag{6}$$

where $\mu > 0$ and $k_1 > 0$ are material parameters with the dimension of stress, and $k_2 > 0$, $\rho \in [0, 1]$ are dimensionless material parameters. The function $\overline{\Psi}$ is sufficiently general to capture the basic anisotropic features of tissue responses of human atherosclerotic plaques, as observed in the experiments of Holzapfel et al. (2004). Note that for the limit $\rho = 1$ we recover the constitutive equation proposed in Holzapfel et al. (2000a), while for the limit $\rho = 0$ we obtain an (isotropic) constitutive equation similar to that documented in Demiray (1972). In addition, we assume that the anisotropic term in (6) contributes when either $\bar{I}_4 > 1$ or $\bar{I}_6 > 1$ or both. Subsequently we use the same form of strain-energy function (6) for each tissue component but with a different set of material constants and different direction vectors $\mathbf{M}$ and $\mathbf{M}'$. For specific material parameters and the associated error measures see Holzapfel et al. (2005), in particular Table 1 therein, and for an extension of the model to consider distributed collagen fibre orientations see Gasser et al. (2005).

Based on the NURBS geometry a mesh with 945, 3150 and 9765 3D-elements was generated. We used eight-node isoparametric hexahedral elements. Only for the elements adjacent to the master contact-surface we used 20-node elements, which form an interface between the smooth contact surface description and the finite element discretization (Stadler and Holzapfel (2004)).

2.3 Stent Model

Geometry and Structure. Parametric design is a useful technique in engineering practice when products are tailored to fit specific customer needs or when numerical optimization is used to generate the 'optimal' design of a product. We propose an efficient strategy of stent parametrization which is able to parameterize the geometry of the stent cells, the geometry of the struts (with width s_{w}, measured in the circumferential direction, and with thickness s_{t}, measured in the radial direction), and the overall dimensions of the stent (i.e. nominal diameter D, number of cells in the axial direction, m, and in the circumferential direction, n; D refers to the stent diameter achieved at any axial position, while the balloon is fully inflated). For the parametrization of the stent cell the developed software requires information about: (i) the cell type, (ii) the geometric quantities to be parameterized (see the upper panels of Fig. 2; each dimension, denoted by lowercase letters, represents a parameter),

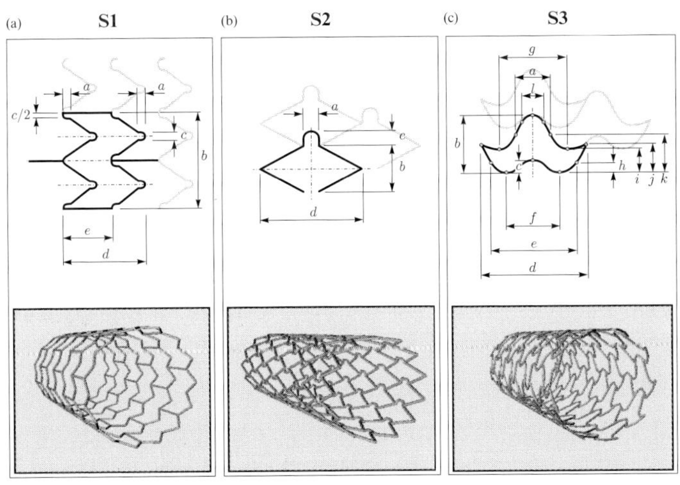

Fig. 2. Three stent geometries described by a number of (geometric) parameters, denoted by lowercase letters (upper panels). The cell types are based on products that are (or were) available commercially: (a) Multi-Link Tetra™ stent (Guidant): S1, (b) NIROYAL™-Elite stent (Boston-Scientific): S2, (c) InFlow-Gold-Flex™ stent (InFlow Dynamics): S3. The lower panels show the generated 3D views of the different stents.

Table 1. Geometrical parameters *a*-*l* of the three cell types S1, S2, S3, as illustrated in Fig. 2. For each cell type two parameters (*b* and *d*) are given, indicated by the symbol •, while all other parameters are functions of these two.

	a	b	c	d	e	f	g	h	i	j	k	l
S1	0.0	•	0.0	•	$0.6d$	–	–	–	–	–	–	–
S2	$0.15d$	•	$0.27b$	•	–	–	–	–	–	–	–	–
S3	$0.3d$	•	$0.2b$	•	$0.77d$	$0.46d$	$0.61d$	$0.17b$	$0.42b$	$0.5b$	$0.69b$	$0.15d$

and (iii) a set of rules describing how the parameters depend on each other. As input data we have only used the two dimensions b and d, while all other dimensions are based upon these two. The lower panels of Fig. 2 show the generated 3D views of the different stents.

We investigate three different types of stent cells that are based on products that are (or were) available commercially. In particular, we employ shapes of stent cells used in products such as (a) the Multi-Link-Tetra™ stent (Guidant), (b) the NIROYAL™ Elite stent (Boston-Scientific) and (c) the InFlow™-Gold-Flex stent (InFlow Dynamics). For subsequent use we will refer to these stent types as S1, S2 and S3, respectively (see Fig. 2).

Material Modeling and Finite Element Discretization. For all three stents we have adopted the material properties for stainless steel ($E = 2.0 \cdot 10^5\,\mathrm{N/mm^2}$, $\nu = 0.3$). The plastic behavior of the stent is described by a yield stress $\sigma_\mathrm{y} = 2.1 \cdot 10^3\,\mathrm{N/mm^2}$. For the finite element formulation we have used a neo-Hookean model for the elastic regime and a J_2-model for the plastic regime. For the discretization of the stents we have used hexahedral finite elements with quadratic shape functions. The parameters defining the different stent cells are summarized in Table 1.

The strut width and the strut thickness were assumed to be $s_\mathrm{w} = 140\,\mu\mathrm{m}$ and $s_\mathrm{t} = 450\,\mu\mathrm{m}$, respectively. For each of the stent types we have assumed four different nominal stent diameters D. The ratio of the metal surface in contact and the arterial wall surface, i.e. the metal-to-artery ratio R_ma at the nominal diameter D, was assumed to be constant; $R_\mathrm{ma} = k s_\mathrm{w} n/(D\pi - k s_\mathrm{w} n) = 0.15$, where k denotes the number of struts for one stent cell in the circumferential direction ($k = 5$ for S1, and $k = 2$ for S2 and S3).

The number of cells in the axial direction is $m = 7$ for all stent types, which gives a stent length of about 26.0 mm. The number of cells (n) in the circumferential direction was computed from the given metal-to-artery ratio, and the geometric parameter b is given as $D\pi/n$. Specific parameters for the three cell types are summarized in Table 2.

2.4 Finite Element Simulation

Stent-artery Interaction. As a rule of thumb, in clinical practice, the nominal diameter D of an expanded stent should be similar to the lumen diameter of the related healthy arterial region, $\overline{L}$ say — in our case 6.0 mm.

Table 2. Specific parameters for the three cell types S1, S2, S3. Nominal stent diameter D, and two geometric parameters b and d in millimeters. Numbers of cells in the axial direction m, and in the circumferential direction n.

	S1				S2				S3			
D	3.0	4.0	5.0	6.0	3.0	4.0	5.0	6.0	3.0	4.0	5.0	6.0
b	4.7	6.28	5.24	4.17	2.35	2.09	2.24	2.09	2.35	2.09	2.24	2.09
d	5.65				3.29				3.29			
m	7				7				7			
n	2	2	3	4	4	6	7	9	4	6	7	9

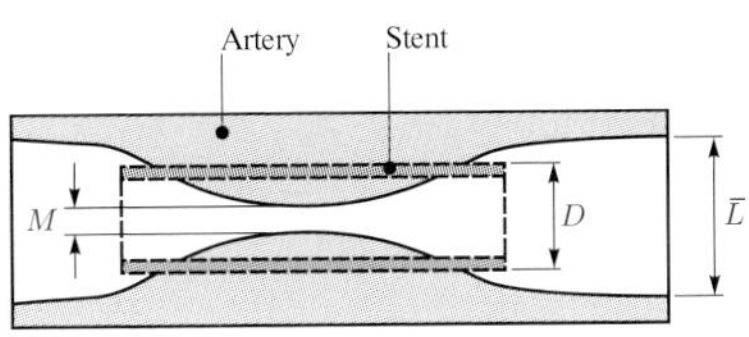

Fig. 3. Mismatch $\Delta M = D - M$ between the smallest lumen diameter M in the stenosis and the diameter D of the expanded stent. The lumen diameter of the healthy arterial region is characterized by $\overline{L}$.

Therefore, during stent expansion, there appears to be a radial mismatch $\Delta M = D - M$ between the smallest lumen diameter M in the stenosis (in our case $M = 1.4\,\text{mm}$) and the expanded nominal stent diameter D. For an illustration of the geometrical situation see Fig. 3. Hence, according to Table 2, from the diameters D we get four different values of the mismatch ΔM, namely for 1.6, 2.6, 3.6, 4.6 (in millimeters). Consequently, a value of $\Delta M = 4.6\,\text{mm}$ reestablishes the original lumen of $\overline{L} = 6.0\,\text{mm}$, provided that no recoil occurs. Note that the stent may also be potentially subject to migration depending on the radial mismatch.

In our simulation we ignored the balloon and applied the (deformation dependent) pressure load directly on the stent struts. For the interaction between the stent and the arterial wall we have employed a contact algorithm, which is based on subdivision surfaces for the interpolation (Stadler and Holzapfel (2004)). For details of the application of boundary conditions for the artery and the stent see Holzapfel et al. (2005).

Loading Procedure. In order to simulate the physiological loading situation two loads are applied on the artery: (i) mean arterial pressure of 100 mmHg (i.e. 13.3 kPa) on the entire inner surface of the artery by (displacement dependent) surface loads, (ii) axial pre-stretch of 1.052 for all nodes located on one end face of the artery by displacement boundary conditions (all nodes on the other end face have fixed z-displacements). After applying these physiological loads the displacement boundary conditions on one end face of the artery are replaced by equivalent reaction forces. This replacement leads to a physically more realistic simulation during subsequent loading.

To expand the stent-artery system, we apply displacement dependent surface loads on the inner side of the stent struts. After full expansion, the pressure loads on the struts are removed gradually. Throughout this loading process, the blood pressure of 100 mmHg on the inner wall of the artery remains.

3 Results

As a representative example, Fig. 4 shows the numerical results in the form of circumferential Cauchy stress distributions. The cutting planes indicate stresses before (Fig. 4(a)) and after stenting (Fig. 4(b)) at locations where

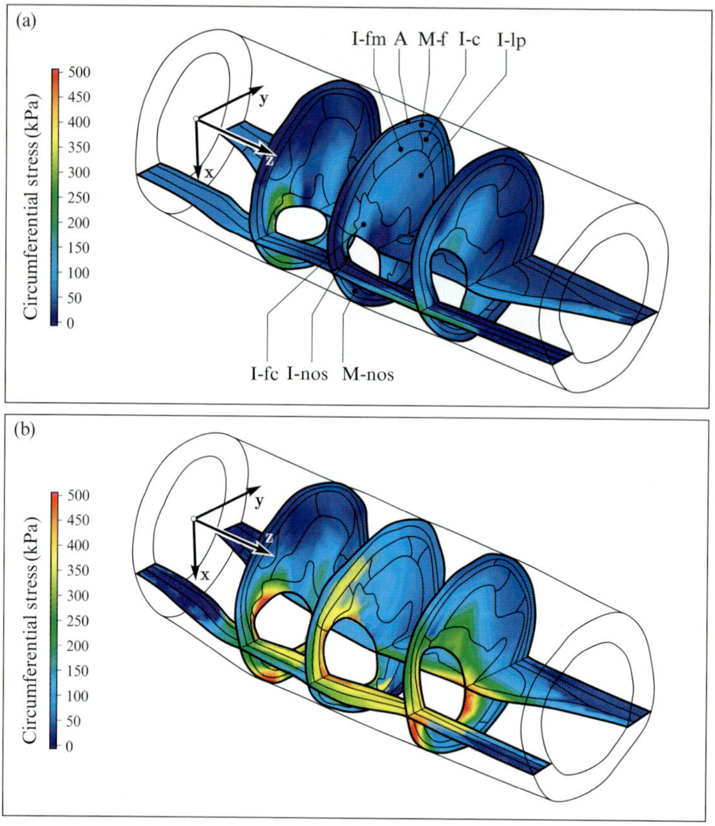

Fig. 4. Circumferential Cauchy stress distributions in the arterial wall before (a), and after stenting for stent S1 at $\Delta M = 3.6$ (b). The only load applied in both configurations is the mean arterial pressure of 100 mmHg. Stresses are projected onto cutting planes at $x = 2.0$ mm, and $z = 12.0$ mm, $z = 20.0$ mm, $z = 18.0$ mm. The cutting planes were selected at locations where changes in stress due to stenting are most pronounced.

changes in stress due to stenting are most pronounced. For the image shown in Fig. 4(b) the stent S1 was used with $\Delta M = 3.6$. As can be seen, stenting induces large stress concentrations in the nondiseased area, while the diseased area remains mainly unchanged. Within the diseased part, the fibrous cap (I-fc) becomes remarkably stressed. High stress in this plaque component may lead to tissue failure and to an increased risk of thrombus formation (Holzapfel et al. (2002)).

We characterize the mechanical effect after deployment and expansion of the stent up to the diameter D by varying the strut thicknesses for the entire stent, and the strut thicknesses for the end cells of the stent only. This parameter study is performed for the four values of mismatch. The smallest value $\Delta M = 1.6\,\text{mm}$ is such that the diameter D of the expanded stent is smaller than the lumen diameter of the healthy arterial region $\overline{L}$, while for the largest value $\Delta M = 4.6\,\text{mm}$ the diameter D is larger than $\overline{L}$, and the stent over-stretches the artery significantly.

3.1 Influence of Mismatch ΔM

We have computed the influence of the mismatch ΔM on the intimal pressure concentration caused by the stent struts (D_1), the stress change within the arterial wall caused by the stent (D_2), and the lumen gain during stenting (LG). For the stent geometries S1, S2, S3 the finite element results are illustrated in Figs 5(a),(b) by solid lines with different thicknesses, and denoted as 'original strut thickness' (abbreviated by 'orig st'). We indicate a particular value ΔM for a specific stent by a circle, which is filled or partly filled. For example, a circle filled up by a quarter indicates the mismatch $\Delta M = 1.6\,\text{mm}$.

3.2 Influence of Strut Thickness

Reduction of Strut Thickness for the Entire Stent. The study documented in Section 3.1 was repeated with a half of the stent strut thickness s_t for all stent types S1, S2 and S3, i.e. $s_t = 225\,\mu\text{m}$. The finite element meshes and all other geometrical data are as above. The results are illustrated in Figs 5(a),(b) by dashed lines with different thicknesses, and denoted as 'half strut thickness' (abbreviated by 'half st').

Reduction of Strut Thickness for the End Cells of the Stent. For all investigated stent designs we parameterize the strut thickness s_t of the cells at both ends of the stent, i.e. for region d. In particular, for these end regions we have reduced s_t to 50% of its original value ($s_t = 225\,\mu\text{m}$), and have repeated the study with the finite element meshes and all other geometrical data, as used in Section 3.1. Since the effect of this modification will mainly occur around the stent ends, we have only considered the regions around the stent ends for the computation of D_1, D_2, LG, according to eqs (1), (3), (4), respectively. In particular, the finite elements considered for the analysis of

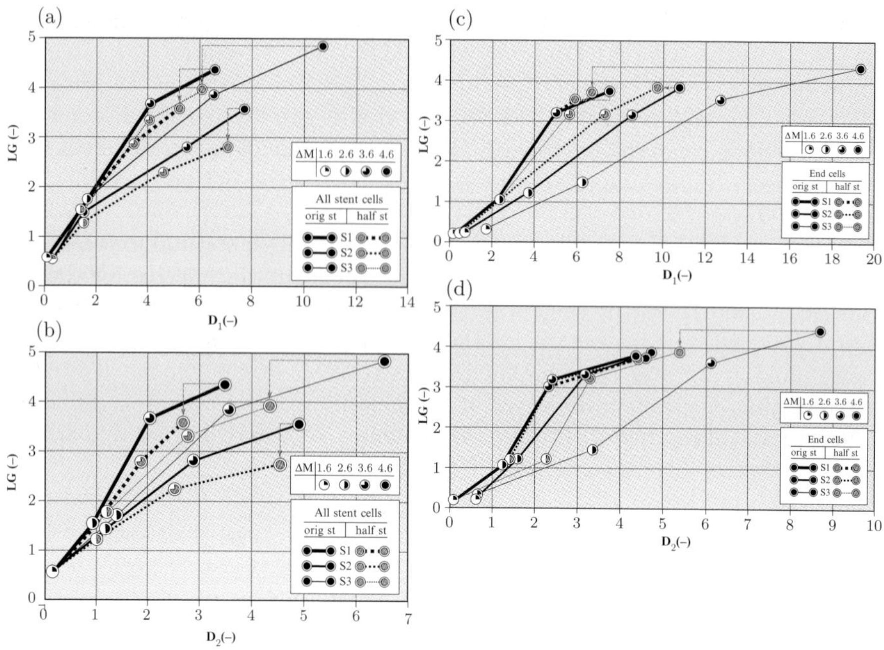

Fig. 5. Influence of mismatch $\varDelta M = D - M$ and the different stent geometries S1, S2, S3 on the three scalar indicators D_1, D_2, LG. A particular value of mismatch for a specific stent is indicated by a circle, which is filled or partly filled. Solid lines indicate the 'original strut thickness' (orig st): in (a) and (b) dashed lines indicate the results obtained by reducing the strut thickness for the entire stent by a half (half st), while in (c) and (d) dashed lines indicate the results obtained by reducing the strut thickness for the end cells of the stent by a half (half st). Only the finite elements inside the dotted frames, as defined in Section D-D of Fig. 1, were considered for the computation of the indicators. For each stent type, arrows indicate the change from 'orig st' to 'half st'-data at $\varDelta M = 4.6$.

the edge effects are from the inside of the dotted frames defined in Section D-D of Fig. 1. The numerical results are shown in Figs 5(c),(d), where solid lines indicate the 'original strut thickness' ('orig st'), and dashed lines the results obtained by reducing the strut thickness for the end cells of the stent by a half (abbreviated by 'half st').

4 Discussion and Conclusion

Three-dimensional finite element simulations of such complex boundary-value problems produce, in general, large amounts of stress and strain data, which is not a straightforward task to interpret. For this reason we have introduced three scalar indicators D_1, D_2 and LG, which provide a measure of the quality of a completed angioplasty procedure by comparing the mechanical situation before the treatment with that after.

General observations of the numerical results obtained, as indicated in Fig. 5, are as follows: there seems to be a global tendency for D_1 and D_2 to increase faster in the region of larger values of LG than in the region of smaller values. Referring to Figs 5(a),(b), for all values of mismatch ΔM, the use of stent S3 (InFlow™-Gold-Flex) leads to the largest and S1 (Multi-Link-Tetra™) to the lowest values of D_1 and D_2. It is interesting to relate this result to the clinical study documented in Kastrati et al. (2001b), which has shown a high restenosis rate of 50.3% for stent S3, and a low restenosis rate of 20.0% for stent S1. In addition, stent S1 gives almost the same lumen gain as S3, but much less change in the mechanical environment of the arterial wall expressed through D_1 and D_2. For example, for $\Delta M = 4.6\,\mathrm{mm}$, stent S1 induces only about half the stress as that induced by stent S3, while S1 achieves approximately 90% of LG of S3.

The dashed lines in Figs 5(a),(b) indicate the results obtained by reducing the strut thickness for the entire stent by a half. For all stent types, the effect of this change becomes more apparent with increasing ΔM. The largest mechanical change in the arterial wall is observed for stent S3 at $\Delta M = 4.6$. Remarkably, by reducing the strut thicknesses of the entire stent S3 by 50% it performs similarly to stent S1, at least for the LG-D_1 ratio. The least advantageous effect of changing the thickness of the struts is achieved for stent S1 at $\Delta M = 3.6$; D_1 and D_2 are almost unchanged, while LG decreases significantly, which is an obvious disadvantage.

The studies of Kastrati et al. (2001a) and Pache et al. (2003) point out that the stent with thinner struts elicits less angiographic restenosis than the thicker-strut stent. From Figs 5(a),(b) we learn that thinner struts lead to a decrease in the values D_1 and D_2, and hence to a decrease in the intimal pressure concentration and the stress change within the arterial wall due to the stent, which produce less local damage. However, we also learn from Figs 5(a),(b) that stents with thinner struts also lead to less lumen gain LG since thinner struts reduce the overall stiffness of the stent.

Evidence is emerging that the abrupt compliance mismatch, which occur at the junction between both ends of the stent and the host arterial wall, disturbs the vascular hemodynamics and the natural wall stress distribution (Berry et al. (2002), Moreno et al. (2005)). In particular, the stress distributions show peaks close to both ends of the stent due to the phenomenon of 'dog-boning' of the stent, analyzed numerically in, for example, Holzapfel et al. (2002). The phenomenon of 'dog-boning' might be avoided by (i) using a shaped balloon, (ii) varying the geometry of the stent cells over the stent length, or (iii) varying the thickness of the struts in certain areas of the stent. In Berry et al. (2002), approach (ii) was chosen to design compliance matching stents. Although approach (iii) seems to be difficult for producing appropriate stent devices (Auricchio et al. (2001)), some manufacturers have started to create stents with variable thickness struts. One example is the VTS™(Variable Thickness

Strut) technology of Guidant, which is designed to provide enhanced flexibility, deliverability, and surface area coverage while maintaining radial strength. Therefore, in our study we have chosen approach (iii). Figures 5(c),(d) show the influence of the mismatch and the different stent geometries on D_1, D_2, LG when the strut thickness for the end cells of the stent is reduced by a half (dashed lines). It turns out that the largest improvement regarding D_1 and D_2 is achieved for stent S3, and the smallest improvement is achieved for stent S1.

Based on the stenosis investigated it is revealed that the best results for all three indicators can be achieved with stent S1, which has a similar structure to the Multi-Link-Tetra™ stent. In addition, the parametric studies reveal that certain modifications for one stent lead to more advantages than for another. For example, reducing the strut thickness by 50% in the end cells drastically reduces the values of D_1 and D_2 for stent S3 by almost maintaining the value of the lumen gain LG, while it has almost no effect for stent S1.

It should be mentioned that the results may be (completely) different for other types of arteries. Although general applicability of the results is not possible, a general application of the proposed methodology for assessing parametric stent designs computationally is, however, possible. The proposed computational methodology has the potential to provide a scientific basis for optimizing treatment procedures and stent geometries and materials, to help stent designers examine new stent designs 'virtually', and to assist clinicians in choosing the most suitable stent for a particular stenosis.

Acknowledgements. Financial support for this research was provided by the *Austrian Science Foundation* under *START-Award Y74-TEC*.

References

Auricchio, F., Loreto, M. D., and Sacco, E. (2001). Finite-element analysis of a stenotic revascularization through a stent insertion. *Comput. Meth. Biomech. Biomed. Eng.* 4:249–264.

Berry, J. L., Manoach, E., Mekkaoui, C., Rolland, P. H., Moore Jr., J. E., and Rachev, A. (2002). Hemodynamics and wall mechanics of a compliance matching stent: *in vitro* and *in vivo* analysis. *J. Vasc. Interv. Radiol.* 13: 97–105.

Betriu, A., Masotti, M., Serra, A., Alonso, J., Fernandez-Aviles, F., Gimeno, F., Colman, T., Zueco, J., Delcan, J. L., Garcia, E., and Calabuig, J. (1999). Randomized comparison of coronary stent implantation and balloon angioplasty in the treatment of de novo coronary artery lesions (START): a four-year follow-up. *J. Am. Coll. Cardiol.* 34:1498–1506.

Degertekin, M., Regar, E., Tanabe, K., Smits, P. C., van der Giessen, W. J., Carlier, S. G., de Feyter, P., Vos, J., Foley, D. P., Ligthart, J. M.,

Popma, J. J., and Serruys, P. W. (2003). Sirolimus-eluting stent for treatment of complex in-stent restenosis: the first clinical experience. *J. Am. Coll. Cardiol.* 41:184–189.

Demiray, H. (1972). A note on the elasticity of soft biological tissues. *J. Biomech.* 5:309–311.

Gasser, T. C., Holzapfel, G. A., and Ogden, R. W. (2005). Hyperelastic modelling of arterial layers with distributed collagen fiber orientations. *J. R. Soc. Interface.* in press.

Holzapfel, G. A., and Ogden, R. W., eds. (2003). *Biomechanics of Soft Tissue in Cardiovascular Systems.* Wien – New York: Springer-Verlag.

Holzapfel, G. A., Gasser, T. C., and Ogden, R. W. (2000a). A new constitutive framework for arterial wall mechanics and a comparative study of material models. *J. Elasticity* 61:1–48.

Holzapfel, G. A., Schulze-Bauer, C. A. J., and Stadler, M. (2000b). Mechanics of angioplasty: Wall, balloon and stent. In Casey, J., and Bao, G., eds., *Mechanics in Biology.* New York: The American Society of Mechanical Engineers (ASME). AMD-Vol. 242/BED-Vol. 46, pp. 141–156.

Holzapfel, G. A., Stadler, M., and Schulze-Bauer, C. A. J. (2002). A layer-specific three-dimensional model for the simulation of balloon angioplasty using Magnetic Resonance imaging and mechanical testing. *Ann. Biomed. Eng.* 30:753–767.

Holzapfel, G. A., Sommer, G., and Regitnig, P. (2004). Anisotropic mechanical properties of tissue components in human atherosclerotic plaques. *J. Biomech. Eng.* 126:657–665.

Holzapfel, G. A., Stadler, M., and Gasser, T. C. (2005). Changes in the mechanical environment of stenotic arteries during interaction with stents: Computational assessment of parametric stent design. *J. Biomech. Eng.* 127:166–180.

Humphrey, J. D. (2002). *Cardiovascular Solid Mechanics. Cells, Tissues, and Organs.* New York: Springer-Verlag.

Kastrati, A., Mehilli, J., Dirschinger, J., Dotzer, F., Schühlen, H., Neumann, F.-J., Fleckenstein, M., Pfafferott, C., Seyfarth, M., and Schömig, A. (2001a). Intracoronary stenting and angiographic results: strut thickness effect on restenosis outcome (ISAR-STEREO) trial. *Circulation* 103: 2816–2821.

Kastrati, A., Mehilli, J., Dirschinger, J., Pache, J., Ulm, K., Schühlen, H., Seyfarth, M., Schmitt, C., Blasini, R., Neumann, F., and Schömig, A. (2001b). Restenosis after coronary placement of various stent types. *Am. J. Cardiol.* 87:34–39.

Lally, C., and Prendergast, P. J. (2005). Simulation of in-stent restenosis for the design of cardiovascular stents. In Holzapfel, G. A., and Ogden, R. W., eds., *Mechanics of Biological Tissue*, 255–267. Heidelberg: Springer-Verlag.

Moreno, M. R., Bedoya, J., Meyer, C., and Moore Jr., J. E. (2005). Computational modeling of stented arteries: Considerations for evolving stent designs. In Holzapfel, G. A., and Ogden, R. W., eds., *Mechanics of Biological Tissue*, 241–253. Heidelberg: Springer-Verlag.

Pache, J., Kastrati, A., Mehilli, J., Schühlen, H., Dotzer, F., Hausleiter, J., Fleckenstein, M., Neumann, F. J., Sattelberger, U., Schmitt, C., Muller, M., Dirschinger, J., and Schömig, A. (2003). Intracoronary stenting and angiographic results: strut thickness effect on restenosis outcome (ISAR-STEREO-2) trial. *J. Am. Coll. Cardiol.* 41:1283–1288.

Stadler, M., and Holzapfel, G. A. (2004). Subdivision schemes for smooth contact surfaces of arbitrary mesh topology in 3D. *Int. J. Numer. Meth. Eng.* 60:1161–1195.

Stary, H. C. (2003). *Atlas of Atherosclerosis: Progression and Regression.* Boca Raton, London, New York, Washington, D.C.: The Parthenon Publishing Group Limited, 2nd edition.

Computational Modeling of Stented Arteries: Considerations for Evolving Stent Designs

M.R. Moreno, J. Bedoya, C. Meyer, J.E. Moore Jr.

Texas A&M University, Department of Biomedical Engineering, USA
michael.moreno@tamu.edu, josejulian@tamu.edu, clarkmeyer@tamu.edu
jmoorejr@iemail.tamu.edu

Stents are small tubular structures that are inserted into the diseased regions of atherosclerotic arteries via a catheter. These devices are designed to prop the constricted passageway open. Clinical failures of stents occur mainly due to restenosis, which may be related to the mechanical conditions that they induce in the artery. Normal vessel response to stenting includes thrombus formation, inflammation, cell proliferation, and geometric remodeling, all of which depend on biomechanical factors. Recently developed drug-eluting stents target smooth muscle cell proliferation, and have been shown to reduce restenosis rates. The development of more effective drug-eluting and bare metal stent designs can be achieved with the continued development of realistic models. Computational modeling of stented arteries is a powerful tool that provides a means to examine issues such as efficiency of drug delivery and the stress field induced in the artery wall. Early models revealed the stress concentrations near the ends of a stent that result from the mismatch in compliance between the relatively rigid stent and the more compliant artery. This led to the design of a stent that provides gradual transitions in compliance at the ends of the stent (Compliance Matching Stent, or CMS).

Computational models have shown that the CMS reduces stress concentrations. Computational models of the drug transfer into the artery wall from drug-eluting stents demonstrate the importance of even vessel coverage in the deployment process. Experimental modeling has demonstrated that important events such as thrombus formation and re-endothelialization are mediated by flow patterns that depend on stent geometry. The temporal nature of these events *in vivo* has led to the development of the Hybrid Dynamic Stent (HDS). The HDS incorporates biodegradable components that allow the stent to change geometry on a timeline coincident with vessel responses to stenting. Employing computational models to develop this type of technology will require more sophisticated techniques than those presently in use. Ultimately, stent design goals should include minimizing the initial trauma induced by the therapy and maximizing the efficiency of vessel recovery. Realistic modeling will necessarily incorporate the target lesion, vessel damage resulting from the stenting process, the physiologic response to damage and the mechanical (fluid and solid) environment. With the application of rapid prototyping and the further development of modeling techniques, patient specific therapies derived from modeling efforts could be the treatment option of the future.

1 Clinical Relevance

Atherosclerosis, the leading cause of death in developed countries, is characterized by a narrowing of the lumen in a diseased artery due to the accumulation of cholesterol, lipids, calcium and cellular by-products. This narrowing or stenosis can constrict blood flow, inhibiting the nourishment of distal tissues. Cardiac arrests and strokes may occur when arteries become critically stenosed, or when there is a total, acute blockage due to rupture of the plaque.

Angioplasty is a minimally invasive procedure that employs a catheter based delivery system to introduce a collapsed balloon into the afflicted region of the artery. Once in place, the balloon is inflated to force the constricted passageway open, allowing the free flow of blood to distal tissues. While this procedure can be effective, it is not uncommon for the artery to recoil once the balloon is removed. A stent is a permanent scaffold that can be left behind to prop the stenotic artery open. These vascular prosthetic devices can be made from metals such as stainless steel, tantalum, chromium cobalt, and Nitinol, or other biocompatible materials. Similar to balloon angioplasty, stents are also delivered using minimally invasive, catheter-based systems. Some stents are collapsed over a balloon and expand when the balloon is inflated (balloon-expandable stents), while others are elastically compressed into a surrounding sheath and released at the affected site (self-expanding stents). In either case the deployed stent must expand to a diameter that is larger than the adjacent healthy artery diameter in order to remain in place. The blockage and artery wall, therefore, undergo a significant deformation of several times the original diameter in this process. Herein lies an important caveat of stent therapy; the artery wall stresses induced by the oversized stent are considerable, non-physiologic, and lead to a variety of tissue responses that may compromise the success of the procedure.

The response of the artery to stent implantation consists of at least four distinct phases, each with its own time course (Fig. 1; from Edelman and Rogers (1998)). Immediately following stent deployment, thrombus formation on the foreign material and denuded endothelium is the primary concern. While acute closure due to thrombosis has been successfully combated with aggressive systemic drug therapies, some platelet deposition will still form and serve as the substrate for later adaptation processes. About a week after the procedure, inflammation and smooth muscle cell proliferation become the dominant issues. The recovery of the endothelial lining occurs within this time frame as well. A quicker re-establishment of the endothelium can help mediate the delivery and incorporation of blood-borne inflammatory cells, and

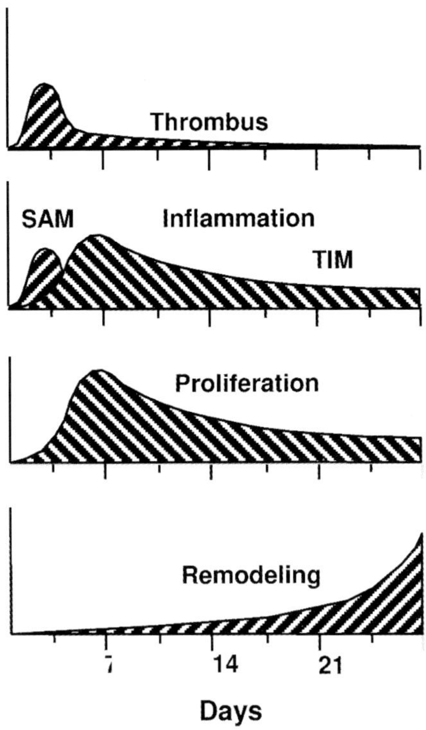

Fig. 1. Data from Edelman and Rogers (1998) illustrating the relative intensity of different types of vessel response vs. time. In inflammation, the first hill represents the heightened activity level of surface adherent monocytes (SAM) while the second hill represents the heightened activity level of tissue infiltrating monocytes (TIM).

the migration of smooth muscle cells from the middle layers of the artery wall. At later time periods, the vessel seeks to remodel, or adapt its geometry, in a search for a new homeostatic state, despite the presence of anomalous stress distributions. In the end, approximately one-third of all patients who have received first-generation bare metal stents have had a new blockage (restenosis) form in or around the stent within six months due to excessive tissue buildup (neointimal hyperplasia) and/or thrombosis (Kastrati et al. (2001)). Nevertheless, stenting procedures have become increasingly popular, with the market for stents estimated to reach \$5B for 2005 (Leon and Bakhai (2003)).

The relatively high clinical failure rate has led researchers and stent manufacturers to search for ways to combat restenosis. The current trend in stent technology is to coat the stent with drugs that target the inhibition of smooth muscle cell proliferation. Drugs such as Sirolimus (Rapamycin) and Paclitaxel (Taxol) have shown great promise, specifically in the reduction of restenosis rates to approximately 10% (Morice et al. (2002), Popma et al. (2004)). These drugs (and others) have been successfully engineered into a relatively new class of stents called 'Drug-Eluting Stents' (DES). Typically, the drug is embedded in a polymer that is coated onto the stent. Additional polymer coatings may be applied to provide more refined control over the release of the drug. The sophistication of DES is evident in their cost, which is approximately three times that of bare metal stents. The present incarnation of a DES is analogous to a 'barbed pill', whereby the therapy is inflicting damage while simultaneously attempting to treat the damage inflicted. While DES have proven to be effective at reducing restenosis, the platforms used should still be optimized to reduce the initial trauma imposed by the treatment and facilitate the recovery process.

In parallel to the development of DES, there has also been some interest in understanding the role of stent design and biomechanics in the restenosis process. Stent implantation provokes dramatic changes in blood flow patterns which range from large-scale flow disturbances created through the diameter of the stented artery due to compliance mismatch (Berry et al. (2002)), and flow anomalies near the artery wall due to strut spacing (Robaina et al. (2003)). In addition, stent implantation also alters the stress field in the artery wall and this has been shown to depend on stent design (Bedoya et al. (2005)). Each phase of the restenosis process is either initiated or mediated by biomechanical actions. Thrombus formation begins with platelet deposition, a process that is flow mediated. Inflammation is initiated by the mechanical injury to the artery wall (Edelman and Rogers (1998)), and also relies on the delivery of blood-borne cells to the artery wall. Smooth muscle cell proliferation, also initiated by the mechanical injury, is part of the artery's effort to re-establish a 'normal' stress environment (Edelman and Rogers (1998)). A greater understanding of the role of biomechanics of restenosis would aid in designing stents that minimize the provocation of the artery wall. As explained below, the nature of the changes in flow patterns and artery wall stress depends strongly on the stent design (strut configuration, strut spacing, etc.). Indeed, clinical data (Table 1) show clearly that stent design is a major risk factor for restenosis (Kastrati et al. (2001)). In a study of more than 4,500 patients whose stent implantations were initially successful, binary restenosis (more than 50% reduction in diameter as determined angiographically) was shown to vary from 20% in some stents to nearly 40% in others. It is important to note that this subset of their data only reflects balloon-expanded stainless steel stents, i.e. the deployment method and material were identical.

2 Computational Modeling of Stented Arteries, Stent Optimization and Drug Delivery

Many of the currently applied principal design criteria for stents arise from concerns over ease of use and deliverability into small, tortuous passages. The

Table 1. Evidence provided by Kastrati et al. (2001) that suggests that restenosis can be modulated by stent design considerations. All stents listed above were constructed from stainless steel and were balloon expanded.

Stent	Restenosis
Guidant Multi-Link	20.0%
Jomed Jostent	25.8%
J&J Palmaz-Schatz	29.0%
PURA-A	30.9%
Inflow Steel	37.3%
NIR	37.8%

increasingly competitive market requires manufacturers to avoid sacrificing these issues for any other concerns. One of the principal issues is the expansion ratio, or the ratio of the crimped to the deployed diameter. This must be as large as possible so that the stent will be easily deliverable to its intended site. Finite element modeling that includes non-recoverable (plastic) deformation can be used to simulate the expansion of balloon-expandable stents. This has been used to investigate the expansion and subsequent recoil of different stent designs (Dumoulin and Cochelin (2000), Migliavacca et al. (2002)), as well as their longitudinal flexibility (Petrini et al. (2004)). Self expanding stents can also be analyzed with computational approaches, but require the inclusion of potentially highly complex material behaviors. Nitinol, a common material for self-expanding stents, can withstand more than 10% strain in its elastic range, but may undergo a phase transformation in the process (Duerig et al. (2000)). This may lead to a highly non-linear stress-strain curve that includes hysteresis. It is also important to recognize that stents must be able to withstand at least 400 million cardiac cycles (equivalent of approximately ten years) without exhibiting fatigue-associated failure (Glenn and Lee (1997)). This can be accomplished with both accelerated fatigue testing apparatus and computational modeling, as outlined by the United States Food and Drug Administration (US Food and Drug Administration (2005)). The US Food and Drug Administration (FDA) must approve the fatigue study results prior to clinical use.

Following implantation, there are a number of changes in the blood flow patterns in the immediate vicinity of the stent and the nearby vasculature. Computational flow models have been developed to analyze both local and global flow patterns in stented arteries. Immediately adjacent to the artery wall, the stent struts protrude into the flow stream, resulting in a flow situation akin to forward and backward facing steps. Flow stagnation, including low wall shear stress and potentially high particle residence times, occurs adjacent to the struts in varying degrees depending very sensitively on the stent strut spacing (Frank et al. (2002)). These flow phenomena have been shown to influence platelet adhesion (initiation of thrombus) both on the stent struts and the wall between them (Fig. 2; Robaina et al. (2003)). Platelet adhesion was lowest when the stent strut spacing was small enough so that flow stagnation was continuous between the stent struts. In this case, the near stagnant recirculating flow fails to mix with the mainstream, limiting the number of platelets that move toward the wall. On the other hand, endothelial regrowth was shown to occur most quickly with large strut spacings (Fig. 3; Walsh et al. (2002)). Similarly, closely spaced struts will likely exacerbate the already

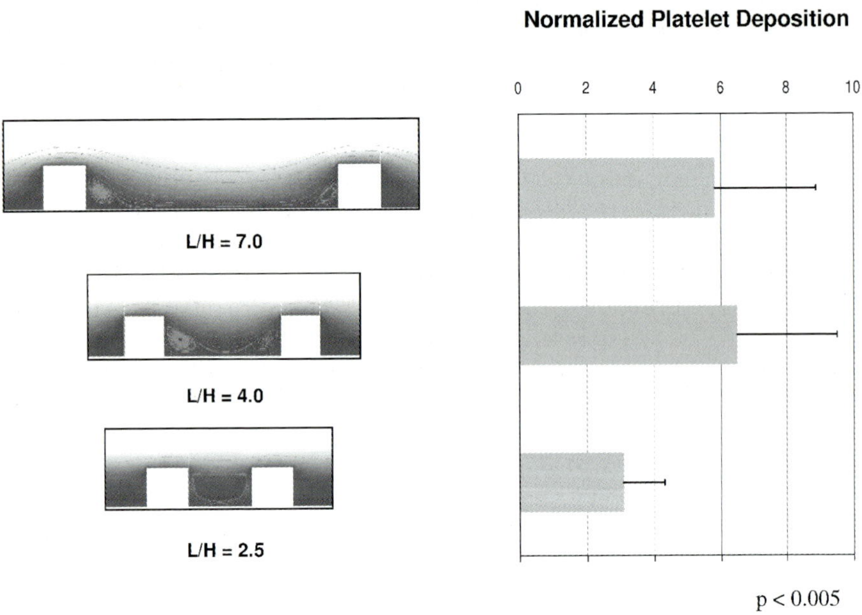

Fig. 2. Platelet deposition as a function of stent strut geometry. The geometry is characterized by a single dimensionless parameter L/H, where L represents the spacing of the struts measured from leading edge to leading edge and H represents the height of the struts. Note that the most narrow strut spacing ($L/H = 2.5$) significantly inhibits platelet deposition. The tight spacing results in the formation of a continuous flow recirculation zone between the struts that makes it difficult for platelets to enter this region. Conversely, the widest strut spacing ($L/H = 7.0$) allows a constant supply of platelets but the higher shear rates inhibit the ability to attach. The mid-level spacing ($L/H = 4.0$) allowed free inflow of platelets and exhibited lower shear rates than the widest strut spacing, traits that collectively resulted in the greatest platelet attachment. Taken from Robaina et al. (2003).

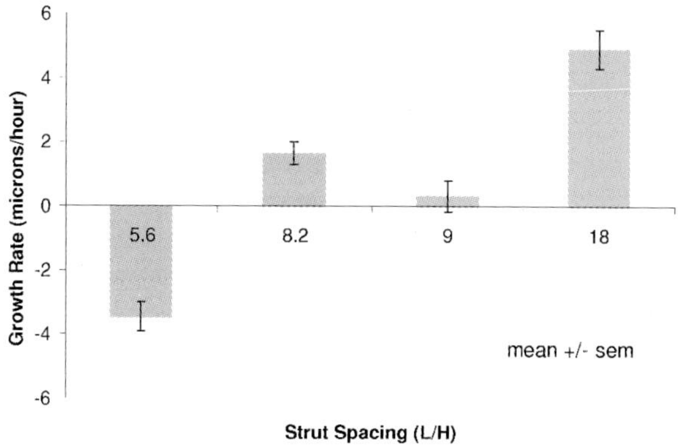

Fig. 3. Endothelial monolayer re-growth as a function of stent geometry under physiologic flow conditions. Note that the widest strut spacing is the most conducive to endothelial re-growth. Also note that the $L/H = 5.6$ case resulted in cell attrition or migration away from the bare area. Taken from Walsh et al. (2002).

high stent-induced stresses in the artery wall. The nature of these competing mechanical factors and tissue responses leads to the proposition that the optimal solution in stent design should attempt to incorporate all of these aspects. The structural rigidity of stents also influences the geometry of the wall to the degree that flow patterns are affected. Changes in local wall curvature, particularly at the ends of the stent, can have a profound effect on wall shear stress (Wentzel et al. (2000)). The pulsatile nature of blood flow proximal and distal to the stent has been shown to be affected by stent implantation in a series of *in vivo* animal experiments (Rolland et al. (1999)). Most stents act as low-pass filters, in that both pulse pressure and pulse flow rate are diminished distal to the stent. This can have implications on perfusion of distal tissues.

Computer modeling of stented artery wall stresses provides a means to identify important design parameters, evaluate the effectiveness of stent design configurations, and ultimately could contribute to the development of patient specific treatment regimes. Since stresses cannot be observed or measured directly, the most effective means of characterizing stress fields in these complex geometries is computer-based finite element modeling. Early models incorporating linearly elastic homogeneous isotropic materials were sufficient to reveal the significant stress concentrations located near the ends of the stent, in the region of compliance mismatch (Berry et al. (2002)). Later studies incorporating more realistic non-linear material properties and finite deformation analysis also revealed similar stress concentrations at the ends of the stent (Fig. 4, from Bedoya et al. (2005)). Using this type of artery model, stent design parameters can be tested and optimized to perform specific functions;

Fig. 4. Stresses in a stented artery model employing non-linear homogeneous isotropic material properties. The color map illustrates the maximum principal stress (Pa) distribution at diastolic pressure on the inner wall of the artery model. The footprint of the stent is clearly visible in this stress map, as are stress concentrations near the ends of the stent, in the region of compliance mismatch. Figure from Bedoya et al. (2005).

e.g., to reduce wall stress. While this artery model incorporates more realistic non-linear material properties, the artery diameter and wall thickness are constant, and the material is homogeneous and isotropic. It is therefore essentially a generic model of a healthy artery. Nevertheless, a comparison of healthy artery models is effective in revealing differences in stent-induced artery wall stress distributions that might be difficult to discern in the presence of large variations in morphology and mechanical properties of diseased arteries. Stent design optimization in a healthy artery platform is a precursor to advanced patient-specific stent optimization. More sophisticated artery models incorporating non-homogeneous anisotropic arteries with multiple lesion materials have been developed (Holzapfel et al. (2000, 2002)). Atherosclerotic lesions vary greatly in material composition and physical properties. Initially, lesions consist mostly of compliant fatty and lipid deposits. In advanced disease, lesions may contain regions of stiff calcification (Holzapfel et al. (2000, 2002), Williamson et al. (2003)). While taking the mechanical properties of the different components into account may be seen as even more realistic, variations in individual plaque geometry and make-up limit the applicability of models with a high degree of specificity. However, this technology provides a foundation upon which to develop patient-specific treatment plans.

One potential shortcoming in current modeling is the lack of inclusion of the potential effects of vessel wall damage and response, which could result in stress redistribution. Typically, the material properties employed represent the behavior of 'healthy' arteries. The stenting process damages the artery, an event that likely affects the stress distribution and the biologic vessel response. The physiologic response of the vessel that results from the damage induced by the stenting process should also be considered in future models. Contributing to this end are models being developed that include vessel response functions (Prendergast and Lally (2004)). In these models, the resulting stress field prompts a response from the vessel simulating neointimal hyperplasia. In this case, an intimal thickening cascade is initiated and modeled with the addition of elements on the inner surface of the artery. The arterial response was modeled by a relatively simple function (unsteady diffusion equation), as this type of modeling is presently in its infancy. Incorporation of the physiologic response of the vessel to the stenting process is an important future direction in the modeling of stented arteries. This would include consideration of both the physiologic response to the stress field induced by the stent, and the physiologic response to the damage inflicted by the stenting process.

Computational modeling has also been used to analyze the effectiveness of drug mass transfer from DES. The efficiency of drug delivery is an important issue, as evidence suggests that often the drugs are effective only in regions very near the delivery points (Hwang et al. (2001), Mongrain et al. (2003)). It is also evident from these studies that the stent struts deploy as uniformly as possible, since regions of the vessel wall far from the stent struts may not receive a therapeutic dosage. This of course depends on the mass transfer characteristics of the drug and local wall properties. Additionally, the areas just proximal and distal to the stent that receive little drug dosage are still susceptible to neointimal hyperplasia. There is clinical evidence that DES are no better than bare metal stents at preventing restenosis that occurs near the ends of the stent, so-called 'edge effects' (Stone et al. (2004), Popma et al. (2004)). While the diffusion of drugs has been modeled, this was done in the absence of other considerations; the mechanical implications were not considered, i.e. the stress distribution was not examined.

3 Role of Modeling in Stent Design

Ideally, as a design tool, stented artery models would allow designers to optimize drug diffusion and mechanical properties. The aims of this highly complex optimization exercise should be to reduce the trauma induced by the stent and improvement of the efficiency of arterial recovery. Towards this first aim, a Compliance Matching Stent (CMS) has been proposed that provides smooth transitions in rigidity at the ends of the stent. This alleviates much of the stress concentration that occurs at the ends of stents where there is normally an abrupt change in mechanical properties. Initial animal experiments on this design have proven encouraging (Berry et al. (2002), Rolland et al. (1999)). Towards the second aim of improving the efficiency of recovery, the effects of stents on blood flow patterns must be considered, in addition to stress reduction. Optimizing stent properties to cope with all of the phases of artery wall reaction will prove challenging. As previously described, the geometric designs that reduce platelet deposition and speed re-endothelialization are mutually exclusive. Specifically, it has been determined that when the stent strut spacing is relatively close, platelet deposition and endothelial regrowth are minimized. Conversely, when the strut spacing is relatively wide, both platelet deposition and the rate of endothelial migration are increased.

Fortunately, the timelines on which these events become critical are not the same. Recognition of this fact has led to the development of the Hybrid Dynamic Stent (HDS, Fig. 5). The HDS incorporates at least one biodegradable component and a permanent scaffold. The initial strut geometry inhibits acute platelet accumulation and, as the degradable components dissolve, the resulting geometry is optimized for re-endothelialization. The initial closely spaced strut geometry should also be effective at holding back intimal flaps, or bits of artery wall material that are partially torn from the wall during stent implantation. Intimal flaps, while dangerous and often quoted as the motivation for designing stents with closely spaced struts, probably heal within a week or two once they are 'tacked up' by the stent. Once this need subsides, a more sparse structure would probably enhance artery healing. In short, the response of the artery wall to stent implantation is highly time-dependent. Incorporating time dependency into stent geometry can exploit the nature of these processes to optimize healing.

The future of interventional technology design (stents and beyond) will benefit greatly from computational modeling tools. The capabilities to model blood flow, artery wall stresses, and drug mass transfer exist currently,

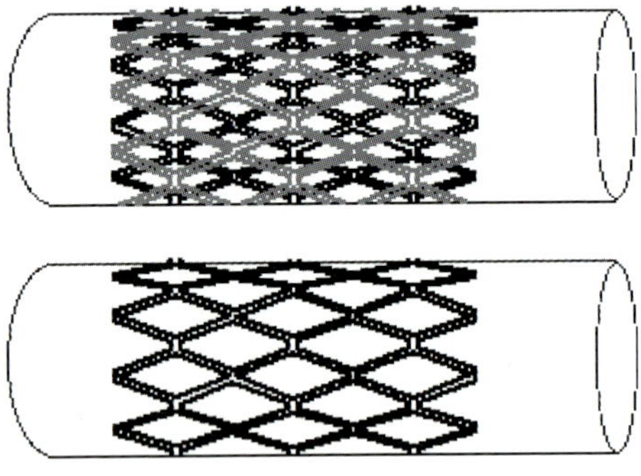

Fig. 5. Conceptual illustration of the Hybrid Dynamic Stent (HDS) depicting the initial (top) and final (bottom) configuration. The dense strut spacing in the initial configuration is designed to inhibit platelet aggregation and tack up intimal flaps. In the final configuration, the wider strut spacing encourages endothelial regrowth and vessel healing while continuing to provide structural support.

although not necessarily in a single computational package. Efforts to model artery wall damage and subsequent reaction are still in early stages of development. As these tools reach greater levels of sophistication, designers will enjoy a more efficient process that results in better patient care. The eventual, ideal stent may be one that combines drug delivery, sound biomechanics, and perhaps time-dependent geometry. The technologies developed and lessons learned from the stenting experience can, and should, be applied to other interventional technologies in other areas of medicine.

References

Bedoya, J., Meyer, C., Moore Jr., J. E., and Moreno, M. R. (2005). Stress quantification in hyperelastic artery models: Tools for improving stent design and reducing restenosis. In *Summer Bioengineering Conference*. Vail, CO.

Berry, J. L., Manoach, E., Mekkaoui, C., Rolland, P. H., Moore Jr., J. E., and Rachev, A. (2002). Hemodynamics and wall mechanics of a compliance matching stent: *in vitro* and *in vivo* analysis. *J. Vasc. Interv. Radiol.* 13: 97–105.

Duerig, T. W., Tolomeo, D. E., and Wholey, M. (2000). An overview of superelastic stent design. *Minim. Invasiv. Ther.* 9:235–246.

Dumoulin, C., and Cochelin, B. (2000). Mechanical behaviour modelling of balloon-expandable stents. *J. Biomech.* 33:1461–1470.

Edelman, E. R., and Rogers, C. R. (1998). Pathobiologic responses to stenting. *Am. J. Cardiol.* 81:4E–6E.

Frank, A. O., Walsh, P. W., and Moore Jr., J. E. (2002). Computational fluid dynamics and stent design. *Artif. Organs* 26:614–621.

Glenn, R., and Lee, J. (1997). Accelerated pulsatile fatigue testing of NI-TI coronary stents. In *Proceedings of the Second International Conference on Shape Memory and Superelastic Technologies*. Pacific Grove, CA.

Holzapfel, G. A., Schulze-Bauer, C. A. J., and Stadler, M. (2000). Mechanics of angioplasty: Wall, balloon and stent. In Casey, J., and Bao, G., eds., *Mechanics in Biology*. New York: The American Society of Mechanical Engineers (ASME). AMD-Vol. 242/BED-Vol. 46, pp. 141–156.

Holzapfel, G. A., Stadler, M., and Schulze-Bauer, C. A. J. (2002). A layer-specific three-dimensional model for the simulation of balloon angioplasty using Magnetic Resonance imaging and mechanical testing. *Ann. Biomed. Eng.* 30:753–767.

Hwang, C.-W., Wu, D., Elazer, R., and Edelman, R. (2001). Physiological transport forces govern drug distribution for stent-based delivery. *Circulation* 104:600–605.

Kastrati, A., Mehilli, J., Dirschinger, J., Pache, J., Ulm, K., Schühlen, H., Seyfarth, M., Schmitt, C., Blasini, R., Neumann, F., and Schömig, A. (2001).

Restenosis after coronary placement of various stent types. *Am. J. Cardiol.* 87:34–39.

Leon, M. B., and Bakhai, A. (2003). Drug-eluting stents and glycoprotein IIb/IIIa inhibitors: Combination therapy for the future. *Am. Heart J.* 146:S13–S17.

Migliavacca, F., Petrini, L., Colombo, M., Auricchio, F., and Pietrabissa, R. (2002). Mechanical behavior of coronary stents investigated through the finite element method. *J. Biomech.* 35:803–811.

Mongrain, R., Nulman-Fleming, N., and Bertrand, O. (2003). 3D numerical simulations of stent-based local drug delivery using realistic stent and vascular wall structures. In *ASME Summer Bioengineering Conference.* Key Biscayne, FL.

Morice, M. C., Serruys, P. W., Sousa, J. E., Fajadet, J., Ban Hayashi, E., Perin, M., Colombo, A., Schuler, G., Berragan, P., Guagliumi, G., Molnar, F., and Falotico, R. (2002). A randomized comparison of a sirolimus-eluting stent with a standard stent for coronary revascularization. *N. Engl. J. Med.* 346:1773–1780.

Petrini, L., Migliavacca, F., Auricchio, F., and Dubini, G. (2004). Numerical investigation of the intravascular coronary stent flexibility. *J. Biomech.* 37:495–501.

Popma, J. J., Leon, M. B., Moses, J. W., Holmes Jr., D. R., Cox, N., Fitzpatrick, M., Douglas, J., Lambert, C., Mooney, M., Yakubov, S., and Kuntz, R. E. (2004). Quantitative assessment of angiographic restenosis after sirolimus-eluting stent implantation in native coronary arteries. *Circulation* 110:3773–3780.

Prendergast, P. J., and Lally, C. (2004). Simulation of in-stent restenosis and its relationship with cardiovascular stent design. In *IUTAM Symposium on 'Mechanics of Biological Tissue'.* Graz, Austria.

Robaina, S., Jayachandran, B., He, Y., Frank, A. O., Schoephoerster, R., and Moore Jr., J. E. (2003). Platelet adhesion to simulated stented surfaces under physiologic flow conditions. *J. Endovas. Ther.* 10:978–986.

Rolland, P. H., Charifi, A. B., Verrier, C., Bodard, H., Piquet, A. F. P., Moulin, G., and Bartoli, J. M. (1999). Hemodynamics and wall mechanics after stent placement in swine iliac arteries: Comparative results from six stent designs. *Radiology* 213:229–246.

Stone, G. W., Ellis, S. G., Cox, D. A., Hermiller, J., O'Shaughnessy, C., Tift Mann, J., Turco, M., Caputo, R., Bergin, P., Greenberg, J., Popma, J. J., and Russell, M. E. (2004). A polymer-based, paclitaxel-eluting stent in patients with coronary artery disease. *N. Engl. J. Med.* 350:221–231.

US Food and Drug Administration. (2005). http://www.fda.gov/cdrh/ode/guidance/1545.html#7b.

Walsh, P. W., Berkani, F., Frank, A., and Moore Jr., J. E. (2002). Endothelial cell regrowth is affected by stent geometry under physiologic flow conditions. In *International Symposium on Endovascular Therapy.* Miami Beach, FL.

Wentzel, J. J., Whelan, D. M., van der Giessen, W. J., van Beusekom, H. M. M., Andhyiswara, I., Serruys, P. W., Slager, C. J., and Krams, R. (2000). Coronary stent implantation changes 3D vessel geometry and 3D shear stress distribution. *J. Biomech.* 33:1287–1295.

Williamson, S. D., Lam, Y., Younis, H. F., Huang, H., Patel, S., Kaazempur-Mofrad, M. R., and Kamm, R. D. (2003). On the sensitivity of wall stresses in diseased arteries to variable material properties. *J. Biomech. Eng.* 125:147–155.

Simulation of In-stent Restenosis for the Design of Cardiovascular Stents

C. Lally,[1] P.J. Prendergast[2]

[1] Dublin City University, School of Mechanical and Manufacturing Engineering, Ireland
triona.lally@dcu.ie
[2] Trinity College, Centre for Bioengineering, Department of Mechanical Engineering, Ireland
pprender@tcd.ie

In-stent restenosis is a problem that has its origins in the high stresses generated in the vessel wall by deployment of a stent. In this paper, finite element modelling is used to quantify the artery stresses after stent deployment. It is predicted that tensile stresses as high as 10 MPa can be generated. Since it is believed that damage (or injury) is the stimulus for restenosis, fatigue tests were performed on vascular tissue *in vitro* (Lally (2004)). The resulting S-N (Wohler) curve was used to compute a remaining life damage variable and thereby establish a non-linear relationship between stress and damage. Cells were assumed to proliferate in response to this damage and to migrate to the inner lumen and create new material. This was incorporated into an iterative procedure to simulate in-stent restenosis. The predictions of in-stent restenosis correlate with clinical observations, thereby suggesting that this approach can be used as a pre-clinical assessment procedure for stents.

1 Introduction

Living tissues adapt to their mechanical environment (Fung (1993)); they adapt both during the life of the individual (ontological adaptation) and over the course of evolution (phylogenetic adaptation) (Prendergast (2002)). Ontological adaptation has long been recognised in bone (Roux (1912)) and more recently it has been recognised in soft connective tissues, including arterial tissue (Humphrey (2002)).

The effects of tissue adaptation and remodelling for the performance of orthopaedics implants has received considerable attention (Huiskes (1993)). Similar remodelling has been observed when cardiovascular stents are used to maintain patency in stenotic arteries. Specifically, it has been found that arteries respond to the presence of a stent by re-forming the blockage – this effect has been termed in-stent *restenosis* to distinguish it from the initial formation of the stenosis. This paper presents a model to simulate restenosis and uses it to explore whether or not the stresses in the vessel wall induced by different stents may be responsible for the different restenosis rates found with different stent designs.

Modelling of the adaptation of tissues to changes in stresses requires the identification of the stimulus for adaptation, and the writing of an evolution equation giving the relationship between the stimulus and the change in the tissue properties. In bone remodelling studies, two classes of stimuli have been proposed:

(i) a strain stimulus giving strain-adaptive remodelling;
(ii) a microdamage stimulus giving damage-adaptive remodelling.

In previous work by our group, the damage-adaptive approach was developed for bone (Prendergast and Taylor (1994)). In this paper, a similar approach is applied to arterial soft tissue to develop a method to simulate in-stent restenosis using the finite element (FE) method. The restenosis model uses a damage stimulus related to the stress levels induced within the tissue by stent deployment. The relationship between stress and damage was determined from fatigue tests on harvested porcine coronary tissue. In the first section of this paper, stresses induced within the artery wall by three different stent designs are studied; the NIR stent (Boston Scientific), S7 stent (Medtronic AVE) and a 'simple' stent design.

2 Materials and Methods

The finite element software used was MSC Marc/Mentat (Santa Ana, CA, USA).

2.1 Simulation of Stent Deployment

Model Geometry. The 3D geometry of the repeating units of the fully expanded 3.5 mm diameter NIR and S7 stents was determined using the coordinate measurement technique described previously (Prendergast et al. (2003)). The thickness of the stent wires was 0.1 mm. Using the repeating unit geometry of each stent, solid models of the full stents were generated in the planar state, and a 'simple' stent design was also developed and used to compare with the two clinically-used stents. The solid models were used to generate a finite element mesh, which was transformed into the final cylindrical shape by transferring the nodal coordinates from Cartesian to cylindrical coordinates (Fig. 1). Note that the NIR and S7 designs differed greatly; the NIR is a slotted tube laser-cut stent with seven crowns whereas the S7 is a modular stent with a circular cross-section, 10 crowns and welded joints. The 'simple' stent

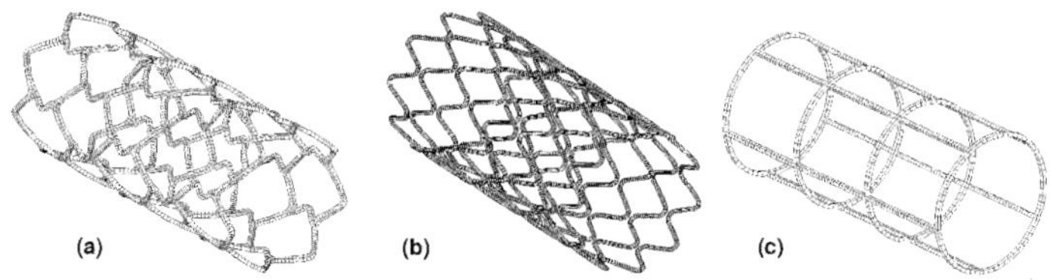

Fig. 1. FE meshes of the expanded stents: (a) NIR, (b) S7 and (c) the 'simple' stent.

design has five crowns. All three stents had an outer diameter of 3.5 mm and a wire thickness of 0.1 mm.

The vessel was modelled as a cylindrical stenotic vessel with a lumen diameter of 2 mm; the inner diameter of the healthy vessel was 3 mm, and the outer diameter of the stenosed vessel was 4 mm. Three elements were assigned across the thickness of the healthy vessel wall, three elements through the plaque, 91 along the length (6.5 mm) and 140 around the circumference.

The adaptive meshing capability within Marc/Mentat was used in the models for the arterial tissue to generate a refined mesh at contact areas.

Material Properties. The arterial tissue and stenotic plaque were modelled using isotropic five-parameter third-order Mooney-Rivlin hyperelastic constitutive equations (Mooney (1940)). Although arterial tissue has been found to be anisotropic (Humphrey (2002)) and therefore is most accurately described by an anisotropic constitutive model, such as that developed by Holzapfel et al. (2000), the isotropic Mooney-Rivlin model has been found to adequately describe the non-linear stress-strain relationship of elastic arterial tissue in this application (Lally and Prendergast (2003)).

The arterial tissue material model was determined by fitting to data from uniaxial and equibiaxial tension tests of human femoral arterial tissue. More details of the determination of the experimental data and this hyperelastic material model are given in Prendergast et al. (2003). The hyperelastic constitutive model used to represent the plaque tissue in the vessel was determined by fitting to published data for human calcified plaques obtained by Loree et al. (1994). Details of these material models are given in Lally et al. (2005). The stent material was 316L stainless steel ($E = 200\,\text{GPa}$, $\nu = 0.3$).

Boundary Conditions. The loading and restraint conditions were applied to the stent/artery construct in two steps. It involved use of the feature that exists in Marc/Mentat (and many other finite element codes as well) that allows elements to be activated and deactivated during an analysis. In the first step, the stent elements were deactivated and the vessel was expanded to a diameter greater than that of the expanded stent by application of a sufficient internal pressure. In the second step, the elements of the stent were activated and the pressure on the inner lumen of the artery was gradually reduced to a value of 13.3 kPa, corresponding to mean blood pressure of 100 mmHg. Due to the elastic nature of the vessel, it contracts around the stent. Frictionless contact between the stent and the artery was assumed.

The contact algorithm implemented in Marc/Mentat was the direct constraint method. In this procedure, the motion of the bodies is tracked and, when contact occurs, direct constraints are placed on the motion using boundary conditions. Deformable-deformable contact was used to describe the contact between the stent and the artery. Both contact bodies were mathematically defined as analytical (NURB) surfaces.

A longitudinal stretch of 1.2 was applied to the artery in an attempt to simulate the longitudinal tethering observed on coronary arteries *in vivo* (Ogden and Schulze-Bauer (2000)). Axial restraints were applied to one end of the vessel. During the analysis the stent was restrained at one node in the circumferential direction to prevent rigid-body rotations. Because of cyclic symmetry it was not necessary to model the entire stented vessel but rather only segments of the stented artery had to be modelled. Using the cyclic symmetry capability in Marc/Mentat, the nodes on the two cyclically symmetrical faces were coupled or tied. This allows out of plane motion, unlike planar symmetry constraints (planar symmetry constraints cannot accurately model a cyclically repeating segment of a stented artery because (r, z) planes do not remain planar during deformation). The NIR stent was represented by a cyclically repeating one-seventh segment whilst the S7 could only be represented by half-cyclic symmetry due to the position of the welds on the stent, and the 'simple' stent design was represented by a one fifth segment.

The number of elements used to represent the geometry of the three stenting simulations differed greatly as a result of the cyclic symmetry constraint applied. The one-seventh segment of the NIR stent was adequately represented by 125 elements and 470 nodes. For the half model of the S7, 5040 elements and 9021 nodes were necessary to model the circular cross-section of the stent, whilst the 'simple' stent was adequately represented by 62 elements and 252 nodes.

2.2 Restenosis Model

The development of damage within a stented vessel is a function of the applied stresses. Following methods used in mechanobiology (van der Meulen and Huiskes (2002)), an algorithm for simulating in-stent restenosis can be proposed as an iterative process; see Fig. 2.

Since the algorithm assumes damage as a stimulus for cell proliferation, a quantification of damage within porcine coronary arterial tissue was required Fatigue tests on arterial tissue were performed and a stress versus number-of-cycles-to-failure curve (S-N or Wohler curve) was obtained; see Fig. 3. The S-N curve shows that there is a non-linear relationship between the stress level and the fatigue life. Using Miner's law (Miner (1945)) a non-linear relationship

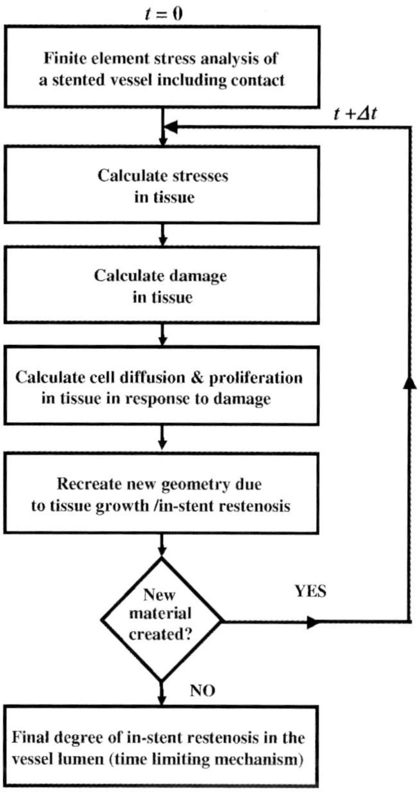

Fig. 2. The iterative process simulating in-stent restenosis; the stresses within the artery wall determine the level of damage in the tissue, and hence the degree of in-stent restenosis.

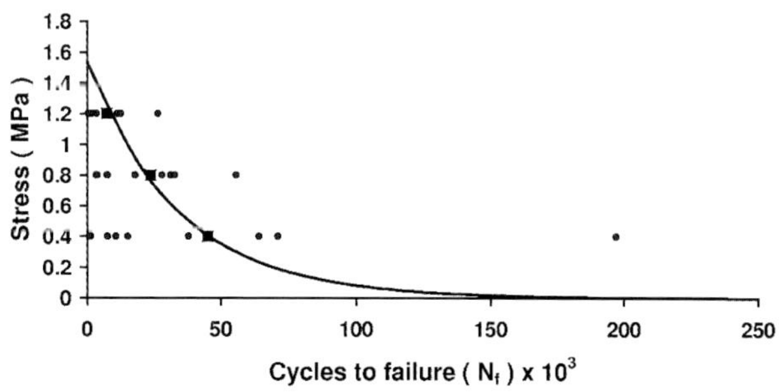

Fig. 3. The S-N curve for porcine coronary arterial tissue. The circles represent test data whilst the squares represent the average number of cycles to failure at each stress level (Brands (2003)).

between stress and the rate of damage accumulation $\dot{D}$ can be derived and rate of damage accumulation within the arterial tissue may be given as

$$\dot{D} = 1/N_{\mathrm{f}}, \tag{1}$$

where N_{f} is the number of cycles to failure and, from the S-N curve,

$$N_{\mathrm{f}} = f(\sigma), \tag{2}$$

where σ is the stress level at which damage is calculated. Using eq. (1), the damage D can be calculated (assuming $D = 0$ at $t = 0$) and it is used as the stimulus for restenosis, as described below.

It may be argued that, since Fig. 3 uses data based on dead tissue, the stimulus D is not relevant in the *in vivo* situation because dead tissue (unlike living tissue) cannot repair the accumulated damage. However, it must be the case that damage would accumulate *in vivo* due to cyclic loading except that it is repaired by repair processes occurring physiologically. The plausibility of this assumption is attested to by the data of Ker et al. (2000) and Schechtman and Bader (2002); both papers show that physiological cyclic stresses in soft tissue (tendon) would cause failure if there were no repair processes.

Since the stress acting in the artery is not uniaxial, the damage stimulus due to both the maximum and minimum principal stresses was calculated and summed. The maximum and minimum principal stresses were chosen as the damage-inducing stimuli since they correspond to the maximum tensile and compressive stresses induced within the artery. It is the elevation of these stresses beyond the baseline tensile and compressive stresses that is deemed to induce damage within the artery, and this then acts as the stimulus for restenosis.

It is the proliferation of cells stimulated by damage D that creates the new volume of material observed as in-stent restenosis. The migration of the cells may be modelled using the diffusion equation (Murray (1989)). If the cell concentration in a region (e.g., a finite element) that is filling with stenosis is denoted n, then the increase in the cell concentration in that region is given by

$$\frac{\mathrm{d}n}{\mathrm{d}t} = \kappa \nabla^2 n + P(D) - K(D), \tag{3}$$

where κ is the diffusion coefficient of cells in arterial tissue, $P(D)$ denotes the proliferation rate as a function of damage and $K(D)$ denotes the apoptosis/death rate of cells; in this case $K(D) = 0$. If a linear association between the damage and proliferation is assumed, then

$$P(D) = cD, \tag{4}$$

where c is an empirical constant.

It is worth remarking that σ could, in principle, be used as a stimulus in the restenosis algorithm; however, our assumption is that D is more directly related to cell proliferation than σ, allowing a linear relationship between the stimulus and the proliferation to be proposed.

Restenosis Algorithm – Implementation in Finite Element Models. The coefficient κ in eq. (3) was calculated from published measurements of neointimal thickness in human coronary arteries (Hoffmann et al. (2002)). Given the area of the neointima and the number of days of growth, the diffusion coefficient for cell migration was estimated from the neointimal growth generated per day (per iteration in the model), i.e. the average neointimal growth was calculated to be $0.016\,\text{mm}^2/\text{day}$ (Lally (2004)). Therefore, κ was set to $0.016\,\text{mm}^2/\text{day}$ in all tissues.

To model the ingrowth of tissue into the lumen of the artery, a finite element mesh of 'deactivated' elements was created inside the vessel with elements that could be activated as restenosis progressed; see Fig. 4. The deactivated grid defined the extent of the potential neointimal growth. The material was initially deactivated in the model by setting the material stiffness low enough that it did not contribute to the structural behaviour of the vessel: $E = 0.05\,\text{MPa}$. Elements of this restenotic tissue were activated if the number of cells within the element reached a critical value. This value was obtained by averaging the calculated cell densities that have been reported for human restenotic specimens at several time points after balloon angioplasty (Schwartz et al. (1996)). The average cell density value was found to be $1.93 \times 10^3\,\text{cells/mm}^2$. It was assumed that cell density is uniform in all directions and therefore the cell density in the three-dimensional models was taken to be the square root of $1.93 \times 10^3\,\text{cells/mm}^2$ cubed giving a critical value of cell concentration (n_{crit}) of $84.79 \times 10^3\,\text{cells/mm}^3$. Activation was achieved by raising the Young's modulus to $1\,\text{MPa}$. This value was chosen

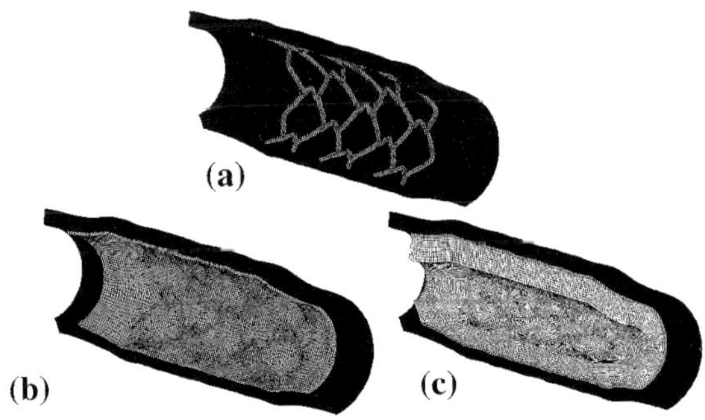

Fig. 4. (a) A segment of the stented vessel illustrating the thrombotic elements (light grey) in (b) and the potentially restenotic tissue growth within the artery lumen surrounding the stent (white) in (c). (Pre-stented artery inner radius, $R_\text{I} = 1.5\,\text{mm}$, artery outer radius, $R_\text{O} = 2\,\text{mm}$, axial stretch applied to the artery, $\lambda_z = 1.2$).

to represent the stiffness of arterial tissue under low loads, as obtained from the uniaxial tensile tests on human femoral and porcine coronary tissue (see Prendergast et al. (2003)).

A thrombotic layer (see Fig. 4), which is a thin layer (0.1 mm thick) of tissue covering the inner lumen of the artery, was assigned as an active material immediately after stent placement to simulate the acute thrombotic response of a vessel to the presence of the stent (Edelman and Rogers (1998)).

The analyses were performed using the coupled thermal-mechanical job specification. This allows the cell diffusion, represented by temperature in the thermal analysis, to be coupled with the stress analysis. A separate routine was written to allow cell generation (modelled as a cell source) to be damage dependant. Following the algorithm laid out in Fig. 2, the finite element analysis was carried out initially to determine the maximum and minimum principal stresses induced within the artery wall. The damage was then calculated using the S-N curve established from the fatigue data at that stress level (Fig. 3), cells were generated according to eq. (4) and migrated and proliferated from areas of damage according to eq. (3). If the cell concentration within any of the deactivated elements reached the critical value of n_{crit} the element was activated. Once elements were activated there was then new material in the model and the stress distribution was altered. Therefore, an iterative process was required (see Section 2.2).

The tissue in the thin thrombotic layer and the arterial tissue were the only sites of new cell generation in the first week of the simulation. For the remaining time of the restenosis process in the simulation, i.e. 93 days, only cells in the thrombotic layer and the activated restenotic material within the lumen of the vessel were allowed to divide and proliferate. This was to simulate the migration of smooth muscle cells from the arterial wall into the lumen of the vessel, whereupon these cells proliferate and diffuse out further from areas of damage, as has been noted in studies of in-stent restenosis (Edelman and Rogers (1998)). Cells could migrate by diffusion through all of the materials at any time during the simulation, including the structurally deactivated material. Cell concentrations in all of the materials were set to zero at the beginning of the simulation. This was to represent the constant concentration of cells everywhere in the tissue before damage was present, which subsequently caused tissue growth.

The restenosis algorithm applied a luminal pressure of 100 mmHg (mean blood pressure) to the inner surface of the lumen. This was achieved by looping through all of the elements at the end of each iteration and determining the innermost surface of the active elements.

As with the FE stenting procedure outlined in Section 2.1, it was possible to represent the restenosis model of the NIR stent by a cyclically repeating one-seventh segment. The mesh densities used for the simulation of restenosis in a NIR stented vessel is the following: stent, 136 elements, 492 nodes; potentially restenotic tissue within the vessel, 13746 elements, 21488 nodes. The cylindrical artery was represented by 3 elements (0.5 mm) in the thickness and

91 in the length (6.5 mm). In the circumference of the cylindrical vessel, 20 elements were to represent the one-seventh segment of the NIR stent design, respectively.

The stent, and the restenotic material surrounding the stent, formed one contact body, whilst the artery formed the other contact body in the contact analysis. Cell movement was enabled across the interface of the two contacting bodies to simulate the migration of cells from the artery into the restenotic material. Since a thermal analysis was used to simulate diffusion, this was analogous to assigning a heat film on the outer contacting surface of each contact body. In fact, for a coupled thermal-stress analysis heat fluxes are automatically created on all the boundaries of the deformable bodies and two film coefficients are assigned to each contact body. One corresponds to free convection and is used whenever the boundary is free. The other is a contact film, and is used whenever the boundary is in contact with another body. Assigning the film coefficient on the outer surface of the artery at the adventitia enabled the cells to diffuse out in every direction and not to be stored in the artery. If the cells were not able to diffuse out at the free outer boundary of the artery a high concentration of cells would have been stored in the artery and would have had the ability to migrate into the intima long after the proliferation within the artery wall had ceased.

As regards material properties, the arterial tissue in the restenotic models was defined using the hyperelastic constitutive equation established for porcine coronary arterial tissue since the fatigue tests to measure damage accumulation were conducted on porcine coronary arterial tissue (see Lally et al. (2004)).

3 Results

3.1 Stress Analysis of Stenting Procedures

The finite element models predicted that all stents restored patency to the stenosed vessels; see Fig. 5. The stresses induced within each stented vessel differed considerably. A larger volume of highly stressed vascular wall was predicted for the NIR and 'simple' stents compared with the S7 stent; see Fig. 6.

3.2 Simulation of In-stent Restenosis

No empirical data exists to relate tissue damage caused by stenting to the number of proliferating smooth muscle cells. Therefore, the constant, c, in eq. (4), was determined for the damage stimulus such that restenosis developed but did not continue to grow indefinitely. The same value of c should be used for all stent designs when comparing their propensity for restenosis. A suitable value of c was determined for this stimulus to be 9.52×10^5 cells/mm^{3}day. The

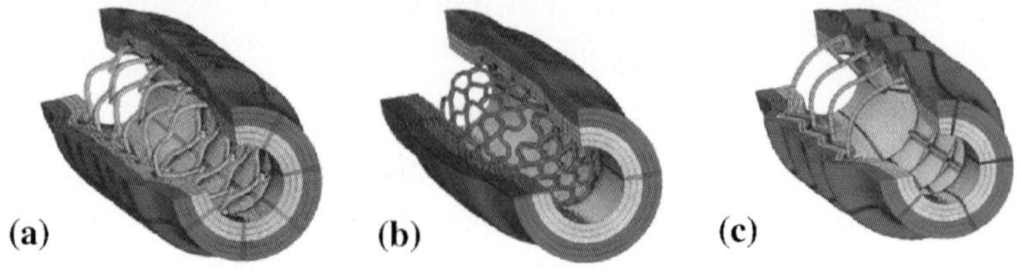

Fig. 5. Deformed geometry of the cylindrical stented vessels with symmetrical cylindrical stenotic plaque (light grey), axial stretch applied ($\lambda_z = 1.2$) and original vessel lumen diameter of 2 mm: (a) NIR stent, (b) S7 stent, (c) 'simple' stent.

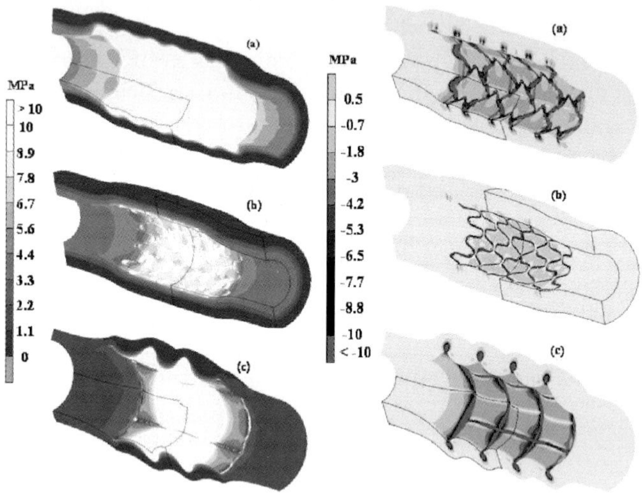

Fig. 6. Maximum and minimum principal stresses induced within segments of idealised cylindrical stenosed vessels stented with the (a) NIR, (b) S7 and (c) 'simple' stents, with axial stretch $\lambda_z = 1.2$ applied.

restenotic growth in the NIR stented vessel was found to be localised around the stent struts and greatest at the central struts and the ends of the stent; see Fig. 7. Restenosis growth was completed at 19 days after stent placement and the percent vessel area stenosis based on the minimum vessel diameter and the outer stent diameter (3.5 mm) was 51%. Therefore, this vessel would be considered restenosed using binary restenosis as the measure of the outcome of stenting (i.e. greater that 50% stenosis).

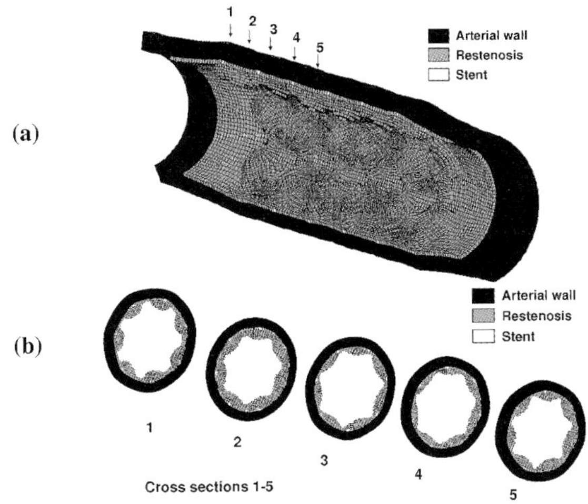

Fig. 7. Simulation of in-stent restenosis for a NIR stent design: (a) the cylindrical vessel stented with a NIR stent showing the predicted pattern of restenotic growth using the restenosis algorithm; (b) cross-sections of the restenotic vessel at locations given in (a). It can be seen from (b) that restenotic growth is predicted to be concentrated around the stent struts and greatest at the stent ends.

4 Discussion

The stenting procedures show that the magnitude of the stresses and the volume of material stressed to high levels in a stented vessel are highly dependent on the stent design. A flexible modular type stent maintains the stresses in the vessel lower than the rigid laser-cut designs (NIR and 'simple' designs). The S7 radially recoils to a greater extent than the NIR or 'simple' stent and in doing so maintains low stresses in the vessel. This difference between rigid and flexible stents has been found to be even more pronounced in a vessel with a localised stenosis (Lally et al. (2005)).

In this paper, a model to simulate restenosis has been developed in the form of a computational algorithm that uses damage as a stimulus variable. A limitation of the model is that an S-N curve obtained from tensile fatigue test data is used to calculate the damage created under the multi-axial stresses experienced *in vivo* in the vessel wall. In the future, if such an algorithm is to be used to give quantitative predictions, data regarding damage accumulation in multi-axial stress states will have to be included.

Another limitation of the model presented here is that the vessel does not include the residual stresses present in an artery in its unloaded state (Fung (1993)). Inclusion of the residual stresses in the artery may have lowered the stresses induced in the artery wall by deployment of the stent. However, since the constant c in the restenosis model has not been obtained from cell studies this model is only useful in its current form as a *comparative* design optimizing tool. It should be noted, therefore, that a great deal of work is still required to fully solve this problem.

This restenosis model has also been applied to the S7 commercial stent design (Lally (2004)). It will now be used in vessels with localised plaque lesions and more realistic vessel geometries as well as for other commercially available stent designs to determine if it can predict the stent with the least propensity for restenosis by comparison to clinical data and to confirm the use of the fatigue damage variable as a reliable stimulus for in-stent restenosis. It may then be used in the future to predict restenosis in new bare-metal and also drug-eluting stents, where drug diffusion can also be modelled using an analysis similar to that of Hose et al. (2004). The long-term goal is to use this approach as a preclinical testing tool to optimise both the biomechanical and pharmacological aspects of new stents.

Acknowledgements. This research was supported by an Applied Research Grant awarded by Enterprise Ireland to Medtronic AVE, Galway, Ireland, and the Centre for Bioengineering, Trinity College, Dublin, Ireland. The authors would like to thank Dr. Finbar Dolan, Medtronic AVE for his support for this research.

References

Brands, J. (2003). Fatigue test of porcine coronary arteries. Technical Report S457818, Technische Universiteit Eindhoven. Internal publication.

Edelman, E. R., and Rogers, C. (1998). Pathobiologic responses to stenting. *Am. J. Cardiol.* 81:4E–6E.

Fung, Y. C. (1993). *Biomechanics. Mechanical Properties of Living Tissues.* New York: Springer-Verlag, 2nd edition.

Hoffmann, R., Mintz, G. S., Haager, P. K., Bozoglu, T., Grube, E., Gross, M., Beythien, C., Mudra, H., vom Dahl, J., and Hanrath, P. (2002). Relation of stent design and stent surface material to subsequent in-stent intimal hyperplasia in coronary arteries determined by intravascular ultrasound. *Am. J. Cardiol.* 89:1360–1364.

Holzapfel, G. A., Gasser, T. C., and Ogden, R. W. (2000). A new constitutive framework for arterial wall mechanics and a comparative study of material models. *J. Elasticity* 61:1–48.

Hose, D. R., Grifftihs, B., Mahmood, S., Gunn, J. A., Sweeney, D., and Lawford, P. V. (2004). Numerical modeling of drug elution from coronary stents. In *Proc. 14th European Society of Biomechanics.* The Netherlands.

Huiskes, R. (1993). Failed innovation in total hip replacement. diagnosis and proposals for a cure. *Acta. Orthop. Scand.* 64:699–716.

Humphrey, J. D. (2002). *Cardiovascular Solid Mechanics. Cells, Tissues, and Organs.* New York: Springer-Verlag.

Ker, R. F., Wang, X. T., and Pike, A. V. L. (2000). Fatigue quality of mammalian tendons. *J. Exp. Biol.* 203:1317–1327.

Lally, C., and Prendergast, P. J. (2003). An investigation into the applicability of a mooney-rivlin constitutive equation for modelling vascular tissue in cardiovascular stenting procedures. In *Proceedings of the International Congress on Computational Biomechanics*, 542–550. Zaragoza, Spain.

Lally, C., Reid, A. J., and Prendergast, P. J. (2004). Elastic behaviour of porcine coronary artery tissue under uniaxial and equibiaxial tension. *Ann. Biomed. Eng.* 32:1355–1364.

Lally, C., Dolan, F., and Prendergast, P. J. (2005). Cardiovascular stent design and vessel stresses: a finite element analysis. *J. Biomech.* in press.

Lally, C. (2004). *An investigation into the influence of cardiovascular stent design on the development of restenosis using the finite element method.* Ph.D. Dissertation, Trinity College Dublin.

Loree, H. M., Grodzinsky, A. J., Park, S. Y., Gibson, L. J., and Lee, R. T. (1994). Static and circumferential tangential modulus of human atherosclerotic tissue. *J. Biomech.* 27:195–204.

van der Meulen, M. C. H., and Huiskes, R. (2002). Why mechanobiology? *J. Biomech.* 35:401–414.

Miner, M. A. (1945). Cumulative damage in fatigue. *J. Appl. Mech.* 12: A159–A164.

Mooney, M. (1940). A theory of large elastic deformation. *J. Appl. Phys.* 11:582–592.

Murray, J. D. (1989). *Mathematical Biology.* Springer-Verlag.

Ogden, R. W., and Schulze-Bauer, C. A. J. (2000). Phenomenological and structural aspects of the mechanical response of arteries. In Casey, J., and Bao, G., eds., *Mechanics in Biology.* New York: The American Society of Mechanical Engineers (ASME). AMD-Vol. 242/BED-Vol. 46, pp. 125–140.

Prendergast, P. J., and Taylor, D. (1994). Prediction of bone adaptation using damage accumulation. *J. Biomech.* 27:1067–1076.

Prendergast, P. J., Lally, C., Daly, S., Reid, A. J., Lee, T. C., Quinn, D., and Dolan, F. (2003). Analysis of prolapse in cardiovascular stents: a constitutive model for vascular tissue and finite element modelling. *J. Biomech. Eng.* 125:692–699.

Prendergast, P. J. (2002). Mechanics applied to skeletal ontogeny and phylogeny. *Meccanica* 37:317–334.

Roux, W. (1912). Anpassungslehre, Histomechanik und Histochemie. Mit Bemerkungen über die Entwicklung und Formgestaltung der Gelenke. *Virchows Arch. Pathologische Anatomie, Klinische Medizin (Berlin)* 209: 168–209.

Schechtman, H., and Bader, D. L. (2002). Fatigue damage of human tendons. *J. Biomech.* 35:347–353.

Schwartz, R. S., Chu, A., Edwards, W. D., Srivatsa, S. S., Simari, R. D., Isner, J. M., and Holmes Jr., D. R. (1996). A proliferation analysis of arterial neointimal hyperplasia: lessons for antiproliferative restenosis therapies. *Int. J. Cardiol.* 53:71–80.

Material Property Alterations with Early Atheroma in an Animal Model

K.B. Chandran[1], H. Kim[1], D.D. McPherson[2]

[1] University of Iowa, Department of Biomedical Engineering, USA
krishnan-chandran@uiowa.edu, hyunggun-kim@uiowa.edu
[2] Northwestern University, Department of Internal Medicine, USA
d-mcpherson@northwestern.edu

Atherosclerosis is a diffuse arterial disease developing over several decades when the arterial wall material properties are continuously altered. An understanding of the nature of such alterations may be exploited for early atheroma detection when pharmacological interventions may be used to arrest or reverse the disease process. As a new approach to evaluate regional alterations in arterial wall material properties *in vivo*, we employed a finite element and optimization analysis to three-dimensional reconstructed segments of femoral and carotid arteries with intravascular ultrasound image data in an animal model. Alterations in material properties predicted by the computational analysis were correlated with histology of the excised segments with induced atheroma lesions. In the femoral arteries, the elastic modulus initially decreased with lipidous lesions and segments with fibrous lesions became stiffer, whereas there was a linear increase in the elastic modulus in the carotid artery compared to that in the normal segments. These variations at different arterial segments could be explained by the phenomenon of arterial remodeling with atheroma formation. This new methodology can be potentially used *in vivo* to detect the extent and location of atheroma formation.

1 Introduction

Atherosclerosis is a diffuse and highly variable disease process that is distinguished by the subintimal accumulation of varying amounts of extracellular lipid, fibrous tissue, smooth muscle cells, and calcium. Atherosclerosis changes arterial wall morphology (Farrar et al. (1978)) and alters its mechanical properties (Kitney et al. (1989), Lee et al. (1992)). This pathologically altered morphology of diseased arterial tissue results in a complex structure that is geometrically irregular, structurally inhomogeneous, anisotropic, incompressible, non-linearly viscoelastic, subject to large strain deformations, and defies straightforward rheological characterization (Born and Richardson (1990), Cheng et al. (1993), Salunke and Topoleski (1997)). Even though the segments. However, the average strain in the lesion-free segments was similar to those with fibrous and advanced fibrous lesions. They were able to prove that elastography has a high sensitivity and specificity for identification of the local mechanical properties of the arterial wall and plaque. Even though the work by the de Korte group demonstrated the variability in the material properties with varying composition of the lesions, information on the three-dimensional regional variations in the arterial wall material properties affected by the atherosclerotic lesion was not presented in this study.

In order to describe the regional alterations in arterial wall material property with atheromatic lesions *in vivo*, we have developed a novel methodology to evaluate three-dimensional regional alterations in arterial wall material properties. We have demonstrated the feasibility of assessment of material property alterations with induced lesions in an animal model. We employed intravascular ultrasound (IVUS) imaging and a three-dimensional reconstruction of vascular segments with and without atheroma formation and performed finite element and optimization analysis to identify regional alterations in arterial wall material property and correlated the same with histology. Our discussion focuses on the benefits of these methodologies to evaluate atheroma progression *in vivo* and compares our findings in the material property alterations between the femoral and carotid arterial beds.

2 Materials and Methods

The details of the materials and methods to identify material property alterations with atheroma are included elsewhere (Chandran et al. (2003), Hamilton et al. (2005), Nagaraj et al. (2005)) and briefly described here.

2.1 Animal Model and Data Acquisition

A Yucatan miniswine atheroma model with similar atherosclerotic characteristics to humans ($n = 12$; weight 20–25 kg) was used. This protocol was approved by the Animal Care and Use Committee of Northwestern University. Following full anesthesia, one femoral artery was exposed with groin incision and one carotid artery with a neck incision. At baseline ($t = 0$ weeks), an intravascular ultrasound (IVUS) catheter was then inserted through the arteriotomy to image the vascular wall in the arterial segment of interest and subsequently the blood pressure signals were recorded using a catheter-tipped transducer. The endothelial cells in these arterial segments were denuded by means of a balloon catheter. Following these procedures, the incisions were closed. Contra-lateral arteries served as control in which no data was obtained and the endothelial cells were not denuded. The animal was placed on a hypercholesterolemic (15% lard, 2% cholesterol) diet regimen one month prior to the denudation and for 8 weeks post denudation (for a total of 3 months), for the development of induced atheroma. After the 8-week period ($t = 8$ weeks), following anesthesia, both sides of the femoral and carotid arteries were exposed. Pressure and imaging data were obtained. Following experimental data acquisition, the animals were euthanized with a bolus of saturated KCl, the arteries excised, and histological processing performed.

2.2 Histology

After sacrifice, the imaged sections of the arteries were tied and filled with 4% buffered neutral formaldehyde at 100 mmHg in order to ensure that histologically the arteries were in near physiological condition. The arteries were marked by a superficial cut into the external media along the longitudinal axis. After fixation the arteries were dehydrated, cut at 2 mm intervals, placed in tomographic sequence, and embedded in paraffin. Each 2 mm block was cut at two levels at 4 μm. One slide of each artery was stained with hematoxylin and eosin and the other with a modified Movat's pentachrome. With this stain, elastic tissue stains black; fibrous tissue, green; cell cytoplasm, red; and nuclei, blue. Digital photomicrographs of the arteries were taken. Magnification was determined by a photographing stage micrometer (100 lines/mm) (final magnifications 51.75$\times$ or 103$\times$). Each artery was marked at 30° intervals around the circumference of the internal elastic lamina starting at a point diametrically opposite from the superficial cut. The marked arterial segments were printed and the intimal and medial thickness measured (perpendicular to the internal elastic lamina) with actual measurements calculated from the magnification. The pathology was determined and coded. The histological coding data and the 3D reconstructed IVUS wall tracings were aligned circumferentially and longitudinally by matching wall thicknesses using custom-designed software. The IVUS images have an axial resolution of 1 mm and hence sampling of the histology data at 2 mm intervals was sufficient to register the histologic data with the IVUS data in the axial direction.

2.3 Finite Element Analysis and Optimization

From the image data, the intimal and adventitial borders in each image were manually segmented and sampled every ten degrees with respect to the catheter center. Putting each cross-section in tomographic sequence resulted in a three-dimensionally reconstructed model of the arterial wall segment of interest. The three-dimensional finite element model with an assumed initial elastic constant was employed to compute the deformation of the vascular segment using the commercial finite element software ANSYS (version 6.0). The wall material was assumed to be homogeneous, and isotropic with an assumed initial linear elastic modulus of 100 kPa (Bergel (1961)) and nearly incompressible with a Poisson ratio of 0.49 (Chuong and Fung (1984)) throughout the wall thickness. A finite element mesh of the arterial structure consisted of 480 wedge elements and 504 nodes and a single layer of elements was used along the arterial wall thickness. Since the predicted nodal displacements were directly compared with the displacements measured from the imaged data later for the design sensitivity analysis, it was appropriate to employ a single layer across the wall thickness. The deformation of the reconstructed vascular geometry was computed from end diastole to end systole with the corresponding pulse pressure measured from the recorded pressure signals applied as the load at the intimal surface. The local minima were computed and the optimization was ended when the difference in error between successive iterations was less than or equal to 0.001. The system identification process using updated regional elastic moduli and regional sensitivity values were continued until the optimum cost was obtained.

The material properties of the arterial segments at baseline were compared with the normal arterial segments (predominantly from contra-lateral arterial segments) at $t = 8$ weeks to identify alterations of material properties with high cholesterol diet. Histology was utilized at $t = 8$ weeks to confirm that all the elements used for the normal comparison were histologically normal. Moreover, comparisons were made between the finite element groups to evaluate material properties with cell injury and high cholesterol diet using the histological data. Normal elements at baseline and those that remained normal after aging to 8 weeks were compared with the elements demonstrating atheroma formation.

2.4 Statistical Analysis

Data were analyzed with the SAS statistical software (version 8.02). Results are presented as mean ± SD, along with the number (n) of cases used in the computations. A two sample t-test was used to compare the animals with predominantly normal segments to those with predominantly abnormal segments. The General Estimating Equation (GEE) was used to estimate mechanical property difference associated with different stages of atheroma development in the denuded arteries (left side) at $t = 8$ weeks (Liang and Zeger (1986), Zeger and Liang (1986)).

3 Results

Resultant wall thickness from the imaged data, elastic modulus distributions (computed from the optimization algorithm) and corresponding histological data for a femoral artery at baseline and following 8 weeks of induced atheroma development are shown in Fig. 1. On the top row, the results of the analysis in arterial segments at baseline ($t = 0$ weeks before endothelial denudation) show uniform distribution of wall thickness and the predicted elastic modulus distribution. These segments were assumed to be without atherosclerotic lesion involvement. In the middle panel, similar data are shown in the arterial segment without denudation at $t = 8$ weeks (Control – contra-lateral) and

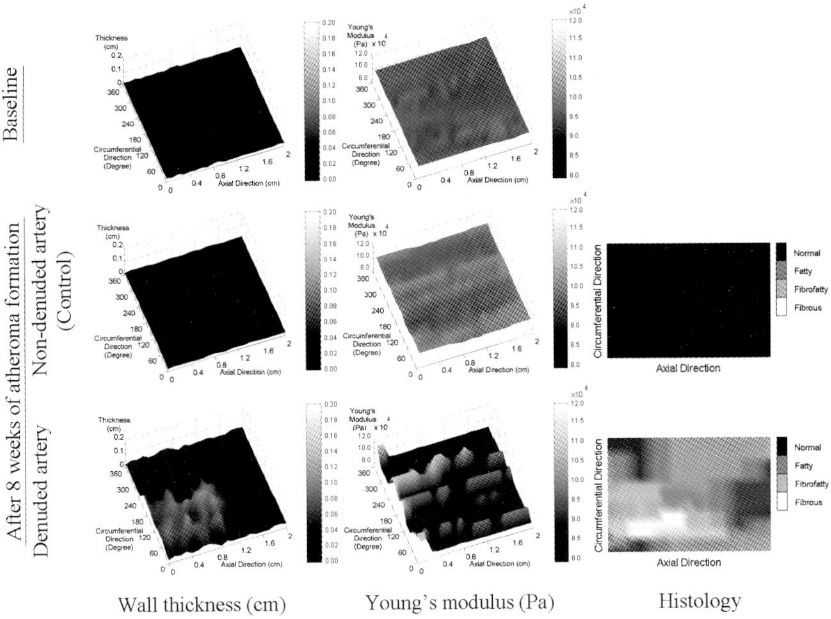

Fig. 1. Wall thickness, elastic modulus and histological distribution in the femoral arterial segment of the Yucatan miniswine model with induced atheroma.

again uniform wall thickness and elastic modulus distribution are observed with histology indicating no lesions. It should be noted that the right side of the artery was not imaged at baseline ($t = 0$ weeks) in order to prevent any possible endothelial injury due to catheter insertion for data acquisition. Hence it was not possible to compare the right side of the artery data at $t = 0$ weeks to that at $t = 8$ weeks.

In contrast, in the bottom row, data are presented for the artery with denuded endothelial cells and high cholesterol diet at $t = 8$ weeks and exhibit regional alterations in wall thickness as well as elastic modulus. The corresponding histological data also show the distribution of fatty, fibrofatty, and fibrous lesions.

For carotid arteries, similar pattern of results was observed, i.e. relatively uniform distribution of elastic modulus and wall thickness for baseline and control ($t = 8$ weeks), but significant regional alteration in both elastic modulus and thickness for the artery with denuded endothelial cells ($t = 8$ weeks).

Figure 2 illustrates average changes in the wall thickness and corresponding elastic modulus of the femoral arterial segments for the baseline ($t = 0$ weeks); control (non-denuded arteries with high cholesterol diet for 8 weeks) and those segments with lesions (denuded arteries with high cholesterol diet for 8 weeks). At 8 weeks of high cholesterol diet alone (Control), there was no significant

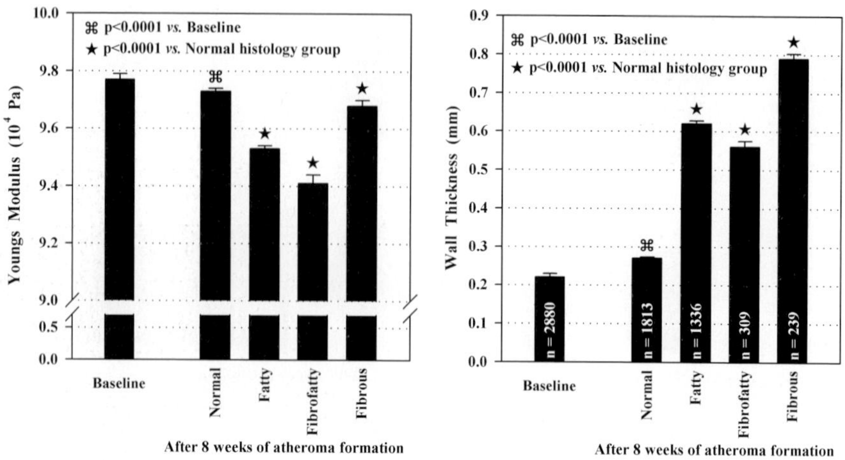

Fig. 2. Elastic modulus and wall thickness comparison for baseline, normal, and atheromatous femoral arterial elements (mean ± sem).

change in the elastic modulus in the finite elements in these segments – Elastic modulus (all values × 10^4 Pa) 9.77 ± 0.02 ($n = 2{,}880$ for $t = 0$ weeks; baseline) vs. 9.73 ± 0.01 ($n = 1{,}813$ for $t = 8$ weeks; non-denuded control) whereas the wall thickness increased – wall thickness (all values in mm) 0.22 ± 0.01 (for $t = 0$ weeks) vs. 0.28 ± 0.003 (for $t = 8$ weeks).

On the other hand, the artery with denuded endothelial cells and high cholesterol diet at 8 weeks showed significant decrease of elastic modulus. The elements in the atherosclerotic segments were classified as fatty, fibrofatty, and fibrous depending upon the histologic grading. The computed elastic modulus was: 9.53 ± 0.01 for fatty elements; 9.41 ± 0.03 for fibrofatty elements; and 9.68 ± 0.02 for fibrous elements (all $p < 0.001$ vs. control). Wall thickness did increase in all the groups – (all values in mm) 0.28 ± 0.003 for control non-denuded, 0.62 ± 0.01 for fatty elements, 0.56 ± 0.02 for fibrofatty elements, and 0.79 ± 0.01 for fibrous elements (all $p < 0.001$ vs. control).

For the carotid arteries in the same animals and at the same time frame after the endothelial injury, the computed average elastic modulus was 9.34 ± 0.36 and measured average wall thickness was 0.41 ± 0.21 for normal segments, whereas the corresponding values were 9.52 ± 0.36 and 0.78 ± 0.30 for segments with early lesions ($p < 0.05$ vs. normal elements).

4 Discussion

It is known that atherosclerotic lesions develop diffusely in arteries and result in a complicated three-dimensional morphology. Any material property alterations with the initiation and development of the disease will also show regional variation. Moreover, material properties will alter with atheroma progression, from early involvement of lipids to advanced calcified plaques as has been demonstrated with the present animal model with induced atheroma for a period of up to 8 weeks after endothelial denudation. Thus, any methodology for assessment of material property alterations *in vivo* should permit better identification of regional vascular involvement in atheroma stages when interventions can be employed to arrest or reverse the disease process. Regional material property distributions in advanced lesions could be used for the mechanical analysis of plaque rupture, a dominant factor in thrombo-embolic complications. By employing a three-dimensional reconstruction of arterial segments in a miniswine model obtained with IVUS imaging and a finite element and optimization analysis, the results demonstrate that the wall thickness increases with early lesions for both femoral and carotid arteries. The alterations of elastic modulus, however, show a different tendency between femoral and carotid arteries. Comparing abnormal segments with normal segments, elastic modulus decreases for femoral arteries, whereas it increases for carotid arteries.

The results for femoral arteries from the present study can be compared with previous studies on arterial segment material property alterations with atheroma formation. Hayashi and Imai (1997) measured the force-deformation characteristics of atherosclerotic and normal arterial specimens obtained from induced atheroma in rabbits with denuded endothelial cells and high cholesterol diet for 12 weeks. Their studies demonstrated that the force deformation characteristics of the specimens with atheromatic plaques were less stiff compared to the normal wall. Their results with induced atheroma of a comparable period to our present study show similar changes with wall material stiffness decreasing with early atheroma. The de Korte group evaluated the application of *in vitro* IVUS elastography studies in diseased human femoral and coronary arteries subject to static intraluminal pressures (de Korte et al. (2000)). Their results showing that strain magnitudes increase with fibrous lesions agree with our results, even though the magnitudes of the alterations are significantly larger in their results compared to ours. The main difference between the two previous studies and the present study is in the methodology. In both the studies, the diseased arterial segments were harvested and a static transmural pressure was applied *in vitro*. Moreover, in both the *in vitro* studies, the arterial segmental deformation measurements with excised specimens obtained *in vitro* ignored the effect of the surrounding structures, whereas the deformation measured in the present work was *in vivo* under actual physiological pulse pressure that includes the effect of those structures. More recently, de Korte et al. (2002) employed IVUS elastography to identify the mechanical effects of atherosclerotic plaque development *in vivo* in an animal model similar to ours. Their results are in agreement with ours. They reported radial strain measurements of about 2% determined from two image frames near end diastole for a pressure differential of 4 mmHg. Computed radial strains in the present study for a pulse pressure of the order of 30–40 mmHg were in the range of 20%. Quantitative comparison of the results between the present study and that of de Korte et al. (2002) is not possible because of the variations in the age of the animals, time of the study, and the protocols employed. The elastic modulus magnitudes computed in the segments without atheroma involvement in the present study agree with the elastic modulus values reported by Bergel (1961) for the physiological range. By the elastic modulus decreasing with early atheroma formation, it is possible that increased local distensibility along with medial thinning would allow for the accommodation of expanding atheroma while maintaining the luminal cross-section. Once later components of atheroma are incorporated, including fibrous and calcific components, then the elastic modulus would increase and no further regional dilation would occur.

Our data demonstrate that, in the carotid arteries, the elastic modulus tends to increase with induced atheroma. These data are in contrast to what we have demonstrated in the femoral artery and what others have demonstrated previously, in which elastic modulus decreased with initial atheroma formation followed by increasing elastic modulus with later atheroma growth. These discrepancies could be explained by the phenomenon of arterial remodeling with atheroma formation. This morphologic phenomenon was initially demonstrated *in vitro* in pathologic specimens by Crawford and Levene (1953) in the iliofemoral arteries. Subsequently, Glagov et al. (1987) demonstrated this phenomenon in coronary arteries *in vitro* in pathologic specimens and McPherson et al. (1991) in our group have demonstrated this phenomenon *in vivo* in coronary arteries. Arterial remodeling is felt to occur in femoral and coronary arteries, but to a much lesser degree in carotid arteries and the aorta (Glagov et al. (1988)). The differences seem to be due to the higher amount of elastic tissues in the carotid arteries and the aorta and, to a lesser extent, the relatively greater amount of smooth muscle in the carotid arteries. In a few carotid arterial segments there was a small decrease in material properties following 8 weeks of atheroma formation. This could be explained by some early, minimal remodeling of the carotid arteries as atheroma develops. However, in the majority of the carotid arterial segments, elastic modulus increased by 8 weeks of atheroma formation – indicating that the degree of arterial remodeling commonly seen in the femoral arteries tends to be much less in the carotid arteries. As atheroma formation at $t = 8$ weeks represents early changes of induced atheroma, any decrease in vascular material properties in the carotid arteries with induced atheroma to 8 weeks would be transient at best and essentially would be gone by the time of lesion evaluation in the clinical setting.

There are a number of factors that may have influenced our results. The exact matching of specimens *in vivo* from $t = 0$ to $t = 8$ weeks was prone to some error. To minimize this, the arteries were marked so that the arterial segment started and ended at similar portions for $t = 0$ and $t = 8$ weeks *in vivo* measurements. When the arteries were evaluated at the terminal experiment, markers were placed to match the segments evaluated *in vivo* to the histological specimens. Some mismatch in segments may have occurred *in vivo* with growth of the artery and mismatch may also have occurred between the IVUS reconstructions and the histological reconstructions. Moreover, we used the non-altered arteries at $t = 8$ weeks as the normal elements. The benefit of this is that the effect of arterial injury on normal elements was discounted. The problem with this analysis, however, is that the effect of different arterial sides was neglected. Side-to-side variability may remain unaccounted for in the data analysis. The IVUS measurements of wall thickness are slightly less accurate when compared to histology. Resolution is approximately 0.1 mm for IVUS and 0.02 mm (for the 51.75× magnification) for histology. Our model was only an 8-week model of atheroma formation. This allowed evaluation of early stages of induced atheroma. Although not examined in this study, changes in material properties with further atheroma development can also be evaluated using the same methodology with the experimental protocols extending to longer duration.

In the mechanical analysis, we assumed the arterial wall to be made of a single material layer and hence ignored the effect of individual layers of intima, media, and adventitia in the deformation due to the applied transmural pressure. Even though the stress-strain characteristics of the arterial wall have been demonstrated to be non-linear (Bergel (1961)), in the physiological range of deformation, we assumed a linear relationship in deformation. The significance of the estimation of elastic modulus distribution using a linear relationship, similar to that presented in this study, lies in the fact that elastic modulus has a straightforward physical meaning allowing for a tangible correlation with histology and hence validation (i.e. modulus would be expected to be high in calcified regions and low in lipid regions). If shown to be accurate, predicted alterations in the elastic modulus would lead to a novel capability of predicting the composition inside the atheroma without having to excise the artery for histology. Our method can be extended to include multiple layers across the wall thickness and a step-wise incremental analysis can also be included in computing the deformation due to the applied pulse pressure. Incorporation of these features may provide more detailed and accurate results for regional material property alterations with atheroma development. However, incorporation of multiple layers in the analysis will require identification of the medial and adventitial borders in order to obtain data on the variations in displacement across the wall thickness.

Once the reliability of our method is determined for predicting regional material property alterations with atheroma formation, this method could potentially be applied to identify regional variations in the diffuse disease with IVUS examination. We would anticipate that regions identified with increased elastic modulus would benefit from interventions such as stent placement and segments with early atheroma and low elastic modulus would benefit from drug therapy to stabilize the arterial smooth muscle tone. With advanced stages of calcified plaques, this method can also potentially be employed to predict plaque rupture that could result in acute thrombus formation. The details of the methodology and the results presented here can be found elsewhere (Hamilton et al. (2005), Nagaraj et al. (2005)).

Acknowledgements. Sponsored in Part by a grant from the National Institutes of Health (HL 62504: National Heart, Lung, and Blood Institute), Iowa Department of Economic Development, and the Feinberg Cardiovascular Research Institute.

References

Bergel, D. H. (1961). The static elastic properties of the arterial wall. *J. Physiol. London* 156:445–457.

Born, G. V., and Richardson, P. D. (1990). *Mechanical properties of human atherosclerotic lesions.* New York: Springer–Verlag.

Buntin, C. M., and Silver, F. H. (1990). Noninvasive assessment of mechanical properties of peripheral arteries. *Ann. Biomed. Eng.* 18:549–566.

Chandran, K. B., Mun, J. H., Choi, K. K., Chen, J. S., Hamilton, A., Nagaraj, A., and McPherson, D. D. (2003). A method for in-vivo analysis for regional arterial wall material property alterations with atherosclerosis: preliminary results. *Med. Eng. Phys.* 25:289–298.

Cheng, G. C., Loree, H. M., Kamm, R. D., Fishbein, M. C., and Lee, R. T. (1993). Distribution of circumferential stress in ruptured and stable atherosclerotic lesions. a structural analysis with histopathological correlation. *Circulation* 87:1179–1187.

Chuong, C. J., and Fung, Y. C. (1984). Compressibility and constitutive equation of arterial wall in radial compression experiments. *J. Biomech.* 17:35–40.

Crawford, T., and Levene, C. I. (1953). Medial thinning in atheroma. *J. Path. Bact.* 56:19–23.

de Korte, C. L., Pasterkamp, G., van der Steen, A. F., Woutman, H. A., and Bom, N. (2000). Characterization of plaque components with intravascular ultrasound elastography in human femoral and coronary arteries in vitro. *Circulation* 102:617–623.

de Korte, C. L., Sierevogel, M. J., Mastik, F., Strijder, C., Schaar, J. A., Velema, E., Pasterkamp, G., Serruys, P. W., and van der Steen, A. F. (2002). Identification of atherosclerotic plaque components with intravascular ultrasound elastography in vivo: a yucatan pig study. *Circulation* 105:1627–1630.

Dobrin, P. B. (1986). Biaxial anisotropy of dog carotid arteriy: estimation of circumferential elastic modulus. *J. Biomech.* 19:351–358.

Farrar, D. J., Green, H. D., Bond, M. G., Wagner, W. D., and Gobbee, R. A. (1978). Aortic pulse wave velocity, elasticity, and composition in a nonhuman primate model of atherosclerosis. *Circ. Res.* 43:52–62.

Farrar, D. J., Green, H. D., Wagner, W. D., and Bond, M. G. (1980). Reduction in pulse wave velocity and improvement of aortic distensibility accompanying regression of atherosclerosis in the rhesus monkey. *Circ. Res.* 47:425–432.

Glagov, S., Weisenberg, E., Zarins, C. K., Stankunavicius, R., and Kolettis, G. J. (1987). Compensatory enlargement of human atherosclerotic coronary arteries. *N. Engl. J. Med.* 316:1371–1375.

Glagov, S., Zarins, C., Giddens, D. P., and Ku, D. N. (1988). Hemodynamics and atherosclerosis. Insights and perspectives gained from studies of human arteries. *Arch. Pathol. Lab. Med.* 112:1018–1031.

Hamilton, A., Kim, H., Nagaraj, A., Mun, J. H., Yan, L. L., Roth, S. I., McPherson, D. D., and Chandran, K. B. (2005). Regional material property alterations in porcine femoral arteries with atheroma development. *J. Biomech.* in press.

Hayashi, K., and Imai, Y. (1997). Tensile property of atheromatous plaque and an analysis of stress in atherosclerotic wall. *J. Biomech.* 30: 573–579.

Hayashi, K., Ide, K., and Matsumoto, T. (1994). Aortic walls in atherosclerotic rabbits – mechanical study. *J. Biomech. Eng.* 116:284–293.

Hayashi, K. (1993). Experimental approaches on measuring the mechanical properties and constitutive laws of arterial walls. *J. Biomech. Eng.* 115: 481–488.

Hudetz, A. G., Mark, G., Kovach, A. G., Kerenyi, T., Fody, L., and Monos, E. (1981). Biomechanical properties of normal and fibrosclerotic human cerebral arteries. *Atherosclerosis* 39:353–365.

Kitney, R. I., Moura, L., and Straughan, K. (1989). 3-D visualization of arterial structures using ultrasound and voxel modelling. *Int. J. Card. Imaging* 4:135–143.

Lee, R. T., Richardson, S. G., Loree, H. M., Grodzinsky, A. J., Gharib, S. A., Schoen, F. J., and Pandian, N. (1992). Prediction of mechanical properties of human atherosclerotic tissue by high-frequency intravascular ultrasound imaging. an in vitro study. *Arterioscler. Thromb.* 12:1–5.

Liang, K. Y., and Zeger, S. L. (1986). Longitudinal data-analysis using generalized linear-models. *Biometrika* 73:13–22.

McPherson, D. D., Sirna, S. J., Hiratzka, L. F., Thorpe, L., Armstrong, M. L., Marcus, M. L., and Kerber, R. E. (1991). Coronary arterial remodeling studied by high-frequency epicardial echocardiography: an early compensatory mechanism in patients with obstructive coronary atherosclerosis. *J. Am. Coll. Cardiol.* 17:79–86.

Nagaraj, A., Kim, H., Hamilton, A., Mun, J. H., Smulevitz, B., Kane, B. J., Yan, L. L., Roth, S. I., McPherson, D. D., and Chandran, K. B. (2005). Porcine carotid arterial material property alterations with induced atheroma: An in vivo study. *Med. Eng. Phys.* 27:147–156.

Pynadath, T. I., and Mukherjee, D. P. (1977). Dynamic mechanical properties of atherosclerotic aorta. a correlation between the cholesterol ester content and the viscoelastic properties of atherosclerotic aorta. *Atherosclerosis* 26:311–318.

Salunke, N. V., and Topoleski, L. D. (1997). Biomechanics of atherosclerotic plaque. *Crit. Rev. Biomed. Eng.* 25:243–285.

Zeger, S. L., and Liang, K. Y. (1986). Longitudinal data-analysis for discrete and continuous outcomes. *Biometrics* 42:121–130.

Microscopic Analysis of Residual Stress and Strain in the Aortic Media Considering Anisotropy of Smooth Muscle Layer

T. Matsumoto,[1] T. Furukawa,[2] K. Nagayama[1]

[1] Nagoya Institute of Technology, Biomechanics Laboratory, Japan
takeo@nitech.ac.jp, k-nagaym@nitech.ac.jp
[2] Nagoya Institute of Technology, Biomechanics Laboratory, Japan (Present affil.: Mitsubishi Heavy Industries, Ltd.)

The aortic media is made of concentric layers whose unit is called a lamellar unit, a pair of elastic lamina (EL) and a smooth muscle-rich layer (SML). We found that the EL was stiffer than the SML. Such heterogeneity may cause residual stress, which is hardly released by macroscopic dissection, but is released in the area close to the sections, causing hills and valleys on the surface. To check this hypothesis, we measured the topography and the stiffness distribution of the sections of the porcine thoracic aortas to find stiff hills and soft valleys. We have estimated residual stress with finite element models considering anisotropy of the SML to find that the values were not smaller than the conventionally estimated ones in all cases examined. Fairly large stresses may still reside in the opened-up aortic media which was believed stress-free.

1 Introduction

It is well known that arterial walls have residual stresses: a ring-like segment of an aorta springs open to form an arc upon radial cutting in many cases. It has been hypothesized that this happens because the radial distribution of the circumferential stress in the aortic media is uniform *in vivo*, and as a result, the segment at no load has compressive residual stress near the inner wall and tensile near the outer, causing the opening-up of the ring-like segment (Fung (1990)). The opened-up configuration should be stress-free if the aortic media is homogeneous. Many studies have been carried out based on this assumption since 1983 when the importance of residual stress was pointed out for the first time (Vaishnav and Vossoughi (1983); Fung (1984)).

However, the wall is heterogeneous at the microscopic level: its media has a layered structure called a lamellar unit which is a pair of elastic lamina (EL) composed of elastin and a smooth muscle-rich layer (SML) (Wolinsky and Glagov (1967)). The elastic modulus of elastin is about 0.6 MPa (Fung (1981)), while that of the smooth muscle is of the order of 0.01 MPa in the relaxed state both at the whole tissue level (Whitemore (1968)) and at the cellular level (Matsumoto et al. (2000)). Thus, the EL might be much stiffer than the SML. In fact, we recently examined the difference in the mechanical properties between the EL and the SML of the porcine thoracic aortas by measuring the change in thickness of the two layers during radial compression, and found that the EL is stiffer than SML (Matsumoto et al. (2004b)).

It has been pointed out that normal stress is maintained at a constant level in various tissues, including artery walls (Wolinsky (1971), Matsumoto and Hayashi (1994)) and left ventricular walls (Grossman et al. (1975)). If the circumferential stress in an *in vivo* state is the same between the two layers, *ex vivo* residual stress of each layer in this direction should be different because the stress-strain relationships differ (Fig. 1). Such residual stress is not released fully by radial cutting, but is released in the area close to the cut surface, causing hills and valleys on the surface due to compressive and tensile stresses, respectively (Fig. 2). To check this hypothesis, we recently measured the surface topography and the distribution of stiffness on the cut surface of the aortas to find that the cut surface exhibits hill and valley pattern, as shown in Fig. 2 (Matsumoto et al. (2004a)).

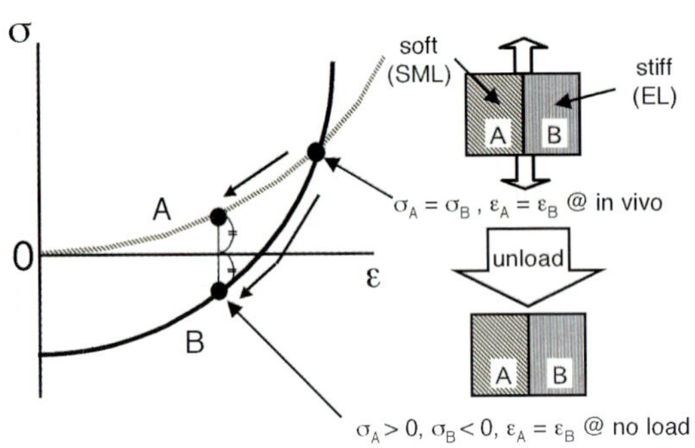

Fig. 1. Schematic diagram on residual stress caused by material heterogeneity (Matsumoto et al. (2004b)).

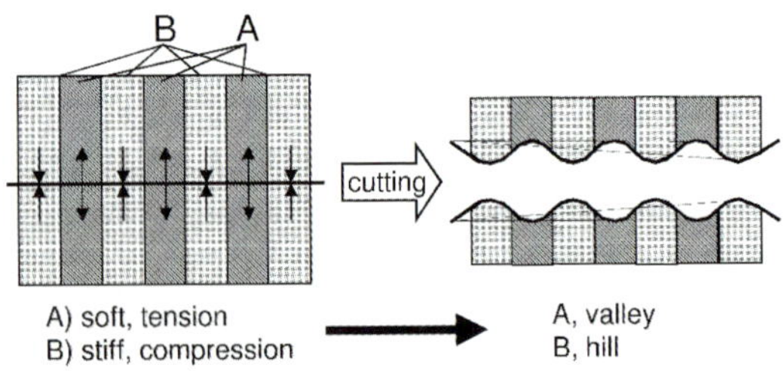

Fig. 2. Hill and valley pattern appearing on the cut surface of a body with microscopic residual stress (Matsumoto et al. (2004b)).

We then estimated the amount of residual stress from a finite element analysis with an isotropic model to find that the stress was very large: it was about 10 times higher than that estimated in the conventional analyses ($\pm 1 - 10$ kPa) (Matsumoto et al. (2004a)). The reason for this discrepancy may partly be attributable to the assumption of isotropy: smooth muscle cells and their intracellular filaments align in the circumferential direction in the SML, and it may therefore be unnatural to assume that the SML is isotropic: the shear modulus of the SML may be much lower than that obtained in the isotropic model.

In this paper, we first introduce the outlines of the radial compression test and the surface topography measurement, and then report the results of our analysis of the relationship between the amount of residual stress and the surface topography, considering anisotropy of the SML.

2 Estimation of Heterogeneity in the Elastic Properties of the Lamellar Unit

2.1 Radial Compression Test

Wall thickness decreases when an artery segment is compressed radially. If the EL is stiffer than the SML, compressive strain is smaller in the EL and larger in the SML than in the whole wall. The ratio of the strain between the EL and the SML is equal to the inverse ratio of the elastic modulus between the two layers.

We used porcine aortas as test models (Matsumoto et al. (2004b)). Three tubular segments of descending thoracic aortas were obtained at a local slaughterhouse. After excess fat and adventitia were carefully removed, fifteen to twenty 9 mm square specimens (2–3 mm thick) were cut out from each segment.

A tensile tester was used for the radial compression test (Fig. 3). The tester was set in a compression mode and jigs for tensile testing were replaced with an indenter and a specimen bath. The indenter has a 20 mm square aluminum plate (2 mm thick) at its tip to compress the specimen. The specimen bath is 60 mm square with the depth of 15 mm. To prevent the slippage of the specimen during compression, a 1 mm long acupuncture needle of 0.2 mm diameter was fixed at the center of its bottom. Each of the specimens was compressed in the radial direction at the rate of 0.5 mm/min in a physiological saline solution at room temperature to various strain levels (from -0.1 to -0.5). The specimen was then immediately fixed with formalin for at least 6 hours while being compressed. The specimen was released after the fixation and histological sections perpendicular to the longitudinal direction were obtained to measure the thickness of the EL and the SML. Specimens adjacent to each compressed specimen were fixed without compression for controls.

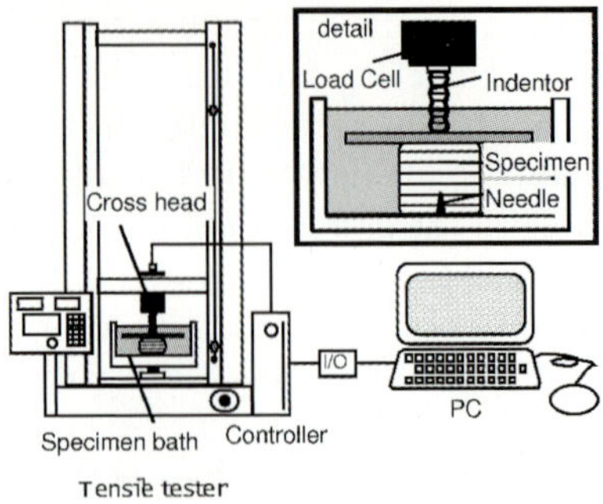

Fig. 3. Experimental set up for the radial compression test (Matsumoto et al. (2004b)).

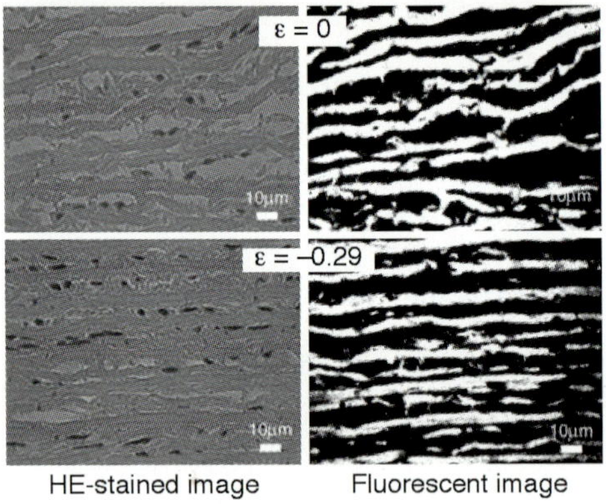

Fig. 4. Examples of histological sections of formalin-fixed specimens subjected to various levels of radial compressive strain (Matsumoto et al. (2004b)).

2.2 Measurement of the Thickness in EL and SML

Histological sections were stained with Hematoxylin-Eosin (HE). When stained with HE, the elastin becomes fluorescent (excitation 480–550 nm, emission > 590 nm), and the EL can be observed clearly, as shown in Fig. 4 (right panels). Eight-bit gray-scale fluorescent images were taken with a digital microscope camera for an image analysis. The image was digitized for the EL and the relative area occupied by the EL (area in white) and the SML (area in black) were obtained. The mean thickness of the EL and the SML were obtained from the relative areas, the scale factor and the number of lamellae. Soft biological tissues fixed with formalin are not completely rigid. The degree of fixation is different among components of the soft tissues. The fixed ELs and SMLs increase their thickness after the compressive force is removed. The thicknesses during compression were obtained by considering the mechanical properties of formalin-fixed elastin and aortic tissues (Matsumoto et al.

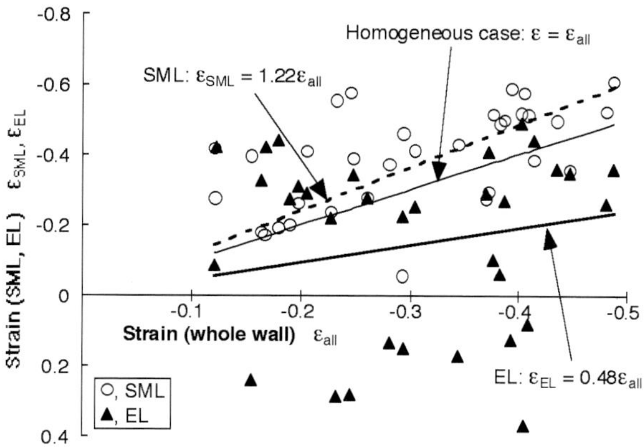

Fig. 5. Difference in strain between EL and SML during radial compression (Matsumoto et al. (2004b)).

(2004b)). Roughly speaking, the initial elastic moduli of the EL, SML, and whole wall estimated to be 180, 80, 100 kPa, respectively for fresh specimen increased to 230, 640, 480 kPa, respectively following fixation. The compressive strain decreased by about 38, 20, and 14% for the EL, SML, and whole wall, respectively.

2.3 Difference in Mechanical Properties between EL and SML

Figure 5 summarizes the change in strain due to radial compression. Although the scatter was large, strain was generally higher in the SML than in the EL. The straight lines are regression lines through the origin for the EL and SML. The ratio of their slope corresponds to the ratio of the compliance of the two layers. The slope was about 2.5 times higher in the SML than in the EL, indicating that the compressive elastic modulus is 2.5 times higher in the elastic lamina than in the smooth muscle-rich layer.

3 Measurement of Surface Topography and Stiffness Distribution

3.1 Scanning Micro Indentation Tester (SMIT)

We developed a scanning micro-indentation tester (SMIT) to measure surface topography and stiffness distribution (Matsumoto et al. (2004a)). A schematic diagram of the SMIT is shown in Fig. 6. It is a tester to measure the surface topography and stiffness distribution of a specimen surface by pressing a cantilever tip against the specimen surface while scanning it as with an atomic force microscope in the contact mode. The cantilever was made from a micro

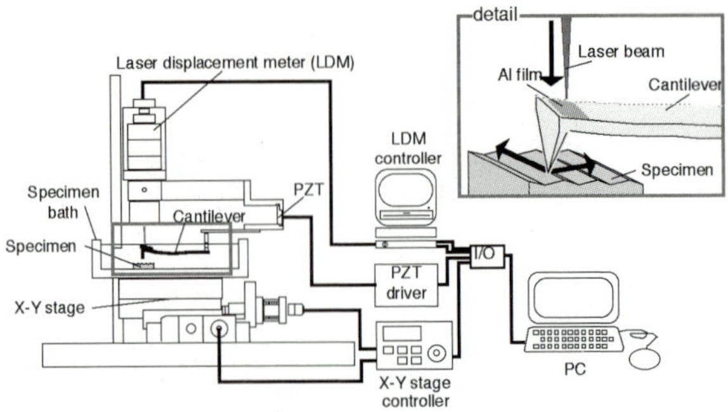

Fig. 6. Schematic diagram of the scanning micro indentation tester (SMIT) (Matsumoto et al. (2004a)).

glass plate of 65 mm × 1 mm × 0.15 mm by pulling it with a pipette puller to make its tip diameter 3–5 μm and bending it with a microforge at right angles at the point 4 mm from the tip. The cantilever was driven by a PZT actuator and the displacement of its tip was measured with a confocal laser displacement meter (LT8100, Keyence Corp., Japan). The measurement was done in a bath filled with a physiological saline solution at room temperature. The specimen bath was set on a motor-driven XY stage to scan the point of measurement. The contact point was determined at the point where the cantilever began deflecting, and a stiffness index α was obtained as the initial slope of the indentation-deflection curve. The index α was converted to elastic modulus E with a formula obtained by calibrating the tester with silicone elastomers with known elastic moduli.

3.2 Indentation Test

Rectangular specimens of 5 mm × 10 mm were excised from porcine thoracic aortas as described in Section 2.1. Macroscopic residual stresses are removed during excision. Each of the specimens was embedded in an agar gel with low gelling temperature (30 $\sim$ 31°C, Nacalai Tesque, Japan) and sliced with a tissue sectioner (Microslicer DTK-1500, Dosaka EM Co., Ltd., Japan) in the saline to obtain the surface perpendicular to the circumferential or axial directions of the aorta. The specimen with its bath was then mounted on the XY stage of the SMIT for the measurement. The indentation test was performed in the area of 100 μm × 100 μm at 2 μm intervals to obtain the surface topography and the stiffness distribution.

3.3 Surface Topography and Stiffness Distribution

The surface of the section shows a hill and valley pattern perpendicular to the radial direction (Fig. 7). The average distances between the peaks and the peak height were $\sim$ 25 μm and $\sim$ 8 μm, respectively. The elastic modulus

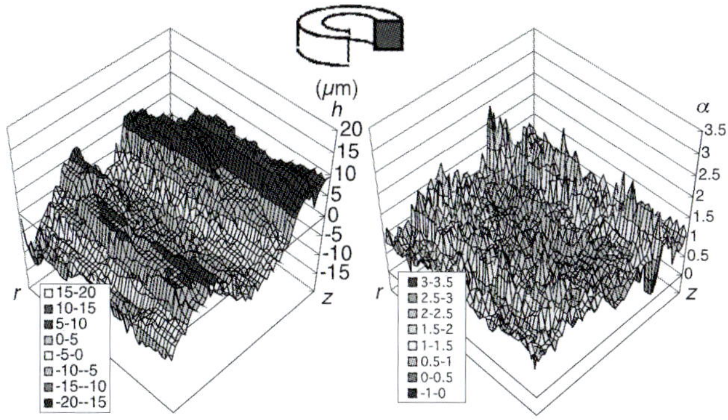

Fig. 7. An example of data obtained with the SMIT for a section of a porcine thoracic aorta ($50\,\mu\text{m} \times 50\,\mu\text{m}$).

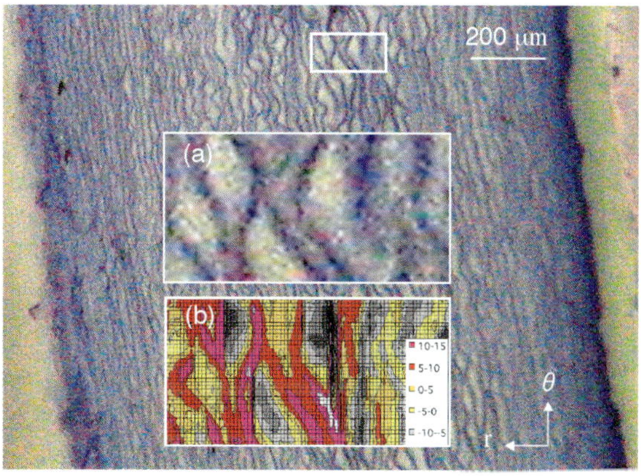

Fig. 8. Elastic lamina stained with gentian violet vs. surface topography. Elastic lamina (a) corresponded well with hills (b). r and θ indicate radial and circumferential directions, respectively.

estimated from a was $\sim$ 180 kPa at the peak and $\sim$ 52 kPa at the bottom. Staining of the measured surface showed that the hills corresponded to the ELs (Fig. 8). These results indicate that the hill is EL and the valley SML, i.e. EL is compressed and SML stretched in the lamellar unit.

4 Estimation of Residual Stress and Strain

4.1 Finite Element (FE) Models

Formation of the hills and valleys following the sectioning was simulated with a 2D FE model, as shown schematically in Fig. 9. A pair of a half of the EL and a half of the SML close to the section perpendicular to the wall circumference

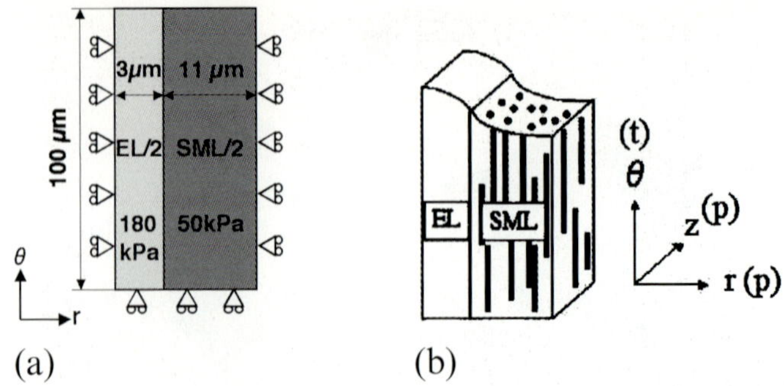

Fig. 9. Finite element models used in the analysis: (a) model dimensions and constraints; (b) transverse isotropy in the anisotropic model.

were modeled with 2D plane strain elements. Each layer was assumed to be homogeneous and linearly elastic. The number of the elements was 300 for the EL and 1100 for the SML. The dimensions of the model were taken based on the measurements: the thicknesses of the EL and the SML were taken as 6 μm and 22 μm, respectively, and the height as 100 μm. The EL was assumed to be isotropic in all analyses, while the SML was assumed to be either isotropic or anisotropic. In the isotropic analysis, the elastic modulus was taken as 180 kPa for the EL and 50 kPa for the SML, and the Poisson's ratio 0.4999. In the anisotropic analysis, transverse isotropy ($r-z$ plane isotropy) was assumed in the SML (Fig. 9(b)). The shear modulus $G_{r\theta}$ of the SML was decreased from the isotropic condition ($G = E/3$). The transverse elastic modulus $E_t = E_\theta$ was taken as 50 kPa. Other moduli were taken by considering incompressibility and Drucker stability (Drucker (1959)). The elastic properties of the EL in the anisotropic analysis were the same as those in the isotropic analysis.

The analysis was performed with ABAQUS/Standard version 5.8 (HKS, Inc.) at the Center for Information and Media Studies, Nagoya Institute of Technology. Deformation and the stress and strain distributions following the slicing, i.e. upon the release of constraints on the top surface, were obtained for four different levels of initial residual stress (σ_{0EL} for the EL and σ_{0SML} for the SML). Residual stresses estimated in the simplified 2D model neglecting the connection between the layers, i.e. $\sigma_{0EL} = -12.6$ kPa and $\sigma_{0SML} = 3.5$ kPa, were used for case 1 (Matsumoto et al. (2004b)). In case n, the residual stresses in the two layers were multiplied by n ($n = 1, ..., 4$).

4.2 Numerical Results

Deformation of the model following cut and resultant distribution of circumferential stress are shown in Fig. 10 for isotropic case 1. The height difference between the hills and valleys (1.2 μm) was much smaller than the measured height difference (8 μm). The height difference Δh increased linearly with the increase in the initial residual stress σ_0 (Fig. 11). The stress that causes

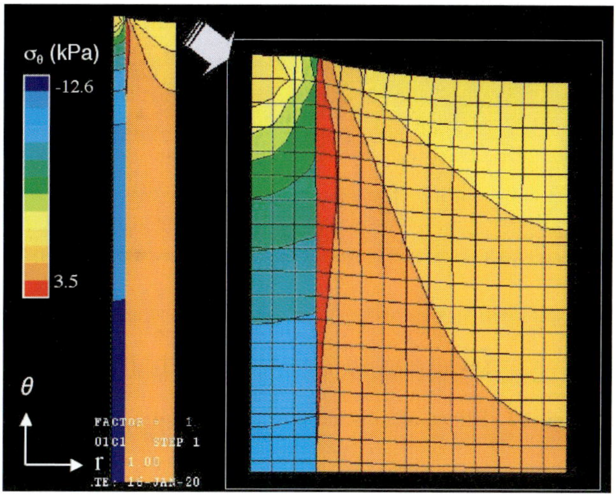

Fig. 10. Stress distribution near the cut surface (isotropic, case 1).

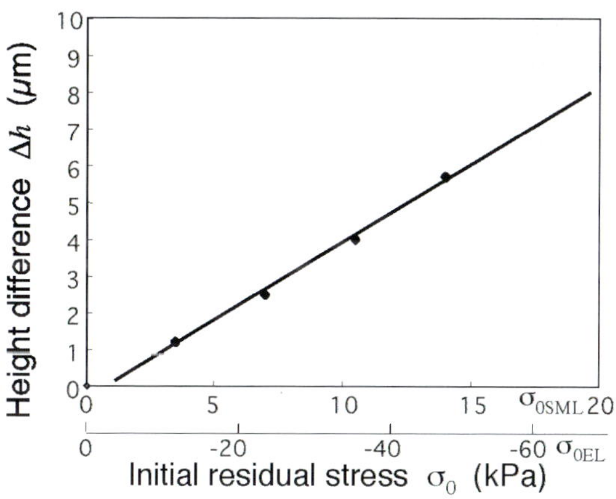

Fig. 11. Effect of the initial residual stress on the height difference.

the measured height difference was considered to be the microscopic residual stress, and was estimated to be −70 kPa for the EL and 20 kPa for the SML from the linear regression line drawn in the figure. The macroscopic residual stress reported in various studies is in the range of 1–10 kPa (Matsumoto and Hayashi (1994)). The microscopic residual stress obtained in the present study is almost one order of magnitude higher than the conventional values.

The height difference increased with the decrease in the shear modulus $G_{r\theta}$ in all cases (Fig. 12). Roughly speaking, the height difference doubled when the shear modulus decreased to 1/3 of the isotropic case, i.e. if the shear modulus is 1/9 of the elastic modulus, the residual stress becomes −35 kPa for the EL and 10 kPa for the SML.

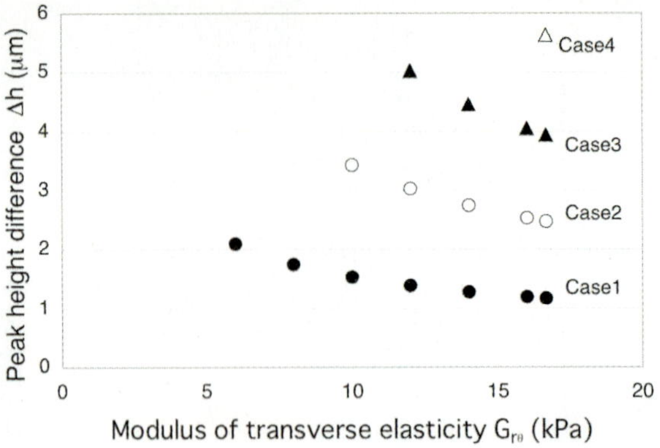

Fig. 12. Effect of anisotropy on the height difference.

5 Discussion

The radial compression test showed that the EL was stiffer than the SML by a factor of 2.5. This factor is fairly small considering that the elastic modulus of elastin is 10–100 times higher than that of the smooth muscle cells. It should be noted that the state of loading used in this test is non-physiological. Elastic recoil may also be an error source. We are now planning to measure the tensile properties of the EL and the SML separately.

This study indicated that fairly large stress might still reside in the opened-up aortic media that was believed to be stress-free. The isotropic analysis showed it was −70 kPa for the EL and 20 kPa for the SML. These are fairly large compared with the macroscopic residual stress reported in various studies (1–10 kPa). Also, these values are as large as 10–30% of the circumferential stress in the physiological condition (Matsumoto and Hayashi (1994)). The values became much lower when anisotropy of the SML is introduced: for a shear modulus that was 1/9 of the elastic modulus, the residual stress became −35 kPa for the EL and 10 kPa for the SML. Our preliminary study on the shear modulus of the whole wall showed that shear modulus was almost 1/10 of the initial elastic modulus measured in the tensile test. This may indicate that the residual stress values obtained in the anisotropic analysis are good estimates as the first step. To obtain better approximations, we need to know multiaxial elastic properties of the EL and the SML, which we are now planning to measure.

The existence of the microscopic residual stress is also confirmed by close examination of histological sections. Figure 13 shows a histological section of a porcine thoracic aorta in the stress-free state. Note that smooth muscle cells (arrows) are stretched while the elastic laminae are corrugated. It is unnatural to assume that the elastic laminae remain wavy in their stress-free state. Thus, this picture may indicate that the smooth muscle cells remain stretched in the unloaded wall, while the elastic laminae are buckled due to compressive stress applied to the EL.

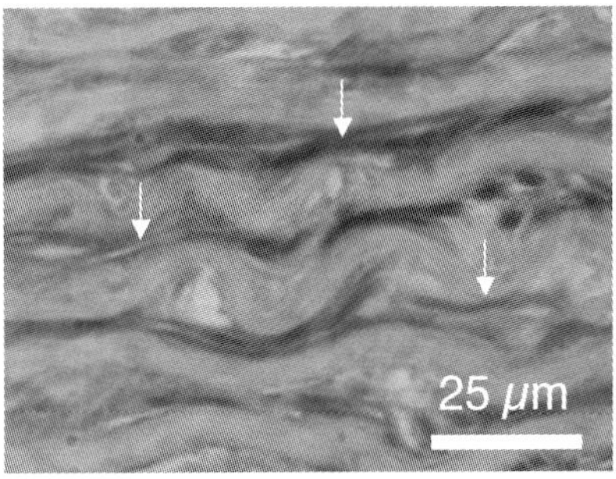

Fig. 13. A histological section stained with Azan of a square segment of porcine thoracic aorta in the stress-free state (Matsumoto et al. (2004b)).

If the corrugation of the EL is buckling, we might be able to estimate the buckling stress. Let us assume that the compressive residual stress in the EL causes buckling and the EL becomes corrugated with the spatial period of p. The critical stress causing the buckling is obtained from the Euler buckling formula for a plate of thickness h and length p, fixed at both ends:

$$\sigma_c = \frac{\pi^2 E h^2}{3p^2}. \tag{1}$$

Histological observation revealed that the spatial period l was around $50\,\mu$m and the thickness $h \sim 4\,\mu$m. The elastic modulus of the EL without bucking can be assumed to be that of the elastin, 0.6 MPa. The buckling stress was thus estimated to be $\sim$ 10 kPa, which is almost comparable to the estimate of the compressive residual stress in the EL.

One of the most important points arising from this study is that stress applied to a tissue is not equal to that applied to the cells within it. Mechanical response of tissues is caused by the cells in the tissues. Thus, it is crucial to know the mechanical environment of cells in the tissues when studying the mechanical response of the biological tissues. The present study indicates that the stress applied to the cells in the tissues could not be estimated from the conventional stress analysis of homogeneous tissues even if the macroscopic residual stress is taken into consideration. We need to know three-dimensional mechanical properties of tissues at a microscopic level and the corresponding distribution of residual stress in various biological tissues. For this purpose, a combination of microscopic measurement and a computational approach is indispensable.

Acknowledgements. We thank Dr. Hiroshi Yamada at Kyushu Institute of Technology for his valuable suggestions on the anisotropic analysis. This work was supported in part by Grant-in-Aid from the Ministry of Education, Science and Culture in Japan (T. Matsumoto, Nos. 15086209 and 16360052) and the Tatematsu Foundation.

References

Drucker, D. C. (1959). A definition of a stable inelastic material. *J. Appl. Mech.* 26:101–106.

Fung, Y. C. (1981). Bio-viscoelastic solids. In *Biomechanics.* Springer-Verlag. 196–214.

Fung, Y. C. (1984). *Biodynamics. Circulation.* New York: Springer-Verlag.

Fung, Y. C. (1990). *Biomechanics. Motion, Flow, Stress, and Growth.* New York: Springer-Verlag.

Grossman, W., Jones, D., and McLaurin, L. P. (1975). Wall stress and patterns of hypertrophy in the human left ventricle. *J. Clin. Invest.* 56:56–64.

Matsumoto, T., and Hayashi, K. (1994). Mechanical and dimensional adaptation of rat aorta to hypertension. *J. Biomech. Eng.* 116:278–283.

Matsumoto, T., Sato, J., Yamamoto, M., and Sato, M. (2000). Smooth muscle cells freshly isolated from rat thoracic aortas are much stiffer than cultured bovine cells: Possible effect of phenotype. *JSME Intern. J. C* 43:867–874.

Matsumoto, T., Goto, T., Furukawa, T., and Sato, M. (2004a). Residual stress and strain in the lamellar unit of the porcine aorta: Experiment and analysis. *J. Biomech.* 37:807–815.

Matsumoto, T., Goto, T., and Sato, M. (2004b). Microscopic residual stress caused by the mechanical heterogeneity in the lamellar unit of the porcine thoracic aortic wall. *JSME Intern. J. A* 47:341–348.

Vaishnav, R. N., and Vossoughi, J. (1983). Estimation of residual strains in aortic segments. In Hall, C. W., ed., *Biomedical Engineering II: Recent Developments.* New York: Pergamon Press. 330–333.

Whitemore, R. L. (1968). *Rheology of the Circulation.* Oxford: Pergamon Press.

Wolinsky, H., and Glagov, S. (1967). A lamellar unit of aortic medial structure and function in mammals. *Circ. Res.* 20:99–111.

Wolinsky, H. (1971). Effects of hypertension and its reversal on the thoracic aorta of male and female rats. *Circ. Res.* 28:622–637.

Parameter Identification in Arteries Using Constraints

J. Stålhand, A. Klarbring

Linköping University, Department of Mechanical Engineering, Sweden
jonst@ikp.liu.se, andkl@ikp.liu.se

To characterize the behaviour of arteries in a reliable way, it is necessary to know both the material parameters and the residual stress remaining in arteries after all external loads have been removed. Parameter identification techniques are often applied to obtain material parameters, while the residual stress is, generally, estimated by dissecting an arterial segment and using the opening-angle method. In a recent study by Stålhand et al. (2004), a method that avoids dissecting arterial segments was proposed. The method is tailored for using clinical *in vivo* data, and estimates the residual strain field simultaneously with the material parameters in a parameter identification process. The method is limited to simple box constraints on the parameters, although more advanced constraints are possible. In this paper, we suggest a way to introduce an *in vitro* observed arterial behaviour as a constraint to the minimization. In addition, we also introduce some motivated constraints that increase the efficiency of the minimization. Finally, to illustrate the potential of the method, we given an example where it is applied to an *in vivo* pressure-radius measurement for a human aorta.

1 Introduction

In a recent study by Stålhand et al. (2004), it is shown that the material parameters and the residual strain field can be simultaneously identified from an *in vivo* measurement on a human aorta by stating a nonlinear minimization problem. The only constraints used therein were simple box constraints. The minimization approach, however, allows for more complex constraints as well, and, in particular, constraints that restrict the behaviour of the arterial response. This type is generally referred to as a *state constraint*. Examples of state constraints in vascular mechanics are the invariance of the axial stretch to the pressure at physiological axial forces (see, for example, Holzapfel et al. (2000), Humphrey (2002), and Schulze-Bauer et al. (2003)), the uniform strain hypothesis proposed by Takamizawa and Hayashi (1987), and the principle of optimal operation proposed by Fung (1997).

This study takes the ideas from Stålhand et al. (2004) a step further and introduces constraints on the behaviour of the solution. For convenience, we choose to include a slight variation of the first example above, and introduce the constraint that the axial force is invariant of the pressure under physiological conditions. The validity of this assumption is found in, for instance, Humphrey (2002). Further, the constraint is equivalent to setting the derivative of the axial force with respect to the pressure to zero. This will put restrictions on the particular form of the constitutive relation, and, in turn, on the strain-energy function, as shown in a paper by Ogden and Schulze-Bauer (2000) for a thin-walled tube with no residual stress. Extending their work to the general case with a thick-walled, residually stressed tube, used to represent arteries, may not be a straightforward task. Therefore, we seek to avoid this problem by relaxing the constraint so that deviation from the preferred behaviour is penalized. This allows for a larger group of strain-energy functions to be used. In addition to the constraint above, we also introduce a physiologically motivated equality constraint for the volume of the arterial wall, and an inequality constraint for the material parameters in the strain-energy function to ensure stability.

To illustrate the potential of the method, it is applied to an *in vivo* measurement on a human aorta. The results show, in our opinion, one of the strengths in the minimization approach. It is possible to characterize the mechanics of the artery from *in vivo* measurements, in particular the material properties and the residual strains without dissection. By utilizing constraints, we can also force the solution to satisfy *in vitro* observed behaviours, and other physically, or physiologically, motivated constraints in a simple way.

2 Method

A model for characterizing the mechanical behaviour of an artery is presented in this section. First, a mechanical model is derived, down to a set of unknown material parameters and a residual strain parameterization, using non-linear continuum mechanics. Second, a minimization algorithm for identifying these parameters from clinical *in vivo* measurements is stated.

The artery is assumed to be a residually-stressed, homogeneous, thick-walled cylinder, with a pressurized inner boundary and a traction free outer boundary. The material is assumed to be hyperelastic and anisotropic. The deformation is taken to be isochoric, and such that the strain only varies in the radial direction. Further, to obtain a parameterization of the residual strain field, a hypothetical stress-free state in the form of a cylindrical sector is introduced. Evidence that these assumptions are reasonable is found in Holzapfel et al. (2000).

2.1 Mechanical Model

Let the stress-free reference configuration $\mathcal{B}_0$ be a cylindrical sector, where a material point is defined by its cylindrical coordinates (R, Θ, Z). Further, let the domain of the reference configuration be given by $R_i \leq R \leq R_o$, $0 \leq \Theta \leq \Theta_0$, $0 \leq Z \leq L$, where R_i, R_o, Θ_0 and L denote the inner and outer radii, sector angle and reference tube length, respectively. Assume that the deformation is such that a material point is mapped to a point in the deformed configuration $\mathcal{B}$ with cylindrical coordinates $(r, \theta, z) = (r(R), \alpha\Theta, \lambda_z Z)$, where α and λ_z are constants. Further, assume that the deformed configuration is a pressurized cylinder, i.e. its domain is given by $r_i \leq r \leq r_o$, $0 \leq \theta < 2\pi$, $0 \leq z \leq \lambda_z L$, where r_i and r_o denote the inner and outer radii, respectively. The deformation gradient is then given by

$$\boldsymbol{F} = \frac{\partial r(R)}{\partial R} \boldsymbol{e}_1 \otimes \boldsymbol{E}_1 + \frac{\alpha r(R)}{R} \boldsymbol{e}_2 \otimes \boldsymbol{E}_2 + \lambda_z \boldsymbol{e}_3 \otimes \boldsymbol{E}_3, \tag{1}$$

where $\boldsymbol{E}_I$ $(I = 1, 2, 3)$, and $\boldsymbol{e}_i$ $(i = 1, 2, 3)$ are the cylindrical base vectors in $\mathcal{B}_0$ and $\mathcal{B}$, respectively. Note that the dependence on the tangential positions for the radial and tangential base vectors above has been suppressed. Here, and in what follows, it is taken that subscripts 1, 2 and 3 refer to the radial, tangential and axial direction, respectively. Using the incompressibility assumption for the artery, the deformed radius can be calculated from eq. (1) as

$$r(R) = \sqrt{\frac{R^2 + D}{\alpha \lambda_z}}, \tag{2}$$

where D is an integration constant; see, for instance, Antmann (1995), Ch. 13, Sec. 3.

If the arterial wall is assumed to be composed of an incompressible hyperelastic material, there exists a strain-energy function Ψ such that the Cauchy stress for is given by

$$\boldsymbol{\sigma} = -p\boldsymbol{I} + \boldsymbol{F} \frac{\partial \Psi(\boldsymbol{E})}{\partial \boldsymbol{E}} \boldsymbol{F}^T, \tag{3}$$

where $\boldsymbol{I}$ is the second–order identity tensor, $\boldsymbol{E}$ is the Green-Lagrange strain tensor

$$\boldsymbol{E} = \frac{1}{2}(\boldsymbol{F}^T \boldsymbol{F} - \boldsymbol{I}), \tag{4}$$

and a superscribed T denotes the transpose of a second-order tensor.

Discarding all body forces and assuming the stress is rotationally symmetric, the equilibrium in terms of the Cauchy stress tensor components is given by

$$\frac{\partial \sigma_{11}}{\partial r} + \frac{\sigma_{11} - \sigma_{22}}{r} = 0; \tag{5}$$

see, for instance, Humphrey (2002). The boundary conditions are taken to be a prescribed pressure P on the inner boundary and a traction free outer boundary, i.e.

$$\begin{cases} \boldsymbol{t} = \boldsymbol{\sigma n} = -P\boldsymbol{n} = P\boldsymbol{e}_1 \text{ at } r = r_i, \\ \boldsymbol{t} = \boldsymbol{\sigma n} = \boldsymbol{0} \qquad\qquad\quad \text{at } r = r_o, \end{cases} \tag{6}$$

where $\boldsymbol{n}$ is the unit outward normal to the boundary. Integrating eq. (5) from r_i to r gives that the radial stress can be computed as

$$\sigma_{11}(r) = \sigma_{11}(r_0) + \int_{r_0}^{r} (\sigma_{22} - \sigma_{11}) \frac{\mathrm{d}r}{r}. \tag{7}$$

In particular, by substituting $r = r_o$ in eq. (7) and applying the boundary conditions in eq. (6), the expression for the pressure is obtained as

$$P = \int_{r_i}^{r_o} (\sigma_{22} - \sigma_{11}) \frac{dr}{r}. \tag{8}$$

The axial force can also be computed via an equilibrium equation in the axial direction. The result is

$$L = 2\pi \int_{r_i}^{r_o} \sigma_{33} r \mathrm{d}r. \tag{9}$$

Some straightforward manipulation of eq. (9) using partial integration and introduction of the *reduced axial force* $L_R = L - \pi r_o^2 P$ lead to

$$L_R = \pi \int_{r_i}^{r_o} (2\sigma_{33} - \sigma_{22} - \sigma_{11})\, r \mathrm{d}r; \tag{10}$$

see, for example, Holzapfel et al. (2000) or Humphrey (2002). Note also that the reduced axial force is the actual force measured in an inflation-extension test *in vitro*, and is, therefore, the force we aim to make invariant of the pressure.

Finally, to complete the mechanical model, a constitutive relation has to be introduced. This is done by specifying the strain-energy function in eq. (3) to be of a general three-dimensional Fung type: $\Psi = \overline{\Psi}(E_{11}, E_{22}, E_{33}, E_{12}, E_{13}, E_{23})$. By using the incompressibility constraint, it can be shown that the integrands in eqs (8) and (10) may be given by a reduced strain-energy function of the form $\Psi = \Psi(E_{22}, E_{33})$ for the considered deformation; see Sect. 4.2.1 in Holzapfel et al. (2000). Since the model will use only the responses given in eqs (8) and (10), it suffices to consider a strain-energy of the reduced form. The reduced strain-energy function is given by

$$\Psi = \frac{c}{2}(e^Q - 1), \tag{11}$$

where $Q = a_1 E_{22}^2 + a_2 E_{33}^2 + 2a_4 E_{22} E_{33}$. Note that the material parameters a_i $(i = 1, 2, 4)$ cannot be chosen arbitrarily if eq. (11) is to be locally convex in the Green-Lagrange strains. In Holzapfel et al. (2000) it is shown that local convexity is guaranteed for $c > 0$ if and only if $a_1 > 0$, $a_2 > 0$ and $a_1 a_2 > a_4^2$. This is a constraint that must be imposed on the parameter identification.

2.2 Parameter Identification

The parameters needed to be determined to characterize the mechanical behaviour of the artery are α, λ_z, R_0 and R_1 for the residual strain, and c, a_1, a_2 and a_4 for the material. This is done by stating a minimization problem. Assume that the n measurements are given pairwise as the inner radius $r_{i,k}$ and the corresponding luminal pressure P_k ($k = 1 \ldots n$). For convenience, the outer radius is replaced by $R_o = R_i + t$, where t is the wall–thickness in the reference configuration. The model parameters are taken to be components of a vector $\boldsymbol{\kappa} = \{R_i, t, \alpha, \lambda_z, c, a_1, a_2, a_4\}$, and the objective function is defined to be

$$\phi(\boldsymbol{\kappa}) = \varphi(\boldsymbol{\kappa}) + \frac{1}{\epsilon} \left(g(\boldsymbol{\kappa})\right)^2 , \tag{12}$$

where $\epsilon > 0$ is a penalty parameter,

$$\varphi(\boldsymbol{\kappa}) = \sum_{k=1}^{n} \frac{1}{2} |P(\boldsymbol{\kappa}, r_{i,k}) - P_k|^2 \tag{13}$$

and

$$g(\boldsymbol{\kappa}) = \sum_{k=2}^{n} |L_R(\boldsymbol{\kappa}, r_{i,k}) - L_R(\boldsymbol{\kappa}, r_{i,k-1})| . \tag{14}$$

The first term in the objective function (12), is a least-squares fitting of the model pressure to the measured pressure. The model pressure is computed by substituting eqs (3) and (11) into eq. (8). The strain measures are computed from eqs (1) and (4) by substituting eq. (2), where the integration constant is given by $D_k = \alpha \lambda_z r_{i,k}^2 - R_0^2$. The second term in the objective function is the sum of the normed differences of all consecutive values of the reduced axial force for the n measurements. The reduced axial force is obtained analogously to the above by substituting eqs (3) and (11) into eq. (9).

As discussed earlier, the idea behind relaxing the axial force constraint is to allow for a larger set of solutions by penalizing the deviation from the preferred behaviour, instead of enforcing it. Any solution that gives an axial force which is not invariant, or close to invariant, of the pressure will have a function $g(\boldsymbol{\kappa}) > 0$ in eq. (14). With a suitable choice of the penalty parameter, these solutions become costly, and are therefore infeasible. Further, note that eq. (14) minimizes the derivative $\mathrm{d}L_R/\mathrm{d}r_i$. However, it is straight-forward to show that this implies a minimization of $\mathrm{d}L_R/\mathrm{d}P$ by noting that $r_i = r_i(P)$ and applying the chain rule.

The measurements are not assumed to have any information regarding the wall thickness, and, as a consequence, there is no knowledge of the arterial wall volume. This is not desirable in the minimization since it allows for the use of an unrestricted amount of material in the arterial wall thickness; if r_o is unrestricted, so is t by eq. (2), and $R_o = R_i + t$. One way to avoid this is to impose a restriction that the volume per unit length of the arterial wall is constant, or

$$V'(\boldsymbol{\kappa}) = \pi \left(r_1^2 - r_0^2\right) = V_0'. \tag{15}$$

The minimization problem now reads

$$\begin{cases} \min\limits_{\boldsymbol{\kappa}} \phi(\boldsymbol{\kappa}) \\ \text{subject to:} \begin{cases} V'(\boldsymbol{\kappa}) = V_0' \\ \boldsymbol{\kappa} \in \mathcal{K}, \end{cases} \end{cases} \tag{16}$$

where $\mathcal{K}$ is the set of admissible model parameters $\boldsymbol{\kappa}$ such that

$$\mathcal{K} = \left\{\boldsymbol{\kappa} : R_0 > 0,\, t > 0,\, 1 \leq \alpha \leq \overline{\alpha},\, \underline{\lambda}_z \leq \lambda_z \leq \overline{\lambda}_z, \\ c > 0,\, a_1 > 0,\, a_2 > 0,\, a_4^2 < a_1 a_2\right\},$$

the overlined and underlined quantities being taken to represent the lower and upper bounds, respectively.

Problem (16) can be solved by a constrained nonlinear minimization algorithm. Such an algorithm is summarized below:

Step 1. Choose an initial parameter vector $\boldsymbol{\kappa}^{(0)}$.
Step 2. Compute a new iterate $\boldsymbol{\kappa}^{(j+1)}$ from $\boldsymbol{\kappa}^{(j)}$.
Step 3. Repeat step 2 with $\boldsymbol{\kappa}^{(j)} = \boldsymbol{\kappa}^{(j+1)}$ until (16) has converged.

3 Materials

To illustrate the potential of the method presented above, it was applied to a clinical *in vivo* measurement of the abdominal aorta from a 47 year-old female made by Sonneson et al. (1994). In their study, the luminal pressure was measured invasively using a catheter, while the inner diameter of the vessel was measured non-invasively using an echo-tracking system. Pressure and diameter were measured simultaneously for multiple cardiac cycles, but we confine ourselves to a representative cardiac cycle, shown in Fig. 1.

The measurement does not include the wall thickness and, therefore, V_0' had to be determined in some other way. Noting that eq. (15) equals the cross-sectional area of the vessel, it is possible to determine V_0' from the literature. In a study by Schulze-Bauer and Holzapfel (2003), the following expression is given for the cross-sectional area of the thoracic aorta as a function of age: $A\,(\mathrm{mm}^2) = 2.169 \times \mathrm{age}\,(\mathrm{yrs}) + 26.06$. If this function is assumed to be correct for the abdominal aorta as well, we obtain $V_0' = 1.28 \times 10^{-5}\ \mathrm{m}^3$ per unit length.

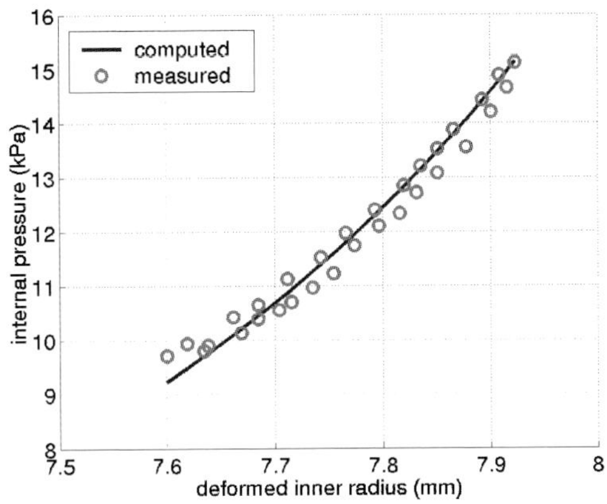

Fig. 1. Pressure-radius data for one cardiac cycle and the model pressure curve.

Table 1. Results from the minimization for the residual strain and material parameters.

R_i(mm)	t(mm)	α(-)	λ_z(-)	c (Pa)	a_1(-)	a_2(-)	a_4(-)
11.04	3.71	1.91	1.23	1407	6.74	5.75	2.39

4 Results

The minimization problem in (16) was solved using a standard Sequential Quadratic Programming method in Matlab (The Mathworks, Natick, USA). The bounds on the admissible set $\mathcal{K}$ were taken to be $\overline{\alpha} = 2.4$, $\underline{\lambda}_z = 1.0$, and $\overline{\lambda}_z = 1.3$. The axial stretch bounds are set to avoid buckling at the lower end, and reflect the maximum axial stretch obtained by Schulze-Bauer et al. (2003) at the upper end. The upper bound $\overline{\alpha} = 2.4$ was set to avoid unphysiological results, although the corresponding opening angle (in this case 210°) is somewhat restrictive compared with values reported by Schulze-Bauer et al. (2003). The penalization parameter ϵ was chosen to 10^{-8} by trial and error. This value obtained a good balance between the fitting to the pressure data and a small variation in the axial force.

The resulting parameters are presented in Table 1. The model fit, the axial force, and the principal stresses are shown in Figs 1, 2 and 3, respectively. The axial force in Fig. 2 is approximately 0.26 N, and has a small maximum variation of about 5%. The axial force is in line with values reported by Schulze-Bauer et al. (2003) for human iliac arteries, and slightly lower than values reported by Schulze-Bauer and Holzapfel (2003) for human aorta. The circumferential stretch (not shown here) is around 1.3, and the axial stretch is

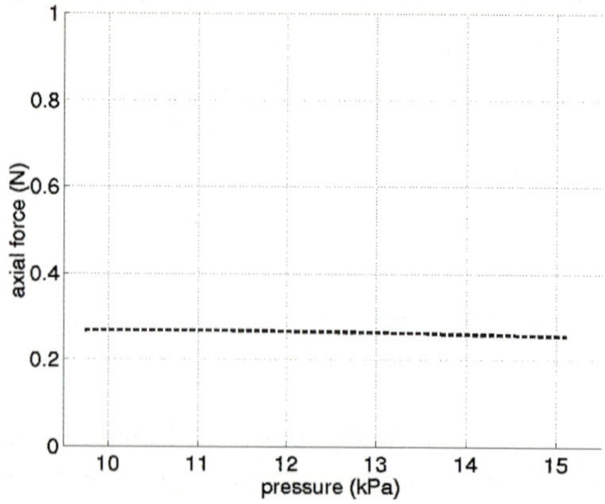

Fig. 2. The axial force as a function of the pressure.

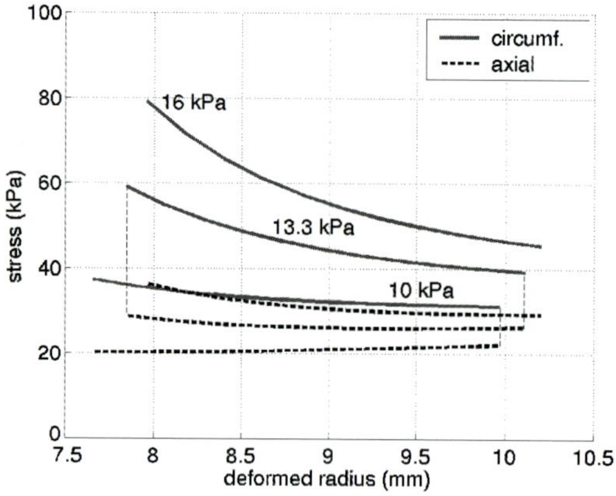

Fig. 3. Circumferential and axial stresses for 10, 13.3 and 16 kPa.

1.23 (from Table 1). These values, and the corresponding stress levels in Fig. 3, are in line with values reported by Schulze-Bauer and Holzapfel (2003).

5 Discussion

The minimization approach as a tool to solve a parameter identification problem is not new in the vascular mechanics field. It is frequently applied to obtain material parameters for different arteries; see, for instance, Holzapfel et al. (2000), Schulze-Bauer and Holzapfel (2003), Schulze-Bauer et al. (2003), Takamizawa and Hayashi (1987), among others. Our research group has recently showed that the minimization approach may be applied to obtain the residual strain parameterization as well (Stålhand et al. (2004)). Since the proposed method therein does not involve any dissection to obtain the residual strain, it offers an embryo of a method, ultimately possible for clinical use. This study takes the ideas in Stålhand et al. (2004) a step further and includes more advanced constraints than simple box constraints. This offers a simple way to include *in vitro* observations in the mechanical model. In addition, it has the desirable side effect of reducing the solution space, and in many cases, thereby, speed up the minimization. These issues are commented on further below.

The reason for introducing the volume constraint is twofold. First, as discussed in Section 2.2, it is physically motivated since it limits the amount of material that can be used for the reference configuration. Second, it also reduces the number of unknown parameters in the minimization by one. Geometrically this means that the m-dimensional solution space is reduced to a space with dimensions $m - 1$ (a hyper–plane). Introducing an equality constraint is equivalent to eliminating one of the unknown parameters. A positive effect of a reduced solution space is that the efficiency of the algorithm increases. From eqs (2) and (15) it can be realized that either R_0 or t is eliminated in this case. Then, why not use the equality constraints to eliminate one, or more, unknown parameters? The answer is that, in general, the equality constraints need not be as simple as in eq. (15), and it may, therefore, be impossible, or very hard, to eliminate unknown parameters. Moreover, elimination of parameters must be used with care since it may lead to ill-conditioning; see Nocedal and Wright (1999).

Even though the material parameters in the constitutive relations are such that the strain-energy function is convex, the objective function in eq. (12) is non-convex. Convergence to the global minimum cannot be guaranteed with a gradient based minimization algorithm under this circumstance. Instead, the gradient based algorithm is likely to be attracted to local minima. An implication of this is that the solution is dependent on the starting point in the minimization, i.e. Step 1 in the algorithm in Section 2.2. This is a typical problem for non-convex minimization, and, in reality, there is no universal remedy for it. The obvious solutions are either to use a global minimization algorithm or to reformulate the objective function such that it becomes convex, or at least convex in a physiologically relevant region. None of these two solutions are straightforward, however. Global minimization algorithms are generally time consuming, and are therefore best suited for smaller problems (Brent (2003)), while a reformulation of the objective function as a convex (or locally convex) function becomes difficult when the residual strain distribution is included. The only option left is to perturb the initial guess and take the solution with the lowest residual, even though it is an unsatisfactory approach. These issues are very important and need to be studied further.

In this study it is assumed that the residual stress of an arterial segment is completely relieved by a single radial cut (once it is dissected from the body).

This is, however, a questionable assumption. In a recent study by Matsumoto et al. (2004) it was shown that the residual stress is only released in the region close to the cutting surface, while more than 50% of the residual stress remains in deeper regions. In addition, if the cut-open state is not assumed stress free, there may exist several residual strain fields that give the same opening angle. This was shown in a paper by Omens et al. (2003) where cross-sectional rings of ventricles from mice were studied. For a given reference configuration, and a given opening angle after a radial cut, they showed that an additional circumferential cut altered the geometry for the new segments drastically, and that it differed substantially between the specimens. The reason is that the specimens have different distribution of the residual stress fields, even though the net effect is the same opening angle after a single cut.

Finally, it should be noted that any active effect from the smooth musculature is neglected in this method. The smooth muscle effect on large elastic arteries, and in particular the aorta, is generally considered small. Therefore, we take the aorta to be a passive structure, although this is an approximation.

In conclusion, we believe that applying minimization methods to obtain both the material parameters and the residual stress is an interesting approach to characterizing the arterial mechanics without dissection, and consequently take the continuum models closer to clinical use. Further, adding state constraints to the minimization process allows us to impose *in vitro* observed behaviour for soft tissues, as well as stability conditions, on the mechanical model in a simple way.

Acknowledgements. The authors wish to thank Prof. T. Länne at the Dept. Medicine and Care, Linköping University for supplying *in vivo* data.

References

Antmann, S. S. (1995). *Nonlinear Problems of Elasticity.* New York: Springer-Verlag.

Brent, R. P. (2003). *Algorithms for Minimization Without Derivatives.* Mineola, New York: Dover Publications.

Fung, Y. C. (1997). *Biomechanics. Circulation.* New York: Springer-Verlag, 2nd edition.

Holzapfel, G. A., Gasser, T. C., and Ogden, R. W. (2000). A new constitutive framework for arterial wall mechanics and a comparative study of material models. *J. Elasticity* 61:1–48.

Humphrey, J. D. (2002). *Cardiovascular Solid Mechanics. Cells, Tissues, and Organs.* New York: Springer-Verlag.

Matsumoto, T., Goto, T., Furukawa, T., and Sato, M. (2004). Residual stress and strain in the lamellar unit of the porcine aorta: experiment and analysis. *J. Biomech.* 37:807–815.

Nocedal, J., and Wright, S. J. (1999). *Numerical Optimization*. New York: Springer-Verlag.

Ogden, R. W., and Schulze-Bauer, C. A. J. (2000). Phenomenological and structural aspects of the mechanical response of arteries. In Casey, J., and Bao, G., eds., *Mechanics in Biology*. New York: The American Society of Mechanical Engineers (ASME). AMD-Vol. 242/BED-Vol. 46, pp. 125–140.

Omens, J. H., McCulloch, A. D., and Criscione, J. C. (2003). Complex distributions of residual stress and strain in the mouse left ventricle: experimental and theoretical models. *Biomech. Model. Mechanobio.* 1:267–277.

Schulze-Bauer, C. A. J., and Holzapfel, G. A. (2003). Determination of constitutive equations for human arteries from clinical data. *J. Biomech.* 36: 165–169.

Schulze-Bauer, C. A. J., Mörth, C., and Holzapfel, G. A. (2003). Passive biaxial mechanical response of aged human iliac arteries. *J. Biomech. Eng.* 125:395–406.

Sonneson, B., Länne, T., Vernersson, E., and Hansen, F. (1994). Sex difference in the mechanical properties of the abdominal aorta in human beings. *J. Vasc. Surg.* 20:959–969.

Stålhand, J., Klarbring, A., and Karlsson, M. (2004). Towards in vivo material identification and stress estimation. *Biomech. Model. Mechanobio.* 2: 169–186.

Takamizawa, K., and Hayashi, K. (1987). Strain energy density function and uniform strain hypothesis for arterial mechanics. *J. Biomech.* 20:7–17.

Collagen Organization and Biomechanics of the Arteries and Aneurysms of the Human Brain

P.B. Canham,[1] R.M. Korol,[1,2] H.M. Finlay,[1] R.R. Hammond,[4]
D.W. Holdsworth,[1] G.G. Ferguson,[5] A.R. Lucas[2,3]

[1] University of Western Ontario, Department of Medical Biophysics, Canada
Peter.Canham@fmd.uwo.ca, rkorol@uwo.ca, hmfinlay@uwo.ca
david.holdsworth@imaging.robarts.ca
[2] John P. Robarts Research Institute, Vascular Biology Research Group, Canada
[3] University of Western Ontario, Department of Microbiology and Immunology, Canada
arl@robarts.ca
[4] University of Western Ontario, Department of Pathology, Canada
rhammond@uwo.ca
[5] University of Western Ontario, Department of Clinical Neurological Sciences, Canada
gary.ferguson@lhsc.on.ca

Saccular aneurysms are balloon-like expansions developing from the branching region of a major brain artery; a small fraction are at risk of rupture with severe clinical consequences. Their wall integrity is related to both collagen fibre strength and orientation. Our approach involved serial section histology and polarized light microscopy, including 3D orientation and birefringence measurements. Results show a marked difference in fibre birefringence across the wall, with strength mainly in the outer layers. Directional and strength measurements on medial gap fibres of major brain artery bifurcations reveal an exceptional structural anisotropy – the apex having a tendon-like backbone.

1 Introduction

Vascular remodelling is an essential process for growth and development in early life, but later manifestations are often pathological with clinical consequences. Such is the case for saccular and fusiform aneurysms, examples in human biology of outward arterial remodelling with the possible consequence of vascular stroke (Stehbens (1972), Ferguson (1989), ISIUA (1998, 2003), Peters et al. (2001), ISAT (2002)). The common form, the saccular brain aneurysm, is a sphere-like expansion from the artery at the region of a major branch, initially appearing as an outward blister from the lumen.

In total there are about 20 arterial branches of the human cerebral circulation (of the many millions of vascular branches, cerebral and non-cerebral) that account for all saccular aneurysms (distribution shown in Fig. 1). Specific branches stand out as the more likely sites, and the anterior cerebral circulation accounts for 87% of all lesions. The incidence of occurrence, including the smallest of 2 mm size, is known from autopsy (in the range of 0.2 to 15%, depending on the study) (Hassler (1961), Stehbens (1972)). Only a small fraction are of clinical significance, those presenting with a minor bleed (which are operable) or fatal rupture, thus indicating the challenge for medical course of action for a discovered lesion – intracranial surgery and clipping the neck of the lesion, endovascular procedure, or nonintervention with follow-up medical imaging. Shown in Fig. 2, prior to treatment, is a large posterior circle aneurysm located at the vertebral-basilar junction, with a manifold of

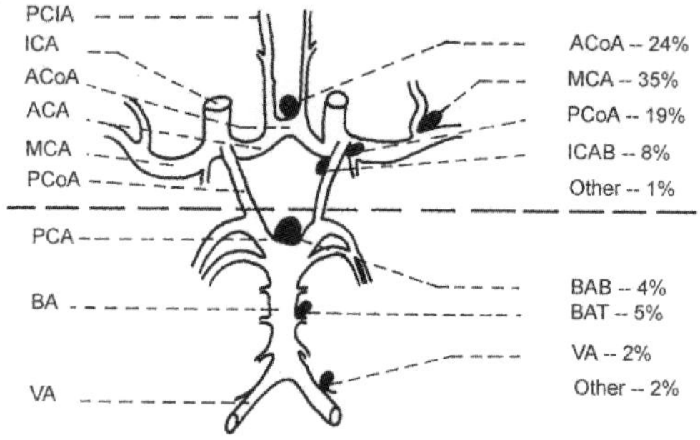

Fig. 1. Map of the frequency of aneurysm occurrence from autopsy studies. (ACA anterior cerebral artery, ACoA anterior communicating artery, BA basilar artery, BAB basilar artery bifurcation, BAT basilar artery trunk, ICA internal carotid artery, ICAB internal carotid artery bifurcation, MCA middle cerebral artery, PCA posterior cerebral artery, PClA pericallosal artery, PCoA posterior communicating artery, VA vertebral artery). (Redrawn from Ferguson (1970).)

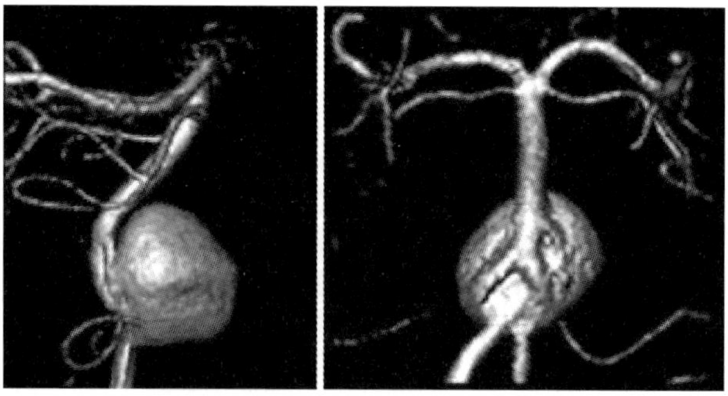

Fig. 2. Volume rendered image from CT angiography of a large fenestration aneurysm, 24 mm across its lumen, at the vertebral artery junction.

arteries at its neck that ruled out clipping. (An endovascular procedure was used with a good outcome.) There are biomechanical challenges for the remodelling collagen fabric of the extracellular matrix (ECM), fibres that provide the structural backbone of the wall, including the requirement of wall integrity, and strength needed to bear the biaxial loading from blood pressure (Canham and Ferguson (1985), Canham et al. (1996), Humphrey and Kyriacou (1996), David and Humphrey (2003)). A measure of the directional organization of the collagen of the aneurysm wall, layer by layer, provides an assessment of how well or poorly structured an aneurysm is to withstand arterial pressure. It also is a measure of how effective the remodelling process has been from the early stages through to a fragile vessel beyond 10 mm in size. Regardless of the underlying biological process, influenced possibly by genetic predisposition (Leblanc et al. (1989)), or localized response to hemodynamic influences (Steinman et al. (2003), Oshima and Torii (2004), Oshima et al. (2005)), the biophysical balance of arterial pressure with wall stress drives the remodelling process and initiates the final event of a bleed into the subarachnoid space.

The main thrust of our work has been on the saccular aneurysm and the arteries and their branches where lesions form. For comparison we are including structural results on a fusiform aneurysm, a differently formed lesion characterized by an asymmetrical expansion of brain artery; although far less common they are another example of pathological outward remodelling, and may be similar in fine structure (Drake and Peerless (1997)).

Collagen fibres of the ECM have the optical property of birefringence, enabling the measurement of two qualities essential to their function, their orientation in three-dimensional space, which defines the direction they can bear load in a tissue, and the strength of birefringence, which correlates with the ultimate tensile strength of its tissue environment. Two attachments to the polarizing microscope enable the measurement of both qualities – the universal stage attachment that reveals the directional alignment in 3D, fibre by fibre, within a tissue section prepared for light microscopy, and the Sénarmont compensator, which enables the measurement of birefringence. We have exploited both instruments in the study of vascular tissue in order to assess strength and tissue anisotropy (Canham et al. (1999), MacDonald et al. (2000)).

2 Methods

2.1 Tissue Preparation and Histology

Brain aneurysms and their contiguous arteries were obtained from the circle of Willis at time of autopsy, and were cannulated and fixed at physiological distending pressure (110 mmHg) with 10% neutral buffered formalin. This is a key step, re-establishing the approximate dimensions of the tissue before

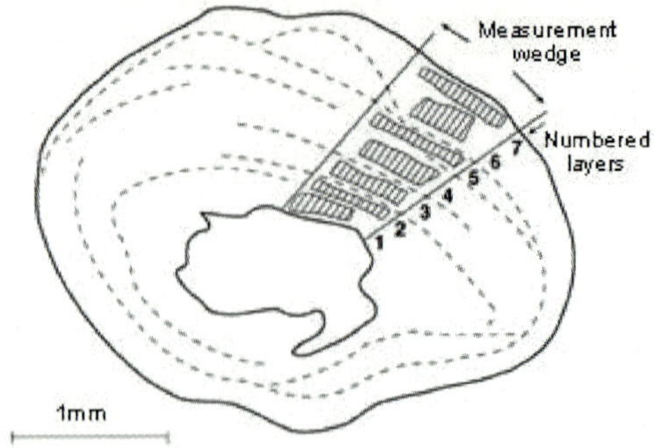

Fig. 3. Schematic aneurysm section with a 'corridor' or wedge of measurements from the lumen surface through to the outer wall. On the assumption of spherical symmetry, the true thickness of each layer can be calculated, and the layer contribution to the directional strength of the lesion calculated for that location.

death, with its in vivo state of approximate strain and the associated alignment of the underlying collagen fabric within the vessel wall (Canham et al. (1996)). For the circumstances when other labs are able to obtain x-ray CT measurements or tissue elasticity measurements on the fresh tissue (Humphrey and Canham (2000)) the fixation is carried out after those measurements (as was the case with the fusiform lesion in this paper).

Refinements of histological techniques for polarizing microscopy led to the preference for 5 μm sectioning along tangential cutting planes. Tangential and near tangential planes reveal the collagen layering at magnified dimensions – like a sheared deck of cards. When a section plane cuts obliquely through the entire wall of an artery or aneurysm, the physical integrity of the wall can be assessed by summing the layers one at a time (concept in Fig. 3), with geometric corrections for true layer thickness (MacDonald et al. (2000)). To maximize the data sets from a saccular aneurysm the tissue is wax-embedded repeatedly, each time at a different location, with serial sectioning at each new orientation cutting through to the lumen and slightly beyond, to permit data retrieval from as many tangential planes as possible. (Pathological lesions, unlike straight normal segments of artery, have limited cylindrical or spherical symmetry and require a rigorous sampling protocol.)

2.2 Birefingence Measurements

The Sénarmont compensator technique has been widely used, both for unstained sections or sections stained for birefringence enhancement with picrosirius red (Whittaker et al. (1988), Nollie et al. (1996)). The brightness of individual fibres of collagen in unstained tissue, or the colour balance in tissue stained with picro-sirius red provides a qualitative assessment of birefringence. By furthering the analysis of Doillon's study (1985) on guinea pig

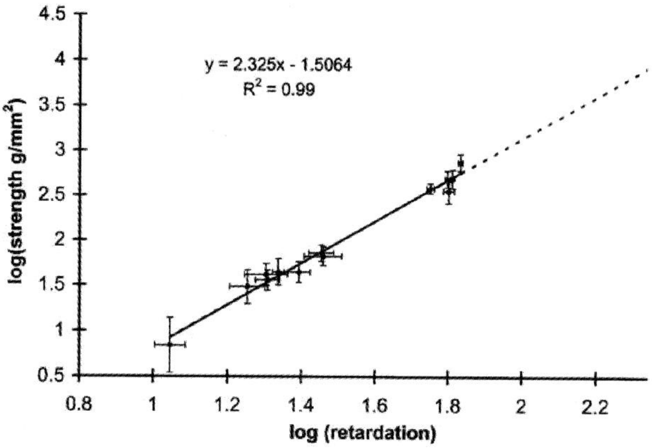

Fig. 4. Graphical presentation of the tabulated data from Doillon et al. (1985): $\sigma_{\max}(\mathrm{g/mm^2}) = 0.031B^{2.33}$ (using authors' units).

wound healing we demonstrated the exponential relationship between tissue ultimate tensile strength ($\sigma_{\max}$, MPa) and fibre birefringence (B,nm): $\sigma_{\max} = 0.000304B^{2.33}$ ($R^2 = 0.99$). The birefringence results on wound healing provide the basis for predicting tissue strength as a function of loading direction in the aneurysm wall, incorporating fibre alignment and birefringence (MacDonald et al. (2000)).

We show the graphical presentation of those data (Fig. 4) in log form to draw attention to the striking linearity of the correlation, providing justification for using the relationship in the extrapolated region as well as interpolation within the data set.

2.3 Fibre Orientation

Fibre orientation was obtained by using the Zeiss polarizing microscope with the four-axis universal stage (similar to the Leitz stage shown in Fig. 5). The stage has been a widely used instrument in physical geology (Emmons (1943)) and incorporated into studies of cardiovascular tissues (e.g., Canham et al. (1991, 1996)). An outstanding value of this instrument is that it enables the 3D alignment measurement of individual fibres directly on a tissue section, region by region as selected by the microscopist. In principle the 5 μm section is treated as a thick transparent section with the fibre singled out by its birefringence. A fibre, or line in 3D space requires two angles to define its orientation – an azimuthal angle in the plane of the microscope stage, and an elevation angle, out of the plane of the stage. Measurements are made at extinction – to within 1° (Smith et al. (1981)). There are limitations with the requirement of a high optical quality of the section and the birefringent fabric must be coherently aligned and thus optically coherent within the zone of measurement (approx. 4 μm × 4 μm).

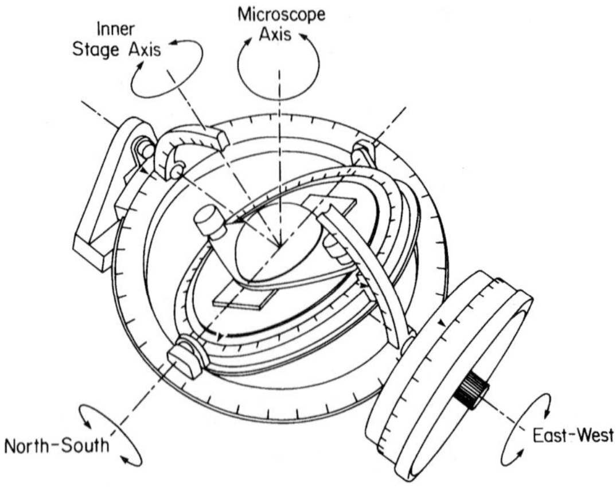

Fig. 5. Diagram of the Leitz universal stage, showing the rotational degrees of freedom to permit tilting the tissue slide. The calibrated arcs provide the measurements of azimuth and elevation angles. (Reprinted from Finlay et al. (1989) with permission of Blackwell Publishing.)

The universal stage, in addition to its measurement capabilities for 3D orientation, is also useful as a qualitative check on the elevation angle of fibres being assessed for strength of birefringence. For ECM fibres at high elevation angles the measurement is more difficult and less accurate. As a consequence we have concentrated on sampling from tangential and near tangential sectioning planes so that all fibre directions are close to the plane of the section and are revealed equally clearly (longitudinal, helical or circumferential directions). Thin fibres within tissues that have not been pressure fixed or fixed at very low pressures (e.g., 30 mmHg transmural pressure) often have a wavy or crimp pattern superimposed on the general alignment of the fibre. For these fibres the optical conditions are not met and measurements are not achievable or have a low accuracy.

Analysis is done by circular statistics (Fisher (1953), Tarling (1983)), and graphical presentations are done using the Lambert equal area projection (Phillips (1971)). The equal area projection is like a map of the north or south pole, showing central zones with correct area and presentation of shape, while the perimeter data are distorted (although areas are quantitatively correct). Description of the use of Lambert projection for vascular collagen fibres has been presented in earlier reports (Finlay et al. (1995), Canham et al. (1996)).

3 Results and Discussion

3.1 Artery Structure

Our goal was to take advantage of the qualitative and quantitative strengths of polarizing microscopy to make advances on the clinical reports on aneurysm biology. Brain arteries are muscular arteries, with a tunica media organized like a vascular compass. Its middle layer of smooth muscle cells and collagen of the ECM is organized within 1 to 2° of being perfectly concentric, and with a remarkably low circular standard deviation of 5.2° (study of 27 brain arteries (Canham et al. (1991)) with similar results on a few coronary arteries). The coherence of the tunica media is demonstrated nicely on tangential sections as shown in Fig. 6, a section that also reveals the exceptionally coherent peripheral collagen fibres of the media. These fibres 'close off' the concentric smooth muscle cells that border against the adventitia. The subendothelium (SE) is a thin layer, increasing in thickness with age and situated between the endothelium and the internal elastic lamina; it consists of matrix fibres and migrated smooth muscle cells from the media. The thicker subendothelial ECM is helically organized, with a structural balance of 'counter-rotating' layers and the alignment changing across the layers in multilayered SE (Finlay et al. (1991)). Over a range of different fixation pressures (vessels fixed without tethering or controlled length) the alignment of the adventitial ECM is increasingly circumferential with increasing pressure (Finlay et al. (1995)). Waviness of the medial fibres only becomes apparent at 0 or very low pressures (30 mmHg). The general lack of waviness for the collagen fibres of the media (at almost all distending pressures) indicates that the type III fibres (or complex of type I/III) of this layer have an elastic modulus substantially lower than type I tendon fibres, and with a much higher tolerance of strain before failure.

Collagen fibre birefringence and its correlation to tissue strength in the wound-healing experiments of Doillon and colleagues (1985) (reported above in Methods) provided an opportunity to test strength mechanics as calculated from microscopical structure against strength measurements of whole arteries (Yamada (1970)). Of particular interest from that review is the higher brain artery strength in the longitudinal direction, 1.3 MPa, compared to 1.0 MPa,

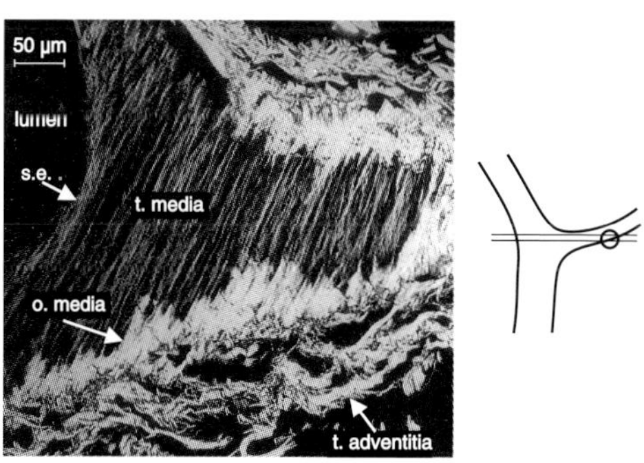

Fig. 6. Circularly polarized light micrograph of a middle cerebral artery branch, showing the parallel finer fibres of the tunica media. At the outer boundary of the concentric smooth muscle cells, similar fibres are merged into a thin very coherent boundary with the adventitia. Schematic on the right indicates the location of the oblique section from the branch vessel. (Reprinted from Rowe et al. (2003) with permission from S. Karger AG, Basel.)

in the circumferential direction. Our microscopically 'predicted' strengths for the same class of arteries were 3.6 and 2.4 MPa, clearly an overestimate but similar in finding the longitudinal direction to be the direction of greater strength (Rowe et al. (2003)). This is despite the histological dominance of the tunica media, which contributes strength entirely in the circumferential direction.

3.2 Branching Regions

Branching regions are widely varied in geometry, including area changes and area ratios, branch angles and vessel curvature (geometric features not unique to brain arteries). The uniqueness of the bifurcation region requires a multi-plane approach (i.e. sectioning in 3 orthogonal planes) – essential for polarized light microscopy, which has a high dependence on the plane of sectioning for the appearance and measurement reliability of individual fibres.

The medial gap remains a controversial feature of brain artery branches. It is unlikely to be the definitive precursor to aneurysm formation because of the gaps' presence in human noncerebral arteries and in many animal vessels (Stehbens (1989)), unlike aneurysms, which are a common pathology only in human brain arteries and in some primates. The average gap width was 0.12 mm (ranging between 0.038 and 0.225 mm) with a length of 0.85 mm. The length is the distance from above the midplane to below at which the serial sections showed the last presence of a muscle cell gap. This length represented approximately 60% of the trunk vessel radius. Within a single series of sections the gap width was consistent and uniform, tapering to zero at its ends. The size of the gap correlated negatively with the angle of the branch, that is the sharpest branching angles had the largest medial gaps ($R = -0.79$, from Canham and Finlay (2004)). This correlation links a clinically accessible measurement from vascular imaging to vessel microstructure that may have significance for vascular pathology.

Within the medial gap is a tendon-like ridge of coherent collagen fibres (Finlay et al. (1998)). The authors reported that 'when viewed in longitudinal sections parallel to the plane of the bifurcation (the conventional view for bifurcations), the coherent alignment of the thin band of collagen was striking.' This unique feature of the apex of bifurcations was discovered through the application of the universal stage.

3.3 Saccular Aneurysm Structure

Reports on the pathology of saccular aneurysms (Forbus (1930), Crawford (1959), Hassler (1961), Crompton (1966)) provide the framework for understanding the lesions' vulnerability to growth and the forces of blood pressure, and for theoretical studies on aneurysm mechanics (Humphrey and Kyriacou (1996), Ujiie et al. (2001), David and Humphrey (2003), Oshima and Torii (2004)). These lesions rupture mostly at the fundus (85%), less frequently at the equatorial region, and rarely at the neck; yet the fundus is generally thicker, conveying the impression of higher strength. Fibre birefringence, rather than fibre direction, has provided an answer to this paradox.

The theme for the continuing polarizing microscopy studies is that either alignment distribution or fibre strength (identified by fibre birefringence) might stand out as the critical factor for aneurysm tissue mechanics (Whittaker et al. (1988), Canham et al. (1996)). Emphasis was put on fibre orientation, chiefly because collagen has a single axis along which it can bear tension, making orientation critical to collagen function and the only measure related to tissue anisotropy. What has emerged is that fibre birefringence is equally critical to collagen function, and that the combination of optical measurements provides an effective indicator of tissue strength mechanics. Results on orientation showed that the small and medium sized aneurysms studied from autopsy are well structured. Even very small lesions appeared well organized with regard to the distribution of fibre directions (example shown in Fig. 7 for a 1.3 mm aneurysm). Shown as well on this figure are examples of measurement corridors, along which each layer of fibres has been sampled as randomly as possible within each identified layer, both for orientation and for birefringence.

Three-dimensional orientation data have provided a strong indication that, wherever sampled, the aneurysm wall (for lesions obtained from autopsy) is well supported from a directional perspective (Canham et al. (1996, 1999), Humphrey and Canham (2000)). Figure 8 shows graphically presented data from two aneurysms. The wedge-shaped positioning of the data on the Lambert projections demonstrate the reasonably close confinement of the data to within the theoretical limits of any tangentially aligned fibres (Canham et al. (1996)). Apparent gaps in data for both projections in Fig. 8 proved to be misleading. The strength along any assigned direction has contributions

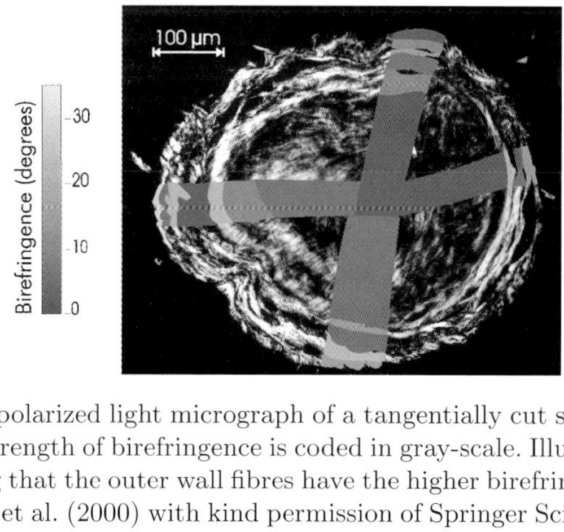

Fig. 7. Circular polarized light micrograph of a tangentially cut section of a 1.3 mm aneurysm. The strength of birefringence is coded in gray-scale. Illustrated is the very consistent finding that the outer wall fibres have the higher birefringence. (Reprinted from MacDonald et al. (2000) with kind permission of Springer Science and Business Media.)

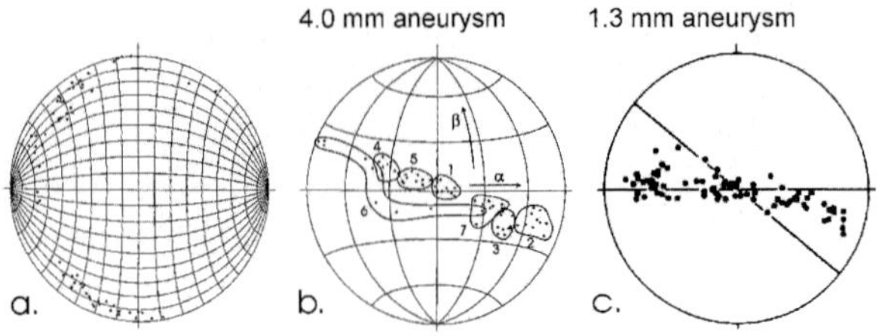

Fig. 8. Lambert projections of the data from two aneurysms. (a) shows the primary data across 7 layers of a near tangential section. The graph lines are at every 10° interval and are similar to lines of longitude and latitude on world maps. Fibres aligned with low elevation angles but with varied azimuth angles are positioned around the perimeter of the projection. In (b) the same data are rotated 90° about the EW axis. Individual points may be traced along vertical lines. Those data in the lower segment rotate to positions above the axis, while those in the upper segment reappear at 180° when they reach the primitive (outer edge) and continue their rotation to positions below the axis. The hand drawn outlines identify groups of data from individual layers. (b) and (c) show rotated data from tangential sections grazing the edge of the lumen of two different aneurysms. (Reprinted from Canham et al. (1996) with permission of Blackwell Publishing.)

from fibres at all angles, particularly those within 60° or less (because of the directional factor of $\cos^2 \theta$) (MacDonald et al. (2000)).

Wall thickness is not an easily accessible measure, despite its importance in aneurysm structure. However, one example of a segment of aneurysm that was extremely thin – 16 to 20 μm (from summing the few 4 μm sections that cut through the wall) – was found to be structurally adequate (although barely) because of the high birefringence of the remaining layers and fibres. This example reinforces the findings from theoretical studies that three equally strong sets of fibres, oriented at 60° from each other, form a virtually isotropic fabric in terms of strength and elasticity.

3.4 Fusiform Aneurysm

Included in the study was a fusiform aneurysm of the basilar artery, discovered as an incidental finding at the time of autopsy (Fig. 9). It was scanned by micro-CT (resolution 40 μm) at the Holdsworth lab (Robarts Research Institute, London, Canada), sent in physiologic saline (kept cool with gel ice pack) by overnight courier to the Humphrey Biomedical Engineering lab (Texas A&M University) for elasticity studies. Multiple surface marker beads for strain assessment were attached to the outer surface, and in-plane surface deformations were obtained using a triplane video system while the lesion was pressure cycled 30 to 150 mmHg (Hsu et al. (1995)). Subsequent to the elasticity testing the lesion was fixed in 10% neutral buffered formalin with a lumenal pressure at 110 mmHg, and returned by courier to the London lab for histology and polarized light studies.

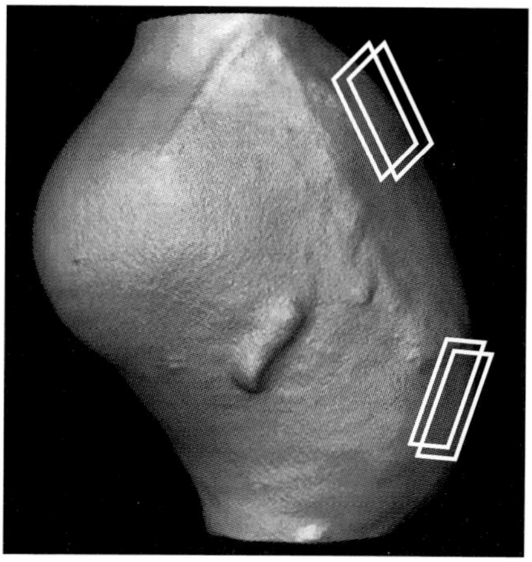

Fig. 9. Volume rendered image of a micro-CT scanned fusiform aneurysm (size 2.5 × 2.0 cm). A few millimetre sized surface patches of calcified areas were evident on 'calcium sensitive' projections. Serial histological 5 μm sections were taken tangentially at two surface regions – upper right and lower right.

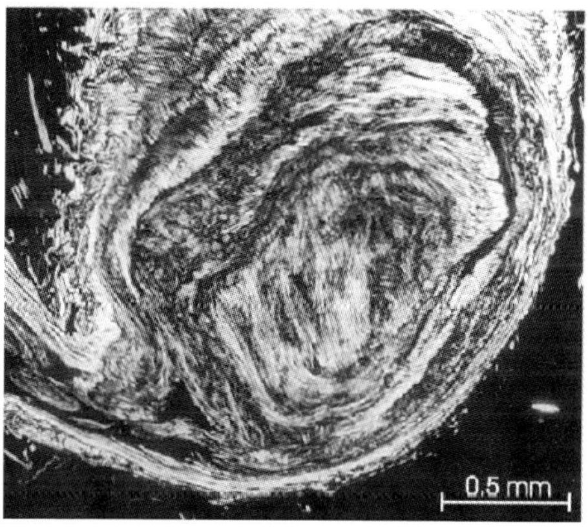

Fig. 10. Circularly polarized light micrograph of a tangential cut section from the upper quadrant of the fusiform aneurysm.

Results of the strain analysis showed that within the resolution of the triplane system (minimum detectable deformation of 8 μm between markers), there was no measurable surface strain, unlike measurable strains obtained from saccular aneurysms. Polarized light micrographs of the aneurysm show a fibrous surface structure that is consistent with an elastically rigid tissue (Fig. 10). The micrograph is from a tangential section taken at a thinner region of the wall, and shows the finely layered high birefringence collagen fibres, lying coherently within each layer. Analysis by universal stage for two tissue sections, one from each of the sampled planes, is shown on Lambert

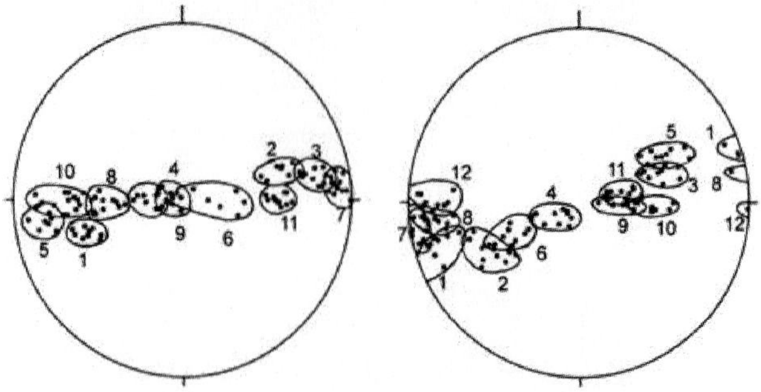

Fig. 11. Lambert projections of sections from two different surface cutting planes on the fusiform aneurysm. The data from each layer are outlined to show the alignment coherence in 3D, which is similar from layer to layer. Data lying perfectly along the equator (in this projection) would have no elevation angle, showing that each section was nearly perfectly tangential. The projection on the left shows less dispersion in terms of elevation, suggesting the lesion had more spherical symmetry at that location. Layers 1, 8 and 12 in the projection on the right have data at two locations near the perimeter. In these projections the radial distance from the centre represents the azimuth direction of the fibre, and the distance from the EW axis is the elevation of the fibre out of the section plane (see Fig. 8).

projections in Fig. 11. Results from the two sections, one with 11 layers, the other 12, indicate a well-balanced complement of fibre directions to mechanically resist biaxial stress. There is no indication with the partial penetration into this thick-walled lesion of inner layers of weakly birefringent fibres, unlike the saccular form of the lesion that invariably has an inner two-thirds of the wall that is virtually non-contributing to wall strength (Canham et al. (1999)).

4 Future Directions

The artery wall and aneurysm tissue involves a composite of ECM proteins – including fibrillary collagens and elastin (Mimata et al. (1997)). Laser-induced fluorescence spectroscopy provides a new approach for assessing vascular remodelling, as illustrated in Figs 12 and 13 (Christov et al. (2005)). A dual methods study combining polarizing microscopy and the laser-induced fluorescence spectroscopy has demonstrated a close correlation between the two methods (Korol et al. (2005)). Data were obtained from rat skin and tail tendon at an early animal age (2-12 weeks) in order to test these methods against each other under the circumstances of tissue remodelling. These studies are setting the stage for using endovascular approaches for investigating vessel wall pathology for advancing atherosclerosis and possibly brain aneurysms.

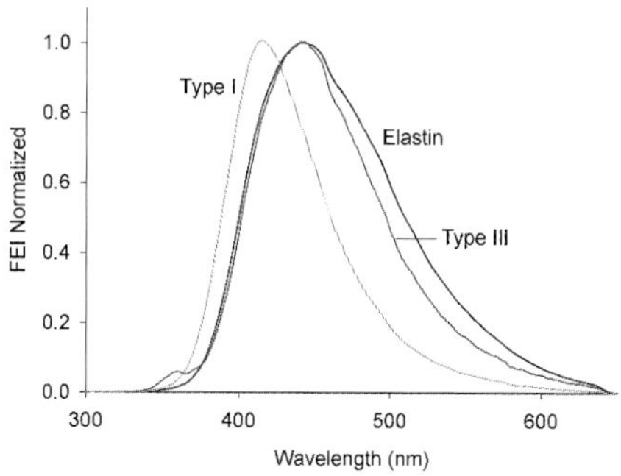

Fig. 12. Fluorescence emission intensity (FEI) spectra of core vascular tissue proteins: elastin (bovine neck ligament), collagen type I (bovine Achilles tendon) and collagen type III (calf skin). Spectra are markedly different with respect to width of the peak and peak location.

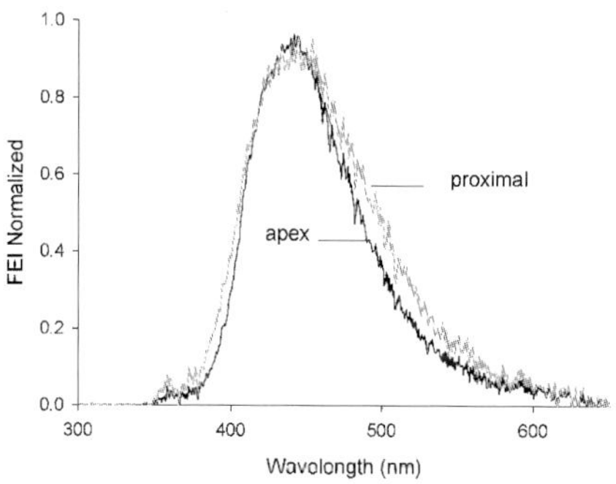

Fig. 13. FEI spectra of middle cerebral bifurcation, proximal versus apex. Spectral differences are visible across the flow impact area (apex), where the 'full width at half maximum' is comparable to collagen type I.

Acknowledgements. The research was funded principally by Canadian Institutes of Health Research grant ♯ MOP38026 with additional support from NIH grant ♯ HL54957 to Dr. Jay Humphrey. The authors wish to recognize the support of the Humphrey lab for the elasticity measurements on the fusiform aneurysm.

References

Canham, P. B., and Ferguson, G. G. (1985). A mathematical model for the mechanics of saccular aneurysms. *J. Neurosurg.* 17:291–295.

Canham, P. B., and Finlay, H. M. (2004). Morphometry of medial gaps of human brain artery branches. *Stroke* 35:1153–1157.

Canham, P. B., Talman, E. A., Finlay, H. M., and Dixon, J. G. (1991). Medial collagen organization in human arteries of the heart and brain by polarized light microscopy. *Conn. Tiss. Res.* 26:121–134.

Canham, P. B., Finlay, H. M., and Tong, S. Y. (1996). Stereological analysis of the layered collagen of human intracranial aneurysms. *J. Microscopy* 183:170–180.

Canham, P. B., Finlay, H. M., Kiernan, J. A., and Ferguson, G. G. (1999). Layered structure of saccular aneurysms assessed by collagen birefringence. *Neurological Research* 21:618–626.

Christov, A., Korol, R., Dai, E., Liu, L., Guan, H., Bernards, M. A., Cavers, P. B., Susko, D., and Lucas, A. R. (2005). *In vivo* optical analysis of quantitative changes in collagen and elastin during arterial remodeling. *Photochem. Photobiol.* in press.

Crawford, T. (1959). Some observations on the pathogenesis and natural history of intracranial aneurysms. *J. Neurol. Neurosurg. Psychiatry* 22: 259–266.

Crompton, M. R. (1966). Mechanism of growth and rupture in cerebral berry aneurysms. *Br. Med. J.* 1:1138–1142.

David, G., and Humphrey, J. D. (2003). Further evidence for the dynamic stability of intracranial saccular aneurysms. *J. Biomech.* 36:1143–1150.

Doillon, C. J., Dunn, M. G., Bender, E., and Silver, F. H. (1985). Collagen fiber formation in repair tissue: development of strength and toughness. *Coll. Relat. Res.* 5:481–492.

Drake, C. G., and Peerless, S. J. (1997). Giant fusiform intracranial aneurysms: review of 120 patients treated surgically from 1965 to 1992. *J. Neurosurg.* 87:141–162.

Emmons, R. C. (1943). The universal stage. *Geol. Soc. Am. Mem.* 8:1–24.

Ferguson, G. G. (1970). *Physical Factors in the Initiation, Growth, and Rupture of Human Intracranial Saccular Aneurysms.* Ph.D. Thesis, The University of Western Ontario, London, Ontario. p. 9.

Ferguson, G. G. (1989). Intracranial arterial aneurysms – a surgical perspective. In Toole, J. S., ed., *Handbook of Clinical Neurology. Vol II: Vascular Disorders (part III.* Amsterdam: Elsevier. 41–87.

Finlay, H. M., Whittaker, P., Hicks, J. G., Taylor, C. P., Park, Y. W., and Canham, P. B. (1989). Spatial orientation of arterial sections determined from aligned vascular smooth muscle. *J. Microscopy* 155:213–226.

Finlay, H. M., Dixon, J. G., and Canham, P. B. (1991). Fabric organization of the subendothelium of the human brain artery by polarized-light microscopy. *Arterioscler. Thromb.* 11:681–690.

Finlay, H. M., McCullough, L., and Canham, P. B. (1995). Three-dimensional collagen organization of human brain arteries at different transmural pressures. *J. Vasc. Res.* 32:301–312.

Finlay, H. M., Whittaker, P., and Canham, P. B. (1998). Collagen organization in the branching region of human brain arteries. *Stroke* 29:1595–1601.

Fisher, R. A. (1953). Dispersion on a sphere. *Proc. R. Soc. Lond. A* 217: 295–305.

Forbus, W. D. (1930). On the origin of miliary aneurysms of the superficial cerebral arteries. *Bull. Johns Hopkins Hosp.* 47:239–284.

Hassler, O. (1961). Morphological studies of the large cerebral arteries, with reference to the aetiology of subarrachniod haemorrhage. *Acta Psychiatr. Neurol. Scand. Suppl.* 154:1–145.

Hsu, F. P. K., Downs, J., Liu, A. M. C., Rigamonti, D., and Humphrey, J. D. (1995). A triplane video-based experimental system for studying axisymmetrically inflated biomembranes. *IEEE Trans. Biomed. Eng.* 42:442–449.

Humphrey, J. D., and Canham, P. B. (2000). Structure, mechanical properties and mechanics of intracranial saccular aneurysms. *J. Elasticity* 61: 49–81.

Humphrey, J. D., and Kyriacou, S. K. (1996). The use of laplace's equation in aneurysm mechanics. *Neurological Research* 18:204–208.

ISAT. (2002). International subarachnoid aneurysm trial of neurosurgical clipping versus endovascular coiling in 2143 patients with ruptured intracranial aneurysms: a randomised trial. *Lancet* 360:1267–1274. International Subarachnoid Aneurysm Trial (ISAT) Collaborative Group.

ISIUA. (1998). Unruptured intracranial aneurysms – risk of rupture and risks of surgical intervention. *New Engl. J. Med.* 339:1725–1733. International Study of Unruptured Intracranial Aneurysm (ISIUA) Investigator.

ISIUA. (2003). Unruptured intracranial aneurysms: natural history, clinical outcome, and risks of surgical and endovascular treatment. *Lancet* 362: 103–110. International Study of Unruptured Intracranial Aneurysm (ISIUA) Investigator.

Korol, R. M., Finlay, H. M., Josseau, M., Lucas, A. R., and Canham, P. B. (2005). Fluorescence spectroscopy and birefringence measurements of molecular changes in maturing rat tail tendon. *Biophys. J.* under revision.

Leblanc, R., Lozano, A. M., van der Rest, M., and Guttmann, R. D. (1989). Absence of collagen deficiency in familial cerebral aneurysms. *J. Neurosurg.* 70:837–840.

MacDonald, D. J., Finlay, H. M., and Canham, P. B. (2000). Directional wall strength in saccular brain aneurysms from polarized light microscopy. *Ann. Biomed. Eng.* 28:533–542.

Mimata, C., Kitaoka, M., Nagahiro, S., Iyama, K., Hori, H., Yoshioka, H., and Ushio, Y. (1997). Differential distribution and expressions of collagens in the cerebral aneurysmal wall. *Acta Neuropath. (Berlin)* 94:197–206.

Nollie, G. J., Sandhu, H. S., Cernovsky, Z. Z., and Canham, P. B. (1996). Regional differences in molecular cross-linking of periodontal ligament collagen of rat incisor, by polarizing microscopy. *Conn. Tiss. Res.* 33:283–289.

Oshima, M., and Torii, R. (2004). Image-based simulation of blood flow and arterial wall interaction for cerebral aneurysms. In *2004 IUTAM Symposium on Mechanics of Biological Tissue.*

Oshima, M., Takagi, K., Hyakawa, M., Torii, R., and Nagano, K. (2005). Effects of vascular morphological changes on cerebral hemodynamics. *J. Biomech.* submitted.

Peters, D. G., Kassam, A. B., Feingold, E., Heidrich-O'Hare, E., Yonas, H., Ferrell, R. E., and Brufsky, A. (2001). Molecular anatomy of an intracranial aneurysm. coordinated expression of genes involved in wound healing and tissue remodeling. *Stroke* 32:1036–1042.

Phillips, W. R. (1971). *Mineral Optics, Principles and Techniques.* San Francisco, California: W. H. Freeman.

Rowe, A. J., Finlay, H. M., and Canham, P. B. (2003). Collagen biomechanics in cerebral arteries and bifurcations assessed by polarizing microscopy. *J. Vasc. Res.* 40:406–415.

Smith, J. F. H., Canham, P. B., and Starkey, J. (1981). Orientation of collagen in the tunica adventitia of the human cerebral artery measured with polarized light and the universal stage. *J. Ultrastruct. Res.* 77: 133–145.

Stehbens, W. E. (1972). *Intracranial arterial aneurysms. Pathology of the Cerebral Blood Vessel.* St. Louis: C. V. Mosby Co.

Stehbens, W. E. (1989). Etiology of intracranial berry aneurysms. *J. Neurosurg.* 70:823–831.

Steinman, D. A., Vorp, D. A., and Ethier, C. R. (2003). Computational modeling of arterial biomechanics: insights into pathogenesis and treatment of vascular disease. *J. Vasc. Res.* 37:1118–1128.

Tarling, D. H. (1983). *Statistical and mathematical analyses. Palaeomagnetism, Principles and Applications in Geology, Geophysics and Archaeology.* London: Chapman & Hall.

Ujiie, H., Tamano, Y., Sasaki, K., and Hori, T. (2001). Is the aspect ratio a reliable index for predicting the rupture of a saccular aneurysm? *J. Neurosurg.* 48:495–502.

Whittaker, P., Schwab, M. E., and Canham, P. B. (1988). The molecular organization of collagen in saccular aneurysms assessed by polarized light microscopy. *Conn. Tiss. Res.* 17:43–54.

Yamada, H. (1970). *Strength of Biological Materials.* Baltimore: Williams & Wilkins.

Image-based Simulation of Blood Flow and Arterial Wall Interaction for Cerebral Aneurysms

M. Oshima,[1] R. Torii,[1] T. Takagi[2]

[1] University of Tokyo, Institute of Industrial Science, Japan
marie@iis.u-tokyo.ac.jp, ryo.torii@imperial.ac.jp
[2] Teikyo University, Department of Neurosurgery, Japan
pktakagi@med.teikyo-u.ac.jp

In order to predict the formation of cerebral aneurysms, it is important to obtain detailed information on hemodynamic and structural quantities such as distributions and magnitudes of wall shear stress and velocity. However, it is difficult to measure them *in vivo* by medical imaging techniques due to the limited resolution and data acquisition capability of currently available systems. In particular, the measurement of such quantities in the cerebral circulation is challenging due to its geometric complexity, the presence of the skull, and small vessels. Since the structural behavior of arterial wall plays an important role, particularly in the formation of aneurysms, a numerical simulation method is developed for investigating the effects of the interaction between arterial walls and blood flow.

1 Introduction

It is known that over 70% of subarachnoid hemorrhages are caused by rupture of a saccular aneurysm, which is a sac-shaped aneurysm that has developed in a relatively large cerebral artery (Moore and Dalley (1999), pp. 886-887). The risk of its rupture is reported to be under 0.1% (The International Study of Unruptured Intracranial Aneurysms Investigators (1998)). However, once the subarachnoid hemorrhage occurs, 30% to 50% of patients die after the first attack (van Gijn and Rinkel (2001)). Thus, it is important to distinguish aneurysms with high risks of rupture from those with low risks as well as where and how they are formed.

Some cardiovascular diseases and cerebral disorders, such as arterioscleroses and cerebral aneurysms, are reported to depend on hemodynamic factors, particularly the blood flow induced wall shear stress (Malek et al. (1999)). Thus, it is important to obtain detailed information of hemodynamic and arterial structure quantities such as the distributions and magnitudes of wall shear stress and of velocity. However, it is difficult to measure them *in vivo* by medical imaging techniques such as MRI (Magnetic Resonance Imaging) or CT (Computed Tomography) due to the limited resolution and data acquisition capabilities of the present medical image systems. In particular, the measurement of hemodynamic and structural quantities in the cerebral circulation is challenging because of its geometrical complexity, the presence of the skull, and small vessels. On the other hand, *in vivo* image-based simulations can provide hemodynamic quantities in a patient-specific manner that cannot be measured experimentally. Thus, *in vivo* image-based simulation has emerged as a powerful tool for cardiovascular (Taylor et al. (1999)) as well as cerebralvascular studies (Cerbral et al. (2002)). The present authors have been developing a patient-specific modeling and numerical simulation system for a better understanding of cerebral hemodynamics (Oshima et al. (2001)). However, our numerical simulation system assumes the arterial wall to be rigid. The formation and growth of aneurysms is associated with both mechanical and physiological factors. In this paper, mechanical factors, in particular the deformations of the vascular walls, are taken into account.

Recent studies (Hisada et al. (2003) and McQueen and Peskin (2001)) reported numerical simulations of the fluid-structure interactions between blood flow and the arterial wall. These studies, however, focus on the heart or abdominal aorta, where the deformations are relatively large. On the other hand, numerical simulations of fluid-structure interaction of cerebral arteries have not been common because the cerebral arteries deform less than the heart and aorta. Even though the deformations are small, they affect the dynamics of the blood flow. Therefore, the present paper examines the effects of structural deformations on cerebral hemodynamics through the numerical simulation of blood flow-arterial wall interaction.

2 Blood Flow-Arterial Wall Interaction

2.1 Blood Flow Mechanics

The present paper analyzes blood flow in a cerebral artery, for which the diameter is 3 mm and the Reynolds number is much lower than 2000. Thus, blood flow can be assumed to be laminar (Karino et al. (1993)). Since the shear rate in a cerebral artery of 3 mm diameter is 400 to 600 s^{-1}, the blood viscosity can be assumed to be Newtonian (Brooks et al. (1970)). The governing equations are the continuity and Navier-Stokes equations for a computational domain Ω,

$$\frac{\partial u_i}{\partial x_i} = 0, \tag{1}$$

$$\rho\left(\frac{\partial u_i}{\partial t} + u_j\frac{\partial u_i}{\partial x_j}\right) = -\frac{\partial P}{\partial x_i} + \mu_f\frac{\partial}{\partial x_j}\left(\frac{\partial u_i}{\partial x_j} + \frac{\partial u_j}{\partial x_i}\right), \tag{2}$$

where u_i is the velocity component in the i-th direction ($i = 1, 2, 3$ denote the x-, y-, and z-directions, respectively), P is the pressure, and μ_f is the viscosity.

The boundary conditions for the governing equations (1) and (2) are given by

$$u_i = g_i(x,t), \quad \text{on } \Gamma_g, \tag{3}$$

$$t_i = h_i(x,t), \quad \text{on } \Gamma_h, \tag{4}$$

$$h_i = \left[-P\delta_{ij} + \mu_f \left(\frac{\partial u_i}{\partial x_j} + \frac{\partial u_j}{\partial x_i} \right) \right] n_j, \tag{5}$$

where g_i is the Dirichlet boundary condition, Γ_g is the Dirichlet boundary, h_i is the Neumann boundary condition, and n_j is the j-th component of a vector normal to the Neumann boundary Γ_h.

Equations (1) and (2) are discretized with the DSD/SST finite element formulation (Tezduyar et al. (1992a) and Stein et al. (2001)). The DSD/SST method is suitable for flow problems with moving boundaries and interfaces. In this formulation, the discretized equation is

$$\begin{aligned}
&\int_{Q_n} w_i \rho \left(\frac{\partial u_i}{\partial t} + u_j \frac{\partial u_i}{\partial x_j} \right) \mathrm{d}Q + \int_{Q_n} \frac{\partial w_i}{\partial x_j} \sigma_{ij} \mathrm{d}Q - \int_{(P_n)_h} w_i h_i \mathrm{d}P \\
&+ \int_{Q_n} q \frac{\partial u_i}{\partial x_i} \mathrm{d}Q + \int_{\Omega} (w_i)_n^+ \left[(u_i)_n^+ - (u_i)_n^- \right] \mathrm{d}\Omega \\
&+ \sum_{e-1}^{(n_{\mathrm{el}})_n} \int_{Q_n^e} \frac{1}{\rho} \left[\tau_{SUPG} \rho \left(\frac{\partial w_i}{\partial t} + u_j \frac{\partial w_i}{\partial x_j} \right) + \tau_{PSPG} \frac{\partial q}{\partial x_i} \right] \\
&\times \left[\rho \left(\frac{\partial u_i}{\partial t} + u_j \frac{\partial u_i}{\partial x_j} \right) - \frac{\partial \sigma_{ij}}{\partial x_j} \right] \mathrm{d}Q = 0,
\end{aligned} \tag{6}$$

where

$$\sigma_{ij} = -P\delta_{ij} + \mu_f \left(\frac{\partial u_i}{\partial x_j} + \frac{\partial u_j}{\partial x_i} \right), \tag{7}$$

and Ω is the spatial domain, Q_n is the slice of the space-time domain between time levels n and $n+1$, and P_n denotes the lateral boundary of the space-time slab. The integrations are performed over the space-time slab Q_n. The functions such as velocity and pressure as well as the test functions are continuous in space but discontinuous in time. Thus, the notations $(\bullet)_n^-$ and $(\bullet)_n^+$ indicate the values as time t_n is approached from before and after

$$(\phi)_n^{\pm} = \lim_{\delta \to 0} \phi|_{t=t_n \pm \delta} \,. \tag{8}$$

In eq. (6), the first five integrals represent the Galerkin formulation. The fifth integral represents the weak continuity of the velocity in time. The sixth term is the stabilization based on the Stream-Upwind/Petrov-Galerkin (PSGS) formulations (Hughes and Brooks (1979) and Tezduyar et al. (1992c)). The parameters τ_{SUPG} and τ_{PSPG} are the stabilization parameters, which are given by

$$\tau_{SUPG} = \left(\frac{1}{\tau_{SUPG12}^2} + \frac{1}{\tau_{SUPG3}^2} \right)^{-1/2}, \tag{9}$$

$$\tau_{SUPG12} = \left(\sum_{a=1}^{n_{\rm en}} \left| \frac{\partial N_a}{\partial t} + u_i \frac{\partial N_a}{\partial x_i} \right| \right)^{-1}, \quad \tau_{SUPG3} = \frac{h_{RGN}^2}{4\nu}, \tag{10}$$

$$h_{RGN} = \left(\sum_{a=1}^{n_{\rm en}} \left| r_i \frac{\partial N_a}{\partial x_i} \right| \right)^{-1}, \quad r_i = \frac{\dfrac{\partial \|u_i\|}{\partial x_i}}{\left\| \dfrac{\partial \|u_i\|}{\partial x_i} \right\|}, \quad \tau_{PSPG} = \tau_{SUPG}, \tag{11}$$

where N_a is the space-time shape function associated with the element node a and $n_{\rm en}$ is the number of element nodes.

2.2 Arterial Wall Mechanics

The governing equations for the structural mechanics are the equilibrium equations

$$\rho^s \left(\frac{{\rm d}^2 y_i}{{\rm d}t^2} - f_i^s \right) - \frac{\partial \sigma_{ij}^s}{\partial x_j} = 0, \tag{12}$$

where y_i are the components of the displacement vector, f_i^s those of the external body force acting on the structure, ρ^s is the density of the material and σ_{ij}^s are the components of the Cauchy stress tensor. The subscript s denotes the structure. We assume the arterial wall to be a linearly elastic material in the paper even though it is known to be non-linear. The Galerkin method is used for finite element discretization based on the virtual work

$$\int_{\Omega_o^s} w_i \rho^s \frac{{\rm d}^2 y_i}{{\rm d}t^2} {\rm d}\Omega + \int_{\Omega_o^s} \delta E_{ij} S_{ij} {\rm d}\Omega = \int_{\Gamma_t^s} w_i t_i {\rm d}\Gamma + \int_{\Omega_t^s} w_i \rho^s f_i {\rm d}\Omega, \tag{13}$$

where S_{ij} and E_{ij} are the components of the second Piola-Kirchoff stress tensor and the Green-Lagrange strain tensor, respectively, and w_i is the test function, which is the virtual displacement. The fluid dynamic forces are included in the traction vector components t_i. Using the Newmark-β method, a non-linear system of equations is obtained at each time step and is written in incremental form. The equations in incremental form are solved using the parameters $\beta = 0.3225$ in order to control the stability and accuracy.

2.3 Mesh Update

Since the fluid domain changes, the fluid mesh is updated using an automatic mesh moving method (Tezduyar et al. (1992b)). In this paper, the motion of the nodes in the fluid domain is governed by the equations of elasticity. The boundary conditions are prescribed as the motion of the interface between the fluid and structure.

2.4 Blood Flow-arterial Wall Coupling

The equations of blood flow (fluid) and arterial wall (structural) mechanics are coupled by a block-iterative coupling method. Since the nodes at the interface between the fluid and structure are defined to be the same, the data for each nodal point at the interface can be exchanged between the fluid and structure. The fluid forces at the interface are treated as external forces for the structure while structural displacements at the interface are treated as Dirichlet boundary conditions for the fluid dynamics in the DSD/SST formulation (Tezduyar et al. (1992a)). In block-iterative coupling of the fluid and structural mechanics equations, the structural response becomes sensitive to small changes in the fluid dynamic forces. Thus, several iterations are performed in each time step in order to obtain convergence. In order to solve the systems of algebraic equations, which result from the discretized equations of both fluid and structural parts, an iterative procedure, the GMRES (Generalized Minimal RESidual) algorithm, is used.

3 Image-based Modeling

3.1 Geometric Modeling Method

The analysis model is constructed in a patient-specific manner using, for example, CT or MRI images. In this paper, CT data are used.

First of all, the iso-surfaces of vascular geometry are extracted from CT horizontal slices using the segmentation method based on the threshold value of the blood vessel. Then, the marching cubes method is applied for surface contraction (Loremson and Cline (1987)). In general, polygonal surfaces that result from iso-surface extraction are over-tessellated. The original over-tessellated surfaces affect the numerical simulation and generally result in excessive wall shear stress distributions due to artifacts. Thus, a smoothing method is developed by combining the polygon reduction technique (Garland and Heckber (1997)) and the smoothing technique (Taubin (1995)). The polygon reduction technique is a surface simplification method that merges adjacent triangular patches by eliminating one vertex and two triangles. On the other hand, the smoothing method based on Taubin's method replaces each vertex with the average of its neighboring vertices without causing shrinkage of the model as in the Gaussian smoothing technique. The polygon reduction technique is advantageous in smoothing the entire surface while Taubin's smoothing method is advantageous in smoothing locally over-tessellated areas. Therefore, the polygon reduction method is applied in order to smooth the entire surface, and then Taubin's smoothing method is applied locally to the over-tessellated areas.

3.2 Analysis Model

The present geometric modeling method is applied in the construction of an analysis model using the CT images of a MCA (Middle Cerebral Artery) aneurysm in a 59 year old female patient. The raw data consist of approximately 100 horizontal slices separated by 0.5 mm in the vertical direction and 0.3 mm of zooming reconstruction. Each slice is a 512×512 pixel 16 bit gray scale image, which results in the resolution of 0.2 mm/pixel.

The present analysis model as shown in Fig. 1. The diameter of the parent MCA is about 3.0 mm. The size of the aneurysm is 6.0 mm in diameter and 3.0 mm in height. The aneurysm has a small bleb as shown in Fig. 1. The arterial wall is assumed to have a uniform thickness of 0.3 mm, which is one-tenth of the diameter of the parent MCA. This assumption is reasonable for the artery but not for the aneurysm. However, the current medical imaging technique does not have enough spatial resolution to obtain the wall thickness. In particular, the wall thickness of the cerebral artery and aneurysm are difficult to measure because of the skull. Thus, we assume a uniform distribution of wall thickness in this analysis as a first step. In the future, we will investigate how the thickness affects the hemodynamics as well as the behavior of the fluid-structure interaction.

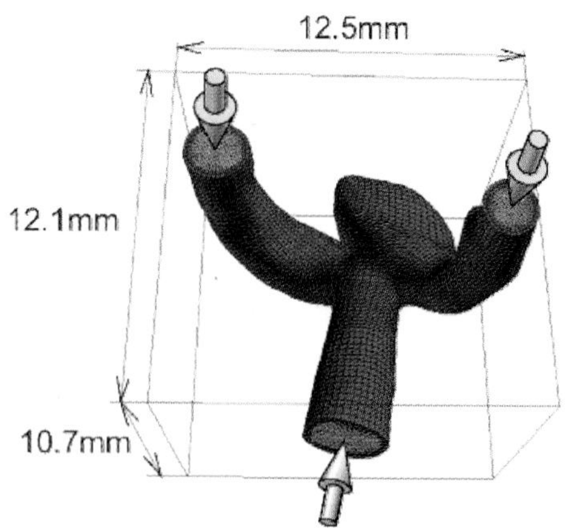

Fig. 1. Analysis model.

4 Numerical Results

4.1 Boundary Conditions

Since blood flow is a time-dependent pulsatile flow, the inflow boundary condition is prescribed as the Womersley velocity profile using the velocity measured by Doppler ultrasound velocimetry. The Womersley velocity profile is a general solution of oscillating flow in a straight tube, given by the axial component equation

$$u_w(r,t) = \frac{2B_o}{\pi R^2}\left[1-\left(\frac{r}{R}\right)^2\right]+\sum_{n=1}^{N}\frac{2B_o}{\pi R^2}\left[\frac{1-\dfrac{J_o\left(\alpha_n \frac{r}{R} i^{3/2}\right)}{J_o\left(\alpha_n i^{3/2}\right)}}{1-\dfrac{2J_1\left(\alpha_n i^{3/2}\right)}{\alpha_n i^{3/2} J_o\left(\alpha_n i^{3/2}\right)}}\right]e^{inwt}, \quad (14)$$

where r is the coordinate in the radial direction, R is the radius of the artery, ω is the angular velocity of pulsation, J_n is a Bessel function of the first kind of order n and $\alpha_n = R\sqrt{n\omega/\nu_f}$ is the Womersley parameter of order n. The coefficients B_n are Fourier coefficients of the flow rate and are determined from

$$Q(t) \approx \sum_{n=0}^{N} B_n e^{inwt}, \quad (15)$$

where $Q(t)$ is the flow rate. In this paper, the Fourier coefficients B_n are determined from the velocity profile measured by Doppler ultrasound velocimetry at the carotid artery of a male in his 20's. The order of the series is 20. The prescribed flow rate in the boundary is described in Fig. 2.

Blood is assumed to be a Newtonian fluid and to have a constant kinematic viscosity of 4.0×10^{-6} m^2/s. The resulting Reynolds number varies from 50 to 500 depending on the inflow velocities at the boundary.

The outflow boundary condition is prescribed as a traction boundary condition given by the Windkessel model. In the Windkessel model of the artery, the pressure is determined by integrating the equation

$$Q = C\frac{\mathrm{d}P}{\mathrm{d}t} + \frac{P}{D}, \quad (16)$$

where Q is the flow rate in the artery, C is the compliance of the artery, and D is the resistance of the distal arterial network. The parameters C and D are determined in an *ad hoc* manner for the pressure to range from 80 mmHg to 120 mmHg, which is the standard range of blood pressure for a healthy human. The transient behavior of the pressure on the inflow boundary is shown in Fig. 2. The pressure on the outflow boundary varies depending on the computed flow rate at each outflow boundary.

In the structural analysis, the boundary conditions are prescribed as no displacement both on the inflow and outflow boundaries.

wall shear stress in the rigid wall case is 300 dyn/cm^2 while that of the elastic wall is about 200 dyn/cm^2, as shown in Figs 4 and 5. The maximum magnitude of the wall shear stress in the rigid wall case occurs around the base of the aneurysm, which is called the neck of the aneurysm. As shown in Figs 4 and 5, the differences in the wall shear stress distribution between the elastic and rigid wall cases are large around the neck area. Significant differences can be observed on the arterial wall of Branch B due to significant arterial wall displacement. Thus, the differences in the wall shear stress distributions between the two cases are associated with deformation. Since the elastic arterial wall tends to expand in a high stress area, the diameter of the artery becomes larger and the wall shear stress decreases.

In order to examine the differences in the wall shear stress distributions, the flow velocities in the two cases are compared. Figures 6(b) and (c) show the velocity vectors in the cutting plane, which is shown in Fig. 6(a). As

(a) Cutting plane at cross section A.

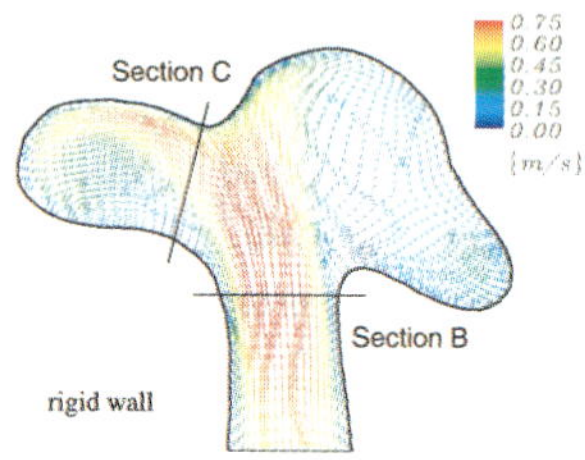

(b) Rigid wall case.

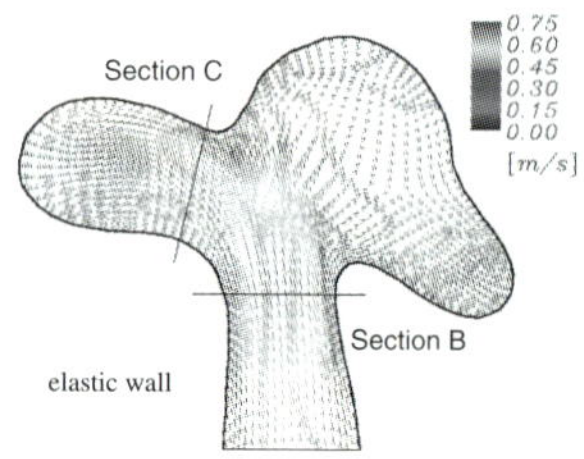

(c) Elastic wall case.

Fig. 6. Velocity distributions at cross section A.

shown in Figs 6(b) and (c), the incoming flow impinges on the neck of the aneurysm. The wall shear stress is affected by the displacement of the arterial wall where the flow impinges. In the elastic wall case, the location where the flow impinges moves away from the inflow, which leads to a reduction in the velocity gradients and also the wall shear stresses.

The magnitude of the velocity around the bifurcation where the flow impinges is strongly affected by the expansion of the proximal artery. As the proximal artery expands in the elastic wall case, as shown in Fig. 6(a), the impinging velocity and wall shear stress become smaller around the bifurcation due to decreases in the axial velocity.

In both rigid and elastic wall cases, the maximum displacement and the maximum wall shear stress occur near the neck of the aneurysm. The incoming flow impinges near the neck of the aneurysm because of the Y-shaped bifurcating artery. Flow impingement on the arterial wall around a bifurcation area was pointed out as an important factor in the formation of an aneurysm by Karino et al. (1993). As seen in the present study, the arterial deformation affects the velocity profile impinging on the bifurcation and the wall shear stress distribution around that area. Therefore, the arterial deformation is an important factor in the formation of the aneurysm.

However, further investigation will be necessary to improve the wall model and the distribution of wall thickness since vessel deformation is affected by distributions of thickness and also wall properties. The present finite element program allows implementation of more complex wall models such as hyperelastic models. Thus, the more sophisticated model will be included in the program in order to obtain more realistic simulations.

Since the boundary conditions will play an important role in more accurate simulations, further investigation will also consider the modeling of the boundary conditions. Since the internal carotid artery, which is located in the upstream region of MCA has large curvature, the magnitude of the secondary flow reaches 10 to 20% of the axial component according to Oshima et al. (2005). Thus, the secondary flow components need to be considered in the inflow boundary conditions in addition to the axial component represented by the Womersley profiles. The effects of the secondary flow will be incorporated into the program using the model, which is a function of Reynolds number and curvature. In this paper, no displacement is prescribed for the boundary conditions of the structural part in the inflow and outflow boundaries. If pressure wave propagation is considered, the boundary condition for the wall requires a more elaborated model. Thus, the wall boundary conditions will also be investigated by taking these aspects into account.

5 Conclusions

This paper has presented a patient-specific modeling and numerical simulation system for blood flow and arterial wall interaction aimed at a better understanding of cerebral hemodynamics. A computational fluid dynamics-based approach combined with patient-specific modeling is able to provide detailed behavior of blood flow and of the arterial wall. The results of a MCA aneurysm in a 59 years old female patient show that the deformation of the arterial wall has a significant influence on cerebral hemodynamics, in particular on the velocity and wall shear stress distributions. A large wall shear stress is observed in the area where the wall displacement is large.

Acknowledgements. The authors would like to thank Professor Tayfun Tezduyar, Rice University, for his kind advice in developing the flow-structure interaction part of the simulation system.

References

Brooks, D. E., Goodwin, J. W., and Seaman, G. V. (1970). Interactions among erythrocytes under shear. *J. Appl. Physiol.* 28:172–177.

Cerbral, J. R., Yim, P. J., Lohner, R., Soto, O., and Choyke, P. L. (2002). Blood flow modeling in carotid arteries using computational fluid dynamics and magnetic resonance imaging. *Academic Radiology* 9:1286–1299.

Garland, M., and Heckber, P. S. (1997). Surface simplification using quadric error metrics. In *Proceedings of ACM SIGRAPH Conference*, 209–216.

Hayashi, K., Hnada, H., Nagasawa, S., Okumura, A., and Moritake, K. (1980). Stiffness and elastic behaviour of human intracranial and extracranial arteries. *J. Biomech.* 13:175–184.

Hisada, T., Watanabe, H., and Sugar, S. (2003). Fluid-structure interaction analysis of human heart by ALE finite element method. In *Proceedings of USNCCM7.* Albuquerque, USA.

Hughes, T. J. R., and Brooks, A. N. (1979). A multi-dimensional upwind scheme with no crosswind diffusion. In Hughes, T. J. R., ed., *Finite Element Methods for Convection Dominated Flows, AMD Vol. 34.* ASME. 19–35.

Karino, T., Takeuchi, S., Kobayashi, N., Motomiya, M., and Mabuchi, S. (1993). Fluid dynamics of cerebrovascular diseases (in japanese). *Neurosurgeons* 12:15–24.

Loremson, W. E., and Cline, H. E. (1987). Marching cubes: A high resolution 3D surface construction algorithm. *Computer Graphics* 18:33–39.

Malek, A. M., Alper, S. L., and Izumo, A. (1999). Hemodynamic shear stress and its role in atherosclerosis. *JAMA* 21:2035–2042.

van Gijn, J., and Rinkel, G. (2001). Subarachnoid hemorrhage: diagnosis, causes and management. *Brain* 124:249–278.

McQueen, D. M., and Peskin, C. S. (2001). Heart simulation by an immersed boundary method with formal second accuracy and reduced numerical viscosity. In *Proceedings of ICTAM 2000.*

Moore, K. L., and Dalley, A. (1999). *Clinically Oriented Anatomy.* Baltimore: Lippincott Williams & Wilkins, 4th edition.

Oshima, M., Torii, R., and Kobayashi, T. (2001). Finite element simulation of blood flow in the cerebral artery. *Comput. Meth. Appl. Mech. Eng.* 191: 661–671.

Oshima, M., Sakai, H., and Torii, R. (2005). Modelling of inflow boundary conditions for image-based simulation of cerebrovascular flow. *Int. J. Numer. Meth. Fluids* 47:603–617.

Stein, K., Benney, R., Tezduyar, T., and Ptovin, J. (2001). Fluid-structure interactions of a cross parachute: Numerical simulation. *Comput. Meth. Appl. Mech. Eng.* 191:673–687.

Taubin, G. (1995). A signal processing approach to fair surface design. In *Proceedings of ACM SIGRAPH Conference*, 351–358.

Taylor, C. A., Draney, M. T., Ku, J. P., Parker, D., Steele, B. N., Wang, K., and Zarins, C. K. (1999). Predictive medicine: Computational techniques in therapeutic decision-making. *Comput. Aided Surg.* 4:231–247.

Tezduyar, T. E., Behr, M., and Liou, J. (1992a). A new strategy for finite element computations involving moving boundaries and interfaces – the deforming-spatial-domain/space-time procedure: I. The concept and the preliminary numerical tests. *Comput. Meth. Appl. Mech. Eng.* 94:339–351.

Tezduyar, T. E., Behr, M., Mittal, S., and Johnson, A. A. (1992b). Computation of unsteady incompressible flows with the finite element methods – spacetime formulations, iterative strategies and massively parallel implementations. In *New Methods in Transient Analysis, PVP.246/AMD-143.* 7–24.

Tezduyar, T. E., Mittal, S., Ray, S. E., and Shih, R. (1992c). Incompressible flow computations with stabilized bilinear and linear equal-order-interpolation velocity-pressure elements. *Comput. Meth. Appl. Mech. Eng.* 95:221–242.

The International Study of Unruptured Intracranial Aneurysms Investigators. (1998). Unruptured intracranial aneurysms – risk of rupture and risks of surgical intervention. *N. Engl. J. Med.* 339:1725–1733.

Part IV

Biological Tissues

A Framework for Soft Tissue and Musculo-skeletal Modelling: Clinical Uses and Future Challenges

J.W. Fernandez, H. Schmid, P.J. Hunter

The University of Auckland, Bioengineering Institute, New Zealand
p.hunter@auckland.ac.nz

A modelling framework is presented to investigate soft tissue and general musculo-skeletal (MSK) problems. This includes geometry development and customization, governing equations and constitutive laws, and an illustration of our methodology for two clinically focused studies: (i) a patella model, and (ii) a Cerebral Palsy (CP) assessment tool. We discuss the motivation behind our approach, challenges, especially with constitutive law formulation, issues with current limitations and future objectives, including the link with the Physiome Project.

1 Motivation

The challenges faced in modelling the mechanical behaviour of soft biological tissues are similar in all organs or organ systems. For example: (i) how to model the anatomy and tissue structure efficiently and in a 'patient-specific' manner; (ii) how to define constitutive laws that capture the material properties of the tissue but are not over-parameterized and have experimentally measurable parameters; (iii) how to link the constitutive parameters to underlying tissue micro-structure so that spatial variation of material properties can be inferred from observation of tissue micro-structure; (iv) how to link the equations of large deformation mechanics to the equations representing other physical processes such as blood flow and electrical activation in the heart and skeletal muscles or air flow in the lungs; (v) how to link tissue properties to cell properties (e.g., for active tension development in muscle or growth processes in all tissues); (vi) how to design computational code that allows the three-dimensional finite deformation mechanics of organs to be solved in a reasonable time on parallel architectures. Here, we address some of these issues in relation to the musculo-skeletal system and, in particular, the modelling of the kinematics and kinetics of the patella and the application of an anatomically-based model of the leg to assessing the kinematics of children with cerebral palsy.

2 Geometry, Development and Customization

The anatomically-based geometries used in our models are a subset from a database (Hunter and Borg (2003), Figs 1a–c) developed as a contribution to the IUPS (International Union Of Physiological Sciences) Physiome Project (Hunter et al. (2002)) and contain such features as muscle fibre orientation fields (Nielsen et al. (1991); LeGrice (1992)), and is illustrated for the rectus femoris (RF) muscle (Fig. 1d). These geometries are derived from the Visible Human (VH) data set (Ackerman (1998)), which has become a common standard for sharing, medical education and virtual simulation (Spitzer and Whitlock (1998)). The soft tissues were manually digitized from 2D images to create 3D data sets for geometric fitting (Figs 1e, f). Numerous researchers have used the VH data sets with anatomically based mechanics investigations including Dong et al. (2002), who segmented individual muscles and described fibre orientations for deformation analysis, and Hirota (2002), who used segmented VH geometry to model the contact mechanics of lower limb flexion. The bones were digitized using a hand-held Polhemus^{TM} Fastscan laser scanner (Applied Research Associates, NZ Ltd, Figs 1g, h). An anatomically accurate physical model (SOMSO, `www.somso.de`), typically found in

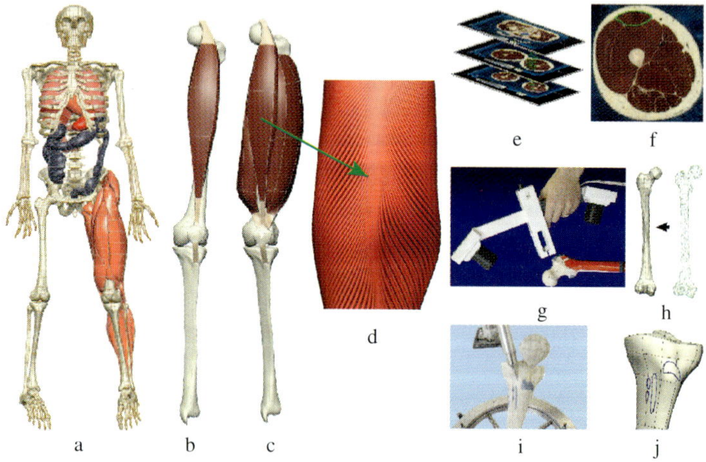

Fig. 1. (a) Geometry from the model database, (b) single rectus femoris and (c) multiple quadriceps subsets with (d) fitted bi-pennate fibre field for rectus femoris; (e) Visible Human axial images; (f) muscle segmentation (Ackerman 1998); (g) Polhemus scanner scanning femur, and (h) filtered digitized data used to fit mesh; (i) FARO arm digitizing tibia nodes and muscle attachment patches, and (j) resulting mesh.

medical school teaching labs was used to obtain the muscle origin and insertion points via a digitizing device, a FARO™ arm (FARO Technologies, Inc. USA, Figs 1i, j). High-order cubic Hermite elements are used for meshing which differ from the usual Lagrangian finite elements in that they preserve both the continuity of the nodal values (C_0 continuity), and their first derivatives (C_1 continuity). This has advantages in describing smooth biological bodies and provides numerical benefits when solving contact problems (Wriggers et al. (2001), Stadler et al. (2003)).

The FE meshes used are 3D volume meshes, and the interpolation or shape functions are determined from the tensor product of 1D interpolation functions. We shall therefore describe the 1D functions and infer their 3D counterparts. The four 1D cubic Hermite interpolation functions are

$$\psi_1^0(\xi) = 1 - 3\xi^2 + 2\xi^3,\ \psi_2^0(\xi) = \xi^2(3 - 2\xi),\ \psi_1^1(\xi) = \xi(\xi - 1)^2,\\ \psi_2^1(\xi) = \xi^2(\xi - 1), \tag{1}$$

where ξ is the local or material coordinate with $\xi \in [0,1]$. In 1D, a field $\mathbf{u}$, which may be either geometry or a field (such as the fibre orientation) is interpolated via

$$\mathbf{u}(\xi) = \psi_1^0(\xi)\mathbf{u}_1 + \psi_2^0(\xi)\mathbf{u}_2 + \psi_1^1(\xi)\frac{\mathrm{d}\mathbf{u}}{\mathrm{d}s}\bigg|_1 L + \psi_2^1(\xi)\frac{\mathrm{d}\mathbf{u}}{\mathrm{d}s}\bigg|_2 L, \tag{2}$$

where $\mathbf{u}(\xi)$ refers to the continuum field evaluated at material point ξ and $\mathbf{u}_n$ refers to the vector of nodal parameters used in the interpolation. Continuity is enforced on arc length based derivatives $\mathrm{d}\mathbf{u}/\mathrm{d}s$ to ensure C_1 continuity in the spatial domain, and L is the arc length scale-factor given by

$$L = \int_0^1 \sqrt{\left(\frac{\mathrm{d}x}{\mathrm{d}\xi}\right)^2 + \left(\frac{\mathrm{d}y}{\mathrm{d}\xi}\right)^2 + \left(\frac{\mathrm{d}z}{\mathrm{d}\xi}\right)^2}\,\mathrm{d}\xi. \tag{3}$$

The full mathematical details of the fitting process used to derive the MSK models from the VH data set, and customize them to patient-specific data using 'host mesh' fitting is presented in Fernandez et al. (2004), and the reader is referred to this as a technical background for this paper.

In summary, the projection of a data point onto a 2D element for geometric fitting, contact mechanics boundary checking or to obtain the 3D material coordinates of a control point within a host mesh is solved via the minimization of a least-squares function,

$$D(\bar{\xi}) = \| \mathbf{u}(\bar{\xi}) - \mathbf{z}_d \|^2 \quad \longrightarrow \quad \frac{\partial D}{\partial \bar{\xi}} = 0, \tag{4}$$

where $\mathbf{z}_d$ are the spatial coordinates of the data point d, and $\mathbf{u}(\bar{\xi})$ is interpolated using the 2D version of eq. (2). Note that for face fitting and contact mechanics $\bar{\xi}=(\xi_1, \xi_2)$ and for host mesh fitting $\bar{\xi}=(\xi_1, \xi_2, \xi_3)$. The following non-linear equations can be solved using the Newton-Raphson procedure. Once the minimum point is obtained ($\bar{\xi}_d$), the geometry or host mesh can be fitted to minimize the objective function, F, for a vector of nodal mesh parameters $\mathbf{u}_n$, where

$$F(\mathbf{u}_n) = \sum_{d=1}^{N} w_d \| \mathbf{u}(\bar{\xi}_d) - \mathbf{z}_d \|^2 + F_s(\mathbf{u}_n), \tag{5}$$

w_d is a weight for each data point, $\bar{\xi}_d$ is obtained from the solution of eq. (4) and $\mathbf{u}(\bar{\xi}_d)$ is evaluated from the interpolation function in 2D (geometric face fitting) or 3D ('host mesh' fitting). $F_s(\mathbf{u}_n)$ is the Sobolev smoothing penalty function and the full 3D form is

$$\begin{aligned} F_s(\mathbf{u}_n) = \int_0^1\!\!\int_0^1\!\!\int_0^1 \Bigg\{ & \gamma_1 \left\| \frac{\partial \Delta \mathbf{u}}{\partial \xi_1} \right\|^2 + \gamma_2 \left\| \frac{\partial \Delta \mathbf{u}}{\partial \xi_2} \right\|^2 + \gamma_3 \left\| \frac{\partial \Delta \mathbf{u}}{\partial \xi_3} \right\|^2 + \gamma_4 \left\| \frac{\partial^2 \Delta \mathbf{u}}{\partial \xi_1^2} \right\|^2 \\ & + \gamma_5 \left\| \frac{\partial^2 \Delta \mathbf{u}}{\partial \xi_2^2} \right\|^2 + \gamma_6 \left\| \frac{\partial^2 \Delta \mathbf{u}}{\partial \xi_3^2} \right\|^2 + \gamma_7 \left\| \frac{\partial^2 \Delta \mathbf{u}}{\partial \xi_1 \partial \xi_2} \right\|^2 + \gamma_8 \left\| \frac{\partial^2 \Delta \mathbf{u}}{\partial \xi_2 \partial \xi_3} \right\|^2 \\ & + \gamma_9 \left\| \frac{\partial^2 \Delta \mathbf{u}}{\partial \xi_3 \partial \xi_1} \right\|^2 + \gamma_{10} \left\| \frac{\partial^3 \Delta \mathbf{u}}{\partial \xi_1 \partial \xi_2 \partial \xi_3} \right\|^2 \Bigg\} \mathrm{d}\xi_1 \mathrm{d}\xi_2 \mathrm{d}\xi_3, \end{aligned} \tag{6}$$

where γ_i, $i = 1, 2, 3$, are the three arc lengths, γ_i, $i = 4, 5, 6$ are the three curvatures in the ξ_1, ξ_2, ξ_3 directions, respectively, and γ_i, $i = 7, 8, 9$ are the three surface area terms for faces (ξ_1, ξ_2), (ξ_2, ξ_3) and (ξ_3, ξ_1), while γ_{10} is related to the volume. Note for 2D fitting, the integral is over a surface and $\gamma_3 = \gamma_6 = \gamma_8 = \gamma_9 = \gamma_{10} = 0$.

To illustrate the above methods a RF muscle is fitted to the VH male data set. This involves starting with a linear mesh (Fig. 2a) which is overlayed with data (Fig. 2b). Note that the data density is less in the ξ_1 direction so Sobolev smoothing on the arc length and curvature is required in that direction (eq. (6)), terms γ_1 and γ_4). Minimizing the objective function (5) leads to the fitted mesh shown in Fig. 2c.

To highlight customization by the free-form deformation technique ('host mesh' fitting), the fitted male RF (slave) is embedded in a host mesh which

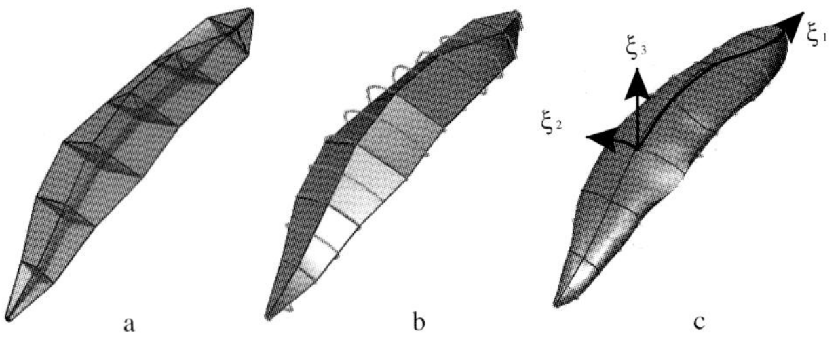

Fig. 2. (a) Initial linear mesh, with (b) data set overlayed, and (c) resulting fitted mesh.

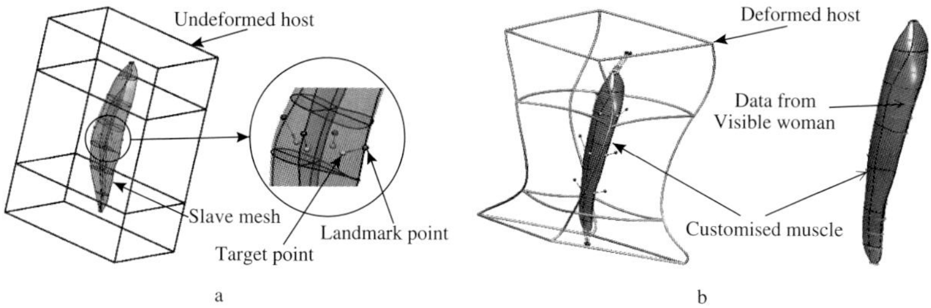

Fig. 3. (a) Undeformed host mesh with slave mesh and landmark points embedded within. The corresponding target points are indicated. (b) Deformed host and customized muscle with data points.

is then morphed to minimize the distance between control points (Fig. 3). The control points are based on easily identifiable markers from MRI or CT images and in this case are from the VH female data set. Hence the male RF is customized to a female RF.

3 Governing Equations

The governing equations are based on the concepts of continuum mechanics (Malvern (1969), Holzapfel (2000)) with implementation relative to our software CMISS[1] detailed in Nash and Hunter (2000). The kinematics of finite deformation deals with a spatial (Eulerian) coordinate system fixed in space and a material (Lagrangian) coordinate system fixed to the continuum. Each point in the continuum may then be defined spatially (x_1, x_2, x_3) and as material coordinates by (X_1, X_2, X_3) assuming that the spatial and material axes are coincident in the undeformed state. $\mathbf{F}$ is the deformation gradient tensor given in coordinate form by

$$F^i_M = \frac{\partial x_i}{\partial X_M}, \tag{7}$$

and used to describe the deformation process. It maps a material vector $\mathrm{d}\mathbf{X}$ in the undeformed configuration to a vector $\mathrm{d}\mathbf{x}$ in the deformed state: $\mathrm{d}\mathbf{x} = \mathbf{F}\mathrm{d}\mathbf{X}$. Under coordinate transformations the three invariants are

$$I_1 = \mathrm{tr}\,\mathbf{C}, \quad I_2 = \tfrac{1}{2}[(\mathrm{tr}\,\mathbf{C})^2 - \mathrm{tr}(\mathbf{C}^2)], \quad I_3 = \det \mathbf{C}, \tag{8}$$

where tr is the trace (the sum of the diagonal terms), det, is the determinant and $\mathbf{C} = \mathbf{F}^T\mathbf{F}$ is the right Cauchy-Green tensor. In our framework, the stresses and strains are referred to a microstructurally based material coordinate system which convects with the deformed body (i.e. remains attached to material particles) consisting of a fibre direction (ν_1), sheet direction (ν_2) and

[1] An interactive computer program developed by the Bioengineering Institute for Continuum Mechanics, Image analysis, Signal processing and System identification.

a sheet-normal axis (ν_3), similar to the studies by Nash (1998) and Stevens (2002). They are orthogonal in the reference state but become curved as the body deforms. This is also useful when describing a constitutive law which refers to microstructural directions, and adding active contractile forces to the fibre direction component. This new coordinate system (denoted ν_α) is derived from base vectors specific to the ν_α coordinate system and related to the standard rectangular Cartesian base vectors ($\mathbf{g}_k^x = \mathbf{i}_k$). These new base vectors are the covariant ($\mathbf{a}_\alpha^\nu$) and contravariant ($\mathbf{a}_\nu^\alpha$) base vectors defined by

$$\mathbf{a}_\alpha^\nu = \frac{\partial x_k}{\partial \nu_\alpha}\mathbf{g}_k^x, \quad \mathbf{a}_\nu^\alpha = \frac{\partial \nu_\alpha}{\partial x_k}\mathbf{g}_x^k. \tag{9}$$

Figure 4 illustrates these vectors in a 2D sense for clarity, where the ν_α coordinate system convects with the deforming body while the spatial coordinate system x_α remains fixed in space. The covariant ($\mathbf{a}_\alpha^\nu$) base vectors are tangent at a point to the ν_α coordinates and the contravariant base vectors are related by $\mathbf{a}_\nu^\alpha \cdot \mathbf{a}_\beta^\nu = \delta_\beta^\alpha$. (i.e. $\mathbf{a}_1^\nu$ and $\mathbf{a}_\nu^2$ are orthogonal and *vice versa*). Note the use of α and β refer to the indices of the microstructural axis vectors and tensors, as compared to i and M in eq. (7) which refer to the indices of the spatial and material coordinate axes respectively. The stress and strain tensors can then be transformed to refer to this microstructural coordinate system through covariant ($a_{\alpha\beta}^\nu$) and contravariant ($a_\nu^{\alpha\beta}$) metric tensors determined from the inner product of base vectors:

$$a_{\alpha\beta}^\nu = \mathbf{a}_\alpha^\nu \cdot \mathbf{a}_\beta^\nu = \frac{\partial x_k}{\partial \nu_\alpha}\frac{\partial x_k}{\partial \nu_\beta}, \quad a_\nu^{\alpha\beta} = \mathbf{a}_\nu^\alpha \cdot \mathbf{a}_\nu^\beta = \frac{\partial \nu_\alpha}{\partial x_k}\frac{\partial \nu_\beta}{\partial x_k}. \tag{10}$$

The base vectors for the ν_α coordinate system can be chosen to be orthogonal in the undeformed state, but after deformation they will in general no longer be orthogonal. This requires the definition of the base vectors ($\mathbf{A}_\alpha^\nu, \mathbf{A}_\nu^\alpha, \mathbf{a}_\alpha^\nu, \mathbf{a}_\nu^\alpha$)

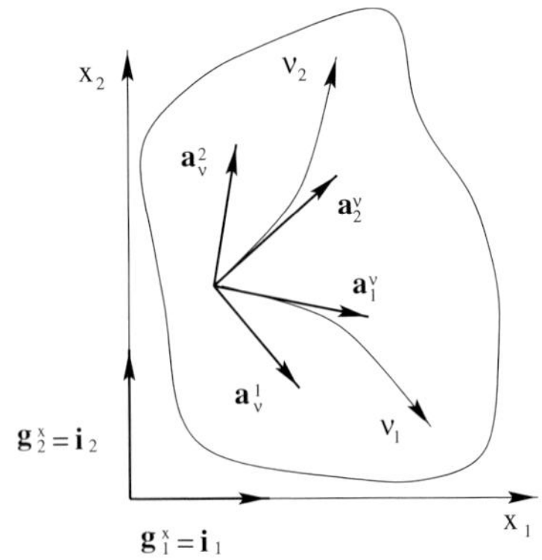

Fig. 4. 2D definition of the microstructural coordinate system (ν_α), and associated covariant ($\mathbf{a}_\alpha^\nu$) and contravariant ($\mathbf{a}_\nu^\alpha$) base vectors with respect to spatial coordinates (x_α).

and metric tensors $(A^{\nu}_{\alpha\beta}, A^{\alpha\beta}_{\nu}, a^{\nu}_{\alpha\beta}, a^{\alpha\beta}_{\nu})$ in both undeformed and deformed states (note that capitals refer to the undeformed state and lower case refers to the deformed state).

The equations of stress equilibrium are derived by considering conservation of mass, linear and angular momentum, giving the weak form

$$\int\limits_{V_0^n} T^{\alpha\beta} F^j_\beta \frac{\partial \delta u_j}{\partial X_\alpha} \mathrm{d}V_0^n = \int\limits_{S_c^n} \mathbf{f}_c \delta u^c_j \mathrm{d}S^n_c, \tag{11}$$

where V_0^n is the undeformed volume, S_c^n the surface in contact for the current configuration of body n, and δu_j is the virtual displacement field in the jth direction approximated by a FE displacement field $\delta u_j = \psi_n^{3D}(\xi_1, \xi_2, \xi_3)\delta u^n_j$, with ψ_n^{3D} the 3D geometric basis functions. Also, f^j_c are the components of the contact force resulting from the penalty implementation of a displacement boundary constraint (Fernandez and Hunter (2005)) and δu^c_j is the relative virtual displacement between the slave and master bodies in contact ($\delta u^c_j = \delta u^S_j - \delta u^M_j$), F^j_β are the components of the deformation gradient $\mathbf{F}$, and $T^{\alpha\beta}$ are the components of the second Piola-Kirchhoff stress tensor, both referred to microstructural axes:

$$T^{\alpha\beta} = \frac{\partial W}{\partial E_{\alpha\beta}} - p a^{\alpha\beta}_{\nu} + T_0(\lambda, [\mathrm{Ca}]) a^{11}_{\nu} \delta^{\alpha}_1 \delta^{\beta}_1. \tag{12}$$

Here, $E_{\alpha\beta}$ are the components of the Green-Lagrange strain tensor (invariant to rigid body rotations) and also referred to a microstructural axes,

$$E_{\alpha\beta} = \frac{1}{2}\left(\frac{\partial x_k}{\partial \nu_\alpha}\frac{\partial x_k}{\partial \nu_\beta} - \frac{\partial X_k}{\partial \nu_\alpha}\frac{\partial X_k}{\partial \nu_\beta}\right) = \frac{1}{2}\left(a^{\nu}_{\alpha\beta} - A^{\nu}_{\alpha\beta}\right), \tag{13}$$

p is the hydrostatic pressure, and W is a strain energy function used to describe the constitutive behaviour. Note that $T_0(\lambda, [\mathrm{Ca}])$ is an active contractile force added to the fibre direction ($\nu = 1$) and is dependent on a fibre extension ratio (λ) and concentration of free–intracellular calcium ([Ca]). A steady-state contraction model had been derived (Bergel and Hunter (1979), Hunter (1995)) based on observations of cardiac tissue and can be included to simulate muscle contractions of tissue and add extra muscle stability during controlled movements such as with an eccentric contraction (active muscle tension with lengthening muscle fibres).

Traditionally, skeletal muscle as well as other soft tissues like myocardium are believed to be very nearly incompressible due to the high amount of water in the cellular space. Modelling soft tissues to be incompressible ($J = \sqrt{I_3} = 1$) simplifies analytical treatment of the balance equations, but to enforce incompressibility in computations usually requires an additional degree of freedom (hydrostatic pressure) to match the kinematic and constraint equation, and the choice of basis function for this variable requires some care (see Oden (1972)). Furthermore, there has been experimental evidence that heart tissue is actually slightly compressible (Waldman et al. (1985)), which can be more readily modelled. Also Oden (1972) states: 'Although in some exact solutions to problems in finite elasticity the assumption of an incompressible material leads to certain simplifications in the analysis, such is not the case in finite element applications.' We therefore shortly mention the conventional approach of the mean-dilatation or hybrid method (two field method, nodal displacement and hydrostatic pressure) and also the more intuitive approach of the so-called penalty method (just nodal displacements). Both are discussed in detail in excellent reviews by Peng and Chang (1997) and Holzapfel (2000), Chapter 8.

The mean dilatation method or hybrid method introduces an additional field into the equilibrium equations with a Lagrange multiplier. Care must be taken with choosing both the right basis functions for the hydrostatic pressure field (see Oden (1972)) as well as a reduced/selective integration scheme for the Gauss integration to prevent the well-known volumetric locking phenomenon. The incompressibility constraint is then enforced by means of a weighted Galerkin integral in an element-wise fashion:

$$\int_{V_0} (\sqrt{I_3} - 1)\psi_n^H \mathrm{d}V_0. \tag{14}$$

The second more intuitive method is a one-field method, where the strain energy function is split into a dilatational and deviatoric part. An in-depth discussion is given in Bonet and Wood (1997) (pp. 126–129), and the treatment of this concept from a continuum point of view is given in Holzapfel (2000) (pp. 227–235, with numerical aspects given in pp. 389–391):

$$W(E_{\alpha\beta}) = W_{\text{dil}} + W_{\text{dev}}. \tag{15}$$

In some constitutional frameworks this split comes for free and it is therefore a natural choice to pursue this method. The dilatational, bulk or volumetric term responsible for describing the response to hydrostatic pressure is then chosen in a way to represent a rather stiff material, mimicking the almost incompressible behaviour. One possible choice is

$$W_{\text{dil}}(J) = \kappa(J-1)^2. \tag{16}$$

For further detail the reader is again referred to Peng and Chang (1997).

The main ideas presented here are that we have introduced the governing equation of static finite elasticity over a volume which is in equilibrium with a contact force in the regions of surface contact (eq. (11)). Both the second Piola-Kirchhoff stress tensor and Green-Lagrange strain tensor are expressed with respect to a microstructural coordinates ν_α rather than the standard rectangular cartesian axis (eqs (12) and (13)). Moreover, the stress tensor is modified with a contribution in the fibre direction from a calcium-tension relation to simulate active contraction. Finally, tissue volume conservation is imposed via a Lagrange multiplier (eq. (14)) or the nearly incompressible approach (eqs (15) and (16)).

4 Constitutive Laws – Strain Energy Functions

Constitutive laws describe the stress-strain relationship for a given material. We assumed in eq. (12) the existence of a scalar potential W. To acknowledge the understanding of the microstructural architecture of, e.g., skeletal or cardiac muscle, the material law has to be transversely isotropic or orthotropic; see Martins et al. (1998) and LeGrice et al. (1995), respectively.

The modeller faces a whole range of challenges when wanting to capture the overall features of the mechanical behaviour. We present the main features for the example of passive myocardium. In general two different sets of issues can be distinguished: the actual material characterization in terms of the constitutive law itself, and the geometrical issues of how to model the geometrical microstructure in all its diversity.

If we only consider the passive material properties and assume that W depends on a set of invariants, eq. (12) can be reformulated as:

$$T^{\alpha\beta} = \frac{\partial W}{\partial E_{\alpha\beta}} = \frac{\partial W}{\partial I_k}\frac{\partial I_k}{\partial E_{\alpha\beta}}. \tag{17}$$

The terms $\partial W/\partial I_k$ are called the response functions, and $\partial I_k/\partial E_{\alpha\beta}$ are called the associated kinematic tensors.

The numerous laws that have been formulated so far are either transversely isotropic or orthotropic where the design was based on either phenomenological observations or on microstructural knowledge. In general these laws do not lead to orthogonal kinematic tensors and orthogonal response functions. It was suggested that this would enhance the precision and speed of inverse material parameter estimations (Criscione et al. (2002)).

We will now present the three main material laws for myocardium which have been proposed over the last decades, all modelling the myocardium to be orthotropic. The so-called Pole-zero law (18 material parameters) is based on the idea that the tissue exhibits an elastic strain limit, which is thought to be easily determined from an extrapolation of experimental data and therefore the number of material parameters that need to be estimated via an inverse method can be reduced. Further reduction was possible by incorporating kinematic constraints (Nash and Hunter (2000)), leading to

$$W(E_{\alpha\beta}) = k_{\alpha\beta}\frac{E_{\alpha\beta}^2}{\left|a_{\alpha\beta} - \left|\frac{1}{2}(E_{\alpha\beta}+E_{\beta\alpha})\right|\right|^{b_{\alpha\beta}}}. \tag{18}$$

The drawback of this law seems to be the high interdependence of the material parameters and the still large number (~ 10) of parameters.

Another model was introduced by Costa et al. (2001) where they extended the idea of an exponential Fung-type law introduced for a transversely isotropic material by Guccione et al. (1991):

$$W(E_{\alpha\beta}) = \tfrac{1}{2}Ce^{Q}, \tag{19}$$

$$Q = c_{ff}E_{ff}^2 + c_{nn}E_{nn}^2 + c_{ss}E_{ss}^2 + c_{fn}(E_{fn}^2 + E_{bf}^2) + c_{fs}(E_{fs}^2 + E_{sf}^2) + c_{ns}(E_{ns}^2 + E_{sn}^2). \quad (20)$$

This has a reduced number (i.e. 7) of material parameters, but simulations have proven that the interdependence of material parameters is still a major problem, i.e. there seem to exist several local minima with greatly varying sets for the material parameters, which give similar final results for the objective function, crucially dependent on the initial guess of the parameters.

Criscione et al. (2002) therefore introduced a novel constitutive model by utilizing the microstructural knowledge of myocardium and introducing so-called strain attributes, α_i, which lead to orthogonal kinematic tensors, $\mathbf{A}_i$:

$$\boldsymbol{\sigma} = \frac{1}{J}\sum_{i=1}^{6}\frac{\partial W}{\partial \alpha_i}(\alpha_1, \alpha_2, \alpha_3, \alpha_4, \alpha_5, \alpha_6)\mathbf{A}_i = \frac{1}{J}\sum_{i=1}^{6} W_{,i}\mathbf{A}_i, \quad (21)$$

where $\boldsymbol{\sigma}$ is the Cauchy stress tensor. Note that this expression uses the Cauchy stress since the $\mathbf{A}_i$ terms refer to the current configuration and they can be calculated via $\mathbf{A}_i = \mathbf{F}(\partial\alpha_i/\partial\mathbf{E})\mathbf{F}^T$. It is work in progress to design the remaining issues of this material law to optimize material parameter robustness.

However, the above effort does not fully model the knowledge of myocardium, i.e. all three preferred directions exhibit a mean fibre direction with a splay around this mean. Ideally, though, one would like to start from the microstructure of myocardium and then use a homogenization technique to obtain the macrostructural behaviour. This would only capture the elastic behaviour and other phenomena like, micromoments (Cosserat theory), viscoelasticity, fluid shifting between cells, plasticity, scars, healing and others will have to be taken into consideration.

The above was solely focusing on describing the issues related to constitutive law formulation itself. It is evident that the best material law description does not suffice to capture the mechanical behaviour, if say the modelled aspect like orthotropy cannot be resolved to a level where the effects of orthotropy actually take effect in the model. For the case of orthotropy it is therefore desirable if, say, a cube of myocardium is modelled that one scans the fibre and sheet distributions with light microscopy. Equivalently it would be necessary to, e.g., use tensor diffusion imaging to obtain data about the fibre direction for *in vivo* myocardium.

Even though it is a fact that material parameters vary spatially throughout the myocardium or in skeletal muscle it is hardly possible to do an inverse parameter estimation for the spatially varying material properties since this problem would not be constrained enough and would therefore again create a whole new set of possible material parameters that lead to the same model answer.

5 Model Applications

The geometries and modelling techniques developed have been illustrated for two clinically focused applications: (i) a patella articulation model and (ii) a Cerebral Palsy (CP) assessment tool. We present a brief overview of these models as an illustration of our goal of developing clinically useful tools.

5.1 Patella Articulation Model

A finite element patella model was developed to predict kinematics, extensor forces, internal stresses and surface contact pressures using the developed database of anatomical geometries, fibre fields and muscle attachment points. Patient-specific features were derived from hospital MRI data using the previously mentioned FFD technique ('host mesh' fitting). Data were used to fit the outer cortical, inner cancellous, cartilage layer and articulating femur profile (Fig. 5). The model was driven via motion capture and the governing equations (Section 3) were solved over the patella and surrounding tissues (muscle and tendon). Flexion simulations for 10°–90° (Figs 6a–h) and the surface contact pressure visualized through a transparent femur (Fig. 6i) are shown. The model agreed well with previously published models for kinematics and extensor forces. Surface contact pressures are shown in Fig. 7. The model can be used to investigate parameter changes for corrective surgeries such as the Maquet procedure (Cohen et al. (2003)) where the tibial tubercle position is changed to increase the effective moment arm and reduce peak stresses. For implementation specifics, results and discussions see Fernandez and Hunter (2005).

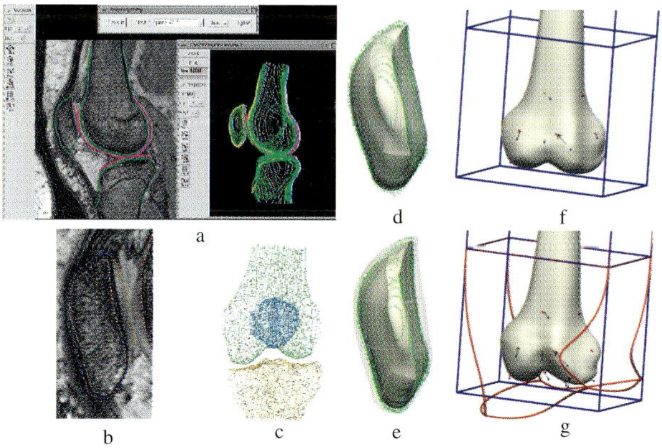

Fig. 5. (a) CMGUI digitizing software screen shot with (b) close-up and (c) 3D data set used for fitting. Saggital views of the fitted outer patella shell (d) and inner cancellous boundary (e). Slave femur mesh with landmark and target points in host mesh (f) and deformed host and customized femur (g).

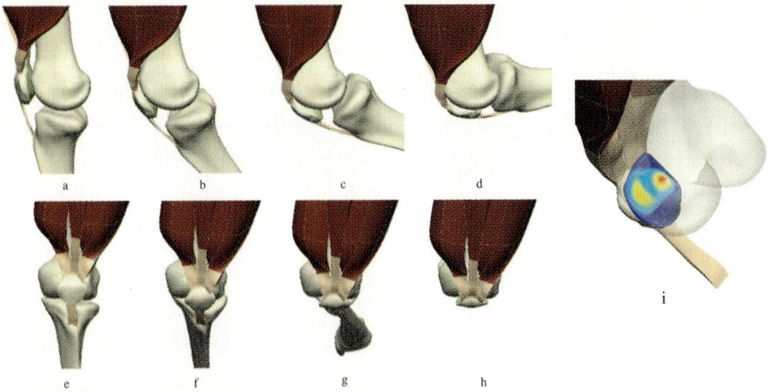

Fig. 6. Saggital view of simulation solutions for (a) 10°, (b) 30°, (c) 60° and (d) 90°. The corresponding anterior views are given in (e)–(h). (i): illustration of the surface contact pressure viewed through a transparent femur.

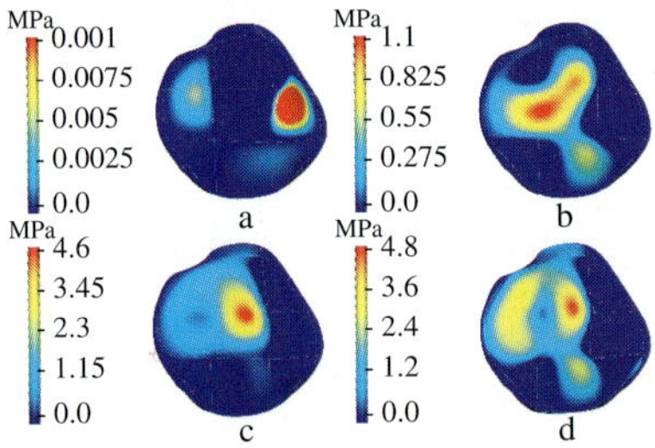

Fig. 7. Surface contact pressure for (a) 10°, (b) 30°, (c) 60° and (d) 90°.

5.2 Cerebral Palsy Tool

A geometric tool for modelling muscle lengths and joint angles was developed to assess the results of corrective muscle lengthening surgeries. The model is driven via motion capture and the user can choose from a subset of muscles in a database. Highlighted is the gastrocnemius (GT) and semimembranosus (SM), two muscles commonly surgically lengthened to improved a subject's gait, embedded in two host meshes (Fig. 8). The model moves muscles via a combination of rigid-body transformations to capture global motion and 'host mesh' fitting to capture local deformations, ensuring that muscles remain attached to bones. A computer movie is also generated of the patient's gait cycle to observe improvements in limb clearance and joint angles. Highlighted are the software outputs for a healthy subject, CP subject and pre- and post-gaits for a subject who has had a GT lengthening (Fig. 9). Using the model, pre- and post-surgery data input can reveal the new arc lengths and joint angles after muscle lengthening surgery. Volume preserving constraints are also used to ensure muscles have the same volume throughout the deformation. For implementation specifics, results and discussions see Fernandez et al. (2005).

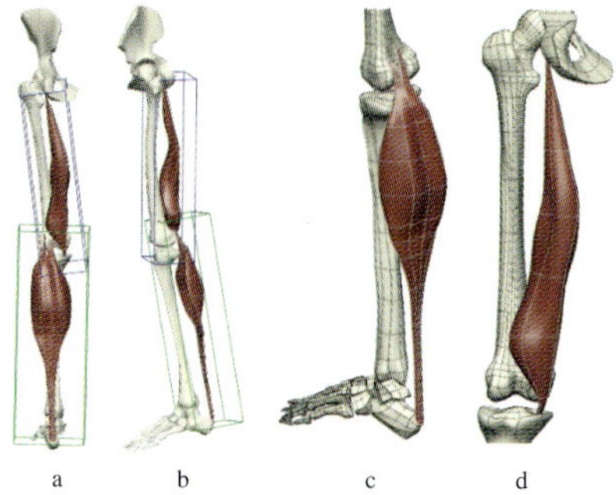

Fig. 8. Geometry set-up with gastrocnemius (GT) and semimembranosus (SM) muscles embedded in individual trilinear host meshes from (a) posterior and (b) lateral views. Closeup of FE discretization for (c) GT and (d) SM used in simulations.

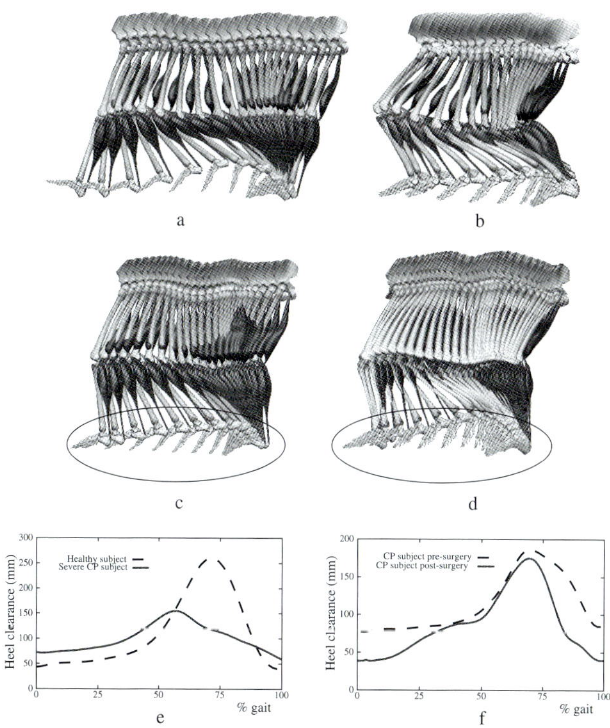

Fig. 9. Software movie outputs for (a) healthy and (b) severe CP subjects. The effects of gastrocnemius lengthening for (c) pre-surgery and (d) post-surgery highlighting improvement in ankle flexion and less dragging of the foot. (e) Shows a comparison between the heel clearance of the (a) healthy and (b) severe CP subjects. (f) Shows a comparison between the (c) pre-surgery and (d) post-surgery effects on heel clearance.

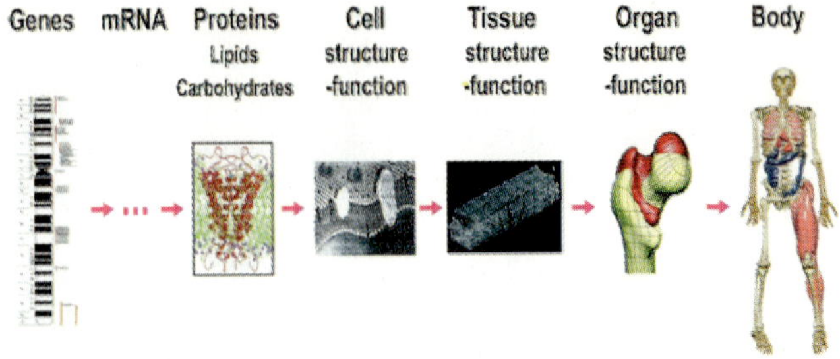

Fig. 10. Diagram showing the links from the gene and protein level across the spatial scales to the organ and whole body level, illustrating the multiscale modelling objective of the Physiome Project.

6 Discussion and Future Work

The future of our work involves linking the current modelling framework at the organ level with the underlying microstructure. Modelling across spatial scales will allow us to extend the scope of the models to investigate the effects of ageing, disease, drugs and a more meaningful interpretation of injury mechanisms. Our previous work in electrocardiology will also provide a good foundation for studying the specific nerve innervations in muscle that lead to muscle contraction. Skin surface EMG (Electromyograms) can be used to deduce nerve locations and firing rates via optimization, leading to a patient-specific activation pattern in muscle. The anatomically-based models of the leg described above are part of a larger project to model the complete musculoskeletal system within the framework of the Physiome Project (Hunter and Borg (2003)). The aim of the Physiome Project is to build models of physiological processes that span multiple spatial scales from genes and proteins to cells, tissues and organs, as illustrated in Fig. 10. The anatomically- and biophysically-based framework that we have presented allows the spatial variation of material properties, including protein expression levels, to be characterized. It also allows links to be made from cell level function described by signal transduction pathways, metabolic pathways and ion channel currents, for example, to be linked to tissue and organ level structure-function relations.

Acknowledgements. The authors would like to acknowledge the important input to this work from the members of the Bioengineering Institute. Financial assistance was received from the Foundation for Research, Science and Technology New Zealand under NERF contract UOAX0232 (PJH), a Bright Future scholarship (JWF) and the International Doctoral Scholarship of the University of Auckland (HS).

References

Ackerman, M. J. (1998). The visible human project. *Proc. IEE-E* 86:504–511.

Bergel, D. H., and Hunter, P. J. (1979). The mechanics of the heart. In Hwang, H. H. C., Gross, D. R., and Patel, D. J., eds., *Quantitative Cardiovascular Studies, Clinical and Research Applications of Engineering Principles, Chapter 4*. Baltimore: University Park Press. 151–213.

Bonet, J., and Wood, R. D. (1997). *Nonlinear Continuum Mechanics for Finite Element Analysis.* Cambridge: Cambridge University Press.

Cohen, Z. A., Henry, J. H., McCarthy, D. M., Mow, V. C., and Ateshian, G. A. (2003). Computer simulations of patellofemoral joint surgery: patient-specific models for tuberosity transfer. *Am. J. Sports Med.* 31:87–98.

Costa, K. D., Holmes, J. W., and McCulloch, A. D. (2001). Modelling cardiac mechanical properties in three dimensions. *Philos. T. Roy. Soc. A* 359: 1233–1250.

Criscione, J. C., McCulloch, A. D., and Hunter, W. C. (2002). Constitutive framework optimized for myocardium and other high-strain, laminar materials with one fiber family. *J. Mech. Phys. Solids* 50:1681–1702.

Dong, F., Clapworthy, G., Krokos, M., and Yao, J. (2002). An anatomy based approach to human muscle modeling amd deformation. *IEEE T. Vis. Comput. Gr.* 8:154–170.

Fernandez, J. W., and Hunter, P. J. (2005). An anatomically based patient-specific finite element model of patella articulation: Towards a diagnostic tool. *Biomech. Model. Mechanobio.* in press.

Fernandez, J. W., Mithraratne, P., Thrupp, S. F., Tawhai, M. T., and Hunter, P. J. (2004). Anatomically based geometric modelling of the musculo-skeletal system and other organs. *Biomech. Model. Mechanobio.* 2:139–155.

Fernandez, J. W., Ho, A., Walt, S., Anderson, I. A., and Hunter, P. J. (2005). A cerebral palsy assessment tool using anatomically based geometries and free-form deformation. *Biomech. Model. Mechanobio.* in press.

Guccione, J. M., McCulloch, A. D., and Waldman, L. K. (1991). Passive material properties of intact ventricular myocardium determined from a cylindrical model. *J. Biomech. Eng.* 113:42–55.

Hirota, G. (2002). *An improved finite element contact model for anatomical simulations.* Ph.D. Dissertation, University of North Carolina, USA.

Holzapfel, G. A. (2000). *Nonlinear Solid Mechanics. A Continuum Approach for Engineering.* Chichester: John Wiley & Sons.

Hunter, P. J., and Borg, T. K. (2003). Integration from proteins to organs: the physiome project. *Nat. Rev. Mol. Cell Bio.* 4:237–243.

Hunter, P. J., Robbins, P., and Noble, D. (2002). The IUPS human physiome project. *Eur. J. Physiol.* 445:1–9.

Hunter, P. J. (1995). Myocardial constitutive laws for continuum models of the heart. In Sideman, S., and Beyer, R., eds., *Molecular and Subcellular Cardiology.* New York: Plenum Press. 303–318.

LeGrice, I. J., Smaill, B. H., Chai, L. Z., Edgar, S. G., Gavin, J. B., and Hunter, P. J. (1995). Laminar structure of the heart: Ventricular myocyte arrangement and connective tissue architecture in the dog. *Am. J. Physiol. Heart Circ. Physiol.* 269:H571–H582.

LeGrice, I. J. (1992). *A finite element model of myocardial structure: Implications for electrical activation in the heart.* Ph.D. Dissertation, University of Auckland, New Zealand.

Malvern, L. E. (1969). *Introduction to the Mechanics of a Continuous Medium.* Englewood Cliffs, New Jersey: Prentice-Hall.

Martins, J. A. C., Pires, E. B., Salvado, R., and Dinis, P. B. (1998). A numerical model of passive and active behavior of skeletal muscles. *Comput. Meth. Appl. Mech. Eng.* 151:419–433.

Nash, M. P., and Hunter, P. J. (2000). Computational mechanics of the heart. *J. Elasticity* 61:113–141.

Nash, M. P. (1998). *Mechanics and material properties of the heart using an anatomically accurate mathematical model.* Ph.D. Dissertation, The University of Auckland, New Zealand.

Nielsen, P. M. F., LeGrice, I. J., Smaill, B. H., and Hunter, P. J. (1991). Mathematical model of geometry and fibrous structure of the heart. *Am. J. Physiol. Cell Physiol.* 260:H1365–H1378.

Oden, J. T. (1972). *Finite Elements of Nonlinear Continua.* New York: McGraw-Hill.

Peng, S., and Chang, W. (1997). A compressible approach in finite element analysis of rubber–elastic materials. *Comput. Struct.* 62:573–593.

Spitzer, V. M., and Whitlock, D. G. (1998). The visible human dataset: The anatomical platform for human simulation. *Anat. Rec.* 253:49–57.

Stadler, M., Holzapfel, G. A., and Korelc, J. (2003). C^n-continuous modeling of smooth contact surfaces using NURBS and applications to 2D problems. *Int. J. Numer. Meth. Eng.* 57:2177–2203.

Stevens, C. (2002). *An anatomically-based computational study of cardiac mechanics and myocardial infarction.* Phd thesis, The University of Auckland, New Zealand.

Waldman, L. K., Fung, Y. C., and Covell, J. W. (1985). Transmural myocardial deformation in the canine left ventricle: Normal in vivo three-dimensional finite strains. *Circ. Res.* 57:152–163.

Wriggers, P., Krstulovic-Opara, L., and Korelc, J. (2001). Smooth C^1-interpolations for two-dimensional frictional contact problems. *Int. J. Numer. Meth. Eng.* 51:1469–1495.

Invariant Formulation for Dispersed Transverse Isotropy in Tissues of the Aortic Outflow Tract

D.R. Einstein[1], A.D. Freed[2], I. Vesely[1]

[1] Children's Hospital Los Angeles, Cardiothoracic Surgery Research, USA
deinst@u.washington.edu, IVesely@chla.usc.edu
[2] NASA's John H. Glenn Research Center at Lewis Field, Bio Science and Technology Branch, USA
Alan.D.Freed@nasa.gov

We present a polyconvex constitutive law for tissues of the aortic outflow tract based on a novel invariant theory for dispersed transverse isotropy. In contrast to traditional structural constitutive equations, fiber splay is treated in closed form and thus can easily handle both two- (2D) and three-dimensional (3D) fiber populations. In the 2D, single preferred-direction case, the model accurately fits biaxial data for aortic valve tissue. Extended to two preferred directions, the model accurately fits available uniaxial data for the aortic sinus and aortic wall. Modification of the fiber stress-strain law requires no re-formulation of the constitutive tangent matrix, making the model flexible for different types of soft tissues, and, in the case of 3D fiber populations, the model may be suitable for describing passive myocardium. Most importantly, the model is computationally expedient in a finite-element analysis.

1 Introduction

The aortic outflow tract consists of aortic valve cusps, aortic sinuses and the wall of the aortic root. Clinically, the outflow tract is the frequent site of pathology in both neonatal and adult populations. Computational analysis is a promising approach for evaluating the mechanics of such pathologies and of their surgical correction with newly proposed implants, bioprostheses or tissue-engineered constructs. However, numerical models of such tissue systems can be computationally intensive. This is particularly true for tissues with marked heterogeneity, such as the aortic valve, and systems that involve fluid-structure interactions. The material models used in these systems must not only be accurate, they must also be computationally efficient.

Micro-structural studies suggest that, in some cardiovascular tissues, collagen fibers are approximately normally distributed about a mean preferred fiber direction. For example, Sacks (2003) showed a correlation between the intensity distribution of the small-angle light scattering of bovine pericardium and both fiber orientation and dispersion. Similarly, Holzapfel et al. (2002) showed that smooth muscle cells in arterial media are statistically oriented around two opposing helical directions. From these cellular orientations, collagen fiber orientations were inferred. An earlier observation of fiber dispersal in bovine pericardium was documented by Zioupos and Barbenel (1994). Structural constitutive equations that account for this dispersion of fibers have been shown to capture the mechanical complexity of these tissues quite well (Sacks (2000), Holzapfel et al. (2002)).

Mathematically, the inclusion of dispersion into constitutive models in the form of probability distribution functions presents little challenge. Computationally, however, the process is cumbersome because the product of the probability density function and the fiber stress-strain rule must be integrated numerically over a semi-circle in the 2D case, as in eq. (1) below, or a hemisphere in the 3D case. Functionally, in the 2D case, this can be represented as

$$\boldsymbol{s} + p\boldsymbol{c}^{-1} = \int_{-\pi/2}^{\pi/2} S_f(\theta, \boldsymbol{c})\, R(\theta)\, \boldsymbol{a}_0(\theta) \otimes \boldsymbol{a}_0(\theta)\, \mathrm{d}\theta, \tag{1}$$

where $\boldsymbol{s}$ is the second Piola-Kirchhoff stress, $\boldsymbol{c}$ is the Lagrangian metric, p is a Lagrange multiplier to force the constraint of incompressibility, $S_f(\theta, \boldsymbol{c})$ is a fiber stress-strain rule, $R(\theta)$ is a probability distribution function, and $\boldsymbol{a}_0(\theta) \otimes \boldsymbol{a}_0(\theta)$ governs the orientation of a fiber family in the 2D plane. A single integral of this kind in the constitutive equation implies two integrals to be evaluated in the constitutive tangent matrix. Likewise, 3D fiber dispersion would require two integrals to be evaluated for stress, and four for the constitutive tangent matrix. In the computational model, these operations are evaluated at every iteration, of every time step, at every Gauss point, for every finite element.

In this paper, we develop an invariant theory for tissues whose collagen fiber populations are statistically distributed. Specifically, the integral in eq. (1) is replaced with a novel closed-form 'splay invariant' that requires a single parameter in the 2D case, and a single operation per iteration. Furthermore, dispersion is represented both in the current configuration and the spatial configuration. Thus, the invariant captures the idea of both fiber stretch and fiber rotation. Two planar constitutive laws are proposed based on this theory. The first models aortic valve tissue as a single fiber family with dispersion. The second models aortic wall and aortic sinus tissue as a two-fiber family with dispersion. To evaluate the first model, we compare its correlative capability against biaxial data for aortic valve tissue. To evaluate the second model, we compare its correlative capability against two sets of orthogonal uniaxial data fitted simultaneously with inverse methods (described elsewhere Einstein et al. (2005)). To illustrate the reduced computational cost of the approach, we perform an illustrative finite element simulation of an aortic valve bioprosthesis. The simulation is performed with both the invariant-based approach presented here, and with a published structural model Billiar and Sacks (2000); Einstein (2002) that is based on the paradigm in eq. (1). The respective CPU times are compared.

2 Theory

In this section, we develop the idea of a scalar invariant based on a 'dispersion tensor', **K** . The specific form of this tensor, $\boldsymbol{\kappa}$, will be derived in local coordinates for simplicity. It will be shown that unlike the integral in eq. (1), **K** admits a closed-form solution. We conclude this section with a tensor approximation to **K** that is both accurate and computationally efficient.

2.1 Invariants

From the classic theory of invariants (Spencer (1972)), there are five invariants that are needed to describe a material with transverse isotropy (i.e. a material with a single fiber family); they are

$$I_I = \operatorname{tr} \boldsymbol{C}, \quad I_{II} = \tfrac{1}{2}\big((\operatorname{tr} \boldsymbol{C})^2 - \operatorname{tr}(\boldsymbol{C}^2)\big), \quad I_{III} = \det \boldsymbol{C} = (\det \boldsymbol{F})^2, \tag{2}$$

$$I_{IV} = \boldsymbol{a}_0 \cdot \boldsymbol{C}\boldsymbol{a}_0 = \boldsymbol{a} \cdot \boldsymbol{a} = \lambda^2, \quad I_V = \boldsymbol{a}_0 \cdot \boldsymbol{C}^2 \boldsymbol{a}_0 = \boldsymbol{a} \cdot \boldsymbol{B}\boldsymbol{a}, \tag{3}$$

where $\boldsymbol{F}$ is the deformation gradient, $\boldsymbol{C} = \boldsymbol{F}^{\mathrm{T}}\boldsymbol{F}$ and $\boldsymbol{B} = \boldsymbol{F}\boldsymbol{F}^{\mathrm{T}}$. The three invariants in eq. (2) account for isotropic effects, while the two invariants in eq. (3) account for anisotropic effects.

Invariants I_{IV} and I_V idealize all fibers in a family as being parallel. We therefore seek an alternative pair of invariants to account for the microscopically observed fiber dispersion in soft tissues. To be clear, these new invariants—call them $I_{\langle IV \rangle}$ and $I_{\langle V \rangle}$—will replace I_{IV} and I_V for dispersed fiber architectures, yet analytically reduce to I_{IV} and I_V in the absence of fiber dispersion. For now, without elaboration, let us define these new invariants as

$$I_{\langle IV \rangle} = \operatorname{tr}(\boldsymbol{F}\boldsymbol{K}\boldsymbol{F}^{\mathrm{T}}) = \operatorname{tr}(\boldsymbol{K}\boldsymbol{C}), \quad I_{\langle V \rangle} = \operatorname{tr}(\boldsymbol{C}\boldsymbol{K}\boldsymbol{C}), \tag{4}$$

where $\boldsymbol{K}(\boldsymbol{X}; t_0)$ is constrained such that $\boldsymbol{K} \to \boldsymbol{a}_0 \otimes \boldsymbol{a}_0$ in the absence of splay. Tensor $\boldsymbol{K}$ is a material constant. The main objective of this paper is to derive an appropriate $\boldsymbol{K}$ for describing the anisotropy caused by fiber dispersion in tissues of the aortic root and valve.

2.2 Local Dispersion Tensor

In order to derive the dispersion tensor, it is useful to switch from the global coordinate system (X, Y, Z) with base vectors $\{\boldsymbol{e}_X, \boldsymbol{e}_Y, \boldsymbol{e}_Z\}$ to a local or intrinsic coordinate system $(1, 2, 3)$ with base vectors $\{\boldsymbol{e}_1, \boldsymbol{e}_2, \boldsymbol{e}_3\}$. These local coordinates are selected so that the unit vector $\boldsymbol{e}_1$ in the 1-direction lies coaxial with the mean direction $\boldsymbol{a}_0$ of fiber orientation, while the unit vectors in the 2- and 3-directions lie in the transverse plane. Because both the intrinsic and global coordinate systems are considered to be rectangular Cartesian, there exists a unique, orthogonal, rotation matrix $\boldsymbol{Q}$ such that

$$\boldsymbol{e}_X = \boldsymbol{Q}\boldsymbol{e}_1, \quad \boldsymbol{e}_Y = \boldsymbol{Q}\boldsymbol{e}_2, \quad \boldsymbol{e}_Z = \boldsymbol{Q}\boldsymbol{e}_3, \tag{5}$$

where $\boldsymbol{Q}^{\mathrm{T}}\boldsymbol{Q} = \boldsymbol{I}$ with $\det \boldsymbol{Q} = 1$. Matrix $\boldsymbol{Q}$ represents a rigid-body rotation.

In the spirit of Lanir (1983), we propose the following definition for our local dispersion tensor. We postulate the existence of a material-constant tensor field that we denote as $\boldsymbol{\kappa}$, which serves as a relative (i.e. normalized) stiffness matrix associated with the anisotropic facets of material geometry. For 2D splay, this field is given by

$$\boldsymbol{\kappa}(\varsigma) = \int_{-\pi/2}^{\pi/2} \frac{\exp\left(-\theta^2/2\varsigma^2\right)}{\varsigma\sqrt{2\pi}\,\mathrm{erf}\left(\pi/2\sqrt{2}\varsigma\right)}\, \boldsymbol{e}_f(\theta) \otimes \boldsymbol{e}_f(\theta)\, \mathrm{d}\theta, \tag{6}$$

and for 3D splay with transverse isotropy it is given by

$$\boldsymbol{\kappa}(\varsigma) = \int_{0}^{2\pi}\int_{-\pi/2}^{\pi/2} \frac{\exp\left(-\theta^2/2\varsigma^2\right)}{\varsigma(2\pi)^{3/2}\,\mathrm{erf}\left(\pi/2\sqrt{2}\varsigma\right)}\, \boldsymbol{e}_f(\theta,\phi) \otimes \boldsymbol{e}_f(\theta,\phi)\, \mathrm{d}\theta\, \mathrm{d}\phi, \tag{7}$$

where the global anisotropic material stiffness $\boldsymbol{K}$ introduced in eq. (4) relates to its intrinsic counterpart $\boldsymbol{\kappa}$ via

$$\boldsymbol{K} = \boldsymbol{Q}\,\boldsymbol{\kappa}\,\boldsymbol{Q}^{\mathrm{T}}. \tag{8}$$

Both of these $\boldsymbol{\kappa}$ matrices obey the required property $\lim_{\varsigma\to 0} \boldsymbol{Q}\,\boldsymbol{\kappa}(\varsigma)\boldsymbol{Q}^{\mathrm{T}} = \boldsymbol{a}_0 \otimes \boldsymbol{a}_0$, and both of the $\boldsymbol{\kappa}$ matrices are symmetric. This implies that their affiliated $\boldsymbol{K}$ matrices are symmetric as well. It should be noted that eqs (6) and (7) are the double or triple integrals of the product of a Gaussian and an orientation tensor scaled by the error function. In contrast to eq. (1), which is a material law, eqs (6) and (7) are material properties. In addition, by adopting a Gaussian distribution to describe fiber splay, the corresponding stiffness matrix $\boldsymbol{\kappa}$ can be computed analytically. In our local coordinate system, for the 2D case, the orientation tensor is given as

$$\boldsymbol{e}_f \otimes \boldsymbol{e}_f = \begin{bmatrix} \cos^2\theta & \sin\theta\,\cos\theta & 0 \\ \sin\theta\,\cos\theta & \sin^2\theta & 0 \\ 0 & 0 & 0 \end{bmatrix}, \tag{9}$$

which leads to

$$\boldsymbol{\kappa}(\varsigma) = \begin{bmatrix} \kappa_{11} & 0 & 0 \\ 0 & \kappa_{22} & 0 \\ 0 & 0 & 0 \end{bmatrix}, \tag{10}$$

where the two analytic non-zero stiffness components have values of

$$\kappa_{11} = \frac{1}{2} + \frac{\mathrm{e}^{-2\varsigma^2}\left[\operatorname{erf}\left(\frac{\pi + i\,4\varsigma^2}{2\sqrt{2}\varsigma}\right) + \operatorname{erf}\left(\frac{\pi - i\,4\varsigma^2}{2\sqrt{2}\varsigma}\right)\right]}{4\operatorname{erf}\left(\frac{\pi}{2\sqrt{2}\varsigma}\right)}, \tag{11}$$

and

$$\kappa_{22} = \frac{1}{2} - \frac{\mathrm{e}^{-2\varsigma^2}\left[\operatorname{erf}\left(\frac{\pi + i\,4\varsigma^2}{2\sqrt{2}\varsigma}\right) + \operatorname{erf}\left(\frac{\pi - i\,4\varsigma^2}{2\sqrt{2}\varsigma}\right)\right]}{4\operatorname{erf}\left(\frac{\pi}{2\sqrt{2}\varsigma}\right)}, \tag{12}$$

where $i = \sqrt{-1}$ is the unit imaginary number. A similar argument can be made for the three-dimensional case but is omitted here for brevity. In both the 2D and the 3D cases, the ability to analytically solve for the anisotropic stiffness $\boldsymbol{\kappa}$, or equivalently $\boldsymbol{K}$, means that this theory will be efficient when implemented into finite element codes.

2.3 An Approximation

Because the error function, $\operatorname{erf}(z)$, is not found in common math software libraries, we introduce a simple approximation to the analytic results in eqs (11) and (12), namely

$$\boldsymbol{\kappa}(\varsigma) \approx \begin{bmatrix} \frac{1}{2}\left(1 + \mathrm{e}^{-2\varsigma^2}\right) & 0 & 0 \\ 0 & \frac{f}{2}\left(1 - \mathrm{e}^{-2\varsigma^2}\right) & 0 \\ 0 & 0 & \frac{1-f}{2}\left(1 - \mathrm{e}^{-2\varsigma^2}\right) \end{bmatrix}, \quad 0 \le f \le 1, \tag{13}$$

which is in keeping with the constraint that $\operatorname{tr}\boldsymbol{\kappa} = 1$. Parameters $f = 0$ and $f = 1$ apply to 2D splay with the normal to the membrane being in the 2- and 3-directions, respectively, while $f = {}^1\!/_2$ applies for 3D splay with transverse isotropy. Splay will be orthotropic whenever $f \neq {}^1\!/_2$; specifically, we conjecture, but do not prove, that there will be an elliptic symmetry in the transverse plane. The analytic solution in eqs (11) and (12) and the approximate solution in eq. (13) for $\boldsymbol{\kappa}$ are contrasted in Fig. 1 for 2D splay. Equation (13) compares similarly for the 3D case (not shown).

3 Elasticity for Finite Elements

Soft tissues are generally considered to be incompressible or nearly incompressible. In a finite element analysis that involves incompressibility, a standard displacement-based interpolation method leads to ill-conditioning of the numerics (Malkus and Hughes (1978)). Following Flory (1961), we seek to decouple pressure from displacement. This decoupling is compatible with two-field displacement-pressure interpolations that avoid volumetric locking in particular, and numerical ill-conditioning in general.

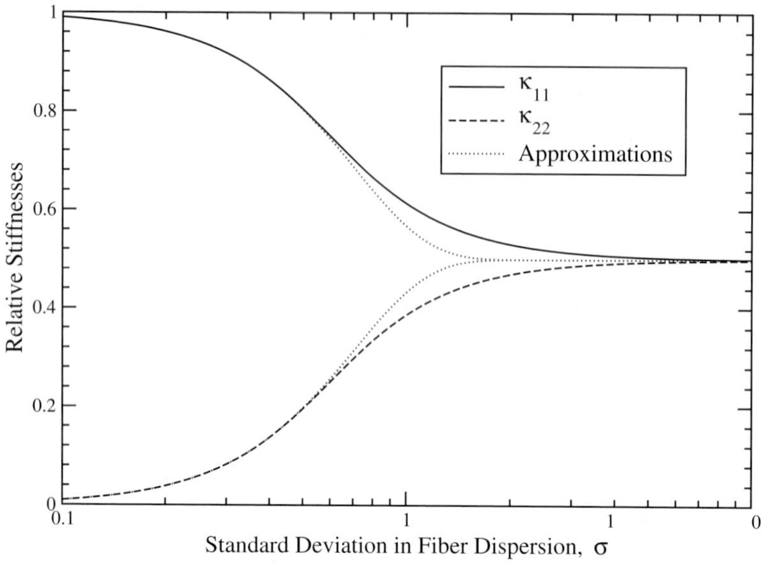

Fig. 1. Plots of relative stiffness vs. fiber dispersion parameter ς for 2D splay, as determined by eq. (13) with $f = 1$ for the approximate solutions, and by eqs (11) and (12) for the analytic solutions.

Adopting the approach and notation of (Simo and Hughes (1998), pp. 358–364), we define

$$J = \det \boldsymbol{F} = \frac{\rho_0}{\rho}, \quad \bar{\boldsymbol{F}} = J^{-1/3}\boldsymbol{F}, \quad \bar{\boldsymbol{C}} = \bar{\boldsymbol{F}}^{\mathrm{T}}\bar{\boldsymbol{F}}, \quad \bar{\boldsymbol{B}} = \bar{\boldsymbol{F}}\bar{\boldsymbol{F}}^{\mathrm{T}}, \tag{14}$$

so that $\det \bar{\boldsymbol{F}} = 1$, and therefore, $\det \bar{\boldsymbol{B}} = \det \bar{\boldsymbol{C}} = 1$. The strain-energy density W is hypothesized to decouple as

$$\varrho_0 W(\boldsymbol{C}) = \widetilde{\mathcal{W}}(J) + \overline{\mathcal{W}}(\bar{\boldsymbol{C}}) + \widehat{\mathcal{W}}(\boldsymbol{K}, \bar{\boldsymbol{C}}), \tag{15}$$

where $\widetilde{\mathcal{W}}$, $\overline{\mathcal{W}}$ and $\widehat{\mathcal{W}}$ are the dilational, distortional-isotropic and distortional-anisotropic strain energies, respectively.

The above definitions allow the general constitutive equation for elasticity,

$$\boldsymbol{S} = 2\varrho_0 \frac{\partial W(T, \boldsymbol{C})}{\partial \boldsymbol{C}}, \tag{16}$$

to be recast as

$$\boldsymbol{S} = J\frac{\partial\, \widetilde{\mathcal{W}}(J)}{\partial\, J}\boldsymbol{C}^{-1} + 2J^{-2/3}\left(\mathrm{DEV}\left[\frac{\partial\, \overline{\mathcal{W}}(\bar{\boldsymbol{C}})}{\partial\, \bar{\boldsymbol{C}}}\right] + \mathrm{DEV}\left[\frac{\partial\, \widehat{\mathcal{W}}(\boldsymbol{K}, \bar{\boldsymbol{C}})}{\partial\, \bar{\boldsymbol{C}}}\right]\right), \tag{17}$$

where

$$\mathrm{DEV}[\bullet] = (\bullet) - \tfrac{1}{3}\,\mathrm{tr}\big((\bullet)\boldsymbol{C}\big)\boldsymbol{C}^{-1} \quad \text{and} \tag{18}$$

is the Lagrangian deviatoric operator. Tensors $\bar{\boldsymbol{F}}$ and $J\boldsymbol{I}$ are the deviatoric (volume preserving) and dilational (volume changing) parts of the deformation gradient $\boldsymbol{F}$, respectively.

3.1 A Simple Model

In a mixed pressure/displacement u/p finite-element scheme (Sussman and Bathe (1987)), the volumetric strain energy is given by

$$\widetilde{\mathcal{W}}(J) = \tfrac{1}{2}\kappa\big(J-1\big)^2. \tag{19}$$

Seeking an isotropic contribution to the deviatoric strain energy, we assign

$$2\,\overline{\mathcal{W}}(\bar{\boldsymbol{C}}) = \tfrac{1}{4}\mu\big(\mathrm{tr}\,\bar{\boldsymbol{C}} + \mathrm{tr}\,\bar{\boldsymbol{C}}^{-1} - 6\big), \tag{20}$$

where μ is the shear modulus. Equation (20) is advocated over the more common neo-Hookean model because it is a convex function whose gradient produces a second-order accurate approximation of true strain.

For an anisotropic contribution to the deviatoric strain energy, going back to the precept that energy is the area under a force/displacement curve, we advocate that

$$\widehat{\mathcal{W}}(\boldsymbol{K}, \bar{\boldsymbol{C}}) = \int_1^{[\mathrm{tr}(\boldsymbol{K}\bar{\boldsymbol{C}})]^{1/2}} \sigma(\lambda)\,\mathrm{d}\lambda, \tag{21}$$

where the fiber stress σ is allowed to be an arbitrary function of fiber stretch λ; it is generally nonlinear in biological tissues. The upper limit of integration is the fourth invariant, as it pertains to the deviatoric part of the deformed state. In order for this strain-energy function to be convex, it is necessary that $E_t(\lambda) > \sigma_e(\lambda)$ for all $\lambda > 0$, wherein

$$E_t(\lambda) = \frac{\mathrm{d}\sigma(\lambda)}{\mathrm{d}\lambda}, \quad \sigma_e(\lambda) = \frac{\sigma(\lambda)}{\lambda}, \tag{22}$$

which are the fiber tangent modulus and engineering stress, respectively, with the fiber stretch λ being quantified by

$$\lambda = \big[\mathrm{tr}(\boldsymbol{K}\bar{\boldsymbol{C}})\big]^{1/2} = \big[\mathrm{tr}(\bar{\boldsymbol{F}}\boldsymbol{K}\bar{\boldsymbol{F}}^{\mathrm{T}})\big]^{1/2}. \tag{23}$$

A physiologically based material model for $\sigma(\lambda)$ has recently been derived by Freed and Doehring (2005) that applies to crimped collagen fibers.

Substituting the strain energies of eqs (19)–(21) into eq. (17) produces an elastic constitutive model suitable for soft tissues, such as aortic valve, possessing a single family of dispersed fibers:

$$\boldsymbol{S} = \tfrac{1}{2}\kappa J\big(J - J^{-1}\big)\,\boldsymbol{C}^{-1} + \tfrac{1}{4}\mu J^{-2/3}\,\mathrm{DEV}\big[\big(\boldsymbol{I} - \bar{\boldsymbol{C}}^{-2}\big)\big] + \sigma_e(\lambda)J^{-2/3}\,\mathrm{DEV}[\boldsymbol{K}]. \tag{24}$$

Similarly, a two-fiber family constitutive model suitable for aortic sinus and aortic wall is

$$\begin{aligned} \boldsymbol{S} = &\ \tfrac{1}{2}\kappa J\big(J - J^{-1}\big)\,\boldsymbol{C}^{-1} + \tfrac{1}{4}\mu J^{-2/3}\,\mathrm{DEV}\big[\big(\boldsymbol{I} - \bar{\boldsymbol{C}}^{-2}\big)\big] \\ &+\sigma_{e1}(\lambda_1)J^{-2/3}\,\mathrm{DEV}[\boldsymbol{K}_1] + \sigma_{e2}(\lambda_2)J^{-2/3}\,\mathrm{DEV}[\boldsymbol{K}_2], \end{aligned} \qquad (25)$$

where $\boldsymbol{K}_1$ and $\boldsymbol{K}_2$ are rotated from local coordinates by their respective transformation tensors $\boldsymbol{Q}_1$ and $\boldsymbol{Q}_2$. In the case of aortic sinus and aortic wall, the two mean fiber directions are separated by an angle θ, following the example of Holzapfel et al. (2002).

3.2 Tangent Moduli

The relationship between $\boldsymbol{S}$ and $\boldsymbol{C}$ in eq. (24) is nonlinear. To obtain a finite-element solution with an iterative Newton-type solution process, that relationship must be linearized with respect to an incremental displacement. This involves the specification of a tangent modulus $\boldsymbol{M} = 2\partial \boldsymbol{S}_n/\partial \boldsymbol{C}_n = 4\varrho_0 \partial^2 W/\partial \boldsymbol{C}_n \partial \boldsymbol{C}_n$.

To obtain this tangent, stress $\boldsymbol{S}$ is linearized over some time interval $[t_n, t_{n+1}]$ such that $\boldsymbol{S}_{n+1} = \boldsymbol{S}_n + \boldsymbol{M}_n \Delta\boldsymbol{E}$ with $\Delta\boldsymbol{E} = \boldsymbol{E}_{n+1} - \boldsymbol{E}_n = \frac{1}{2}\big(\boldsymbol{C}_{n+1} - \boldsymbol{C}_n\big) = \boldsymbol{F}_n^{\mathrm{T}}\hat{\boldsymbol{E}}(\boldsymbol{x}_n; t_n, t_{n+1})\boldsymbol{F}_n$. Said differently, the tangent modulus corresponds to the slope affiliated with a forward Euler integration step, and since it depends on step number n it needs to be re-evaluated at each step along the solution path.

In a u/p formulation (Sussman and Bathe (1987)), the pressure tangent modulus corresponding to eq. (19) is given by

$$\widetilde{\boldsymbol{M}} = \kappa J\Big[J\,\boldsymbol{C}^{-1}\otimes\boldsymbol{C}^{-1} + (1-J)\big(2\boldsymbol{C}^{-1}\boxtimes\boldsymbol{C}^{-1} - \boldsymbol{C}^{-1}\otimes\boldsymbol{C}^{-1}\big)\Big]. \qquad (26)$$

The isotropic tangent modulus corresponding to eq. (20) is determined to be

$$\begin{aligned} \overline{\boldsymbol{M}} = &\ \mu\Big\{J^{2/3}\Big[\boldsymbol{C}^{-1}\boxtimes\boldsymbol{C}^{-2} - \tfrac{1}{3}\big(\boldsymbol{C}^{-1}\otimes\boldsymbol{C}^{-2} + \boldsymbol{C}^{-2}\otimes\boldsymbol{C}^{-1}\big) + \tfrac{1}{9}\big(\mathrm{tr}\,\boldsymbol{C}^{-1}\big)\,\boldsymbol{C}^{-1}\otimes\boldsymbol{C}^{-1}\Big] \\ &+\tfrac{1}{6}\Big[\big(J^{-2/3}\,\mathrm{tr}\,\boldsymbol{C} - J^{2/3}\,\mathrm{tr}\,\boldsymbol{C}^{-1}\big)\big(\boldsymbol{C}^{-1}\boxtimes\boldsymbol{C}^{-1} + \tfrac{1}{3}\boldsymbol{C}^{-1}\otimes\boldsymbol{C}^{-1}\big) \\ &-\boldsymbol{C}^{-1}\otimes\big(J^{-2/3}\boldsymbol{I} - J^{2/3}\boldsymbol{C}^{-2}\big) - \big(J^{-2/3}\boldsymbol{I} - J^{2/3}\boldsymbol{C}^{-2}\big)\otimes\boldsymbol{C}^{-1}\Big]\Big\}. \end{aligned} \qquad (27)$$

Lastly, the anisotropic tangent modulus corresponding to eq. (21) is found to be given by

$$\begin{aligned} \widehat{\boldsymbol{M}} = &\ J^{-2/3}\Big\{\big(E_t(\lambda) - \sigma_e(\lambda)\big)\Big[J^{-2/3}\lambda^{-2}\,\boldsymbol{K}\otimes\boldsymbol{K} \\ &-\tfrac{1}{3}\big(\boldsymbol{K}\otimes\boldsymbol{C}^{-1} + \boldsymbol{C}^{-1}\otimes\boldsymbol{K}\big) + \tfrac{1}{9}J^{2/3}\lambda^2\,\boldsymbol{C}^{-1}\otimes\boldsymbol{C}^{-1}\Big] \\ &+\tfrac{2}{3}\sigma_e(\lambda)\Big[J^{2/3}\lambda^2\big(\boldsymbol{C}^{-1}\boxtimes\boldsymbol{C}^{-1} + \tfrac{1}{3}\boldsymbol{C}^{-1}\otimes\boldsymbol{C}^{-1}\big) - \boldsymbol{K}\otimes\boldsymbol{C}^{-1} - \boldsymbol{C}^{-1}\otimes\boldsymbol{K}\Big]\Big\}. \end{aligned} \qquad (28)$$

For the two-fiber family case, eq. (28) is repeated twice, once for each dispersed family.

4 Examples

Because the modulus $E_t(\lambda)$ and stress $\sigma_e(\lambda)$ are generic scalar components of the anisotropic tangent modulus in eq. (28), we are free to adopt any reasonable fiber stress-strain rule. In other words, modification of the fiber stress-strain law requires no re-formulation of the constitutive tangent matrix, making the model very flexible for different types of soft tissues.

In this article, we adopt the micro-structurally based collagen fiber stress-strain rule based on the physiology of crimped collagen fibers that was recently proposed by Freed and Doehring (2005). This is an algorithm for the elastic response of crimped collagen fibers based on the observation that fibril crimp has a three-dimensional structure at the μm scale whose geometry can be approximated as a cylindrical helix. For pre-failure analysis, the model is defined in terms of three physiologic parameters, namely the initial normalized wavelength of the crimp H_0/r_0, the initial normalized amplitude of the crimp R_0/r_0, and the elastic modulus of the collagen fiber in the linear region E_f. Parameters H_0 and R_0 are normalized with respect to fibril radius r_0.

4.1 Fit to Biaxial Aortic Valve Tissue

For aortic valve biaxial data, parameter values were estimated as $H_0/r_0 = 14.4$, $R_0/r_0 = 2.19$, $E_f = 10.6\,\text{MPa}$, and $\varsigma = 0.768$. Overall, the fit of this constitutive model was excellent (see Fig. 2), with a residual error of 0.37.

4.2 Aortic Valve Finite Element Implementations

Equation (24) and the crimped collagen model from Freed and Doehring (2005) were implemented into Adina™ (Adina R&D Inc., Watertown, MA), where a two-field pressure/displacement interpolation is employed (Sussman and Bathe (1987)). Using this compute engine, numeric experiments were performed to demonstrate the effect that splay has on the predicted state of stress. The results are shown in Fig. 3. Values for the material constants used to construct this figure were the same as those used in Fig. 2, except for the splay variable ς. Equibiaxial extensions were imposed for deformations up to $F_{11} = F_{22} = 1.4$.

Additionally, we simulated a dynamic *invacuo* bioprosthestic aortic valve model. A total of 5400 C^1 brick elements were used. Principal fiber angles were circumferential. Fixed displacement and rotation boundary conditions were applied to three sides of each leaflet. Pressure, applied on the ventricular face, was ramped from 0 to 0.5 kPa (3.5 mmHg) then decreased linearly to −10 kPa (80 mmHg). The identical simulation, adopting the same fiber model from Freed and Doehring (2005), was run with the fully integrated constitutive model in eq. (1). Material constants were the same except for the splay variable, which was set to 0.192, consistent with previous estimations.

Figure 4 shows the fully opened and fully closed configurations of solutions based on eq. (24) (see subfigures A and B), and on eqs (1) (see subfigures C

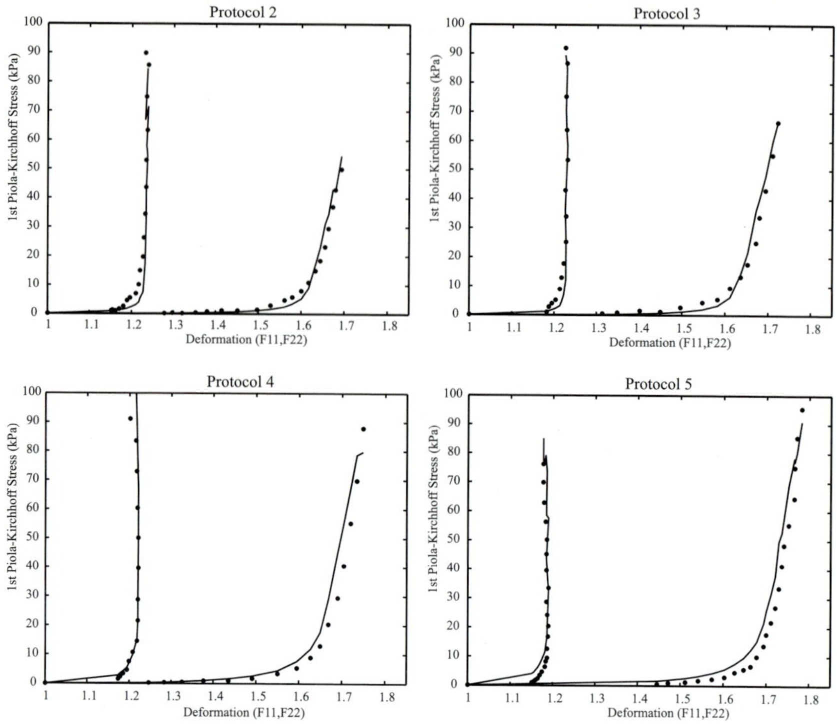

Fig. 2. Constitutive model fit to biaxial data from fresh aortic-valve tissue: fiber crimp stress-strain rule (left – fiber direction, right – cross-fiber direction). Dots are the original data. Solid lines are our constitutive model eq. (24) using the crimped collagen model from Freed and Doehring (2005).

and D). The deformed configurations and maximum principal stresses between the two models are comparable. CPU times were 6424 seconds for the invariant formulation and 16968 seconds for the integrated statistical model. Both simulations were run on a Linux machine with a 2.4 GHz Pentium IV processor and 2 GB of memory. Thus, our invariant formulation for fiber dispersion ran 2.64 times faster than the simulation based on integrated splay.

4.3 Aortic Wall and Sinus Finite Element Implementations

Aortic sinus presents an unusual topological challenge owing to the fact that it has a hole in its center. This configuration effectively precludes biaxial testing. Thus, orthogonal strips of tissue were tested uniaxially and eq. (25) was fit by performing an inverse finite element analysis of two simultaneous data sets – one for each direction. For these simulations the exponential fiber crimp model

$$\sigma_e(\lambda) = A\lambda^{-1}\left[e^{B(\lambda^2-1)/2} - 1\right], \quad E_t(\lambda) = AB\lambda e^{B(\lambda^2-1)/2}. \tag{29}$$

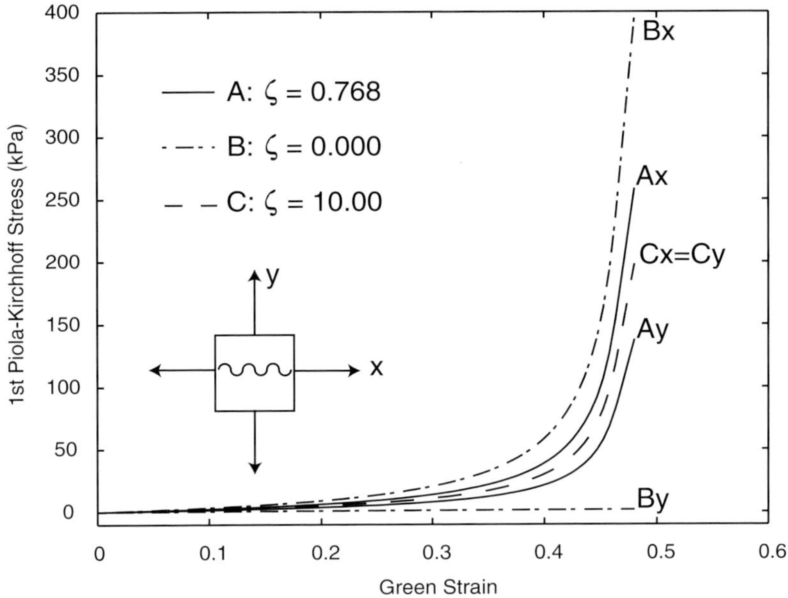

Fig. 3. The effect of splay on equibiaxial stress/strain response. All parameters were held fixed except for ς. The constants of condition A were the same as those used to obtain Fig. 2, condition B is the limiting case of transverse isotropy (i.e. no splay), while condition C is the limiting case of planar isotropy.

was adopted. The inverse approach was based on an iterative sampling of the parameter space via the response surface method and is described elsewhere Einstein et al. (2005). Briefly, minimization proceeds along a quadratic response surface constructed from a set of forward solutions at points in the parameter space selected by design of experiments criteria. The minimum of one set of experiments, or of one response surface, becomes the starting point for the next set of experiments. This cycle of forward solutions at selected points followed by the minimization of the objective function with respect to the response surface is repeated until the prescribed tolerance in the residual is reached. With this approach, parameter values were estimated at: $\mu = 37\,\text{kPa}$, $\theta = 24.9$ deg, $A = 37.67$, $B = 55.52$ and $\varsigma = 1.09$. The two fiber families are offset by an angle θ with respect to the axis of the aorta. Overall, the fit of this constitutive model was excellent (see Fig. 5), with a residual error of 0.51.

5 Conclusion

We have proposed an efficient, invariant-based alternative to structural constitutive equations that accounts for statistical dispersion of fibers. In contrast to

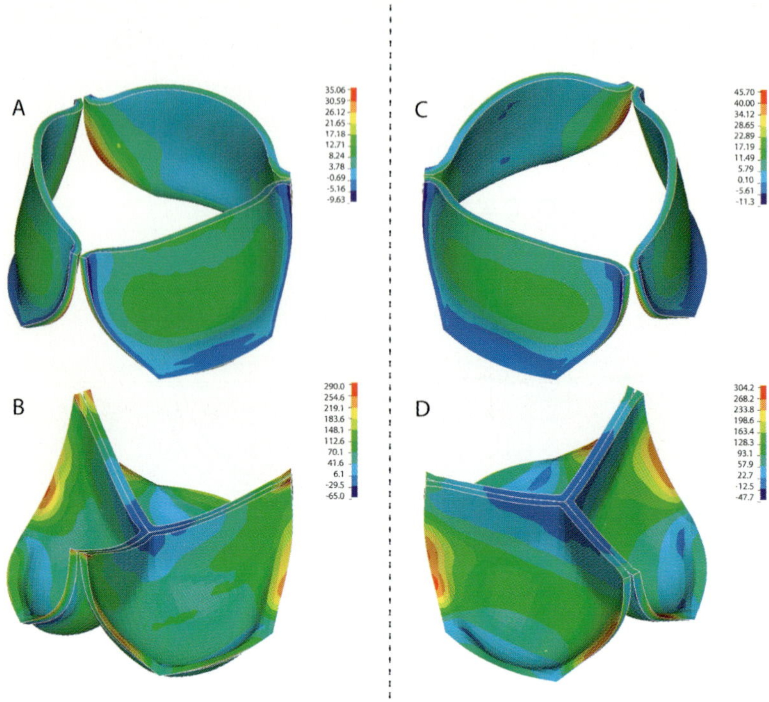

Fig. 4. A comparison of two bioprosthetic valve simulations: one based on the invariant formulation (A and B), the other based on classical integrated splay (C and D). The invariant formulation was 2.64 times faster.

existing models, our new invariant theory easily handles a 3D fiber population with a single mean preferred direction. The invariant theory is based on a novel closed-form 'splay invariant' that requires a single parameter in the 2D case, and two parameters in the 3D case. The proposed model is polyconvex, and fits biaxial data for aortic valve tissue better than existing aortic-valve models Billiar and Sacks (2000). A modification in the fiber stress-strain law requires no re-formulation of the constitutive tangent matrix, making the model flexible for different types of soft tissues. Most importantly, the model is computationally expedient in a finite element analysis.

Acknowledgements. The authors take this opportunity to thank Prof. Michael Sacks at the University of Pittsburgh for providing us with his biaxial experimental data, and to Dr. Todd Doehring at the Cleveland Clinic and Mr. Dimitri Deserranno at the National Center for Microgravity Research for many delightful discussions on this and related topics.

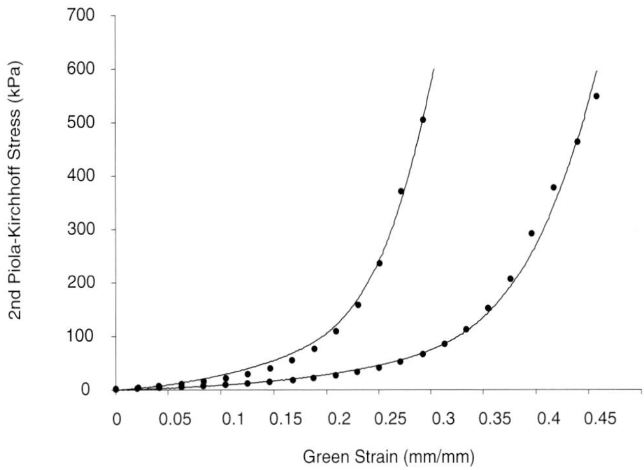

Fig. 5. Equation (25) and the exponential collagen model of eq. 29 fit to two simultaneous orthogonal data sets. An inverse finite element scheme was used to estimate tissue parameters.

References

Billiar, K. L., and Sacks, M. S. (2000). Angioscopic findings during coronary angioplasty of coronary occlusions.biaxial mechanical properties of the natural and glutaraldehyde-treated aortic valve cusp – Part II: A structural constitutive model. *J. Biomech. Eng.* 122:327–335.

Einstein, D. R., Freed, A. D., Stander, N., Fata, B., and Vesely, I. (2005). Inverse parameter fitting of biological tissues: A response surface approach. submitted.

Einstein, D. R. (2002). *Nonlinear Acoustic Analysis of the Mitral Valve.* Ph.D. Dissertation, University of Washington, Seattle.

Flory, P. J. (1961). Thermodynamic relations for high elastic materials. *Trans. Faraday Soc.* 57:829–838.

Freed, A. D., and Doehring, T. C. (2005). Elastic model for crimped collagen fibrils. *J. Biomech. Eng.* in press.

Holzapfel, G. A., Gasser, T. C., and Stadler, M. (2002). A structural model for the viscoelastic behavior of arterial walls: Continuum formulation and finite element analysis. *Eur. J. Mech. A/Solids* 21:441–463.

Lanir, Y. (1983). Constitutive equations for fibrous connective tissues. *J. Biomech.* 16:1–12.

Malkus, D. S., and Hughes, T. J. R. (1978). Mixed finite element methods – reduced and selective integration techniques: A unification of concept. *Comput. Meth. Appl. Mech. Eng.* 15:63–81.

Sacks, M. S. (2000). Biaxial mechanical evaluation of planar biological materials. *J. Elasticity* 61:199–246.

Sacks, M. S. (2003). Incorporation of experimentally-derived fiber orientation into a structural consitutive model for planar collagenous tissues. *J. Biomech. Eng.* 125:280–287.

Simo, J. C., and Hughes, T. J. R. (1998). *Computational Inelasticity.* New York: Springer-Verlag.

Spencer, A. J. M. (1972). *Deformations of Fibre-reinforced Materials.* Oxford: Clarendon Press.

Sussman, T., and Bathe, K.-J. (1987). A finite element formulation for nonlinear incompressible elastic and inelastic analysis. *Comput. Struct.* 26: 357–409.

Zioupos, P., and Barbenel, J. C. (1994). Mechanics of native bovine pericardium. i. the multiangular behaviour of strength and stiffness of the tissue. *Biomaterials* 15:366–373.

Mathematical Modelling of Cardiac Mechanoenergetics

J. Engelbrecht[1], M. Vendelin[1,2]

[1] Tallinn Technical University, Centre for Nonlinear Studies, Estonia
je@ioc.ee
[2] Université J. Fourier, Laboratory of Fundamental and Applied Bioenergetics, France markov@ioc.ee

It has been shown how to bridge the physiological descriptions to continuum mechanics using the formalism of internal variables. Due to the complicated microstructure of cardiac tissues, these internal variables are switched in successively forming a hierarchy. The stress developed in the myocardium is separated into active and passive parts. The passive stress depends upon the elastic properties of fibres, the active stress, according to the Huxley model, depends on triggering the myofibrils due to Ca^{2+} signals. Overall, the process includes a three-level hierarchy of internal variables. The test problems, such as isometric contraction, Adenosin-Tri-Phosphate (ATP) consumption, etc. have shown good matches with experimental results. The numerical calculations based on FEM are presented for the left ventricle, modelled geometrically as a spheroid.

1 Introduction

The complexity of living tissues and cells is a challenge for modelling various biophysical and biochemical processes. The progress of *in silico* modelling shows the strength of a bio-mathematical approach supported by large-scale computing techniques. There are several reviews which reflect the recent results in this fast progressing research; see Kohl et al. (2000), Kolston (2000), Humphrey (2003), van Leeuwen and Aerts (2003), etc. The most promising results are obtained by integrating macro- and microbehaviour of tissues. Here macrobehaviour is understood for an organ or tissue as a whole and microbehaviour for the behaviour and processes of its constituents, including the processes in cells. Such a matching of physiology and continuum theory leads to a comprehensive description of tissues and their assemblies with predictive power and a wide range of *in silico* experiments. Those experiments should, however, always be validated against physical experiments *in vivo* and *in vitro*.

In this paper we focus on cardiac mechanoenergetics. Models of cardiac mechanics and electrical activation have been developed by several groups (reviewed by Hunter et al. (2003)). The primary aim is to combine stress analysis in cardiac muscles with cell energetics. It is shown how to bridge the physiological descriptions to continuum mechanics. In this context special attention is focused on the hierarchy of processes within the cell resulting in active stress. The concept of internal variables is used (Maugin (1990), Maugin and Muschik (1994)). Due to the complicated hierarchical (step-by-step) processes in cells and microstructured tissues, the basic concept of internal variables is generalized into the *concept of hierarchical internal variables.* This concept is demonstrated for the cardiac contraction based on the Huxley model. That means a sequence of internal variables: Ca^{2+} signal, the number of activated cross-bridges, and the number of force-producing cross-bridges. Such a successive modelling permits the build up of a mathematical model for cardiac contraction that can be tested against various physiological experiments (Vendelin et al. (2000)). Numerical calculations using the finite-element method for the left ventricle (LV) demonstrate the effectiveness of such an approach (Vendelin et al. (2002)). The geometry of the LV is taken as a spheroid.

The paper is organized as follows. Section 2 gives a brief overview of the concept of internal variables with the generalization to model hierarchies. In Section 3, the basic features of cardiac modelling are presented and formulated within the framework of hierarchical internal variables. Section 4 involves the discussion, including numerical results and conclusions.

2 The Concept of Internal Variables

The concept of internal variables has its origin in thermodynamics and chemical systems. Contemporary understanding has been reviewed by Maugin and Muschik (1994). It rests upon the assumption that the thermodynamic state is determined not only by observable variables χ (like strain) but also by internal variables α hidden to the external observers. Observable variables are internally governed by a balance law with a kinetic energy. Internal variables, however, do not possess inertia and are governed by kinetic equations. A typical case of an internal parameter in the mechanics of solids is the damage parameter.

The formalism of internal variables in a nutshell is the following (for more detailed description, see Maugin (1990), Maugin and Muschik (1994)). The dependent variables, for example stress σ, must be simultaneously a function of both observable and internal variables:

$$\sigma = \sigma(\chi, \alpha). \tag{1}$$

This must be complemented by an evolution law, such as

$$\dot{\alpha} = f(\chi, \alpha) + g(\chi, \alpha)\dot{\chi}, \tag{2}$$

describing the temporal evolution of the variable α. As usual, stress (i.e. an inertial variable) is derived from a free energy function ψ. In addition to that, a dissipation potential $\mathcal{D} > 0$ is postulated. Then it is possible to show (see Maugin (1990)) that the governing eq. (2) for α is derived as

$$\frac{\delta\,\psi}{\delta\,\alpha} + \frac{\partial\,\mathcal{D}}{\partial\,\dot{\alpha}} = 0, \tag{3}$$

where $\delta/\delta\alpha$ denotes the Euler-Lagrange derivative. This concept has also been used by Maugin and Engelbrecht (1994) for description of nerve pulse dynamics where the ion current is dependent on internal variables.

Biological processes, however, are complex and beside the structural hierarchy are often characterized by several embedded microprocesses, including also the cellular level. In this case we have proposed (Engelbrecht et al. (2000)) the concept of hierarchical internal variables. In general terms, the idea of building up the mathematical model is the following.

(i) a constitutive equation for a dependent variable σ depends on the observable variable χ and the *first-level* internal variable α:

$$\sigma = \sigma(\chi, \alpha); \tag{4}$$

(ii) the evolution law for α is

$$\dot{\alpha} = f_1(\chi, \alpha, \beta), \tag{5}$$

where β is the next, *second-level*, internal variable influencing σ only through dynamics of the first-level internal variable α;

(iii) the evolution law for β is

$$\dot{\beta} = f_2(\chi, \alpha, \beta, \gamma), \tag{6}$$

where γ is again the next, now the *third-level*, internal variable that influences σ only through dynamics of the second-level internal variable β;

(iv) the evolution law for γ is

$$\dot{\gamma} = f_3(\chi, \alpha, \beta, \gamma, ...), \tag{7}$$

etc.

Internal variables $\alpha, \beta, \gamma, ...$ form a hierarchy reflecting the hierarchical processes in the medium (tissue). Note that here we have dropped the gradients and did not discuss the entropy fluxes.

3 Cardiac Contraction

3.1 Physiological Background

Based on fundamental treatises on cardiac performance (Glass et al. (1991), Zipes and Jalife (1995)) we focus our attention on the mechanical behaviour of the heart. In terms of continuum mechanics, the ventricles are thick-walled shells with complex geometry made of muscle fibres with essential variation of their orientation. A single fibre is made up by the bunches of myofibrils with a surrounding sarcotubular system. The myofibrils convert metabolic energy into mechanical energy and the sarcotubular system governs the Ca^{2+} ions needed for activation. A myofibril, in its turn, is composed of repeating units of myosin and actin filaments (sarcomeres). The excitation of a muscle is triggered by an action potential from the conducting system. This potential releases Ca^{2+} ions that activate the troponin molecules at the actin filament so that they will be able to attach to the heads of myosin molecules. These heads – the cross-bridges – swivel and cause sliding of filaments against each other, i.e. cause contraction of the muscle.

The transformation of the whole actomyosin complex is driven by the free energy of ATP hydrolysis to ADP and inorganic phosphate Pi.

The early phenomenological models tried to describe the relationships between observed macroscopic data. Starting from the Huxley (1957) models, contemporary modelling has put much more emphasis on mechanoenergetics (Glass et al. (1991), van Campen et al. (1994), Humphrey (2003)). Here we follow the ideas of Vendelin et al. (2000), casting them in the formalism of internal variables.

3.2 Mathematical Modelling of Tissue Properties

We assume that the total (Cauchy) stress in the cardiac muscle can be split into two parts,

$$\boldsymbol{\sigma} = \boldsymbol{\sigma}_p + \boldsymbol{\sigma}_a, \tag{8}$$

where $\boldsymbol{\sigma}_p$ and $\boldsymbol{\sigma}_a$ denote passive and active stress, respectively. The passive stress results from the elastic deformation of the tissue and can be calculated as

$$\boldsymbol{\sigma}_p = \partial\psi/\partial\boldsymbol{\varepsilon}^e, \tag{9}$$

where $\boldsymbol{\varepsilon}^e$ is the strain. The strain tensor is taken in its full form corresponding to large deformations. The active stress $\boldsymbol{\sigma}_a$ is generated in myofibrils by activation and is directed parallel to the fibre orientation. Hence

$$\boldsymbol{\sigma}_a = \sigma_a \boldsymbol{\varepsilon}_1 \otimes \boldsymbol{\varepsilon}_1, \tag{10}$$

where $\boldsymbol{\varepsilon}_1$ is the unit vector identifying the orientation. The mechanism for generating σ_a involves internal variables, as demonstrated below.

The force developed by the actinmyosin complex depends on the distance z between an attached cross-bridge and the nearest actin site. Actin and myosin produce the mechanical force through cyclic interaction using the free energy of ATP hydrolysis to ADP and Pi. The kinetic scheme of such an interaction

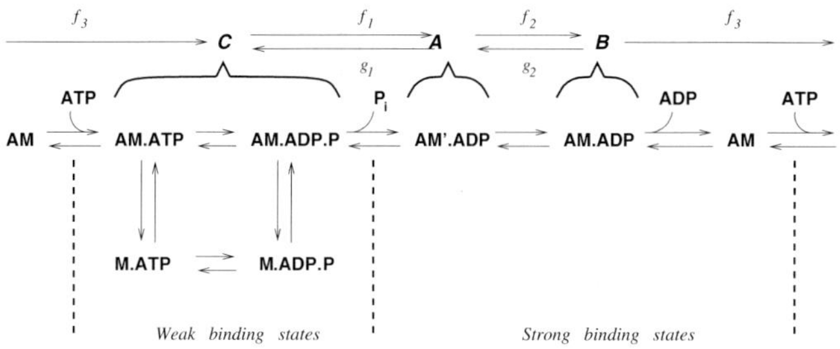

Fig. 1. The kinetic scheme of actin (A) and myosin (M) interaction (lower part) and its simplification as used in the model (upper part). States which include only M in the scheme present the states where myosin head is not attached to actin filament. Adapted from Vendelin et al. (2000).

and the simplified version used in the model are shown in Fig. 1. In the model, there are two states through the cycle producing force. Denoting them by A and B, we may calculate the corresponding forces by

$$F_A = K_A\, z, \quad F_B = K_B\, z, \tag{11}$$

where K_A, K_B are elastic constants. Further we take $K_A = K_B = K$. Assuming the uniform distribution of cross bridges in z over an interval d, we find the active stress

$$\sigma_a = \frac{m l_s K}{2d} \left(\int_{-d/2}^{d/2} z n_A(z) \mathrm{d}z + \int_{-d/2}^{d/2} z n_B(z) \mathrm{d}z \right), \tag{12}$$

where m is the number of cross-bridges per unit volume and $n_A(z), n_B(z)$ are relative numbers of cross-bridges producing force (i.e. being in states A and B). The variables, n_A and n_B, are the *first-level* internal variables. They are governed by the (coupled) kinetic equations (Hill (1974))

$$\frac{\partial n_A}{\partial t} + w \frac{\partial n_A}{\partial z} = f_1 n_C + g_2 n_B - (g_1 + f_2) n_A, \tag{13}$$

$$\frac{\partial n_B}{\partial t} + w \frac{\partial n_B}{\partial z} = f_1 n_A - (g_2 + f_3) n_B, \tag{14}$$

where w is the speed of lengthening, f_1, f_2, f_3, g_1, g_2 are kinetic constants between the states (see Vendelin et al. (2000)) and n_C is the number of cross-bridges that do not produce force. Clearly, the summation of all activated cross-bridges gives

$$A = n_A + n_B + n_C. \tag{15}$$

Now, A is the next in the hierarchy, i.e. the *second-level* internal variable, the change in which affects the observable σ_a only through n_A, n_B. The internal variable A (the activation parameter) has its own kinetic equation, i.e.

$$\frac{\mathrm{d}A}{\mathrm{d}t} = c_1(l_1)[Ca^{2+}](1 - A) - c_2(l_s)A, \tag{16}$$

with $c_1(l_s), c_2(l_s)$ as certain parameters. This kinetic equation involves $[Ca^{2+}]$ – the *third-level* internal variable. This variable is governed by its own kinetic equation:

$$\frac{\mathrm{d}[Ca^{2+}]}{\mathrm{d}t} = f\left([Ca^{2+}]\right). \tag{17}$$

As is clear from equations presented, the observable σ_a is influenced by three levels of internal variables. In addition to finding σ_a, it is possible to compute the ATP consumption rate V_{ATP} in the tissue using the kinetic scheme of actomyosin interaction (Fig. 1):

$$V_{\mathrm{ATP}} = \frac{1}{d} \int_{-d/2}^{d/2} f_3(z) n_B(z,t) \mathrm{d}z. \tag{18}$$

Thus, through the model we are able to relate the mechanical function of the muscle (stress and strain) to biochemical energy consumption (ATP consumption).

The model was tested against several experimental results that are described in detail in Vendelin et al. (2000). In short, the following tests were performed: (a) the relationship between ATP consumption and specific area in the stress-strain diagram is linear, with contractile efficiency close to the measured one; (b) the computed isometric active stress during a beat replicates the measured stress in isosarcometric contractions at different sarcomere length values; (c) the contraction duration is smaller in the isotonic case if compared with the isometric case, in agreement with the isotonic contraction experiment results; (d) the end-systolic point in the stress-strain diagram in isotonic contraction lies close to the end-systolic line computed for the isometric case. The model was able to predict the following properties of the muscle: (a) the shortening velocity-afterload relationship at afterloads higher than 2.5 kPa; (b) the drop of ATP consumption by the cross-bridges during a cycle by about 40% if the muscle is released at the time of peak force.

3.3 Mathematical Modelling of the Left Ventricle

Using the material properties described above, we developed a mathematical model of the left ventricle (Vendelin et al. (2002)). The model computes the deformation of the ventricle, local strains, passive and active stress, and ATP consumption in the ventricular wall. In short, the following assumptions were adopted when constructing the model. First, in the reference state of the model, defined as the state with zero transmural pressure, the endocardial and epicardial surfaces are represented by truncated focal ellipsoids (Streeter and Hanna (1973)), leaving a thick wall between them. The residual stresses in the unloaded state were ignored (Hunter et al. (2003)). Second, the calculations are based on the law of conservation of momentum. We neglected inertial and gravitational effects in our simulations. Third, the hemodynamic coupling of the left ventricle to the aorta is described by a three-element aortic input impedance (Bovendeerd et al. (1992)). Axial displacement of the nodes in the basal surface and circumferential displacement of subepicardial basal ring are suppressed. A uniform left ventricular pressure is applied to the entire endocardial surface. Epicardial pressure is assumed to be zero during a cardiac cycle. Finally, the governing equations were discretized using the finite element method in conjunction with Galerkin's method.

Using this model we studied the influence of fiber orientation in the left ventricular (LV) wall on the ejection fraction, efficiency and heterogeneity of the distributions of developed fiber stress, strain and ATP consumption (Vendelin et al. (2002)). The fiber orientation was quantified by two angles: the helix fiber angle, describing the crossover of fibers between base and apex of the heart, and the transverse angle, describing the crossover of fibers between inner and outer layers of the cardiac wall. For simplicity, the influence of the laminar structure of the myocardium on the distributions of stress and strain in the left ventricular wall was not considered. The computed variances of sarcomere length (VarSL), developed stress (VarDS) and ATP consumption (VarATP) have several minima at different transmural courses of helix fiber angle. Intriguingly, we identified only one region in the design space used with high ejection fraction, high efficiency of the LV and relatively small VarSL, VarDS and VarATP. This region corresponds to the physiological distribution of the helix fiber angle in the LV wall. Transmural fiber angle can be predicted by minimizing VarSL and VarDS, but not VarATP. If VarATP is minimized then the transverse fiber angle is considerably underestimated. The reasons for such differences in estimation are not clear yet. However, our results suggest that the ATP consumption distribution does not regulate the fiber orientation in the heart.

The model was tested against several experiments (Vendelin et al. (2002)). First, the computed increase in the equatorial wall thickness, outer equatorial ventricular radius and outer ventricular length during systole were close to the values measured by Olsen et al. (1981). Second, torsion of the apex during systole was found to be very sensitive to the model parameter values and somewhat different from the measured values (see Vendelin et al. (2002) for results and discussion). Third, the relationship between the pressure developed and the oxygen consumption of the ventricle is reproduced accurately.

An example solution is depicted in Figs 2 and 3. As is shown in the figures, the model can predict mechanical and energetic properties of the left

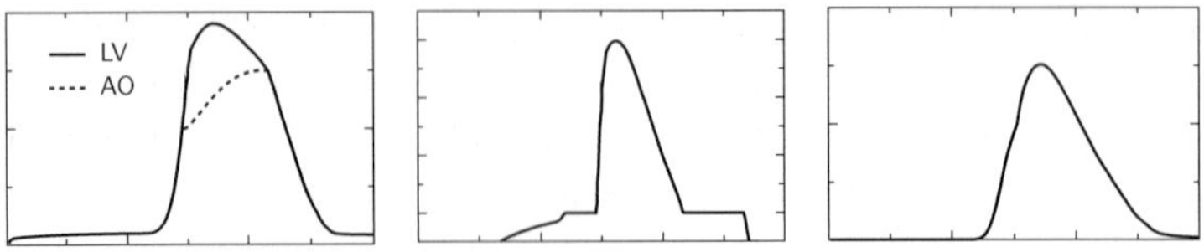

Fig. 2. Developed pressure in left ventricle (LV) and pressure in aorta (AO) (left), blood flow from LV (center), and ATP consumption by LV (right) computed by the model.

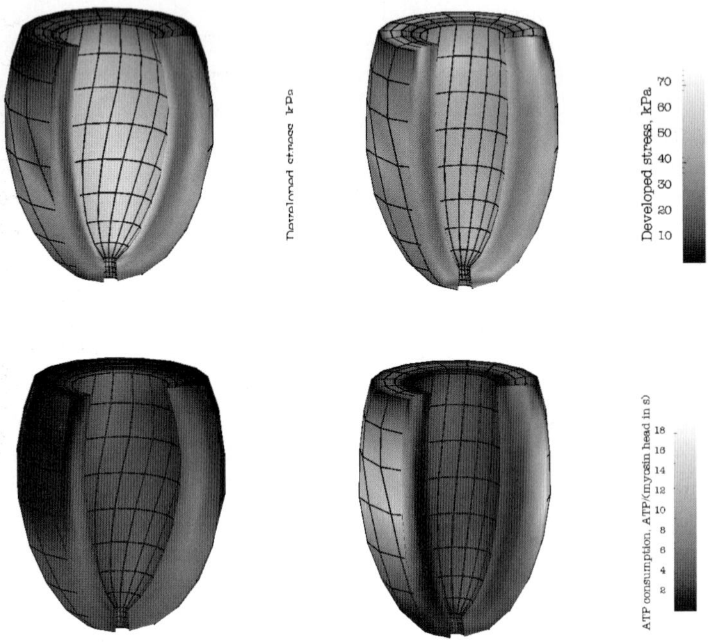

Fig. 3. Developed stress (the top row) and ATP consumption rate (the bottom row) during a systole computed by the model. The time-moments are 0.3 s (the left column) and 0.4 s (the right column).

ventricle. Now the model has to be tested against available experimental data on the distribution of oxygen consumption in the ventricle as well as regional deformation.

4 Discussion

The calculations of contraction for the idealized spheroidal LV are performed using the model described briefly in Section 3 and FEM. Several test problems were solved and then compared with experimental results (Vendelin et al. (2000)). The tests included isotonic and physiologic contractions, ATP consumption, the quick-release experiment, etc. The model is in good agreement with the classical measurements of the SSA and oxygen consumption dependency.

Acknowledgements. This work was supported in part by Marie Curie Fellowship of the European Community programme 'Improving Human Research Potential and the Socio-economic Knowledge Base' (M. Vendelin, contract N HPMF-CT-2002-01914). The authors also acknowledge the support from the Estonian Science Foundation.

References

Bovendeerd, P. H. M., Arts, T., Huyghe, J. M., van Campen, D. H., and Reneman, R. S. (1992). Dependence of local left ventricular wall mechanics on myocardial fiber orientation: A model study. *J. Biomech.* 25:1129–1140.

Engelbrecht, J., Vendelin, M., and Maugin, G. A. (2000). Hierarchical internal variables reflecting microstructural properties: Application to cardiac muscle contraction. *J. Non-Equilib. Thermodyn.* 25:119–130.

Glass, L., Hunter, P., and McCulloch, A., eds. (1991). *Theory of Heart. Biomechanics, Biophysics, and Nonlinear Dynamics of Cardiac Function.* New York: Springer.

Hill, T. L. (1974). Theoretical formalism for the sliding filament model of contraction of striated muscle. part I. *Prog. Biophys. Molec. Biol.* 28:267–340.

Humphrey, J. D. (2003). Continuum biomechanics of soft biological tissues. *Proc. R. Soc. Lond. A* 459:3–46.

Hunter, P. J., Pullan, A. J., and Smaill, B. H. (2003). Modeling total heart function. *Ann. Rev. Biomed. Eng.* 5:147–177.

Huxley, A. F. (1957). Muscle structure and theories of contraction. *Prog. Biophys. and Biophys. Chem.* 7:257–318.

Kohl, P., Noble, D., Winslow, R. L., and Hunter, P. (2000). Computational modelling of biological systems: tools and visions. *Philos. Trans. R. Soc. Lond. A* 358:579–610.

Kolston, P. J. (2000). Finite-element modelling: a new toll for a biologist. *Philos. Trans. R. Soc. Lond. A* 358:611–631.

Maugin, G. A., and Engelbrecht, J. (1994). A thermodynamical viewpoint on nerve pulse dynamics. *J. Non-Equilib. Thermodyn.* 19:9–23.

Maugin, G. A., and Muschik, W. (1994). Thermodynamics with internal variables. *J. Non-Equilib. Thermodyn.* 19:217–249, 250–289. in two parts.

Maugin, G. A. (1990). Internal variables and dissipative structures. *J. Non-Equilib. Thermodyn.* 15:173–192.

van Campen, D. H., Huyghe, J. M., Boverndeerd, P. H. M., and Arts, T. (1994). Biomechanics of heart muscle. *Eur. J. Mech. A/Solids* 13 (suppl):19–41.

van Leeuwen, J., and Aerts, P. (2003). Modelling in biomechanics. *Philos. Trans. R. Soc. Lond. B* 358:1425–1603.

Olsen, C. O., Rankin, J. S., Arentzen, C. E., Ring, W. S., McHale, P. A., and Anderson, R. W. (1981). The deformational characteristics of the left ventricle in the conscious dog. *Circ. Res.* 49:843–855.

Streeter, D. D., and Hanna, W. T. (1973). Engineering mechanics for successive states in canine left ventricular myocardium. I. cavity and wall geometry. *Circ. Res.* 33(6):639–655.

Vendelin, M., Bovendeerd, P. H. M., Arts, T., Engelbrecht, J., and van Campen, D. H. (2000). Cardiac mechanoenergetics replicated by cross-bridge model. *Ann. Biomed. Eng.* 28:629–640.

Vendelin, M., Bovendeerd, P. H. M., Engelbrecht, J., and Arts, T. (2002). Optimizing ventricular fibers: uniform strain or stress, but not ATP consumption, leads to high efficiency. *Am. J. Physiol. Heart Circ. Physiol.* 283:H1072–H1081.

Zipes, D. P., and Jalife, J., eds. (1995). *Cardiac Electrophysiology: From Cell to Bedside.* Philadelphia: Saunders.

Creep and Relaxation in Ligament: Theory, Methods and Experiment

A.L. Oza,[1] R. Vanderby,[1,2,3] R.S. Lakes[2,3]

[1] University of Wisconsin, Department of Orthopedics and Rehabilitation, USA
oza@surgery.wisc.edu, vanderby@surgery.wisc.edu
[2] University of Wisconsin, Department of Biomedical Engineering, USA
[3] University of Wisconsin, Department of Engineering Physics, USA
lakes@engr.wisc.edu

Ligament is a viscoelastic material. Although its time dependent behavior has been traditionally modeled as independent of load level, recent studies in ligament have shown that the creep and relaxation are stress and strain dependent, i. e. it is nonlinearly viscoelastic. Nonlinear superposition integrals were developed to model these behaviors and derive an interrelationship between the creep and relaxation parameters. It has been shown herein that this interrelationship can predict relaxation from creep data and *vice versa*. Unlike the new phenomenological models developed in this study, quasilinear viscoelasticity (QLV) cannot interrelate creep and relaxation. Furthermore, the new models can simulate ligament behavior with fewer parameters than QLV. Models and interrelations are sufficiently general that they have broader utility; e.g., an application to aluminum alloy at an elevated temperature is presented.

1 Introduction

1.1 Viscoelasticity

When a material undergoes time dependent deformation under load and time dependent recovery from deformation with unloading, that material is said to be 'viscoelastic'. Materials that exhibit viscoelasticity are plastics, wood, natural and synthetic fibers, concrete, metals (at higher temperatures), biological tissues, etc. Common characteristics of a viscoelastic material are creep and relaxation. Creep is change in strain over time under application of constant stress. Similarly, relaxation is change in stress over time under application of constant strain.

1.2 Constitutive Equations

Creep compliance and relaxation modulus in linearly viscoelastic materials are dependent only on time and independent of stress and strain respectively, as shown in eq. (1). In linearly viscoelastic materials, the constitutive equation relating time dependent strain $\varepsilon(t)$ and time dependent stress $\sigma(t)$ is a Boltzmann integral

$$\varepsilon(t) = \int_0^t J(t-\tau)\frac{\mathrm{d}\sigma(\tau)}{\mathrm{d}\tau}\mathrm{d}\tau, \quad \sigma(t) = \int_0^t E(t-\tau)\frac{\mathrm{d}\varepsilon(\tau)}{\mathrm{d}\tau}\mathrm{d}\tau, \tag{1}$$

where $J(t)$ is the creep compliance and $E(t)$ is the relaxation modulus. $J(t)$ is obtained experimentally by applying a step stress and observing the resulting time dependent creep strain. $E(t)$ is obtained experimentally by applying a step strain and observing the resulting time dependent stress in relaxation. Linearity is verified by tests at several stress or strain levels.

In nonlinearly viscoelastic materials creep compliance is a function of both time and stress, and relaxation modulus is a function of both strain and time. Nonlinear superposition integrals, as in eq. (2), can describe these nonlinearities (Lakes (1998)):

$$\varepsilon(t) = \int_0^t J(t-\tau,\sigma(\tau))\frac{\mathrm{d}\sigma(\tau)}{\mathrm{d}\tau}\mathrm{d}\tau, \quad \sigma(t) = \int_0^t E(t-\tau,\varepsilon(\tau))\frac{\mathrm{d}\varepsilon(\tau)}{\mathrm{d}\tau}\mathrm{d}\tau. \tag{2}$$

The nonlinear creep function $J(t,\sigma)$ is determined experimentally by series of creep tests at different stress levels, and the nonlinear relaxation function $E(t,\varepsilon)$ is determined experimentally by series of creep tests at different strain levels. Quasilinear viscoelasticity (QLV) is a special form of the nonlinear superposition (2). In QLV, time and stress dependence are separable so the creep compliance (or relaxation modulus) is written as a product of time dependent and stress (or strain) dependent functions.

1.3 Ligament

Ligaments are soft connective tissues that connect bone to bone and primarily consist of collagen, water, elastin and proteoglycans. Ligaments contain closely packed collagen fiber bundles that are axially oriented to provide strength and stiffness in tension. Figure 1 shows the results of a typical stretch to failure test at constant strain rate. Although these data reveal a nonlinear stress-strain behavior as well as ligament ultimate strength, this common test modality mixes strain dependence with time dependence since both strain and time increase during the test. Strain dependence and time dependence may be decoupled by conducting a series of creep tests at different stress levels or a series of relaxation tests at different strain levels. Then, an isochronal

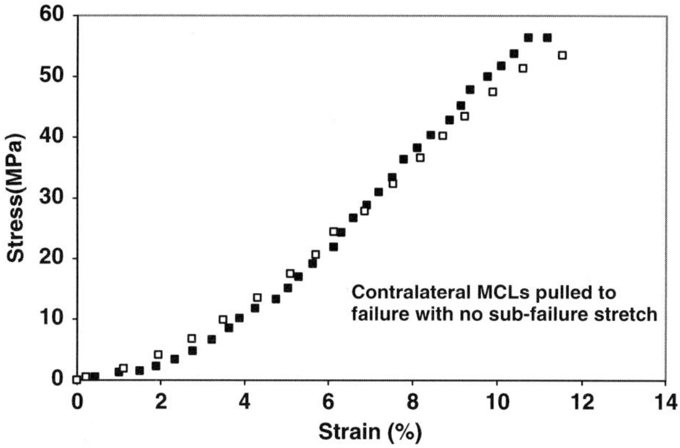

Fig. 1. Stress-strain curves of contralateral rat medial collateral ligaments (Provenzano et al. (2002b)) loaded at a constant strain rate until failure. (□) represents a ligament without a prior stretch and (■) represents a stretched ligament after sub-failure stretch.

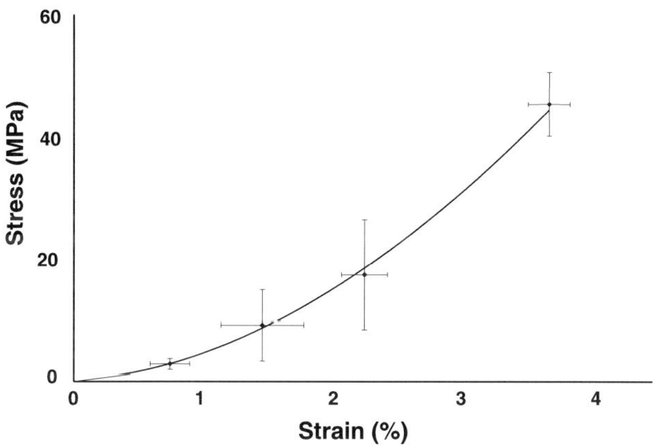

Fig. 2. Isochronal curve at time 2.4 s obtained from a series of stress relaxation test performed on 15 MCLs. It shows the nonlinear strain-stiffening behavior that is a part of the viscoelastic properties found in ligaments. Error bars represent standard deviations in the data and the solid line is a curve fit (Hingorani et al. (2004)).

stress-strain curve can be obtained from these results by plotting stress and strain at a particular time from these tests, as shown in Fig. 2. The isochronal tensile behavior of these tissues is also nonlinear with a compliant region at low loading that becomes a stiffer region (Fig. 2). One causal mechanism for this nonlinearity observed in the ligament is its fibril arrangement, as shown in Fig. 3.

Under unloaded conditions, the microstructure has a wavy appearance, also known as crimp (Viidik (1968)). Collagen fibers and fascicles have varying

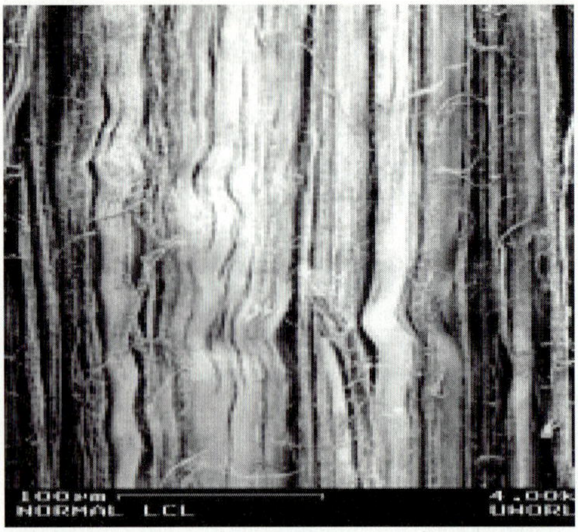

Fig. 3. Scanning electron micrograph showing the wavy appearance of collagen fibers in a normal rat lateral collateral ligament at 250× (Provenzano et al. (2002a)).

amounts of crimp such that increasing tensile deformation recruits additional load bearing fibrils to resist tensile stress. This uncrimping and recruitment of collagen fibers is thought to be responsible for the concave-up 'toe' region, as shown in Fig. 1, where fibers straighten and elongate in a strain-stiffening fashion (Rigby et al. (1959), Abrahams (1967), Viidik (1973), Broom (1978), Lanir (1980)). Rigby et al. (1959) and Diamant et al. (1972) showed that the crimp pattern disappears above 4% strain in rat tail tendon and the onset of mechanical damage in rat medial collateral ligament (MCL) has been shown by Provenzano et al. (2002b) to begin at 5.1% above preload. The physiological function of a ligament is typically confined to this toe region (Viidik et al. (1982)); neuromuscular control of muscles crossing a joint tends to prevent high ligament loadings.

2 Background

Viscoelasticity is an important part of ligament behavior. Stretching prior to physical activity produces creep, isometric joint contractions produce relaxation, cyclic loadings produce hysteresis. Sometimes the injury is so severe that the ligaments are surgically reconstructed. It is logical then that ligament grafts (tissue or prosthetic) should have viscoelastic properties that mimic normal ligament. This requires a broad understanding of ligament viscoelasticity and robust models to characterize and predict behavior for varied and complex load histories.

Researchers frequently quantify viscoelastic behavior with phenomenological models of intervertebral disc, ligament, tendon, cartilage, muscle, mesentery, cornea, skin, fascia, etc. The general types of viscoelastic models include linear (Viidik (1968)), quasi-linear (Haut and Little (1972)) and nonlinear (Johnson et al. (1996), Puso and Weiss (1998)). By far, the most common ligament phenomenological model has been the QLV approach of Fung (1972). This method has been used to model a number of experiments with ligaments and tendons (e.g., Woo et al. (1981), Woo (1982), Kwan et al. (1993), Johnson et al. (1994), Funk et al. (2000), Sarver et al. (2003), Abramowitsch et al. (2004), Abramowitsch and Woo (2004)). Quasilinear viscoelasticity has also modeled numerous other tissues (Best et al. (1994) for skeletal muscle, Carew et al. (1999) and Carew et al. (2000) for aortic valves, Huyghe et al. (1991) for heart muscle, Woo et al. (1980) for cartilage, Rubin et al. (1998) for facial tissues, Fung (1972) for mesentery, Demiray (1996) for arterial walls, Kim et al. (1998) for pharyngeal tissue, etc.). Quasi-linear viscoelasticity has also been the basis for models that include finite deformations (Johnson et al. (1996), Puso and Weiss (1998)).

In QLV, creep and relaxation functions can be represented by eqs (3) and (4) respectively (Provenzano et al. (2001), so that

$$\varepsilon(\sigma,t) = \int\limits_0^t \bar{J}(t-\tau)\frac{\mathrm{d}\varepsilon}{\mathrm{d}\sigma}\frac{\mathrm{d}\sigma(\tau)}{\mathrm{d}\tau}\mathrm{d}\tau, \qquad \bar{J}(t-\tau)\frac{\mathrm{d}\varepsilon}{\mathrm{d}\sigma} = J(t,\sigma), \tag{3}$$

$$\sigma(\varepsilon,t) = \int\limits_0^t \bar{E}(t-\tau)\frac{\mathrm{d}\sigma}{\mathrm{d}\varepsilon}\frac{\mathrm{d}\varepsilon(\tau)}{\mathrm{d}\tau}\mathrm{d}\tau, \qquad \bar{E}(t-\tau)\frac{\mathrm{d}\sigma}{\mathrm{d}\varepsilon} = E(t,\varepsilon), \tag{4}$$

where $\bar{J}$ is the reduced creep function and $\bar{E}$ is the reduced relaxation function. Thus, QLV is a special case of the single integral nonlinear superposition formulation in eq. (2).

Let the stress history in (3) and strain history in (4) be controlled by a Heaviside step function of magnitude σ_0 and ε_0 in time respectively. Since the derivative of a step function becomes a delta function, creep in eq. (3) and relaxation in eq. (4) can be expressed as

$$\varepsilon(\sigma,t) = \sigma_0\bar{J}(t)f(\sigma), \quad \sigma(\varepsilon,t) = \varepsilon_0\bar{E}(t)g(\varepsilon). \tag{5}$$

The strain in eq. $(5)_1$ is clearly dependent on stress level but its time dependence is separable and does not depend on stress. Similarly, stress in eq. $(5)_2$ is clearly dependent on strain level, but its time dependence is separable and does not depend on strain. In a graphical form, a purely strain dependent elastic nonlinearity manifests itself in the overall height of the relaxation curve while time dependence manifests itself in the shape of the curve. Since, in QLV, time dependence is decoupled from stress dependence in creep and strain dependence in relaxation, creep curves from any level of stress and relaxation curves from any level of strain must have the same time dependent behavior as depicted in Fig. 4 with continuous gray lines.

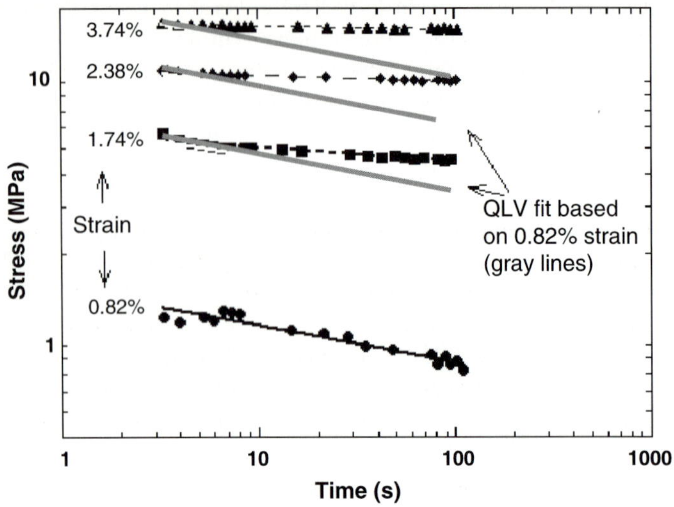

Fig. 4. Experimental stress relaxation (solid points) at multiple levels of strain for a rat ligament. A QLV fit based on 0.82% strain would predict the same rate for all strains, and hence QLV (as indicated by gray lines) does not capture the observed strain rate dependence (Provenzano et al. (2001)).

In addition, relaxation proceeds faster than creep (Graf et al. (1994), Thornton et al. (1997), Provenzano et al. (2001)). Thornton et al. (1997) showed that linear viscoelasticity cannot predict ligament creep from relaxation. Lakes and Vanderby (1999) demonstrated that the observed rate difference between creep and relaxation is consistent with continuum concepts when the stress-strain nonlinearity is of a strain stiffening type. They developed an interrelationship between creep and relaxation, assuming a separable viscoelastic model for creep which gave rise to a non-separable form of relaxation (see Section 5.1). They showed that if QLV is used to empirically fit creep and relaxation data, then the constitutive parameters of the two models are not related by continuum mechanics. Furthermore, the interrelationships between parameters predicted a faster rate of ligament relaxation (than creep), consistently with the Thornton et al. (1997) data.

Researchers using QLV for modeling typically perform a creep or relaxation test at a single stress or strain level, respectively, and a pull to failure test. The creep or relaxation test is to obtain the time dependent kernel in eq. $(5)_1$ or $(5)_2$ and the pull-to-failure test is to obtain the elastic portion of eq. $(5)_1$ or $(5)_2$. This kind of experiment does not allow the experimenter to see if stress level affects the rate of creep or if strain level affects the rate of relaxation, and it does not permit formulation of a robust viscoelastic description. Pioletti and Rakotomanana (2000) considered the QLV time-strain separability hypothesis, but they did not deal directly with the viscoelastic functions. A difference between force at a particular time and final force normalized by the total difference in force was used, which made their data difficult to interpret. With the above limitations of QLV in mind, experiments were designed and conducted to explore the nonlinear viscoelasticity of ligaments.

3 Methods

Fifteen white New Zealand rabbits (weight = 3.2–3.6 kg) were used (Hingorani et al. (2004)). A similar study using medial collateral ligaments (MCLs) from eighteen rats was also performed in our laboratory (Provenzano et al. (2001)). In both studies ligaments were harvested immediately after death of the animal and stored at −73°C prior to testing.

3.1 Rabbit Testing

On the day of testing, tissues were kept at −6°C for 4 hours and then warmed to room temperature immediately prior to testing. All the extraneous tissue was carefully removed and bone blocks of about 2 cm per side were cut from the femoral condyle and tibial plateau. Bone blocks were fixed in plastic molds with polyester resin in an anatomical position. During this time, tissue was hydrated at room temperature with saline. The width and the thickness of the MCL were measured using vernier calipers, with a caliper accuracy of 0.01 mm. The specimen was then placed in the testing machine (MTS 858 BIONIX Test System; 200 N load cell) containing a hydration bath. Bone blocks were arranged to depict a joint with −70° of flexion. The rabbit femur-MCL-tibia complex might not provide a totally uniaxial state of strain but this angle of knee flexion provides a natural anatomic position for the rabbit MCL that best approximates uniform loading (Thornton et al. (2002)) and is ideal for biomechanical evaluation of this tissue (Thornton et al. (2001a), Thornton et al. (2001b)). Ligaments were marked near their insertions with Verhoff stain. These optical markers were used with video analysis to calculate ligament strain since use of optical markers at the ends of the viewing area increases the digitization for strain. The gage length was measured at 0.5 N and the bath was then filled with phosphate-buffered saline (PBS), so that the ligament was completely submerged.

One knee was tested in stress relaxation at varying strain levels up to 5%. A preload of 0.5 N was applied and then an idealized step input for strain elongated the ligament. The rise time was 0.2 s. This strain was maintained for 100 s, after which the tissue grips were returned to their starting position and the ligament recovered for 1000 s while remaining hydrated. Data were analyzed after ten times the rise time in loading. This procedure, accepted in the viscoelasticity community (Turner (1973)), ensures that the results are (relatively) independent of the details of the ramp portion leading up to constant stress. The above protocol was repeated for different strain levels. Five different strain levels were applied in a random order.

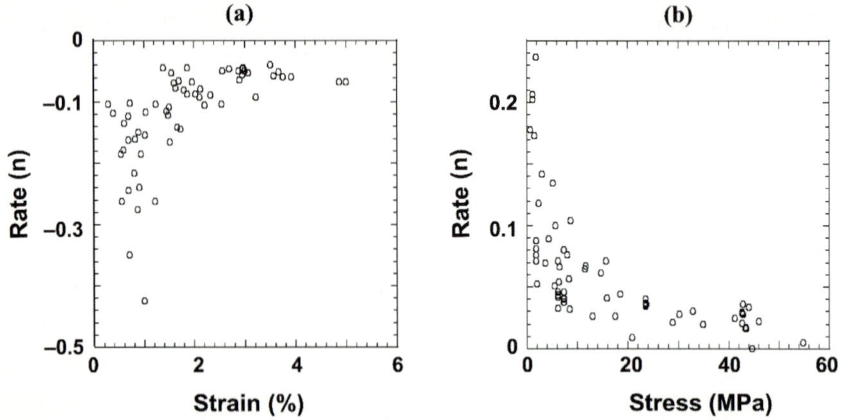

Fig. 5. The rate of stress relaxation in rabbit medial collateral ligament is seen to be nonlinear with respect to strain. Rate means the exponent in a power law in time and is the slope on a log-log plot: graph (a) shows the rate of stress relaxation (n) changes by an order of magnitude throughout the low load region in MCL; graph (b) shows the rate of creep (n) in a rabbit medial collateral ligament to be nonlinear with respect to stress level and changes by an order of magnitude through the low load region in rabbit MCL (Hingorani et al. (2004)).

Creep testing was done similarly to stress relaxation except that tensile force was prescribed as the step function in time. Load values corresponding to the peak force from the contralateral relaxation test were used in the creep tests. All creep tests were performed at stress levels up to ~55 MPa. Examples of the creep and relaxation rate changes with stress and strain respectively in rabbit MCL are shown in Figs 5(a) and 5(b), respectively. Rate means the exponent in a power law in time and is the slope on a log-log plot. Deformation was recorded using a Sony CCD-IRIS camera, a VCR, and a viewing monitor, which displayed synchronized displacement data with force data acquired using Labtech Notebook data acquisition software (Laboratory Technologies Corp., Wilmington, MA). Video images of the tests were digitized to calculate strain for a particular load and time in the tissue with N.I.H. image software. The x–y coordinate center of each marker was used to calculate the distance between the Verhoff markers, and hence ligament displacement, after loading.

4 Interrelation Using Nonlinear Superposition

The objective of this study was to develop a robust, interrelated, nonlinearly viscoelastic models for creep and relaxation for ligaments. That is, the models must fit the experimentally observed load dependence in creep and relaxation and interrelate the parameters used in eq. (2). This single integral constitutive model allows the creep function to be stress dependent and the relaxation function to be strain dependent.

Time dependent strain $\varepsilon(t)$ due to constant stress can be written as a sum of immediate and delayed Heaviside step functions $H(t)$ in time:

$$\varepsilon(t) = \varepsilon(0)H(t) + \sum_{i=0}^{N} \Delta\varepsilon_i H(t - t_i). \tag{6}$$

Decomposition of creep in eq. (6) is shown in the top part of Fig. 6(a). Each step strain in the summation gives rise to a relaxing component of stress in view of the definition of the relaxation function. Nonlinearity is accommodated in this analysis since the relaxation function E depends on the strain level:

$$\sigma = \varepsilon(0)E(t, \varepsilon(0)) + \sum_{i=0}^{N} \Delta\varepsilon_i E(t - t_i, \varepsilon(t_i)). \tag{7}$$

Equation (7) is represented graphically in the bottom part of Fig. 6(b). The single-integral nonlinear response considered assumes that there is no interaction between the incremental step components (Lakes and Vanderby (1999)). Dividing eq. (7) by σ and using the definition of creep compliance yields

$$1 = J(0, \sigma)E(t, \varepsilon(0)) + \sum_{i=0}^{N} \Delta J_i E(t - t_i, \varepsilon(t_i)). \tag{8}$$

Equation (8) can be written as

$$1 = J(0, \sigma)E(t, \varepsilon(0)) + \int_0^t E(t - \tau, \varepsilon(\tau)) \frac{\partial J(\tau, \sigma)}{\partial \tau} \mathrm{d}\tau. \tag{9}$$

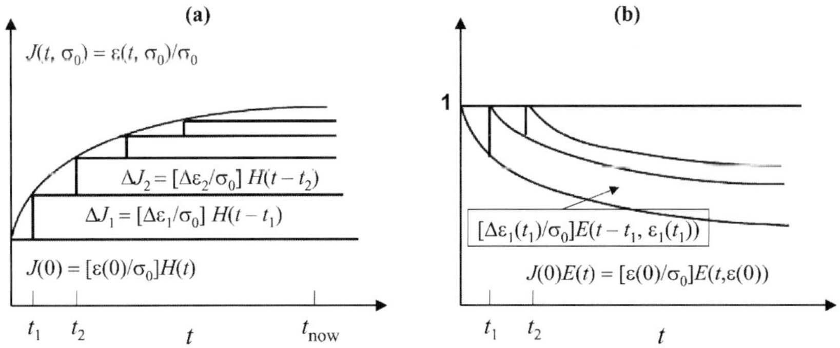

Fig. 6. (a) Decomposition of creep function $J(t, \sigma_0)$ as a sum of immediate and delayed Heaviside step functions $H(t - t_i)$ in time t; (b) constant stress σ_0 gives rise to creep expressed as a sum of relaxing components each of which comes from a step function in the decomposition of the creep curve (adapted from Lakes (1998)).

This is the Stieltjes integral, which will be used in the following section to solve explicit problems.

The analysis is exact with the nonlinear superposition constitutive form assumed. To obtain explicit interrelations, several explicit time dependent functions are assumed in the following section. Experimental studies in our lab show the power law in time,

$$J(t,\sigma) = (g_1 t^n + g_2 \sigma t^{2n}), \tag{10}$$

to be adequate for ligament behavior (Manley et al. (2003)). This equation contains three independent parameters (g_1, g_2, n). It is both more general than QLV (since it accommodates the observed stress dependence of creep rate) and simpler than QLV (since it does not contain an artificial time constant). For a single stress, as considered in QLV, only two parameters, a compliance and a rate, are needed here, compared with three parameters in QLV. In addition, the parameters do not have to be chosen outside the data window, as is the case for QLV time parameters. Power laws have the limitation that the modulus tends to infinity as time tends to zero, but this asymptotic behavior is not obtrusive as experimental data are available for only a limited window of the time domain (Oza et al. (2003)).

5 Explicit Analytical Interrelations

5.1 Interrelation 1 (Lakes and Vanderby (1999))

Lakes and Vanderby (1999) considered the interrelationship between creep and relaxation for a hypothetical material, in which time dependence and strain dependence were separable in creep (as in QLV). The nonlinear creep function was assumed to be a power law in time, written as

$$J(t,\sigma) = (g_1 + g_2\sigma + g_3\sigma^2 + \ldots)t^n. \tag{11}$$

The above interrelation analysis gave rise to the non-separable form of relaxation

$$E(t,\varepsilon) = (f_1 t^{-n} + f_2 \varepsilon t^{-2n} + f_3 \varepsilon^2 t^{-3n} + \ldots). \tag{12}$$

Equations (11) and (12) are inserted into (9) (Stieltjes integral) to develop a relation between different coefficients, which leads to

$$f_1 = \frac{1}{g_1}, \quad f_2 = \frac{-g_2}{g_1^3}, \quad f_3 = \frac{(2g_2^2 - g_1 g_3)}{g_1^5}. \tag{13}$$

Results of the interrelation and comparison of a quadratic versus cubic model are given in Fig. 7.

Although the creep behavior was assumed separable, the corresponding relaxation behavior was not separable. So, even if a material obeys QLV in creep, it would not obey QLV in relaxation and therefore it is inconsistent to assume QLV for both creep and relaxation.

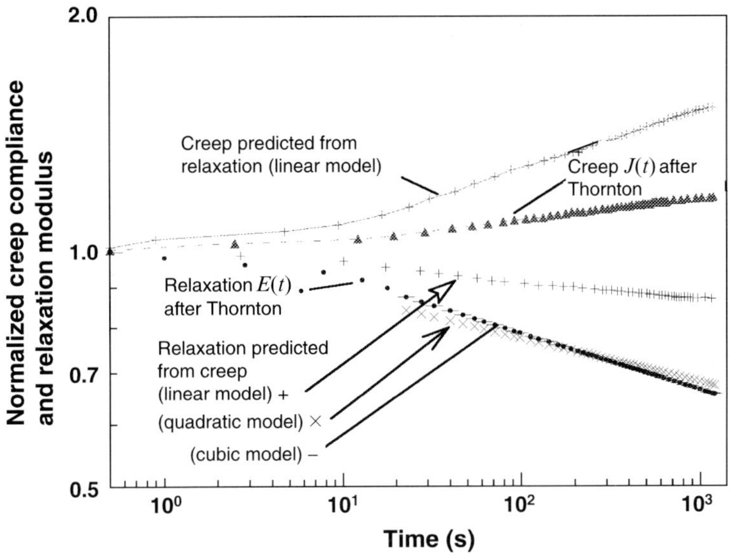

Fig. 7. Relaxation predicted from creep using a linearly viscoelastic (Laplace transform) approach, +, does not agree with observed relaxation, •. Normalized data of Thornton et al. (1997) for ligament creep $J(t)$ and relaxation $E(t)$, •, replotted on a log-log scale, after Lakes and Vanderby (1999). Use of a nonlinear, non-separable constitutive approach with a quadratic model, ×, or a cubic model, $*$, offers better predictive power (Lakes and Vanderby (1999)).

5.2 Interrelation 2 (Oza et al. (2003))

Lakes and Vanderby (1999) assumed a separable form of creep that gave rise to a non-separable form of relaxation, as seen in Section 5.1. To develop a more robust model, Oza et al. (2003) assumed a non-separable form of creep given as

$$J(t,\sigma) = (g_1 t^n + g_2 \sigma t^m), \tag{14}$$

and a non-separable form of relaxation as

$$E(t,\varepsilon) = (f_1 t^{-n} + f_2 \varepsilon t^{-q}). \tag{15}$$

Inserting eqs (14) and (15) into the Stieltjes integral and solving the equation gives the relations

$$f_1 = \frac{1}{g_1}\frac{\sin(n\pi)}{n\pi}, \quad f_2 = \frac{-m f_1 g_2 \Gamma(-n+1)\Gamma(m)}{n g_1^2 \Gamma(-2n+m+1)\Gamma(n)}, \quad q = 3n - m. \tag{16}$$

So, once the coefficients of creep compliance (g_1, g_2, n, m) are known by curve fitting, eq. (16) can be used to obtain the coefficients of relaxation modulus (f_1, f_2, q). These coefficients of relaxation modulus are inserted into eq. (15) to predict the relaxation response. Results of the interrelation developed using this method are given in Fig. 8(a) and 8(b).

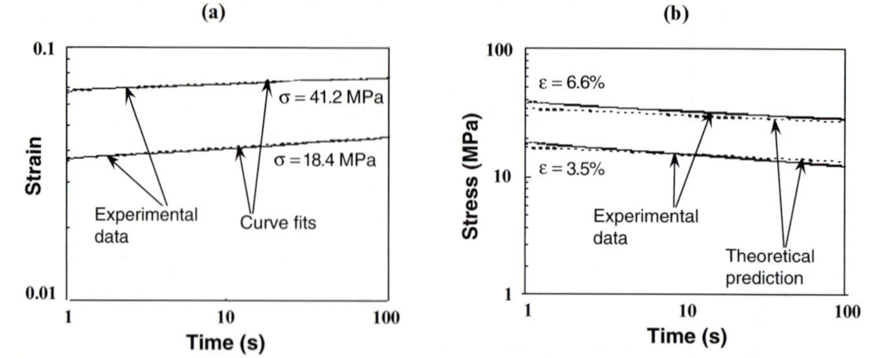

Fig. 8. (a) Curve fitting of creep of rabbit medial collateral ligament at 18.4 MPa and 41.2 MPa using eq. (15). Dense points represent the experimental data and dashed line is the curve fit; (b) prediction of relaxation from creep. Dense points are experimental data points of relaxation at 3.5% and 6.6% and dashed line is the theoretical prediction represented by eq. (16) (Oza et al. (2003)).

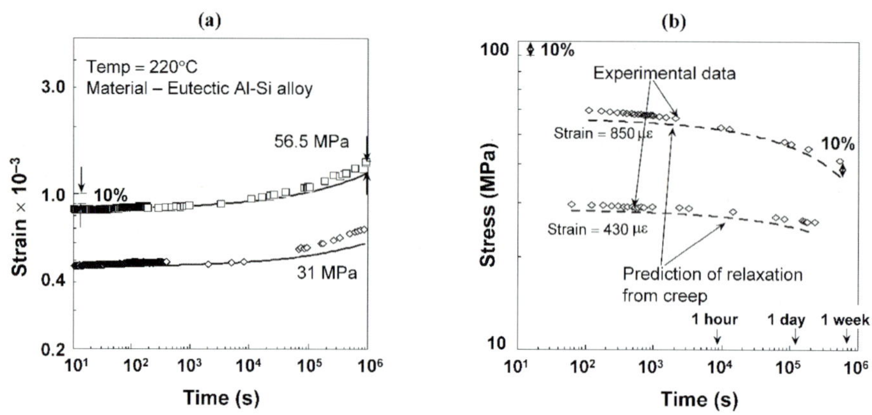

Fig. 9. (a) Two creep curves (31 MPa and 56 MPa) at a constant temperature of 220°C are plotted against time; (b) prediction of relaxation from creep; comparison with experimental relaxation of alloy. The two corresponding strain levels for relaxation are ε_1 (430×10^{-6}) and ε_2 (850×10^{-6}). The points in the figure give the experimental curve, while the dashed lines are theoretical predictions from creep data (Oza et al. (2004)).

5.3 Application of Interrelation to Metals (Oza et al. (2004))

The interrelation technique developed was also applied to metals to demonstrate its broader applicability in other materials. Results lie within 10% of the experimental data, as shown in Fig. 9.

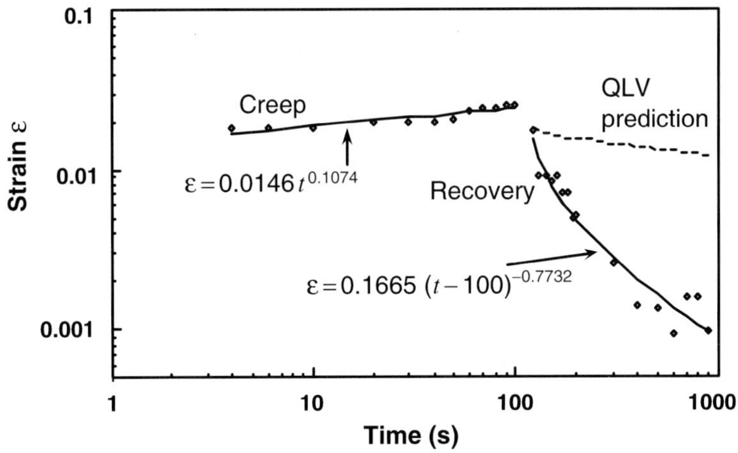

Fig. 10. Rate of recovery for the same specimen is much faster than the rate of creep. QLV predicts the same rate.

6 Recovery

Recovery is the response of a tissue after creep or relaxation when the applied load or strain, respectively, is removed. Preliminary studies (Jensen et al. (2004)) in our laboratory have shown that the rate of tissue recovery after creep in rat medial collateral ligaments is much faster than the rate of creep. Since QLV does not include stress dependence in creep, it would predict the recovery at the same rate as creep. This is shown in Fig. 10.

From a physiological standpoint, these data suggest that MCLs are slower to stretch out and faster to recover to their normal lengths. QLV fails to predict recovery but the single integral nonlinear superposition described in Section 4 remains to be tested.

7 Discussion

The effect of composition and microstructure on ligament viscoelastic behavior is not entirely known and data are sometimes contradictory. Elliott et al. (2003) and Robinson et al. (2004a,b) have shown that glycosaminoglycans contribute to the viscoelastic behavior of soft tissues in their study on mouse tail tendon. They found greater stress relaxation in a group with reduced proteoglycans and their associated glycosaminoglycans (GAGs). Proteoglycans are usually associated with tissue hydration, so presumably these tissues had greater viscosity with lower water content. The presence of fluid in ligament tissue has been consistently shown to affect the mechanical response

(Atkinson (1997), Chimich et al. (1992), Thielke et al. (1995)). Water content in tendon decreases with static and cyclic loading (Hannafin and Arnoczky (1994)). Chimich et al. (1992) showed that ligaments with higher water content demonstrated greater relaxation than ligaments with lower water content, in contrast to the above mouse tail studies. A decrease in relaxation rate here could be interpreted as the result of larger strains causing more fluid to be driven out of the ligament, which causes tissue to be more elastic and less viscous than the tissue subjected to lower strains. However, a recent study by Liao and Vesely (2004) tested stress relaxation on fifty mitral valve chordae tendineae from eight hearts. These chordae tendianeae were of three types: marginal, basal and strut. The water and the elastin content was found to be the same for all three types and collagen content of the marginal chordae was only slightly less than for the basal and strut chordae. The total concentration of glycosaminoglycans decreased in the order of marginal, basal and strut. Strut chordae underwent faster relaxation than basal chordae, which in turn relaxed faster than the marginal chordae. Since water content, collagen, and elastin for all three types were almost the same, the greater number of proteoglycans appear responsible for reduced relaxation. These authors speculate that a greater number of proteoglycan linkages may prevent the slippage of fibrils with respect to each other and thus, reduce the rate of relaxation.

Thornton et al. (1997) speculated that differences between rates of stress relaxation and creep were micro-structural. During creep, collagen fibers are progressively recruited, but this does not occur during stress relaxation. The authors state that this explains the observed decrease in rate of creep with increasing load. As larger loads are applied to the ligament, more fibers are recruited leaving fewer fibers to be progressively recruited after initial loading and therefore decreasing the creep response.

The goal of this work is to illuminate certain aspects of nonlinear viscoelasticity in ligament, which are not revealed by previous studies that do not test the viscoelastic behavior over multiple load levels. Experimental data (Provenzano et al. (2001), Hingorani et al. (2004)) strongly support the concept that creep and relaxation in ligaments are dependent on time and on the level of stress or strain respectively, particularly within the lower levels of physiologic loading. A more general viscoelastic formulation than the QLV is required to model this aspect of ligament behavior. The rates of creep and relaxation approach asymptotic values at higher levels of stresses and strains, respectively, which correspond to the linear region of the stress-strain curve. The separable QLV formulation is adequate to describe the rate of creep or relaxation for stresses or strains, respectively, in this subregion (Provenzano et al. (2001)). Most normal ligament loading occurs at low strains in daily life making nonlinearities in the rate of creep and relaxation physiologically relevant.

Creep and relaxation are interrelated for primary creep described by a sum of power-law terms in time, within the framework of single integral nonlinear superposition. The model in Section 5.1 developed by Lakes and Vanderby (1999) (with a separable form of creep and non-separable form of relaxation) shows that a separable form in creep gives rise to a non-separable form in relaxation. Models in the subsequent sections use non-separable forms of both creep and relaxation to develop an interrelation. The model in Section 5.3 has been applied to interrelate creep and relaxation in aluminum alloys for 6 decades of time in seconds. The prediction of relaxation from creep has been shown to lie within 10% of the experimental data. This ability to predict relaxation from creep is extremely useful considering the time requirements associated with viscoelastic testing. This method of interrelation appears to have broad utility and can be applied to other kinds of nonlinearly viscoelastic materials.

When considering the results of this study the limitations in this work must be taken into consideration. One of the potential limitations of the study is that no preconditioning was done prior to the creep and relaxation tests. But, the following observation suggests that the tissue is relatively insensitive to preconditioning. Repeated creep tests separated by 1000 s of recovery of the MCL at the same load and repeated relaxation tests separated by 1000 s of the MCL at the same strain respectively gave reproducible results (Hingorani et al. (2004)). Second, during video analysis the strain obtained from optical markers is an average strain over the entire length of the ligament between the two ligament insertions. Hence, our strain measure provides an average over the regional variations within the tissue. But, as mentioned in the methods section, using optical markers at the ends of the viewing area increases the digitization for strain. Third, nonlinear superposition integrals represented in eq. (2) are not the most general approach for a nonlinear viscoelastic material. The superposition approach developed in Section 5 is amenable to higher-order expansion of the kernel. This provides a wider window of applicability, but at the cost of additional complexity. This limitation, however, is more than compensated for by the time and cost that may be saved by using analytical interrelations. The single integral form used here may not suffice regardless of the kernel and in that case a multiple integral constitutive equation (Findley et al. (1976)) must be used. Experimental result for creep recovery (Fig. 10) suggests that this may be required for sufficiently complex loadings. Multiple integral constitutive laws may be regarded as general but the experimental program required to characterize the kernel functions rapidly grows in complexity for higher-order terms. Therefore single integral formulations are preferred provided they suffice to model the data and offer sufficient predictive power for loadings of interest. The method for interpreting experiment in multiple integral theory is well developed, and has been performed in polymers (Morgan and Ward (1971)).

The simplicity implicit in the phenomenological viscoelastic model shown herein and analytical interrelationship can then be used to model more robustly many complex viscoelastic materials and to reduce the number of tests required to characterize both creep and relaxation.

Acknowledgements. This research was supported by NSF via Grant CMS 9907977.

References

Abrahams, M. (1967). Mechanical behaviour of tendon in vitro. A preliminary report. *Med. Biol. Eng.* 5:433–443.

Abramowitsch, S. D., and Woo, S. L. (2004). An improved method to analyze the stress relaxation of ligaments following a finite ramp time based on quasi-linear viscoelastic theory. *J. Biomech. Eng.* 126:92–97.

Abramowitsch, S. D., Woo, S. L., Clineff, T. D., and Debski, R. E. (2004). An evaluation of quasi-linear viscoelastic properties of healing medial collateral ligament in a goat model. *Ann. Biomed. Eng.* 32:329–335.

Atkinson, T. S. (1997). A microstructural poroelastic model for patellar tendon. In *Proceedings ASME Bioengineering Conference*. Sunriver, OR.

Best, T. M., McElhaney, J., Garrett Jr., W. E., and Myers, B. S. (1994). Characterization of passive response of live skeletal muscle using quasilinear viscoelastic theory of viscoelasticity. *J. Biomech.* 27:413–419.

Broom, N. D. (1978). Simultaneous morphological and stress-strain studies of the fibrous components in wet heart valve leaflet tissue. *Conn. Tiss. Res.* 6:37–50.

Carew, E. O., Talman, E. A., Boughner, D. R., and Vesely, I. (1999). Quasilinear viscoelastic theory applied to internal shearing of porcine aortic leaflets. *J. Biomech. Eng.* 121:386–392.

Carew, E. O., Barber, J. E., and Vesely, I. (2000). Role of preconditioning and recovery time in repeated testing of aortic valve tissues: Validation through quasilinear viscoelastic theory. *Ann. Biomed. Eng.* 28:1093–1100.

Chimich, D., Shrive, N., Frank, C., Marchuk, L., and Bray, R. (1992). Water content alters viscoelastic behavior of the normal adolescent rabbit medial collateral ligament. *J. Biomech.* 25:831–837.

Demiray, H. (1996). A quasi-linear relation for arterial wall materials. *J. Biomech.* 29:1011–1014.

Diamant, J., Keller, A., Baer, E., Litt, M., and Arridge, R. G. C. (1972). Collagen: Ultrastructure and its relation to mechanical properties as a function of ageing. *Proc. R. Soc. Lond. B* 180:293–315.

Elliott, D. M., Robinson, P. S., Gimbel, J. A., Sarver, J. J., Abboud, J. A., Iozzo, R. V., and Soslowsky, L. J. (2003). Effect of altered matrix proteins on quasilinear viscoelastic properties in transgenic mouse tail tendons. *Ann. Biomed. Eng.* 31:599–605.

Findley, W. N., Lai, J. S., and Onaran, K. (1976). *Creep and Relaxation of Nonlinear Viscoelastic Materials.* Amsterdam: North Holland.

Fung, Y. C. (1972). Stress strain history relations of soft tissues in simple elongation. In Fung, Y. C., Perrone, N., and Anliker, M., eds., *Biomechanics: Its Foundations and Objectives.* New Jersey: Prentice-Hall, Englewood Cliffs.

Funk, J. R., Hall, G. W., Crandall, J. R., and Pilkey, W. D. (2000). Linear and quasilinear viscoelastic characterization of ankle ligaments. *J. Biomech. Eng.* 122:15–22.

Graf, B., Vanderby, R., Ulm, M., Rogalski, R., and Thielke, R. (1994). The effect of preconditioning on the viscoelastic response of primate patellar tendon. *Arthroscopy* 10:90–96.

Hannafin, J. A., and Arnoczky, S. P. (1994). Effect of cyclic and static tensile loading on the water content and solute diffusion in canine flexor tendons: and in vitro study. *J. Orthop. Res.* 12:350–356.

Haut, R. C., and Little, R. W. (1972). A constitutive equation for collagen fibers. *J. Biomech.* 5:423–430.

Hingorani, R., Provenzano, P. P., Lakes, R. S., Escarcega, A., and Vanderby Jr., R. (2004). Nonlinear viscoelasticity in rabbit medial collateral ligament. *Ann. Biomed. Eng.* 32:306–312.

Huyghe, J. M., van Campen, D. H., Arts, T., and Heethaar, R. M. (1991). The constitutive behaviour of passive heart muscle tissue. A quasi-linear viscoelastic formulation. *J. Biomech.* 24:841–849.

Jensen, K. T., Dwyer, K. W., Lakes, R. S., and Vanderby, R. (2004). The rate of viscoelastic recovery is faster than the rate of creep. In *50th Annual Meeting of the Orthopaedic Research Society.* San Francisco, CA, paper 0046.

Johnson, G. A., Tramaglini, D. M., Levine, R. E., Ohne, K., Choi, N. Y., and Y. Woo, S. L. (1994). Tensile and viscoelastic properties of human patellar tendon. *J. Orthop. Res.* 12:796–803.

Johnson, G. A., Livesay, G. A., Y. Woo, S. L., and Rajagopal, K. R. (1996). A single integral finite strain viscoelastic model of ligaments and tendons. *J. Biomech. Eng.* 118:221–226.

Kwan, M. K., C. Lin, T. H., and Y. Woo, S. L. (1993). On the viscoelastic properties of the anteromedial bundle of the anterior cruciate ligament. *J. Biomech.* 26:447–442.

Lakes, R. S., and Vanderby, R. (1999). Interrelation of creep and relaxation: a modeling approach for ligament. *J. Biomech. Eng.* 121:612–615.

Lakes, R. S. (1998). *Viscoelastic Solids.* Boca Raton, Florida: CRC Press.

Lanir, Y. (1980). A microstructure model for the rheology of mammalian tendon. *J. Biomech. Eng.* 102:332–339.

Liao, J., and Vesely, I. (2004). Relationship between collagen fibrils, glycosaminoglycans, and stress relaxation in mitral valve chordae tendineae. *Ann. Biomed. Eng.* 32:977–983.

Manley, E., Provenzano, P., Heisey, D., Lakes, R. S., and Vanderby, R. (2003). Required test duration for group comparisons in ligament viscoelasticity: A statistical approach. *Biorheology* 40:441–450.

Morgan, C. J., and Ward, I. M. (1971). The temperature dependence of nonlinear creep and recovery in oriented polypropylene. *J. Mech. Phys. Solids* 19:164–178.

Oza, A., Vanderby, R., and Lakes, R. S. (2003). Interrelation of creep and relaxation for nonlinearly viscoelastic materials: application to ligament and metal. *Rheol. Acta* 42:557–568.

Oza, A., Jaglinski, T., Vanderby, R., and Lakes, R. S. (2004). Application of nonlinear superposition to creep and relaxation of commercial die-casting aluminum alloys. *Mech. Time-Depend. Mat.* 8:385–402.

Pioletti, D. P., and Rakotomanana, L. R. (2000). On the independence of time and strain effects in the stress relaxation of ligaments and tendons. *J. Biomech.* 33:1729–1732.

Provenzano, P., Lakes, R. S., Keenan, T., and Vanderby, R. (2001). Non-linear ligament viscoelasticity. *Ann. Biomed. Eng.* 28:908–914.

Provenzano, P., Hayashi, K., Kunz, D. N., Markel, M. D., and Vanderby, R. (2002a). Healing of subfailure ligament injury: comparison between immature and mature ligaments in a rat model. *J. Orthop. Res.* 20:975–983.

Provenzano, P., Heisey, D., Hayashi, K., Lakes, R. S., and Vanderby, R. (2002b). Subfailure damage in ligament: A structural and cellular evaluation. *J. Appl. Physiol.* 92:362–371.

Puso, M. A., and Weiss, J. A. (1998). Finite element implementation of anisotropic quasi-linear viscoelasticity using a discrete spectrum approximation. *J. Biomech. Eng.* 120:62–70.

Rigby, B. J., Hirai, N., Spikes, J. D., and Eyring, H. (1959). The mechanical properties of rat tail tendon. *J. Gen. Physiol.* 43:265–289.

Robinson, P. S., Lin, T. W., Reynolds, P. R., Derwin, K. A., Iozzo, R. V., and Soslowsky, L. J. (2004a). Strain-rate sensitive mechanical properties of tendon fascicles from mice with genetically engineered alterations in collagen and decorin. *J. Biomech. Eng.* 126:252–257.

Robinson, P. S., Lin, T. W., Reynolds, P. R., Jawad, A. F., Iozzo, R. V., and Soslowsky, L. J. (2004b). Investigating tendon-fascicle structure- function relationships in a transgenic-age mouse model using multiple regression models. *Ann. Biomed. Eng.* 32:924–931.

Rubin, M. B., Bodner, S. R., and Binur, N. S. (1998). An elastic-viscoplastic model for excised facial tissues. *J. Biomech. Eng.* 120:686–689.

Sarver, J. J., Robinson, P. S., and Elliott, D. M. (2003). Methods of quasilinear viscoelastic modeling of soft tissue: Application to incremental stress-relaxation experiments. *J. Biomech. Eng.* 125:754–758.

Thielke, R. J., Vanderby, R., and Grood, E. S. (1995). Volumetric changes in ligaments under tension. In *Proceedings ASME Bioengineering Conference.* Breckenridge, CO.

Thornton, G. M., Oliynyk, A., Frank, C. B., and Shrive, N. G. (1997). Ligament creep cannot be predicted from stress relaxation at low stresses: a biomechanical study of the rabbit medial collateral ligament. *J. Orthop. Res.* 15:652–656.

Thornton, G. M., Frank, C. B., and Shrive, N. G. (2001a). Ligament creep behavior can be predicted from stress relaxation by fiber recruitment. *J. Rheol.* 45:493–507.

Thornton, G. M., Shrive, N. G., and Frank, C. B. (2001b). Altering ligament water content affects ligament pre-stress and creep behavior. *J. Orthop. Res.* 19:845–851.

Thornton, G. M., Shrive, N. G., and Frank, C. B. (2002). Ligament creep recruits fibres at low stresses and can lead to modulus-reducing fibre damage a higher creep stresses: a study in rabbit medial collateral ligament model. *J. Orthop. Res.* 20:967–974.

Turner, S. (1973). Creep in glassy polymers. In Howard, R. H., ed., *The Physics of Glassy Polymers.* Chichester: John Whiley & Sons.

Viidik, A., Danielsen, C. C., and Oxlund, H. (1982). On fundamental and phenomenological models, structure and mechanical properties of collagen, elastin and gycolsaminoglycan complexes. *Biorheology* 19:437–451.

Viidik, A. (1968). A rheological model for uncalcified parallel-fibered collagenous tissue. *J. Biomech.* 1:3–11.

Viidik, A. (1973). Functional properties of collagenous tissues. *Int. Rev. Conn. Tiss. Res.* 6:127–215.

Woo, S. L., Simon, B. R., Kuei, S. C., and Akeson, W. H. (1980). Quasilinear viscoelastic properties of normal articular cartilage. *J. Biomech. Eng.* 102:85–90.

Woo, S. L., Gomez, M. A., and Akeson, W. H. (1981). The time and history-dependent viscoelastic properties of the canine medial collateral ligament. *J. Biomech. Eng.* 103:293–298.

Woo, S. L. (1982). Mechanical properties of tendons and ligaments I. quasi-static and nonlinear viscoelastic properties. *Biorheology* 19:384–396.

Viscoelastic Constitutive Law Based on the Time Scale of the Mechanical Phenomena

D.P. Pioletti

Ecole Polytechnique Federale de Lausanne (EPFL), Laboratory of Orthopedic Research, Switzerland
dominique.pioletti@epfl.ch

Experimental characterizations of biological soft tissues show that the stress-strain curves are sensitive to the strain rate at which they are performed and relaxation phenomena that are present. Hence, the strain rate is an important variable in the understanding of the mechanical behavior of soft tissues. Many of the constitutive models that have been proposed fail to properly consider this variable for large deformations. The development of a conceptual framework to test different viscoelastic constitutive laws is presented. This framework has the advantage that it satisfies *a priori* thermodynamical restrictions and is valid for large deformations. In addition, the different mechanical contributions are separated according to the time scale of their effects. The constitutive law obtained takes into account the non-linear stress-strain curves with the strain rate as an explicit variable. As an illustration of its ability to model the immediate, short-time memory and long-time memory contributions, the framework is used to identify mechanical tests performed on human ligaments. Good correlations are found between experimental and theoretical stress-strain curves. The identification results then in a realistic viscoelastic constitutive law. The viscoelastic law developed is proposed to model biological tissues exhibiting an important viscoelastic behavior.

1 Mechanical Behaviors According to their Time Scale

The viscoelastic properties of soft tissues play an essential role in the kinematics and biomechanics of the human body. Intervertebral discs, as shock absorbers, can fulfill their function thanks to their viscoelastic properties (see, e.g., Palmer and Lotz (2004)). Ligament or tendon stiffnesses vary accordingly to the strain rate (see, e.g., Danto and Woo (1993) and Pioletti et al. (1999)), allowing an optimal mechanical behavior between physiological joint laxity and stability to be achieved. From a more general point of view, the mechanical behavior of soft tissues can be distinguished according to the time scale of its effects. Uniaxial tension tests performed on a human anterior cruciate ligament (ACL) are displayed in Figs 1, 2 and 3 in Section 3.

It would be interesting to describe with the same constitutive law the contribution of the immediate, the short-time memory and the long-time memory. This approach would facilitate not only the comparison of viscoelastic properties between tissues but also the implementation of the constitutive law in a numerical model. To make the constitutive law as general as possible, it should be valid for large deformations and should satisfy the principles of thermodynamics. We present here the development of a general constitutive law describing, in the same conceptual framework, the immediate, the short-time and the long-time memory contributions. The experimental data are obtained from a PhD thesis (Pioletti (1997)).

2 Time-scale Based Viscoelastic Constitutive Law

We start our description by considering the most general form for a constitutive law (Truesdell and Noll (2004)), namely

$$\boldsymbol{S}(t) = \boldsymbol{S}_e(\boldsymbol{C}(t)) + \Im_{s=0}^{\infty}\{\boldsymbol{G}(t-s);\boldsymbol{C}(t))\}, \tag{1}$$

where $\boldsymbol{S}$ is the second Piola-Kirchhoff stress tensor, $\boldsymbol{C}$ is the right Cauchy-Green deformation tensor, $\Im$ is a functional representing the history of $\boldsymbol{G}(t-s) = \boldsymbol{C}(t-s) - \boldsymbol{C}(t)$, and $\boldsymbol{S}_e(\boldsymbol{C}(t))$ is an equilibrium term. The stress $\boldsymbol{S}$ and deformation tensor $\boldsymbol{C}$ are two symmetric second-order tensors.

The key point in the development of a constitutive law based on the time scale of the mechanical behaviors is to transform eq. (1) by adjusting its time interval (Pioletti and Rakotomanana (2000)) to give

$$\boldsymbol{S}(t) = \boldsymbol{S}_e(\boldsymbol{C}(t)) + \Im_{s=0}^{\delta}\{\boldsymbol{G}(t-s);\boldsymbol{C}(t))\} + \Im_{s=\delta}^{\infty}\{\boldsymbol{G}(t-s);\boldsymbol{C}(t))\}, \tag{2}$$

where $\delta \simeq 0$. The first term on the right hand-side of eq. (2) is the immediate contribution (elastic behavior) since it supports the contribution of the deformation at the actual time t. The second term in eq. (2) represents the short-time memory effect while the third term represents the long-time memory effect. It has been shown that eq. (2) can take the form (Pioletti and Rakotomanana (2000))

$$\boldsymbol{S}(t) = \boldsymbol{S}_e(\boldsymbol{C}(t)) + \boldsymbol{S}_v(\dot{\boldsymbol{C}}(t);\boldsymbol{C}(t)) + \int_{\delta}^{\infty} \Sigma(\boldsymbol{G}(t-s), s; \boldsymbol{C}(t))\mathrm{d}s. \tag{3}$$

The development of eq. (3) has the advantage of taking into account the different mechanical behaviors in one framework and hence facilitates the identification process. Moreover, as this law uses objective stress and strain tensors, the frame invariance of the law is automatically satisfied. To be consistent with thermodynamical principles, the general constitutive law (3) is expressed in term of elastic and dissipative potentials (see, e.g., Germain (1986) and Truesdell and Noll (2004)). Explicitly,

$$\boldsymbol{S}_e(t) = 2\rho_0 \frac{\partial W_e}{\partial \boldsymbol{C}} \quad \text{(immediate contribution)}, \tag{4}$$

$$\boldsymbol{S}_v(t) = 2\rho_0 \frac{\partial W_v}{\partial \dot{\boldsymbol{C}}} \quad \text{(short-time memory)}, \tag{5}$$

$$(\boldsymbol{S} - 2\rho_0 \mathbf{D}\aleph_{s=\delta}^{\infty}\{\boldsymbol{C}(t-s)\}) : \frac{\dot{\boldsymbol{C}}}{2} \geq 0 \quad \text{for all } \boldsymbol{C}, \dot{\boldsymbol{C}} \quad \text{(Long-time memory)}, \tag{6}$$

where $\mathbf{D}$ is a differential operator acting on the functional $\aleph$. For the long-time memory, in the particular case where exponential relaxation functions are used, it can be demonstrated that the thermodynamic principles are satisfied (Rabotnov (1977)).

3 Identification of the Time-scale Based Viscoelastic Constitutive Law

Different elastic and viscous potentials can be tested in the general framework proposed until a satisfactory identification is obtained. In the present work, the identification is performed with uniaxial tension tests performed on human ACL (Pioletti (1997)). In order to obtain a tractable identification, the specimens are considered as isotropic, homogeneous and incompressible.

The potential for the immediate contribution is taken from Veronda and Westmann (1970) as

$$W_e = \frac{\alpha}{2} exp[\beta(I_1 - 3)] - \frac{\alpha\beta}{2}(I_2 - 3), \tag{7}$$

where α and β are two elastic (immediate) parameters, $I_1 = \mathrm{tr}\boldsymbol{C}$ and $I_2 = [(\mathrm{tr}\boldsymbol{C})^2 - \mathrm{tr}(\boldsymbol{C}^2)]/2$ are two invariants of the deformation tensor. Identification with the tension test performed at the lowest strain rate allowed us to determine the elastic parameters (see Fig. 1).

The potential for the short term memory is taken from Pioletti et al. (1998). This is

$$W_v = \frac{\eta'}{4} tr((\dot{\boldsymbol{C}})^2)(I_1 - 3), \tag{8}$$

where η' is the short time memory parameter. Identification with tension tests performed at four different strain rates allowed us to determine the short term memory parameter (see Fig. 2).

Finally, for the long time memory, a normalized exponential Prony series is used (Pioletti and Rakotomanana (2000)), so that

$$\int_{\delta}^{\infty} \Sigma(\boldsymbol{G}(t-s), s; \boldsymbol{C}(t))\mathrm{d}s = \int_{\delta}^{t} \boldsymbol{S}_e(\boldsymbol{C}(t-s))\dot{M}(s)\mathrm{d}s, \tag{9}$$

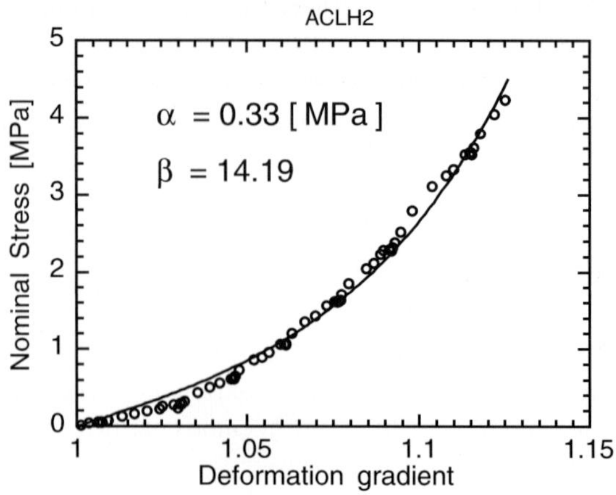

Fig. 1. Experimental uniaxial tension test (dots) on human ACL and analytical results using the elastic potential (7).

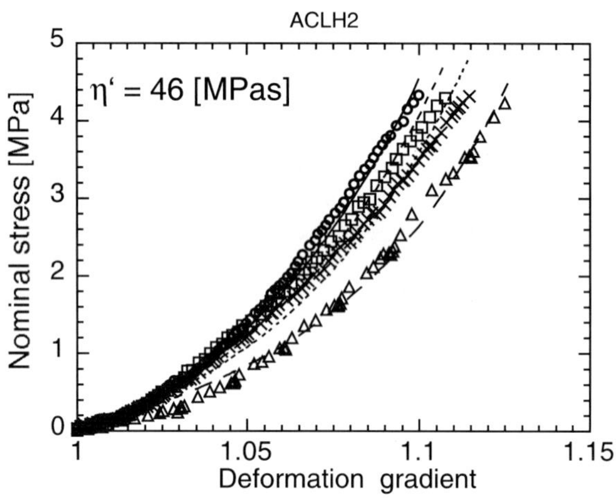

Fig. 2. Experimental uniaxial tension tests (points) performed at 4 different strain rates on human ACL and analytical results (curves) using the viscous potential (8).

with

$$M(s) = \left(\sum_{k=1}^{3} a_k \exp(-s/\tau_k) \right) \bigg/ \left(\sum_{k=1}^{3} a_k \right) . \tag{10}$$

This approach has been shown to satisfy thermodynamical principles as well as the principle of fading memory (Coleman and Noll (1961)). A possible set of six parameters a_k and τ_k is identified with stress relaxation experiments (see Fig. 3). It has to be noted that the choice of the six parameters is not

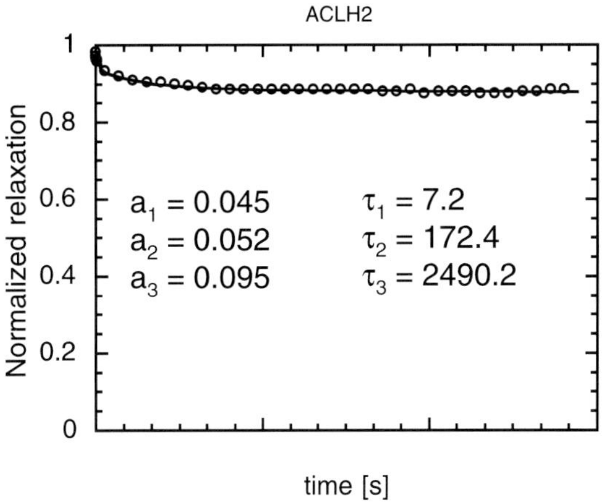

Fig. 3. Experimental stress relaxation test (points) and analytical results (curves) using the Prony series (9) and (10).

unique and alternative sets of six parameters might equally well reproduce the curve.

In summary, the constitutive law developed takes into account immediate, short-time memory and long-time memory contributions and has the form

$$S_e(t) = -pC^{-1} + \alpha\beta(2\exp[\beta(I_1 - 3)] - I_1)I + \alpha\beta C, \quad (11)$$

$$S_v(t) = \eta'(I_1 - 3)\dot{C}, \quad (12)$$

$$\int_\delta^\infty \Sigma(G(t-s), s; C(t))\mathrm{d}s = \int_\delta^t S_e(C(t-s))\dot{M}(s)\mathrm{d}s. \quad (13)$$

The presence of the term $-pC^{-1}$ is due to the kinematic constraint of incompressibility used.

4 Advantages and Limitations of the Time-scale Based Constitutive Law

In the constitutive law presented, the main limitation is the isotropic assumption. It is well known that soft tissues such as ligaments present a transverse isotropic symmetry. However, the developed constitutive law based on time scale eq. (2) is general enough to incorporate different symmetries. The limitation to isotropy followed in this study was more due to the one-dimensional experimental identification. Indeed, performing mechanical tests in the transversal direction of ligaments or tendons is a challenge. As highlighted in the identification process, the separation of the different mechanical behaviors according to their time scale is an advantageous process. It also facilitates the incorporation of resulting constitutive laws in numerical codes.

It has to be mentioned that the approach followed is phenomenological. This kind of description does then not furnish insight to the nature or origin of the viscoelastic phenomena in soft tissues. For such a description, poroelastic descriptions may be more appropriate. But it has to be clarified that finally any descriptions will be considered as phenomenological compared to a description taking more phenomena into account.

Probably the major advantage of the proposed description is that it satisfies *a priori* objectivity and thermodynamical requirements. The description developed could be used to model the viscoelastic behavior of different tissues such as bone or intervertebral discs. This represents the next step of our work.

References

Coleman, B. D., and Noll, W. (1961). Foundations of linear viscoelasticity. *Rev. Modern Phys.* 3:239–249.

Danto, M. I., and Woo, S. L.-Y. (1993). The mechanical properties of skeletally mature rabbit anterior cruciate ligament and patellar tendon over a range of strain rates. *J. Orthop. Res.* 11:58–67.

Germain, P. (1986). *Mecanique – Tome 1-2.* Ellipses.

Palmer, E. I., and Lotz, J. C. (2004). The compressive creep properties of normal and degenerated murine intervertebral discs. *J. Orthop. Res.* 22:164–169.

Pioletti, D. P., and Rakotomanana, L. R. (2000). Non-linear viscoelastic laws for soft biological tissues. *Eur. J. Mech. A/Solids* 19:749–759.

Pioletti, D. P., Rakotomanana, L. R., Benvenuti, J.-F., and Leyvraz, P.-F. (1998). Viscoelastic constitutive law in large deformations: Application to human ligaments and tendons. *J. Biomech.* 31:753–757.

Pioletti, D. P., Rakotomanna, L. R., and Leyvraz, P. F. (1999). Strain rate effect on the mechanical behavior of the anterior cruciate ligament-bone complex. *Med. Eng. Phys.* 21:95–100.

Pioletti, D. P. (1997). *Viscoelastic properties of soft tissues: application to knee ligaments and tendons.* Ph.D. Dissertation, Department of Physics, Ecole Polytechnique Federale de Lausanne.

Rabotnov, Y. C. (1977). *Elements of Hereditary Solid Mechanics.* Mir.

Truesdell, C., and Noll, W. (2004). In Antman, S. S., ed., *The Non-linear Field Theories of Mechanics.* Berlin: Springer-Verlag. 3rd edition.

Veronda, D. R., and Westmann, R. A. (1970). Mechanical consideration of skin-finite deformation. *J. Biomech.* 3:111–124.

A Coupled FE Analysis of the Intervertebral Disc Based on a Multiphasic TPM Formulation

W. Ehlers, B. Markert, N. Karajan, A. Acartürk

University of Stuttgart, Institute of Applied Mechanics, Germany
ehlers@mechbau.uni-stuttgart.de, markert@mechbau.uni-stuttgart.de
karajan@mechbau.uni-stuttgart.de, acar@mechbau.uni-stuttgart.de

In general, charged hydrated soft biological tissues exhibit distinct dissipative properties resulting from hydraulic and osmotic phenomena, which are strongly coupled with the intrinsic viscoelastic behaviour of the extracellular matrix. As naturally grown materials, their inner structures and thus, the associated physical properties, are inhomogeneously distributed over the 3D anatomic shape of the tissue. In addition, anisotropic properties may occur due to embedded collagen fibers and orientation-dependent hydraulic and electrolytic conductivities. From a biological perspective, a combination of all these difficulties can be encountered in the human intervertebral disc (IVD). On the microscale, the IVD is a complex arrangement of mostly ionized water (interstitial fluid), collagen arranged in fiber bundles (anulus fibrosus) as well as proteoglycans and proteins forming a charged solid meshwork. Accordingly, a macroscopic mixture approach provides a convenient strategy for the description. In particular, the Theory of Porous Media (TPM) frames a well-founded theoretical basis for the consistent treatment of multiphasic continua with internal interactions. Moreover, proceeding from the effective stress principle, the intrinsic, viscoelastic and anisotropic properties of the extracellular matrix can be easily included. An efficient numerical implementation of the governing strongly coupled multi-field formulation can be realized within the FE tool PANDAS, which is especially designed for the treatment of porous media problems. The overall applicability of the TPM macro model to soft biological tissues will be shown by a 3D simulation of the IVD under physiological loading conditions, which reveals the essential biomechanical functioning of the disc.

1 Introduction

The second European survey on working conditions conducted by Paoli (1996) reported an overall average of 30% of the people in Europe suffering from low back pain. This significant number causes not only a substantial increase in health-care costs, but also an increased research interest in numerical simulation techniques aimed at understanding the complex behaviour of the spine and especially the IVD, aiding the design of new implants. The Finite Element Method (FEM) in combination with an elaborated model has the potential to be the best choice.

The biomechanical challenge lies within the structural complexity of the spine itself arising from a great individuality of the patient data with an irregular and hard to determine 3D anatomical shape. Currently, most models that describe the IVD as a soft biological tissue are restricted to single-phase materials, which are not capable of describing the electro-chemomechanical behaviour of charged hydrated tissues; see, e.g., Eberlein et al. (2001, 2004) for references. Moreover, the available models are often limited to small strains or to finite strains but purely elastic skeleton deformations without considering any of the inhomogeneities that occur; see Klisch and Lotz (1999) and Elliott and Setton (2000). In fact, these types of naturally grown materials undergo large viscoelastic deformations, where the extracellular matrix exhibits so-called intrinsic viscoelastic properties (Iatridis et al. (1996), Ehlers and Markert (2001)). These flow-independent viscoelastic effects are strongly coupled with the dissipative phenomena resulting from the interstitial fluid flow and the electrochemical (osmotic and electrostatic) swelling mechanisms. Furthermore, their inner structure and thus, the associated physical properties are distributed inhomogeneously over the 3D anatomic shape of the tissue. In addition, anisotropic properties may occur due to embedded collagen fibers and orientation-dependent hydraulic and electrolytic conductivities. Another difficulty arises in the application of the numerical model. It is often difficult to obtain reliable experimental data to fit the material parameters needed and to define realistic boundary conditions for boundary-value problems (BVP).

The goal of this contribution is to present a consistent theoretical framework based on the Theory of Porous Media (TPM) (see Bowen (1980), Mow et al. (1989), de Boer (2000) and Ehlers (2002)), where the constitutive assumptions, necessary to cover all relevant tissue properties, are merged in a modular manner. This actually allows for a concise representation of the resulting system of coupled partial differential equations, which can be efficiently treated within the FEM and thus, enables the numerical solution of even large 3D problems at suitable computational costs. Therein, special attention is given to the inhomogeneous fiber reinforcement that arises as well as to the consideration of swelling phenomena.

2 Tissue Properties

2.1 Structural Composition of the IVD

The intervertebral disc is a very sophisticated cushion-like structure embedded in between two adjacent vertebrae constituting 20–30% of the entire spine height. The structural integration of the IVD within the spine provides the capability of load transmission between vertebral bodies, shock absorption and limited vertebral motion.

On the macroscale, the intervertebral disc is composed of three significant substructures. In the center of the disc, a gel structure is found, known as the nucleus pulposus (NP), which exhibits a fluid-like behaviour and covers about 30–50% of the cross sectional area (Iatridis et al. (1996)). The portion of the IVD that becomes gradually prominent within the periphery of the NP forms the outer boundary, known as the anulus fibrosus (AF). At this, the AF appears as a laminate structure surrounding the NP. Marchand and Ahmed (1990) observed that the anulus consists of 15 to 26 layers of discontinuous concentric lamellae, each having embedded string-like structures, known as the collagen fibers. This natural fiber reinforcement runs obliquely around the AF, whereas the direction of the fibers is alternating within successive layers. The upper and lower faces of the IVD are covered with a layer of hyaline cartilage, known as the cartilaginous end plates (CEP).

2.2 Porous Multi-component Microstructure

On the microscale, the biochemical composition of the IVD is rather complex. Generally, the intervertebral disc is composed of a charged hydrated extracellular matrix consisting of proteoglycans (PGs), glycosaminoglycans (GAGs) and embedded collagen fibers as well as an ionized interstitial fluid including dissolved ions and cations. Besides these free movable solutes, there are further chemically active negative charges attached to the long chained molecules (i.e. PGs and GAGs) within the solid matrix. According to Ayad and Weiss (1987), the water content is highest within the NP, reaching approximately 70–90%, and decreases towards the outer periphery of the AF. The same distribution holds for the PG and GAG content, which reaches the highest percentage within the NP, namely 10–20% of the overall dry weight. Hence, the fixed charges density (FCD) is also highest in the center of the IVD, which corresponds to measurements performed by Iatridis et al. (2003) and Urban and Holm (1986). However, a different distribution is observed for the ubiquitous collagen. Out of the several different types of collagen, there are two main groups to be found within the IVD. Firstly, this is collagen of type I, which is well organized and forms parallel fiber bundles within the AF, having a diameter of approximately 2–10 μm. In contrast, collagen fibers of type II are of a much smaller diameter, ranging from 20–200 nm, and form a loose network of fibrils mostly present in the nucleus pulposus, see Mow and Hayes (1997) and references therein. However, there is no sharp border in the occurrence of the two types of collagen between the NP and AF, but rather a smooth transition.

2.3 Inhomogeneities Within the IVD

The intervertebral disc is not uniform in many of its aspects. Firstly, there is the inhomogeneous distribution of collagen type I, which yields a varying alignment within the lamellae of the AF. This variation can be expressed in terms of a fiber angle φ_0 between the fiber and the axial direction of the spine. Holzapfel et al. (2005) measured an anterior fiber angle $\varphi_0 = 66.8°$ as well as a posterior angle of $\varphi_0 = 43.4°$ and discovered an almost linear relationship between the two angles in circumferential direction. However, this linearity stays constant in the radial direction. Furthermore, Marchand and Ahmed (1990) found a decrease of the embedded collagen fibers of type I as one moves towards the NP. Secondly, there is a variation concerning the discontinuities in the lamellar structure of the AF. A detailed study was performed by Tsuji et al. (1993), elaborating the structural variation of the anterior and posterior parts of the AF in terms of the number of continuous and discontinuous fiber bundles. At this, all anterior parts consist of mostly complete fiber bundles, whereas the posterior part is composed of a very complex structure having several incomplete and discontinuous lamellar bundles.

The differences in continuity as well as in the number of embedded fiber bundles finally leads to an inhomogeneous mechanical behaviour of the lamellae, when pulled in fiber direction. Hence, it is the superimposed effect of the structural inhomogeneities that finally predicts the stress-strain relationship of the fibers at any point in the AF. In a comprehensive study, Holzapfel et al. (2005) measured these mechanical variations by performing tensile tests in the fiber direction on single lamellae taken from different regions of the AF. The results show a non-linear exponential stress-strain behaviour of the collagen fibers, where the overall stiffness of fibers found within the periphery of the AF is about three times greater than the values found close to the NP. Moreover, there is also a variation in the circumferential direction, thus classifying the posterior part as being slightly softer than the anterior parts. This behaviour is in good accordance with the qualitative findings of Skaggs et al. (1994) and Ebara et al. (1996).

Another inhomogeneity of the IVD concerns the hydraulic permeability, which is a measure of the ease of fluid flow through a porous medium. For the case of a non-degenerated IVD, Gu et al. (1999) found direction-dependent hydraulic permeabilities yielding different values in the axial, circumferential and radial directions. Therein, an average *Darcy* permeability of about 2×10^{-11} m/s of IVD tissue corresponds to the permeability of a silt seal unit within an embankment and reveals the approximate impermeability of the IVD. However, in the course of disc degeneration, this direction dependency vanishes.

2.4 Coupled Dissipative Behaviour

Overall, there are three superimposed effects characterizing the time-dependent deformation behaviour of the IVD. In particular, these are the intrinsic viscoelastic extracellular matrix, the interstitial fluid flow and a swelling phenomena. According to Hayes and Bodine (1978), Mak (1986) and Iatridis et al. (1996) as well as Ehlers and Markert (2001), the extracellular matrix within soft biological tissues generally behaves viscoelastically.

Moreover, an applied load causes a creep deformation due to the viscous pore-fluid flow within the porous IVD. Since the fluid flow is driven by the overall pore-pressure gradient, it is also coupled to the osmotic pressure, which again is coupled with the deformation process through the FCD.

3 Theory of Porous Media (TPM)

3.1 Basic Concept and Kinematics

The TPM is a macroscopic continuum theory, which is based on the theory of mixtures extended by the concept of volume fractions (see, for example, Bowen (1980), Mow et al. (1989), de Boer (2000) and Ehlers (2002)). Proceeding from a binary mixture consisting of solid and fluid constituents φ^α denoted by $\alpha = \{S, F\}$, the solid phase is extended by incorporating the volume-free fixed negative charges φ^{fc} (see Fig. 1). Furthermore, the interstitial fluid φ^F is composed of water as a liquid solvent φ^L and mobile electrolytes (Na^+Cl^-) as solutes φ^+ and φ^-. In this context, it is assumed that all constituents φ^α are in a state of ideal disarrangement. Following this, the prescription of a real or a virtual averaging process over a representative elementary volume (REV) leads to a model of superimposed and interacting continua, i.e. the homogenized or 'smeared' model.

Particularly in the case of soft biological tissues, the tiny mobile ions are assumed to diffuse rapidly through the liquid and by themselves do not give rise to concentration gradients. Hence, the intervertebral disc is always immediately in electrochemical equilibrium, so that the arising osmotic properties can be sufficiently described without considering the ion concentrations (or electrochemical potentials) as additional unknowns (degrees of freedom) of the process (Lanir (1987)). Moreover, the volume fractions $n^\alpha = \mathrm{d}v^\alpha/\mathrm{d}v$ (partial volume v^α per bulk volume v) serve as describing physical quantities consistent with the saturation constraint

$$\sum_\alpha n^\alpha = n^S + n^F = 1 \quad \text{with} \quad n^{fc} \approx 0. \tag{1}$$

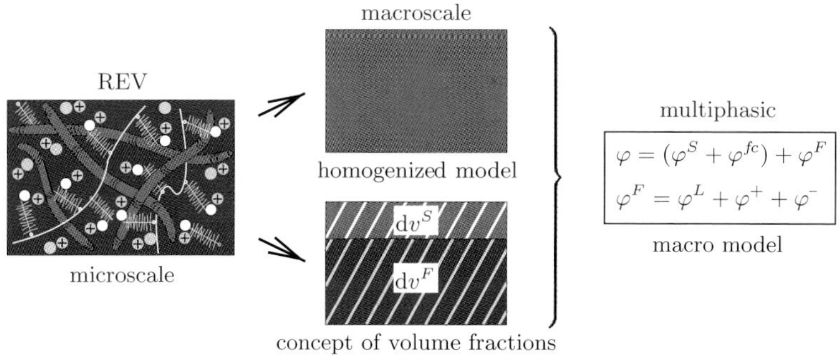

Fig. 1. Multicomponent TPM model of charged hydrated biological tissues.

Exploiting the definition of the volume fractions n^α, a partial density $\rho^\alpha = \mathrm{d}m^\alpha/\mathrm{d}v$ (constituent mass m^α) and a material or realistic density $\rho^{\alpha R} = \mathrm{d}m^\alpha/\mathrm{d}v^\alpha$ can be defined for each constituent φ^α, which are related to each other via $\rho^\alpha = n^\alpha \rho^{\alpha R}$. Note in passing that the property of material incompressibility (defined by $\rho^{\alpha R}$ = constant) does not lead to a macroscopic incompressibility, as the partial density ρ^α can still change through changes in the volume fraction n^α.

In the framework of superimposed continua, there is each spatial point $\mathbf{x}$ of the current configuration at any time t simultaneously occupied by material points P^α of all constituents. Following this, all particles proceed from different reference positions $\mathbf{X}_\alpha$, which leads to individual motion and velocity fields for each constituent, reading $\mathbf{x} = \boldsymbol{\chi}_\alpha(\mathbf{X}_\alpha, t)$ and $\overset{\prime}{\mathbf{x}}_\alpha = \mathrm{d}\boldsymbol{\chi}_\alpha(\mathbf{X}_\alpha, t)/\mathrm{d}t$, respectively. Therein, $(\,\cdot\,)'_\alpha$ denotes the material time derivative following the motion of φ^α. Describing multiphasic materials, it is convenient to express the solid in a *Lagrange*an setting via the displacement vector $\mathbf{u}_S = \mathbf{x} - \mathbf{X}_S$ and the overall fluid in a modified *Euler*ian setting relative to the deforming solid skeleton through the seepage velocity $\mathbf{w}_F = \overset{\prime}{\mathbf{x}}_F - \overset{\prime}{\mathbf{x}}_S$. Furthermore, the fixed charges are assumed to move with the solid skeleton described by the solid deformation gradient given through $\mathbf{F}_S = \partial\mathbf{x}/\partial\mathbf{X}_S$.

3.2 Balance Relations

The model under consideration is assumed to be composed of materially incompressible constituents, which are incapable of an internal mass production or exchange, thereby excluding biochemical reactions and growth processes. Moreover, assuming quasi-static conditions and excluding thermal effects yields the corresponding volume, concentration and momentum balances of the constituents φ^α (see Ehlers et al. (2002)):

$$
\begin{aligned}
0 &= (n^\alpha)'_\alpha + n^\alpha \operatorname{div} \overset{\prime}{\mathbf{x}}_\alpha, \\
0 &= (n^F c_m^{fc})'_S + n^F c_m^{fc} \operatorname{div} \overset{\prime}{\mathbf{x}}_S, \\
\mathbf{0} &= \operatorname{div} \mathbf{T}^\alpha + \rho^\alpha \mathbf{b} + \hat{\mathbf{p}}^\alpha.
\end{aligned}
\tag{2}
$$

Herein, c_m^{fc} is the molar concentration of the fixed charges, $\mathbf{T}^\alpha$ is the partial *Cauchy* stress tensor and $\rho^\alpha \mathbf{b}$ is the body force. Furthermore, $\hat{\mathbf{p}}^\alpha$ represents the momentum production term satisfying the conservation constraint $\hat{\mathbf{p}}^S + \hat{\mathbf{p}}^F = \mathbf{0}$. Combining the specific volume and momentum balances for the solid and the fluid phase yields the respective balances of the overall mixture as

$$
\begin{aligned}
0 &= \operatorname{div} [(\mathbf{u}_S)'_S + n^F \mathbf{w}_F], \\
\mathbf{0} &= \operatorname{div} (\mathbf{T}^S + \mathbf{T}^F) + (n^S \rho^{SR} + n^F \rho^{FR})\, \mathbf{b},
\end{aligned}
\tag{3}
$$

where $\mathbf{T} := \mathbf{T}^S + \mathbf{T}^F$ defines the overall stress of the model. Moreover, the specific volume balance of the solid skeleton as well as the concentration balance of the fixed charges can be analytically integrated from initial quantities $(\,\cdot\,)_{0S}$ leading to

$$n^S = n^S_{0S} \det \mathbf{F}_S^{-1},$$

$$c^{fc}_m = c^{fc}_{0S}\, n^F_{0S}\, (\det \mathbf{F}_S - n^S_{0S})^{-1}.labelDoblare-eq:fcd \tag{4}$$

3.3 Constitutive Equations

In order to close the materially independent set of equations for the extended two phase model, constitutive assumptions for the partial *Cauchy* stress tensors $\mathbf{T}^\alpha$ as well as the interaction term $\hat{\mathbf{p}}^F = -\hat{\mathbf{p}}^S$ need to be defined. From general thermodynamical considerations (Ehlers (1993)), these assumptions read

$$\mathbf{T}^\alpha = -n^\alpha\, p\, \mathbf{I} + \mathbf{T}^\alpha_E, \quad \hat{\mathbf{p}}^F = p \operatorname{grad} n^F + \hat{\mathbf{p}}^F_E. \tag{5}$$

Therein, $\mathbf{I}$ is the identity tensor and $(\,\cdot\,)_E$ denotes the so-called extra terms governed by the solid deformation state and the fluid viscosity. As is usual in hydraulics, the fluid extra stress can be neglected, i.e. $\mathbf{T}^F_E \approx \mathbf{0}$, which can be deduced from a dimensional analysis. Furthermore, p can be interpreted as the entire pore pressure, which introduces a third independent scalar field variable. Thus, an extra equation needs to be generated, which is achieved by inserting the constitutive assumption $\hat{\mathbf{p}}^F_E = -(n^F)^2\, \gamma^{FR} (\mathbf{K}^F)^{-1}\, \mathbf{w}_F$ into the specific fluid momentum balance. This leads to *Darcy*'s well-known filter law

$$n^F \mathbf{w}_F = \frac{1}{\gamma^{FR}}\, \mathbf{K}^F\, \mathbf{i} \quad \text{with} \quad \mathbf{i} = -(\operatorname{grad} p - \rho^{FR}\, \mathbf{b}), \tag{6}$$

relating the hydraulic gradient $\mathbf{i}$ to the filter velocity $n^F \mathbf{w}_F$. Herein, $\mathbf{K}^F$ is the *Darcy* permeability tensor and $\gamma^{FR} = \rho^{FR} g$ is the effective fluid weight ($g = |\,\mathbf{b}\,|$ is the gravitation constant).

However, the entire pore pressure $p = \mathcal{P} + \pi$ introduced is constitutively decomposed into hydraulic and osmotic contributions, respectively. Herein, in the sense of continuum thermodynamics, $\mathcal{P}$ is understood as a *Lagrange*an multiplier, whereas π is directly related to the deformation of the solid skeleton via the molar concentration change c^{fc}_m. Due to the assumption of Lanir (1987), chemical equilibrium expressed by the *Donnan* equation is not limited to the domain boundary. Following this and exploiting the electroneutrality condition for the internal and external solutions (Huyghe and Janssen (1997)) yields the local colloid osmotic pressure (*Donnan* osmosis) expressed through

$$\pi = 2R\Theta \left[\sqrt{\bar{c}^2_m + (c^{fc}_m/2)^2} - \bar{c}_m \right], \tag{7}$$

where R is the universal gas constant, Θ is the absolute temperature and $\bar{c}_m$, in this context, denotes the molar concentration of the monovalent external salt solution.

The final constitutive assumption concerns the solid *Cauchy* extra stress tensor $\mathbf{T}_E^S$ describing the purely mechanical contribution of the extracellular matrix. Proceeding from the evaluation of the entropy or dissipation inequality, the following relation holds:

$$\boldsymbol{\tau}_E^S = 2\,\mathbf{F}_S\,\frac{\partial W^S}{\partial \mathbf{C}_S}\,\mathbf{F}_S^T \qquad \text{with} \qquad \mathbf{T}_E^S = \det \mathbf{F}_S^{-1}\,\boldsymbol{\tau}_E^S. \tag{8}$$

Therein, $\boldsymbol{\tau}_E^S$ is the extra solid *Kirchhoff* stress, while $\mathbf{C}_S = \mathbf{F}_S^T\mathbf{F}_S$ is the right *Cauchy-Green* deformation tensor of the solid skeleton. Hence, it is possible to capture the intrinsic viscoelastic and anisotropic effects in terms of an admissible Ansatz for the solid strain-energy function W^S. This is achieved by a constitutive split of W^S into purely isotropic (PGs, GAGs, etc.) and anisotropic (oriented collagen of type I) contributions (Holzapfel and Weizsäcker (1998)), viz.

$$W^S = W^S_{\mathrm{ISO}} + W^S_{\mathrm{ANISO}} \qquad \text{with} \qquad W^S_{\mathrm{ISO}} = W^S_{\mathrm{EQ}} + W^S_{\mathrm{NEQ}}\,. \tag{9}$$

Herein, the isotropic part can be further split into equilibrium (EQ) and non-equilibrium (NEQ) contributions yielding a finite viscoelastic material law based on a generalized *Maxwell* model, as is described in Ehlers et al. (2002) and Markert (2005). However, the focus of this paper lies in the description of the osmotic swelling pressure, the influence of the collagen fibers and their inhomogeneous distribution. Therefore, only a purely elastic compressible *neo-Hooke*an material law is used for W^S_{ISO}, reading

$$W^S_{\mathrm{ISO}} = W^S_{\mathrm{EQ}} = \tfrac{1}{2}\,\mu_0^S\,(\mathrm{I}_1 - 3) + U^S(J_S), \tag{10}$$

where μ_0^S is the classical ground-state shear modulus, $\mathrm{I}_1 = \mathrm{tr}\,(\mathbf{C}_S)$ is the first principal invariant of $\mathbf{C}_S$, and $J_S = \det \mathbf{F}_S$ is the solid *Jacobi*an. In accordance with Eipper (1998), the volumetric response function $U^S(J_S)$ characterizes the macroscopic compressibility of the porous material and is given by

$$U^S(J_S) = \frac{\Lambda_0^S}{\gamma\left(\gamma - 1 + \dfrac{1}{(1-n_{0S}^S)^2}\right)} \times \left(J_S^\gamma - 1 - \gamma\,\ln\frac{J_S - n_{0S}^S}{1 - n_{0S}^S} + \gamma\,n_{0S}^S\,\frac{J_S - 1}{1 - n_{0S}^S}\right) \tag{11}$$

with the convexity constraint $\gamma \geq 1$ and the second *Lamé* constant Λ_0^S.

Moreover, the anisotropic contribution has to describe a locally orthotropic fiber reinforcement, since the fibers within neighbouring layers are mechanically equivalent. Here, an Ansatz according to Holzapfel et al. (2000) is chosen as

$$W^S_{\mathrm{ANISO}} = \frac{\beta_1}{2\,\beta_2}\{\,\exp\,[\,\beta_2(\mathbf{a}_0\cdot\mathbf{C}_S\,\mathbf{a}_0-1)^2\,] \\ +\exp\,[\,\beta_2(\mathbf{b}_0\cdot\mathbf{C}_S\,\mathbf{b}_0-1)^2\,]-2\}\,, \tag{12}$$

where β_1 and β_2 denote material parameters, and $\mathbf{a}_0$ and $\mathbf{b}_0$ represent the unit fiber vectors defined in the undeformed state. Note in passing that the mixed invariants $\mathrm{I}_4 = \mathbf{a}_0\cdot\mathbf{C}_S\,\mathbf{a}_0$ and $\mathrm{I}_6 = \mathbf{b}_0\cdot\mathbf{C}_S\,\mathbf{b}_0$ describe the squared stretches in the fiber directions and, hence, have a clear physical interpretation (see Spencer (1984)). Like in rope theory, the collagen fibers are assumed to carry loads only under tension. Therefore, the restriction $\{\mathrm{I}_4,\mathrm{I}_6\}\geq 1$ must hold, which simultaneously satisfies restrictions stemming from polyconvexity considerations (Schröder and Neff (2003), Holzapfel et al. (2004)).

4 Numerical Modelling and Example

For the numerical treatment within the FEM, weak forms of the governing multi-field eqs (3) are required. Eliminating the seepage velocity by use of the *Darcy* filter law (6) yields a concise representation of the respective balance relations by means of the primary variables $\mathbf{u}_S$ and p. Subsequently, weighted by independent test functions and integrated over the spatial domain Ω with the surface $\partial\Omega$, one finds the weak forms of the mixture volume (MV) and the mixture momentum (MM) balances:

$$\mathcal{G}_{\mathrm{MV}} \equiv \int_\Omega \delta p\,\mathrm{div}\,(\mathbf{u}_S)'_S\,\mathrm{d}v + \int_\Omega \mathrm{grad}\,\delta p\cdot\frac{k^F}{\gamma^{FR}}\,(\mathrm{grad}\,p-\rho^{FR}\,\mathbf{b})\,\mathrm{d}v \\ +\int_{\partial\Omega}\delta p\,\bar{q}\,\mathrm{d}a = 0,$$

$$\mathcal{G}_{\mathrm{MM}} \equiv \int_\Omega (\mathbf{T}^S_E - p\,\mathbf{I})\cdot\mathrm{grad}\,\delta\mathbf{u}_S\,\mathrm{d}v - \int_\Omega (\rho^S+\rho^F)\,\mathbf{b}\cdot\delta\mathbf{u}_S\,\mathrm{d}v \\ -\int_{\partial\Omega}\bar{\mathbf{t}}\cdot\delta\mathbf{u}_S\,\mathrm{d}a = 0. \tag{13}$$

Therein, $\delta\mathbf{u}_S$ and δp are the corresponding test functions, $\bar{q}$ denotes the efflux of the interstitial fluid over the surface $\partial\Omega$, and $\bar{\mathbf{t}} = (\mathbf{T}^S_E - p\,\mathbf{I})\,\mathbf{n}$ is the external load vector acting on the entire mixture, where $\mathbf{n}$ is the outward oriented unit surface normal. Note in passing that the spatial discretization has to be carried out using stable mixed finite elements, such as the *Taylor-Hood* elements with a quadratic approximation of $\mathbf{u}_S$ and a linear approximation of p.

Furthermore, the numerical approximation of eq. (12) demands discrete fiber vectors $\mathbf{a}_0$ and $\mathbf{b}_0$ to be computed initially at every *Gauss* point ($\mathcal{GP}$) of the mixed finite elements representing the AF. Due to the irregular geometry of the IVD, it is rather inconvenient to define a closed mathematical distribution function. Hence, the fiber vectors are defined with an algorithm based on tangential planes, which are computed at points resulting from a $\mathcal{GP}$ projection onto the outer surface of the IVD. As depicted in Fig. 2, the fibers are then distributed with a circumferentially varying fiber angle φ_0. This is achieved by a linear regression over the polar angle α with respect to the center of gravity (COG), as is described in Eberlein et al. (2004) and Holzapfel et al. (2005).

Moreover, the overall stiffness variation of the collagen fibers has to be considered locally at every $\mathcal{GP}$. At this point, the anisotropic stress contribution stemming from eq. (12) is fitted to trilinear material laws of four distinct regions, which are given by Eberlein et al. (2004). In particular, these are the anterior inner (AI) and outer (AO) as well as the posterolateral inner (PI) and outer (PO) regions (see Fig. 3). Furthermore, a linear distribution of the respective material parameters β_1 and β_2 is assumed according to Fig. 3, which can be explained by changes in the morphological microstructure of the lamellae discussed by Tsuji et al. (1993).

The initial FCD distribution is modelled in a simplified fashion using a constant concentration of $c_{0S}^{fc} = 0.10\,\text{mol/l}$ within the AF and of $c_{0S}^{fc} = 0.26\,\text{mol/l}$ within the NP. Accordingly, eq. (7) together with a constant concentration of

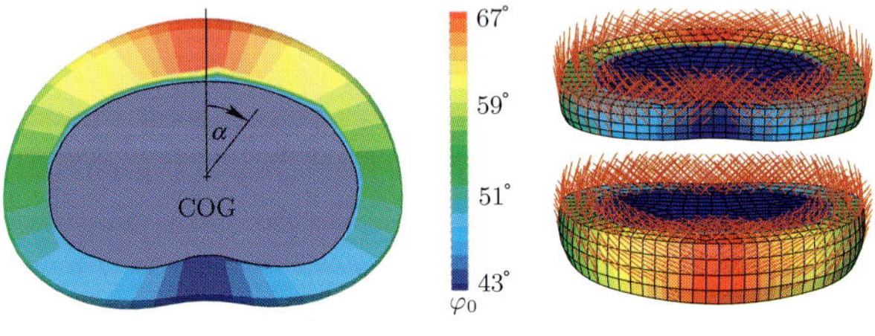

Fig. 2. Distribution of the varying fiber angle φ_0 depending on the polar angle α.

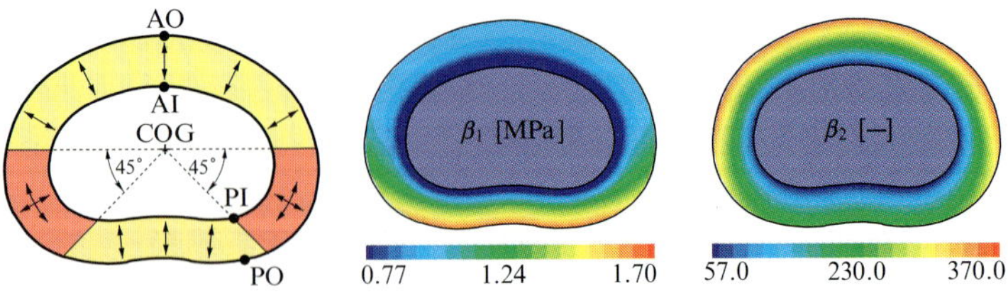

Fig. 3. Assumed distribution of the varying fiber stiffness due to the lamellar microstructure.

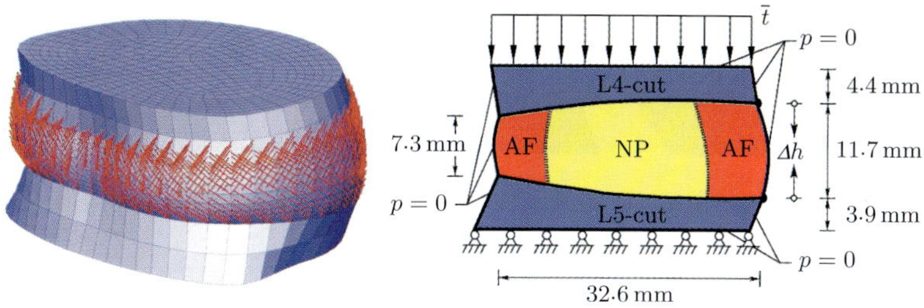

Fig. 4. Uniaxial compression test, where $\bar{t} = 0.33\,\mathrm{N/mm^2}$ is linearly applied within 1 s.

$\bar{c}_m = 0.15\,\mathrm{mol/l}$ of the external solution causes an initial osmotic pressure of $\pi_0 = 0.04\,\mathrm{MPa}$ and of $\pi_0 = 0.24\,\mathrm{MPa}$ in the respective regions. Proceeding from an overall stress-free reference configuration ($\mathbf{T} = \mathbf{0}$), eq. $(5)_1$ yields a prestressed solid skeleton, i. e. $\mathbf{T}^S_{E0} = \pi_0\,\mathbf{I}$.

In order to compare the influence of the above described inhomogeneities, eqs (13) are implemented in the FE tool PANDAS[1]. The corresponding boundary-value problem (BVP) concerns a simple axial compression test of an L4-L5 motion segment with removed spinal processes, where a physiological load of 400 N is applied, thus leading to an external load of $\bar{t} = 0.33\,\mathrm{N/mm^2}$ (see Fig. 4). Since the focus lies on the IVD, the vertebrae L4 and L5 are horizontally cut in parallel having boundary conditions as depicted in Fig. 4. The simplified motion segment is discretized using 7275 20-noded hexahedral elements, each equipped with 27 $\mathcal{GP}$s for the numerical quadrature.

A total number of six simulations is performed using material parameters from the related literature; see, e.g., Gu et al. (1999), Iatridis et al. (2003) and Eberlein et al. (2004) among others. Overall, the findings presented in Fig. 5 are in good accordance with the experimental results reported in the related literature, e.g., Brown et al. (1957), Nachemson et al. (1979) and Wilke et al. (1999). Moreover, the influences of the osmotic pressure and the inhomogeneities concerning the collagen fibers become apparent regarding the entire pore pressure p and the relative change Δh in height, which is computed at two anterior points within the sagittal plane of the disc (Fig. 4). In particular, neglecting the attached fixed charges (i.e. $c_m^{fc} = 0 \rightarrow \pi = 0$) leads to an underestimate of the entire pore pressure, which somehow represents a degenerated disc having a reduced FCD. Hence, the anulus fibers are not fully utilized, yielding an overall softer response and thus a non-physiological bearing mechanism of the IVD. Furthermore, the differences in Δh ranging from 0.25 mm to 0.44 mm reveal the necessity of modelling the fiber inhomogeneities. Note in passing

[1]Porous media Adaptive Nonlinear finite element solver based on Differential Algebraic Systems: `http://www.get-pandas.com`

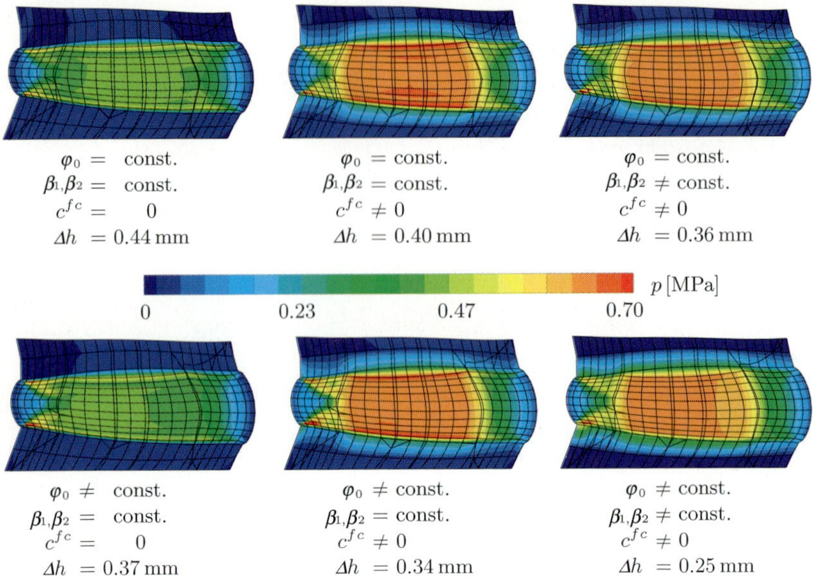

Fig. 5. Results for six simulations carried out with a different degree of inhomogeneity.

that this influence is altered concerning the height change as well as the disc bulge when other regions of the disc are considered.

5 Conclusions

The results presented above show that the model under consideration is capable of describing the complex inhomogeneous and coupled behaviour of the IVD. Herein, the different characterizing moduli, namely inhomogeneous fibers, permeability, FCD or intrinsic viscoelasticity, are easily activated or deactivated, which yields an efficient tool to aid the study of their influence on a variety of effects, such as the interstitial fluid flow or the disc bulge. As a next step, the theoretically introduced material parameters must be determined by independent experiments involving more complex deformation modes, such as superimposed flexion, lateral bending and torsion. Finally, together with an appropriate damage model, it is then possible to describe further degeneration effects and explore mechanisms leading to disc herniation.

References

Ayad, S., and Weiss, J. B. (1987). Biochemistry of the intervertebral disc. In Jayson, M. I. V., ed., *The Lumbar Spine and Back Pain*. Churchill Livingstone, 3rd edition. 100–137.

Bowen, R. M. (1980). Incompressible pourous media models by use of theory of mixture. *Int. J. Eng. Sci.* 18:1129–1148.

Brown, T., Hansen, R. J., and Yorra, A. J. (1957). Some mechanical tests on the lumbosacral spine with particular reference to intervertebral discs. *J. Bone Joint Surg.* 39:1135–1164.

Ebara, S., Iatridis, J. C., Setton, L. A., Foster, R. J., Mow, V. C., and Weidenbaum, M. (1996). Tensile properties of nondegenerate human lumbar anulus fibrosus. *Spine* 21:452–461.

Eberlein, R., Holzapfel, G. A., and Schulze-Bauer, C. A. J. (2001). An anisotropic model for annulus tissue and enhanced finite element analyses of intact lumbar disc bodies. *Comput. Meth. Biomech. Biomed. Eng.* 4:209–230.

Eberlein, R., Holzapfel, G. A., and Fröhlich, M. (2004). Multi-segment FEA of the human lumbar spine including the heterogeneity of the annulus fibrosus. *Comput. Mech.* 34:147Ű–163.

Ehlers, W., and Markert, B. (2001). A linear viscoelastic biphasic model for soft tissues based on the theory of porous media. *J. Biomech. Eng.* 123:418–424.

Ehlers, W., Markert, B., and Acartuerk, A. (2002). Large strain viscoelastic swelling of charged hydrated porous media. In Auriault, J.-L., and al., eds., *Poromechanics II, Proceedings of the Second Biot Conference on Poromechanics*, 185–191. The Netherlands: Lisse.

Ehlers, W. (1993). Constitutive equations for granular materials in geomechanical context. In Hutter, K., ed., *Continuum Mechanics in Environmental Sciences and Geophysics*. Wien: Springer-Verlag. 313–402. CISM Courses and Lectures No. 337.

Ehlers, W. (2002). Foundations of multiphasic and porous materials. In Ehlers, W., and Bluhm, J., eds., *Porous Media: Theory, Experiments and Numerical Applications*, 3–86. Berlin: Springer-Verlag.

Eipper, G. (1998). *Theorie und Numerik finiter elastischer Deformationen in fluidgesättigten Porösen Medien.* Ph.D. Dissertation, Universität Stuttgart. Bericht Nr. II-1 aus dem Institut für Mechanik (Bauwesen)

Elliott, D. M., and Setton, L. A. (2000). A linear material model for fiber-induced anisotropy of the anulus fibrosus. *J. Biomech. Eng.* 122:173–179.

Gu, W. Y., Mao, X. G., Foster, R. J., Weidenbaum, M., Mow, V. C., and Rawlins, B. A. (1999). The anisotropic hydraulic permeability of human lumbar annulus fibrosus. Influence of age, degeneration, direction, and water content. *Spine* 24:2449–2455.

Hayes, W. C., and Bodine, A. J. (1978). Flow-independent viscoelastic properties of articular cartilage matrix. *J. Biomech.* 11:407–419.

Holzapfel, G. A., and Weizsäcker, H. W. (1998). Biomechanical behavior of the arterial wall and its numerical characterization. *Comp. Biol. Med.* 28:377–392.

Holzapfel, G. A., Gasser, T. C., and Ogden, R. W. (2000). A new constitutive framework for arterial wall mechanics and a comparative study of material models. *J. Elasticity* 61:1–48.

Holzapfel, G. A., Gasser, T. C., and Ogden, R. W. (2004). Comparison of a multi-layer structural model for arterial walls with a Fung-type model, and issues of material stability. *J. Biomech. Eng.* 126:264–275.

Holzapfel, G. A., Schulze-Bauer, C. A. J., Feigl, G., and Regitnig, P. (2005). Single lamellar mechanics of the human lumbar anulus fibrosus. *Biomech. Model. Mechanobio.* 3:125–140.

Huyghe, J. M., and Janssen, J. D. (1997). Quadriphasic mechanics of swelling incompressible porous media. *Int. J. Eng. Sci.* 35:793–802.

Iatridis, J. C., Weidenbaum, M., Setton, L., and Mow, V. (1996). Is the nucleus pulposus a solid or a fluid? mechanical behaviors of the human intervertebral disc. *Spine* 21:1174–1184.

Iatridis, J. C., Laible, J. P., and Krag, M. H. (2003). Influence of fixed charge density magnitude and distribution on the intervertebral disc: Applications of a poroelastic and chemical electric (PEACE) model. *J. Biomech. Eng.* 125:12–24.

Klisch, S. M., and Lotz, J. C. (1999). Application of a fiber-reinforced continuum theory to multiple deformations of the annulus fibrosus. *J. Biomech.* 32:1027–1036.

Lanir, Y. (1987). Biorheology and flux in swelling tissue. I Biocomponent theory for small deformation including concentration effect. *Biorheology* 24:173–187.

Mak, A. F. (1986). The apparent viscoelastic behaviour of articular cartilage – the contributions from the intrinsic matrix viscoplasticity and interstitial fluid flows. *J. Biomech. Eng.* 108:123–130.

Marchand, F., and Ahmed, A. M. (1990). Investigation of the laminate structure of lumbar disc anulus fibrosus. *Spine* 15:402–410.

Markert, B. (2005). *Porous media viscoelasticity with application to polymeric foams.* Ph.D. Dissertation, Universität Stuttgart. Bericht Nr. II-11 aus dem Institut für Mechanik (Bauwesen).

de Boer, R. (2000). *Theory of Porous Media. Highlights in the Historical Development and Current State.* Heidelberg: Springer-Verlag.

Mow, V. C., and Hayes, W. C. (1997). *Basic Orthopaedic Biomechanics.* New York: Lippincott-Raven.

Mow, V. C., Gibbs, M. C., Lai, W. M., Zhu, W. B., and Athanasiou, K. A. (1989). Biphasic indentation of articular cartilage-II. A numerical algorithm and an experimental study. *J. Biomech.* 22:853–861.

Nachemson, A., Schultz, A., and Berkson, M. (1979). Mechanical properties of human lumbar spine motion segments - influences of age, sex, disc level and degeneration. *Spine* 4:1–8.

Paoli, P. (1996). *Second European Survey on Working Conditions in the European Union.* Dublin, Ireland: European Foundation for the Improvement of Living and Working Conditions.

Schröder, J., and Neff, P. (2003). Invariant formulation of hyperelastic transverse isotropy based on polyconvex free energy functions. *Int. J. Solids Structures* 40:401–445.

Skaggs, D. L., Weidenbaum, M., Iatridis, J. C., Ratcliffe, A., and Mow, V. C. (1994). Regional variation in tensile properties and biomechanical composition of the human lumbar anulus fibrosus. *Spine* 19:1310–1319.

Spencer, A. J. M. (1984). Constitutive theory for strongly anisotropic solids. In Spencer, A. J. M., ed., *Continuum Theory of the Mechanics of Fibre-Reinforced Composites*. Wien: Springer-Verlag. 1–32. CISM Courses and Lectures No. 282, International Centre for Mechanical Sciences.

Tsuji, H., Hirano, N., Ohshima, H., Ishihara, H., Terahata, N., and Motoe, T. (1993). Structural variation of the anterior and posterior anulus fibrosus in the development of human lumbar intervertebral disc. *Spine* 18:204–210.

Urban, J., and Holm, S. (1986). Intervertebral disc nutrition as related to spinal movements and fusion. In Hargens, A. R., ed., *Tissue Nutrition and Viability*. Springer-Verlag. 101–119.

Wilke, H.-J., Neef, P., Caimi, M., Hoogland, T., and Claes, L. E. (1999). New in vivo measurements of pressure in the intervertebral disc in daily life. *Spine* 24:755–762.

Is the Free Energy of Hydrogel the Sum of Elastic Energy and Ionic Energy?

R.W. Roos, J.M. Huyghe, F.P.T. Baaijens

Eindhoven University of Technology, Department of Biomedical Engineering, The Netherlands
r.w.roos@tue.nl, j.m.r.huyghe@tue.nl, f.p.t.baaijens@tue.nl

Equilibrium behavior of hydrogel can be derived from the free energy function W, which depends on both strain and ionic concentration. In common macroscopic analyses (Flory and Rehner Jr. (1943)), the free energy is assumed to be the sum of an elastic energy and an ionic energy. In this article we fit the elastic stiffness and the osmotic coefficient to the results of swelling and consolidation measurements on hydrogel performed by Frijns et al. (2003). With the ionic energy described by the classical Donnan model, the raw data cannot be fitted satisfactorily. Although no experimental indication for physical instability is found, the fit yields a negative elastic stiffness, which implies strain softening, i.e. unstable material behavior. Furthermore, the fitted osmotic coefficient inside the gel is larger than the one in the bathing solution, which is in contrast with Debye-Hückel theory. We therefore propose to use an extra term in the free energy function, which is a function of both strain and ionic concentration.

1 Introduction

Saturated ionized porous media are profusely present in both biology and geology. To learn more about the constitutive behavior of these media, we use hydrogel as a model material. A picture of hydrogel is shown in Fig. 1. Hydrogel is an ionized polymer network with a large water content, and its electromechanical behavior is not fully understood. The main problem is caused by a lack of knowledge of the interplay between its free energy components:

- W_{el}, the energy change caused by deformation of network chains to a more elongated state;
- W_{ion}, the energy change caused by electrostatic interactions between fixed charges and mobile ions;
- W_{mix}, the energy change caused by polymer-solvent mixing.

Typically, the strain of the gel and the apparent density of the components are the state variables of the mixture. But, because the elastic stiffness of the

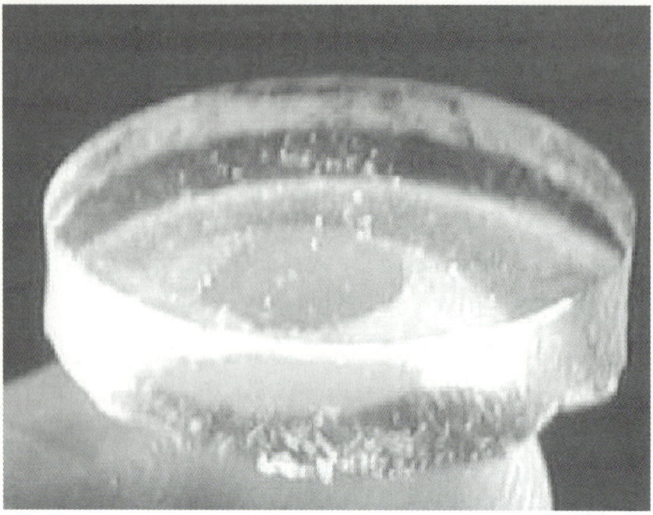

Fig. 1. An example of an Acrylic Acid Acrylamid hydrogel sample.

gel is orders of magnitude lower than the volumetric modulus of water, it is reasonable to assume incompressibility of all mixture components. Hence, we may replace the apparent densities by volume fractions N^β, which is the volume fraction of the component β per unit initial mixture, i.e.

$$N^\beta = \frac{V^\beta}{V_0}, \tag{1}$$

where the superscript β denotes solid (s), fluid (f), cations (+) and anions (-), V^β is the current volume associated with the component β and V_0 is the initial mixture volume.

The strain is described by the deformation gradient tensor (F), which defines the deformation with respect to the initial configuration. It is convenient to choose the stress free state for this initial configuration. In hydrogel the stress free state is not unambiguously defined, but in this paper we use the situation of the hydrogel just after polymerization because it seems reasonable to assume that the elastic stress in the polymer network is then at a minimum. In addition, the volumetric variation with respect to the initial configuration is described by J, the third invariant of the deformation gradient tensor ($\det(F)$). Thus,

$$J = \frac{V}{V_0}, \tag{2}$$

where V is the mixture volume (m^3), and saturation requires that

$$J = N^s + N^f + N^+ + N^-. \tag{3}$$

Okay and Sariisik (1999) assumed that the total free energy (J m^{-3}) is the sum of all these separate parts (W_{el}, W_{ion} and W_{mix}), which is an extension of the Flory-Rehner Hypothesis (Flory and Rehner Jr. (1943)) to ionized porous media. Furthermore, they assumed that W_{el} is only a function of deformation, W_{ion} depends only on the apparent densities of the constituents of the gel, and W_{mix} is a function of the volume fraction of the polymer network N^s:

$$W(J, N^s, N^f, N^+, N^-) = W_{\text{el}}(J) + W_{\text{mix}}(N^s) + W_{\text{ion}}(N^f, N^+, N^-). \quad (4)$$

The state variables in (4) are not independent. It seems likely that the volume contribution of the ions is negligible:

$$N^+, N^- \ll N^f, \quad N^+, N^- \ll N^s. \quad (5)$$

Therefore, the saturation condition (3) simplifies to

$$N^f = J - N^s, \quad (6)$$

where, in turn, the solid volume fraction depends on the deformation and the initial solid fraction N_0^s. Thus,

$$N^s = \frac{N_0^s}{J}. \quad (7)$$

We can further reduce the number of state variables using electro-neutrality, i.e.

$$\frac{N^+}{\overline{V}^+} - \frac{N^-}{\overline{V}^-} + c_0^{fc} = 0, \quad (8)$$

which relates N^- to N^+ and c_0^{fc}, the concentration of 'fixed charge' with respect to the initial mixture volume, and $\overline{V}^\beta$ is the molar volume ($\text{m}^3\,\text{mol}^{-1}$). Because c_0^{fc} is constant in time, the state variables in the general equation for the free energy (4) can be reduced to J and N^+ by using (6)–(8):

$$W(J, N^+) = W_{\text{el}}(J) + W_{\text{mix}}(J) + W_{\text{ion}}(J, N^+). \quad (9)$$

Finally, we can write the sum of $W_{\text{el}}(J)$ and $W_{\text{mix}}(J)$ as $W_{\text{def}}(J)$, which accounts for changes in free energy due to deformation:

$$W(J, N^+) = W_{\text{def}}(J) + W_{\text{ion}}(J, N^+). \quad (10)$$

It is common practice to use Donnan osmosis for W_{ion} (Donnan (1924)):

$$\begin{aligned} W_{\text{ion}}(N^f, N^+, N^-) = {} & \mu_0^f N^f + \mu_0^+ N^+ + \mu_0 N^- - RT\Gamma_g \left(\frac{N^+}{\overline{V}^+} + \frac{N^-}{\overline{V}^-}\right) \ln N^f \\ & + RT\frac{N^+}{\overline{V}^+}\left[\ln\left(\frac{f^+ N^+}{\overline{V}^+}\right) - 1\right] \\ & + RT\frac{N^-}{\overline{V}^-}\left[\ln\left(\frac{f^- N^-}{\overline{V}^-}\right) - 1\right], \end{aligned} \quad (11)$$

where R is the universal gas constant ($\mathrm{J\,mol^{-1}K^{-1}}$), T is the absolute temperature (K) and f^β is the activity coefficient (−) of component β. Γ_g is the osmotic coefficient inside the gel (−), which is a measure for electric screening. Now, it follows that

$$\frac{\partial W_{\text{ion}}}{\partial N^f} = -RT\Gamma_g\left(\frac{N^+}{\overline{V}^+ N^f} + \frac{N^-}{\overline{V}^- N^f}\right) \tag{12}$$

is minus the Donnan osmotic pressure of the gel. The right-hand side of eq. (12) is the classical Van 't Hoff's expression for osmotic pressure. The classical expression for the electrochemical potentials μ^+ and μ^- of the ionic species is recovered from expression (11) by

$$\mu^+ = \frac{F\xi}{\overline{V}^+} + \frac{\partial W_{\text{ion}}}{\partial N^+} = \frac{F\xi}{\overline{V}^+} + \frac{RT}{\overline{V}^+}\ln\left(\frac{f^+N^+}{\overline{V}^+(N^f)^{\Gamma_g}}\right),$$

$$\mu^- = \frac{F\xi}{\overline{V}^-} + \frac{\partial W_{\text{ion}}}{\partial N^-} = \frac{F\xi}{\overline{V}^-} + \frac{RT}{\overline{V}^-}\ln\left(\frac{f^-N^-}{\overline{V}^-(N^f)^{\Gamma_g}}\right), \tag{13}$$

in which ξ is the Donnan electrical potential (V) and F the Faraday constant ($\mathrm{C\,mol^{-1}}$).

In this article, we discuss the results of confined swelling and compression experiments performed by Frijns et al. (2003). In these experiments, an Acrylic Acid Acrylamid copolymer gel was allowed to deform in one direction under the influence of a combination of a mechanical load and an ionic load using a set-up as shown in Fig. 2. While the samples were loaded by the protocol of Table 1, the corresponding height in the equilibrium situation was measured.

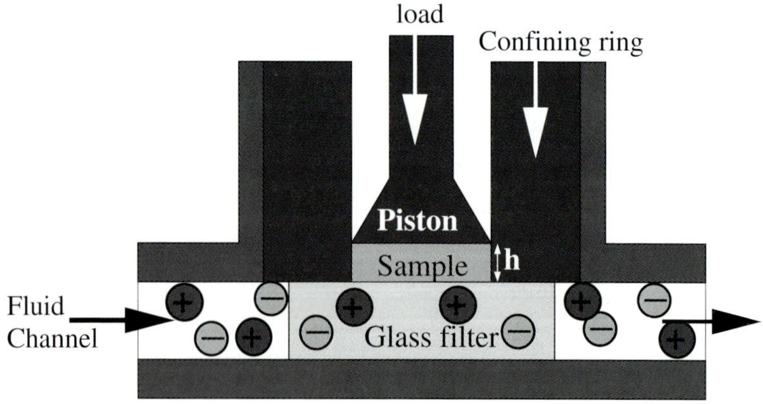

Fig. 2. Schematic representation of the experimental set-up of a confined swelling and consolidation experiment.

Table 1. Experimental protocol.

Stage	Conditioning (1)	Swelling (2)	Consolidation (3)
Mechanical load [kPa]	78	78	195
NaCl concentration $[\text{mol}\,\text{l}^{-1}]$	0.469 ± 0.026	0.159 ± 0.014	0.159 ± 0.014

2 Method

The applied load σ on the sample is the effective stress σ_{eff} minus the hydrostatic pressure inside the gel π_{gel} (Huyghe and Janssen (1997)), i.e.

$$\sigma = \sigma_{\text{eff}} - \pi_{\text{gel}}, \tag{14}$$

where π_{gel} is related to the chemical potential and the derivative of the free energy with respect to the molar volume of the fluid:

$$\begin{aligned} \mu^f_{\text{gel}} &= \left(\frac{\partial W}{\partial N^f}\right)_{J,N^+,N^-} + \pi_{\text{gel}} \\ &\overset{(12)}{=} -RT\Gamma_g\left(\frac{N^+}{\overline{V}^+N^f} + \frac{N^-}{\overline{V}^-N^f}\right) + \pi_{\text{gel}}. \end{aligned} \tag{15}$$

The pressure in the external bath is taken as a reference ($\pi_{\text{bath}} = 0$). Applying Debye-Hückel theory for the external bath yields

$$\mu^f_{\text{bath}} = -2RT\Gamma_b c_b, \tag{16}$$

where c_b and Γ_b are the molar concentration $(\text{mol}\,\text{m}^{-3})$ and the osmotic coefficient $(-)$ of the external bath. In the range from 0.45 M NaCl to 0.15 M NaCl, Γ_b is 0.924 (Robinson and Stokes (1968)) and its variation may be neglected (Houbon (1996), Oomens et al. (1995)). Now, using continuity of the chemical potential across the sample surface (15, 16) yields

$$\pi_{\text{gel}} = RT\Gamma_g\left(\frac{N^+}{\overline{V}^+N^f} + \frac{N^-}{\overline{V}^-N^f}\right) - 2RT\Gamma_b c_b \tag{17}$$

after equilibrating the gel with the external solution. With (6)–(8), this result can be expressed in terms of J and N^+:

$$\pi_{\text{gel}} = RTJ\Gamma_g\left(\frac{2N^+ - c_0^{fc}\overline{V}^+}{\overline{V}^+(J^2 - N_0^s)}\right) - 2RT\Gamma_b c_b. \tag{18}$$

This means that not only the effective stress but also the pressure in the gel depends directly on the deformation via the fixed charge density. Subsequently, this variation of fixed charge density changes the cation and anion concentration inside the gel. Applying continuity across the gel surface of the electrochemical potentials of the cations and anions (13) and using electroneutrality both inside and outside the gel (8), we obtain

$$c_g^+ = \frac{N^+}{\overline{V}^+ N^f} = \frac{1}{2}\left(-c^{fc} + \sqrt{(c^{fc})^2 + 4c_b^2}\right),$$
$$c_g^- = \frac{N^-}{\overline{V}^- N^f} = \frac{1}{2}\left(-c^{fc} - \sqrt{(c^{fc})^2 + 4c_b^2}\right), \tag{19}$$

where c^{fc} is the concentration of fixed charge per unit fluid volume (mol m^{-3}):

$$c^{fc} = \frac{c_0^{fc}}{N^f}, \tag{20}$$

and c_g^+, c_g^- is the cation concentration and the anion concentration respectively inside the gel per unit fluid volume (mol m^{-3}). Although it is known that the screening parameter Γ_g depends on concentration, we assume that this variation is small enough to be neglected. Hence, we take Γ_g as a constant, where its value is obtained by fitting the experiments.

3 Results

The unconstrained diameter increase from polymerization to the equilibrium situation in a 0.45 M NaCl solution (conditioning) is 190 ± 40%. Therefore, the volumetric variation tensor J_1 in this situation is taken to be 2.9 ± 0.4. Because the deformation of the sample during the experiments takes place in one direction only, the relative volume change in other configurations can be derived from the height of the sample. Thus,

$$J = \frac{J_1}{h_1} h, \tag{21}$$

and it may also be replaced in the former equations by the one-dimensional strain ϵ:

$$\epsilon = J - 1. \tag{22}$$

Further, eq. (14) yields an expression for the effective stress:

$$\sigma_{\text{eff}} = \sigma + \pi_{\text{gel}}. \tag{23}$$

Another expression for the effective stress can be derived from the free energy function:

$$\sigma_{\text{eff}} = \left(\frac{\partial W}{\partial \epsilon}\right)_{N^\beta} = \left(\frac{\partial W_{\text{def}}}{\partial \epsilon}\right)_{N^\beta}, \qquad \beta = \text{s}, \text{f}, +, -. \tag{24}$$

Without loss of generality, we linearize the stress-strain dependence within the experimental range as

$$\sigma_{\text{eff}} = H\epsilon + \sigma_{\text{offset}}, \tag{25}$$

where H is the apparent stiffness of the hydrogel (Pa) and σ_{offset} is the offset stress. Their values should be obtained by fitting the experimental data. Now, we can substitute eqs (18) and (25) into (14) and solve the resulting set of equations for all three different loading situations, with H and Γ_g unknown:

$$\sigma = H\epsilon + \sigma_{\text{offset}} + RT(\epsilon+1)\Gamma_g \left(\frac{2N^+ - c_0^{fc}\overline{V}^+}{\overline{V}^+\left[(\epsilon+1)^2 - N_0^s\right]} \right) - 2RT\Gamma_b c_b. \quad (26)$$

This fit is performed for 9 hydrogel samples separately. In these fits, the computed fixed charge density c^{fc} was equal to $-0.301\,\text{mol}\,\text{l}^{-1}$ during the conditioning situation (Frijns et al. (2003)). The results, including the influence of uncertainty in the estimated stress free state, are shown in Table 2. Furthermore, the effective stress is calculated for all different loading situations and is displayed in Fig. 3 as a function of strain. Finally, the strain dependence of the pressure inside the gel is illustrated in Fig. 4.

Table 2. The values of the fitted parameters H and Γ_g for different estimates of stress free configurations.

ϵ	H (kPa)	Γ_g (-)
1.5	-30 ± 12	0.97 ± 0.01
1.9	-34 ± 14	0.97 ± 0.01
2.3	-40 ± 16	0.97 ± 0.01

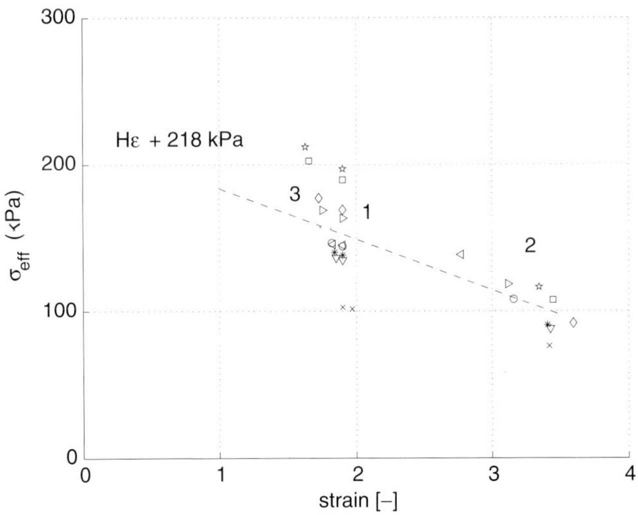

Fig. 3. The effective stress during conditioning (1), swelling (2) and consolidation (3) fitted with expression (25). The fitted curve shows strain softening and a positive offset: $\sigma_{\text{eff}} = H\epsilon + 218$ kPa.

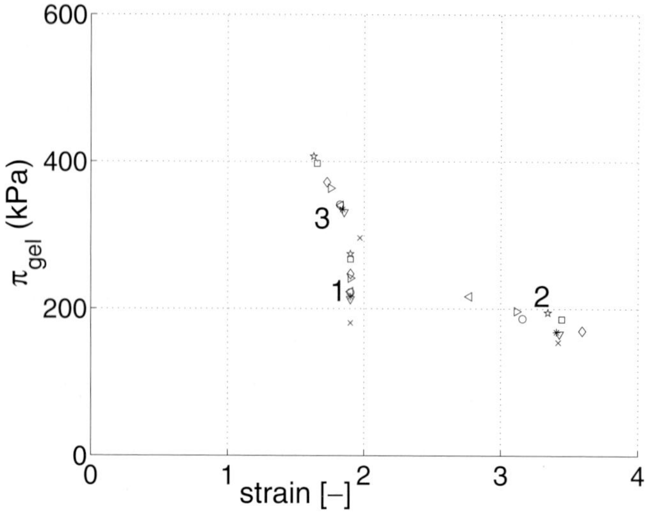

Fig. 4. The gel pressure π_{gel} is different in the conditioning (1), swelling (2) and consolidation (3) situation. According to this graph, π_{gel} is lower in the swollen state than in the conditioning state.

4 Discussion

The relation between equilibrium behavior of hydrogel and sample strain and ionic concentration of the external bath has been studied. From the experiments performed by Frijns et al. (2003), the effective stress has been calculated by subtracting the pressure contribution from the total stress. The effective stress of 1D swelling and consolidation experiments as a function of strain consistently shows the same qualitative picture. All experiments were fitted separately, yielding negative values for the elastic modulus H and an osmotic coefficient Γ_g of 0.97. These fits differ from the values found by Frijns et al. (2003). This difference may be caused by the fact that Frijns neglects the contribution of gel pressure variation as a function of strain.

Figure 4 shows the gel pressure as a function of strain. It shows a lower gel pressure in the swollen state than in the conditioning state, which contradicts physical intuition. The influence of this unphysical pressure behavior can be seen in Fig. 3 as strain softening. Therefore, the question is raised as to whether the mechanical behavior of hydrogel may be represented by the Donnan model. Replacing the linear elastic relationship (25) by a non-linear stress-strain relationship does not improve the fit shown in Fig. 3 because these data consistently exhibit apparent strain softening. Since these data are obtained through subtraction of the osmotic contribution from the raw data of the experiment, the reason for the unphysical strain softening is probably associated with an unrealistic osmosis model.

In various scientific fields, there is discussion about the applicability of the Donnan model for soft hydrogels. From molecular analysis, Vilgis and Wilder (1998) have shown that the elastic modulus depends on the Debye screening parameter, i.e. on ionic concentration. They predict an increasing elastic modulus with increasing salt concentration. This dependence is not accounted for in expression (11) of the free energy W. On the other hand, experiments on cartilage samples performed by Jin and Grodzinsky (2001) show the opposite result. In these experiments, cylindrical cartilage samples equilibrated with different salt solutions were subjected to torsional loading. They found that the shear modulus of the cartilage increases with decreasing salt concentration.

The problems that are brought forward in this paper are inherent to highly charged and highly deformable polymer networks. This property couples ionic pressure to strain behavior of the gel, and it may become even more pronounced for less stiff gel-like materials like living cells.

In conclusion, we propose to adjust the free energy such that the calculated effective stress of raw experimental data always increases with strain. In order to do so, the apparent pressure contribution to the effective stress should be altered. This can be achieved by regarding Γ_g to be dependent on concentration, or by adding an extra free energy term that is a function of both sample deformation and concentration. With the experiments of Frijns, we cannot calculate this adjustment term. Therefore, other experiments on hydrogel with more loading conditions are needed. Options are MRI-measurements of H^+ and Na^+ profiles during swelling, swelling and compression experiments involving more equilibrium states or repeating the experiment of Jin and Grodzinsky (2001) for hydrogel samples.

Acknowledgements. This research is supported by the Technology Foundation STW, Applied Science Division of NWO, and the technology programs of the Ministry of Economic Affairs.

References

Donnan, F. G. (1924). The theory of membrane equilibria. *Chem. Review* 1:73–90.

Flory, P. J., and Rehner Jr., J. (1943). Statistical mechanics of cross-linked polymer networks. *J. Chem. Phys.* 11:512–526.

Frijns, A. J. H., Huyghe, J. M., Kaasschieter, E. F., and Wijlaars, M. W. (2003). Numerical simulation of deformations and electrical potentials in a cartilage substitute. *Biorheology* 40:123–131.

Houbon, G. B. (1996). *Swelling and compression of intervertebral disc tissue.* Ph.D. Dissertation, Eindhoven University of Technology.

Huyghe, J. M., and Janssen, J. D. (1997). Quadriphasic mechanics of swelling incompressible porous media. *Int. J. Eng. Sci.* 35:793–802.

Jin, M., and Grodzinsky, A. J. (2001). Effect of electrostatic interactions between glycosaminoglycans on the shear stiffness of cartilage: A molecular model and experiments. *Macromolecules* 34:8330–8339.

Okay, O., and Sariisik, S. B. (1999). Swelling behavior of poly(acrylamide-co-sodium acrylate) hydrogels in aqueous salt solutions: theory versus experiments. *Eur. Polym. J.* 36:393–399.

Oomens, C. W. J., de Heus, H. J., Huyghe, J. M., Nelissen, L., and Janssen, J. D. (1995). Validation of triphasic mixture theory for a mimic of intervertebral disk tissue. *Biomimetics* 3:171–185.

Robinson, R. A., and Stokes, R. H. (1968). *Electrolyte Solutions.* London: Butterworths.

Vilgis, T. A., and Wilder, J. (1998). Polyelectrolyte networks: elasticity, swelling, and the violation of the Flory-Rehner hypothesis. *Comput. Theor. Polym. S.* 8:61–73.

In Vivo Experiments to Characterize the Mechanical Behavior of the Human Uterine Cervix

E. Mazza[1], A. Nava[1], M. Bauer[2], R. Winter[2], M. Bajka[3], G.A. Holzapfel[4]

[1] Swiss Federal Institute of Technology Zurich, Department of Mechanical Engineering, Switzerland
mazza@imes.mavt.ethz.ch

[2] Medical University Graz, Department of Obstetrics and Gynecology, Austria
margit.bauer@klinikum-graz.at

[3] University Hospital Zurich, Department of Obstetrics and Gynecology, Switzerland michael.bajka@hin.ch

[4] Royal Institute of Technology, School of Engineering Sciences, 100 44 Stockholm, Sweden; Graz University of Technology, Institute for Structural Analysis – Computational Biomechanics, Austria
gh@hallf.kth.se, gh@biomech.tu-graz.ac.at

Experimental results of *in vivo* measurements for characterizing the mechanical behavior of human uterine cervices are documented. Aspiration experiments were performed *in vivo*, before vaginal/abdominal hysterectomy, and *ex vivo*, approximately 1.5 hours after extraction. We defined various parameters to characterize the mechanical response from the aspiration experiments. The reproducibility of the mechanical data from the *in vivo* aspiration experiments has been analyzed. The *in vivo* and *ex vivo* mechanical responses of uterine cervices did not differ considerably: the *ex vivo* situation shows a stronger history dependence in tissue pre-conditioning, whereas the differences in the time dependence of the mechanical response were not significant. This study represents a first step of a clinical application aiming at characterizing the mechanical response of normal cervical tissue at different gestational ages, and to identify the mechanical parameters that characterize pathologic conditions such as cervical insufficiency leading to preterm delivery.

1 Introduction

The identification of the mechanical properties of soft biological tissue is essential to the understanding of their functions. In particular, experimental data obtained from *in vivo* measurements that mimic real loading conditions in a physiological environment may be essential for the design of new diagnostic devices. However, very limited quantitative data are available that describe the mechanics of soft biological tissues *in vivo* because of the severe technical and ethical problems related to the experiments. Qualitative evaluation of the mechanical properties of soft biological tissues through palpation is an established practice in medicine. Specifically, the palpation of the uterine cervix represents a standard procedure, which is applied during regular pelvic examination; see Bishop (1964). Another method of diagnosing mechanical changes is elastography, which complements ultrasound examination by providing qualitative data on differences in stiffness between tissues of internal organs (Ophir et al. (1991), Manduca et al. (2001)).

Quantitative local measurements of the mechanical properties of biological tissues have the potential to provide a method for (more) accurate tissue classification and early detection of diseases. To the authors' knowledge the first quantitative mechanical measurements on human organs *in vivo* were performed by Carter et al. (2001). A number of different procedures for the quasi-static *in vivo* testing of human and animal soft tissues have been developed recently. They are based on indentation, aspiration or shear testing; see, for example, Zheng and Mak (1996), Miller et al. (2000), Kauer et al. (2002), Nasseri et al. (2002), Ottensmeyer (2002), Hendriks et al. (2004), Tonuk and Silver-Thorn (2004).

In the present study an 'aspiration device' (Vuskovic (2001)) was used to characterize the *in vivo* mechanical properties of human uterine cervices (Mazza et al. (2005)). The method based on the aspiration device allows control of kinematic and kinetic boundary conditions so that identical aspiration protocols can be repeated several times on the same tissue sample (Nava et al. (2004b)). Recently, the device was used for intra-operative measurements on human uteri, which served as a basis for the inverse finite element characterization of the associated mechanical properties (Kauer et al. (2002)). The aspiration device allows assessment of the mechanical response of internal organs under sterile conditions without harm to the tested tissue; therefore, this approach has the potential to serve for diagnosis. The present work is the first application of the aspiration device in a clinical study. The scatter of the mechanical data obtained from intra-operative experiments is evaluated in order to verify the feasibility of *in vivo* measurements on human uterine cervices. In this work preliminary experimental results are presented and conclusions are proposed on the basis of the expected sensitivity of the present procedure for the detection of changes in the mechanical behavior of cervices. The experiments were performed at the Department of Obstetrics and Gynecology at the Medical University, Graz, in Austria.

2 Experiments

2.1 Aspiration Device and Method

The aspiration device, an illustration of the working principle and a representative image of a deformed cervical tissue are shown in Fig. 1. The experiment is based upon the pipette aspiration technique of Aoki et al. (1997). The device, designed for quasi-static aspiration tests and for *in vivo* applications, consists of a tube in which an internal pressure can be controlled according to a desired pressure law. The device characteristics were driven by issues such as safety, sterilizability, space limitation and short data acquisition time, which are important when dealing with *in vivo* applications.

The experiment is performed by (i) gently pushing a tube against the tissue to ensure a good initial contact, and (ii) creating a (time variable) vacuum inside the tube so that the tissue is sucked through the aspiration area (with a diameter of 10 mm); see Fig. 1. For an isotropic and homogeneous tissue a complete description of the tissue deformation field can be obtained by monitoring the side-view profile of the tissue during the vacuum change. An optical fibre, which is connected to an external source of light, provides the necessary illumination in the inner part of the tube. The images of the side-view (see Fig. 1) are reflected by a mirror and are captured at a frequency of 25 Hz by a digital camera mounted on the upper part of the device. The captured images are analyzed off-line in order to extract the profiles of the deformed tissue.

The pressure inside the device is controlled by means of a pump, an air reservoir and two valves. For the present application the pressure starts from atmospheric pressure, reaches a minimum of $p_{\min} = 220$ mbar negative relative pressure, stays at that pressure level for about 8 seconds and then goes back to atmospheric pressure. From the analysis of the time histories of pressure and deformation profiles constitutive models can then be determined through iterative finite element calculations. In Nava et al. (2004a,b),

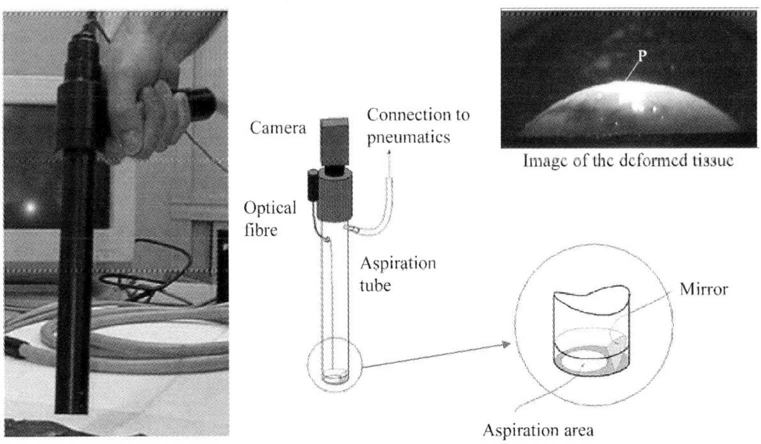

Fig. 1. Aspiration device: an image of the instrument and a sketch illustrating the working principle. On the top right: representative image of a deformed tissue captured by the digital camera. The light half bubble is part of the tissue aspirated into the device. A point P is indicated for later reference.

non-linear viscoelastic models have been used to describe the mechanical response of animal and human organs tested *ex vivo* with the aspiration device.

2.2 Experimental Procedure

All patients were asked to take part in the study and informed consent was given by all participants. All tested cervices were of menopausal women. For each patient the following data were recorded: age, numbers of births, abortions, pathology and hormonal replacement therapy. Data are summarized in Table 1.

In vivo experiments were performed after anesthetization and some presurgical procedures. Specula were inserted to get access to the cervix. Under visualization of the cervix the device was placed at the upper lip (see Fig. 2),

Table 1. Clinical data.

uterus number	age	given births	abortions	pathology	hormonal replacement therapy
1	59	4	0	myoma	NO
2	49	2	3	vaginal prolaps	NO
3	47	3	1	vaginal prolaps	NO
4	77	1	0	carcinoma	YES
5	72	3	0	vaginal prolaps	NO
6	47	1	2	myoma	YES
7	61	4	1	vaginal prolaps	NO
8	69	3	0	endometrial cancer	YES

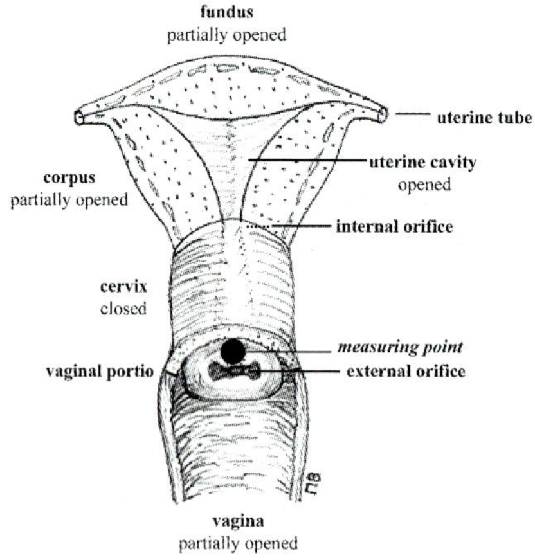

Fig. 2. Anatomy of the uterus and cervix. The testing location is indicated.

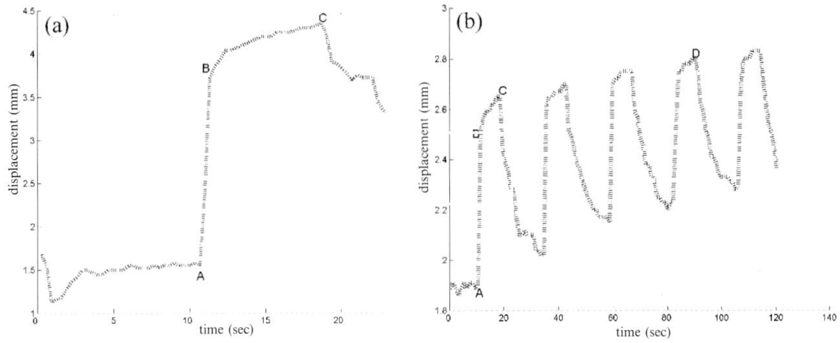

Fig. 3. Representative displacement history of point P (see Fig. 1): one cycle test (a); multiple cycle test (b).

and was then kept in contact with the tissue (on the same spot) to achieve optimal results. All tests for one organ were performed at the same location.

Four of the eight uteri (designated 2, 3, 5, and 7 in Table 1) were tested *ex vivo*, approximately 1.5 hours after the extraction. The uteri were kept in a closed plastic container (in air) inside the operation room at ambient temperature. No solutions were used for tissue conservation. *Ex vivo* tests were performed at the same testing location as for the *in vivo* experiments. The remaining four uteri were not available for *ex vivo* testing because they had to be dissected during the vaginal hysterectomy procedure to complete surgery. In general, the following protocol was adopted for *in vivo* and *ex vivo* testing: (i) a first test with one single loading-unloading cycle, Fig. 3(a); (ii) a second test with the same cycle as in the first test; (iii) a third test with four to five loading-unloading cycles, Fig. 3(b). The resting time between individual cycles was then approximately 10 sec.

For the same organ the tests were always repeated at the same location, i.e. on the upper lip of the cervix (see Fig. 2). For the *in vivo* tests the aspiration device was placed at the anterior lip at the 12 o'clock position. A slight imprint remained on the tissue after performing the aspiration experiment so that the testing location could be identified. In this way the same location could then be selected for *ex vivo* measurements.

2.3 Data Analysis

The deformation profiles obtained from the aspiration experiments were processed for the extraction of the displacement history of the highest point of the profile, i.e. the point P indicated in Fig. 1. Figure 3 shows representative plots of the displacement history of point P for one cycle and for multiple cycles. Specific features of the deformation history are identified such as the locations A, B, C, D, and the displacement history between B and C, used for data analysis and comparison purposes. In particular, the following parameters are defined:

$$d_0 = \text{displacement at B} - \text{displacement at A},$$
$$d_1 = \text{displacement at C} - \text{displacement at A},$$
$$d_4 = \text{displacement at D} - \text{displacement at A}.$$

We propose that the curve between the levels B and C is interpolated by an exponential function f, with the origin shifted to location B, i.e.

$$f(t) = A_0[1 - \exp(-t/\tau)], \quad A_0 = \frac{d_1 - d_0}{1 - \exp(-t_0/\tau)}, \tag{1}$$

where τ is the characteristic time (in sec) of the exponential function (referred to as the 'rising time'), which was determined from the first cycle of the experimental curves, and t_0 is the time between the points B and C ($t_0 = 8$ sec). The four parameters d_0, d_1, d_4 and τ are determined from measured values and characterize the displacement history. Together with the known values of the applied negative pressure history, these parameters enable evaluation and comparison of the different mechanical responses of the organs. For subsequent data analysis it is useful to introduce three additional parameters, i.e. the so-called 'stiffness' η in bar/mm, and two dimensionless quantities, the so-called 'softening' γ, and 'creep' δ, defined to be

$$\eta = \frac{p_{\min}}{d_1}, \quad \gamma = \frac{d_4 - d_1}{d_1}, \quad \delta = \frac{d_1}{d_0}. \tag{2}$$

3 Results and Discussion

3.1 Mechanical Behavior

Figure 4 shows the stiffness parameter η in bar/mm calculated from the data obtained from eight uteri. Several values of η were determined for each cervix by evaluating the data from *in vivo* and *ex vivo* tests (i), (ii) and the first cycle of tests (iii), as described in Section 2.2.

From these data the average stiffness η_{ave} and the related standard deviation σ_η were calculated for each organ, considering *in vivo* as well as *ex vivo* results. The average values for each uterus range from 0.095 (uterus 2) to 0.24 bar/mm (uterus 3), which gives a factor of 2.5 between these values. Such scatter is common for mechanical measurements on soft biological tissue. The standard deviation σ_η for each organ has a maximum of 30% (uterus 6). These results indicate that the scatter between the different organs is larger than the scatter of the measurement results obtained from one single organ.

In order to eliminate the organ to organ variability, the stiffness parameters of each organ were normalized with respect to the organ specific average *in*

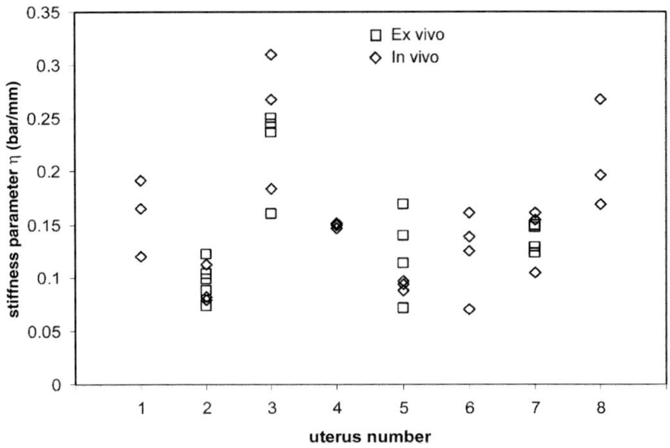

Fig. 4. Stiffness parameter η calculated from the measurements on eight uterine cervices.

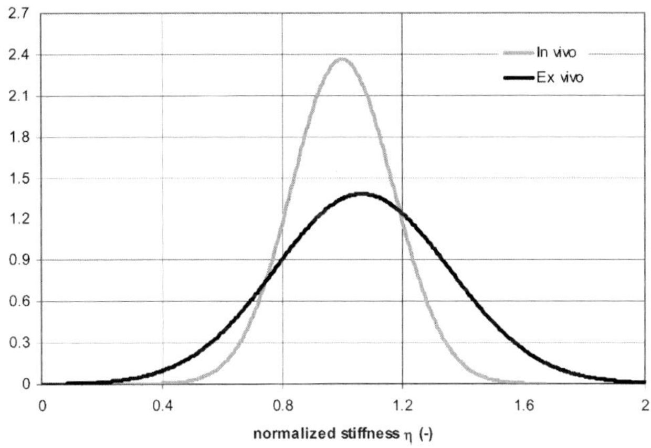

Fig. 5. Gaussian distribution of the stiffness parameter η normalized with respect to the organ specific average *in vivo* stiffness.

vivo stiffness. The resulting normalized distribution of the *in vivo* and *ex vivo* data are compared in Fig. 5. The average of the normalized *in vivo* values is 1 (by definition) and its standard deviation is 19%, whereas the average of the normalized *ex vivo* values is 1.045, with a standard deviation of 27%. These results indicate that there is no significant difference between *in vivo* and *ex vivo* with respect to the stiffness parameter η. Thus, the overall mechanical response of the uterine cervix tissue measured *ex vivo*, approximately 1.5 hours after extraction, seems to be representative for the *in vivo* behavior. On the other hand, history and time dependence of the material response changed to some extent from *in vivo* to *ex vivo*, as described later in this section. Gefen and Margulies (2004) have recently drawn similar conclusions from a comparison of *in vivo* and *ex vivo* indentation experiments on porcine brain: the short-term mechanical response did not differ significantly but to some extent the long-term relaxation behavior did.

Since the measurements on each organ were all performed at the same location, the scatter of the normalized values of Fig. 5 may partly be attributed to the experimental set-up (experimental error). The scatter of *in vivo* data (standard deviation: 19%) is smaller than that of *ex vivo* (standard deviation: 27%). It may be concluded that for the stiffness parameter the experimental conditions for *in vivo* measurements did not increase the scatter of the aspiration test results, this is a prerequisite for a possible application of this procedure for diagnostic purposes in an *in vivo* clinical study.

The softening parameter γ relates the mechanical response in the fourth cycle to the response obtained during the first cycle. Previous work has shown that the increase in displacement with cycles is due to the history dependence of the tissue mechanical behavior (Nava et al. (2004a)). History dependence is associated with the phenomenon 'pre-conditioning'; see, e.g., Humphrey (2002). A comparison of γ obtained from *in vivo* and *ex vivo* data is reported in Fig. 6. The values of γ obtained from *in vivo* experiments range from 0.05 to 0.19, whereas the corresponding values from *ex vivo* tests are between 0.11 (uterus 3) and 0.29 (uterus 2). The variability of γ is comparable with that of the stiffness parameter η (data not shown).

The results of Fig. 6 highlight a significant difference between *in vivo* and *ex vivo* mechanical response: all the points are consistently located at the upper left part of the diagram with *ex vivo* values of γ larger than *in vivo* by a factor

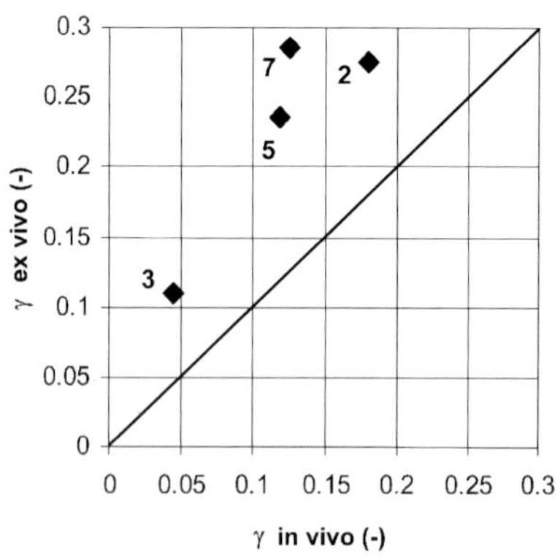

Fig. 6. Comparison of the softening parameter γ from *in vivo* and *ex vivo* measurements performed on the same organs (values shown for *ex vivo* tests of uteri 2, 3, 5 and 7). The solid line represents the location of equal *in vivo* and *ex vivo* values.

of approximately two. This is in agreement with the observations of Gefen and Margulies (2004): pre-conditioning was found to affect the mechanical response of porcine brain tissue to a larger extent *ex vivo* with respect to *in vivo*. Similar observations were also reported by Brown et al. (2003) from *in vivo* and *ex vivo* measurements on porcine abdominal organs.

Finally, the time dependence of the mechanical response is evaluated from the displacement history of the time period with 'constant' pressure during the first loading cycle. Organ to organ variability is eliminated by normalizing the data of each organ with respect to the organ specific average value of the parameter obtained from *in vivo* tests. It turns out that the average value of the normalized creep parameter is smaller for *ex vivo* tests, whereas the corresponding normalized rising time is larger (data not shown). The differences between *in vivo* and *ex vivo* data, however, are small and cannot be regarded as significant for the given scatter of the measured values.

3.2 Constitutive Modeling

No constitutive model has been derived from the experimental data yet. In the present work the tissue behavior has been analyzed in terms of parameters that characterize the overall mechanical response, as measured with the aspiration device . The parameters 'stiffness', 'softening', 'creep', and 'rising time' are phenomenological quantities used to compare the deformation curves obtained from the different experiments. They cannot be considered as material parameters: i.e. δ and τ cannot be directly associated with the creep compliance of the material; no elastic constant can be calculated from η; the softening parameter γ refers to an increased deformation for the same applied negative pressure but it does not necessarily reflect a history-dependent reduction of the elastic moduli. The present approach enables the characteristics of a biological tissue to be featured directly without the need for the lengthy procedure of solving the inverse problem. The parameters proposed in this work can be evaluated in short time (almost 'on-line') with possible advantages for applications for diagnosis.

The design of a three-dimensional continuum model and the identification of the related parameters from the aspiration experiments represent challenging future tasks of the present work. For this purpose finite element procedures and optimization algorithms will be used that were successfully applied in previous work for the determination of parameters of a quasi-linear viscoelastic model from aspiration test results; see, for example, Kauer et al. (2002) and Nava et al. (2004a,b). A three-dimensional constitutive model of the cervix could be applied for the investigation of the organ mechanical behavior under physiological loading conditions or for surgical simulation purposes. In addition, a constitutive model based on the histological tissue structure will provide further insights in the processes that govern tissue biomechanics – for a discussion of, for example, the biomechanics of soft tissue in cardiovascular systems, see the books by Humphrey (2002) and Holzapfel and Ogden (2003).

3.3 Limitations and Applicability for Diagnostic Purposes

The results on the stiffness parameter evaluation show that there is a scatter of measured data (standard deviation up to 27%), which may partly be attributed to experimental uncertainties, and to patient-specific mechanical properties *ex vivo* and *in vivo*. The main source of uncertainty in the aspiration experiment is represented by the force required to bring the device into contact with the tissue. The contact force determines the initial deformation of the tissue and any variation of this force during the measurement influences the time history of deformation. Modifications of the aspiration device are under consideration in order to reduce and/or quantify the influence of the contact force on the measured results. One other source of error is the oscillation of the pressure level around the prescribed minimum value, which is about ±10%. Improvements of the pressure control system are being implemented that will allow a significant reduction of the oscillation.

Although clinical palpation is a well established examination for revealing cervical changes, the method is not indisputable, especially at late gestational ages, when inoculation of bacteria into the cervical canal might cause severe problems. Additionally, palpation is based on individual experiences and detection of very early changes of cervical tissue, which is crucial in order to avoid preterm delivery, might not be achieved. The use of the aspiration test performed during a routine speculum examination may provide 'objective' information on the biomechanical properties of the cervix related to pathologic conditions. The analysis of the present data allows conclusions to be drawn on the expected sensitivity of the present technique. The sensitivity is assessed here from a statistical point of view using the data of the stiffness parameter η, for which a much larger number of values could be determined from the present experiments as compared with the other parameters.

Assuming a variance component model and treating *in vivo* and *ex vivo* data as equally valid, the variability of measured data can be defined as (i) the standard deviation σ_{A} for the eight uteri (which depends on the uterus to uterus variability), and (ii) the standard deviation σ_{B} for each organ (a measure of the errors in the measurement procedure). For the stiffness parameter η these standard deviations were calculated to be $\sigma_{\mathrm{A}} = 32\%$ and $\sigma_{\mathrm{B}} = 22\%$. Based on this result certain changes in the parameter η may successfully be assessed by the experimental procedure. In a clinical study the initial average stiffness value (reference stiffness) of one patient-specific organ can be determined by repeating the measurement several times, for example, five times. Figure 7 shows the detection rate as a function of the change in the stiffness parameter η for one patient by assuming five measurement repetitions. With a false positive detection rate of 15%, the statistical distribution of the present data yields a positive predictive value of 90% for detecting a change by 1/3 of the initial stiffness value, or of 99% for detecting a change by 1/2 of the initial stiffness value. Significant changes in the stiffness of the cervix are expected due to hormonal influences under, for example, hormonal

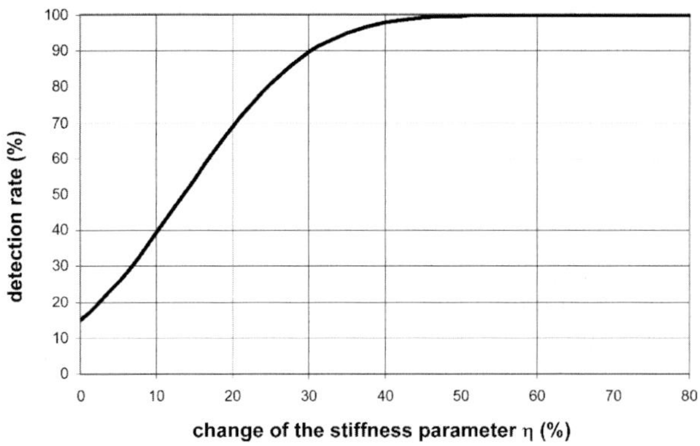

Fig. 7. Detection rate as a function of the change in the stiffness parameter η.

replacement therapy and during pregnancy (Rechberger et al. (1988), (Fung, 1993, pp. 263ff)), which may cause variations by one order of magnitude for the stiffness parameter η.

The present results encourage the application of the aspiration test to cervices of pregnant women in order to establish the mechanical properties of normal cervical tissue at different gestational ages.

4 Conclusions

The main purpose of the present study was to test the reliability and sensitivity of mechanical data obtained from human cervices with respect to a possible clinical application for diagnostic purposes. Future studies will be performed with the goal of using the proposed method for the detection of early cervical changes associated with pathologic conditions.

The investigation of the mechanical behavior of the human uterine cervices was based on the use of an aspiration device. Intra-operative *in vivo* measurements were performed on eight organs without delaying the surgical procedure. The quality of the *in vivo* data is comparable with that obtained *ex vivo*, and with the experience of previous *ex vivo* bench top applications of the aspiration device.

A comparison of measurements of the same organ *in vivo* and *ex vivo* has shown that: (i) the *ex vivo* mechanical response of the uterine cervix tissue, measured approximately 1.5 hours after extraction, did not differ considerably from that observed *in vivo*; (ii) a stronger history dependence in tissue pre-conditioning was identified in the *ex vivo* situation when compared with *in vivo*; (iii) the differences in the time dependence of the mechanical response (parameters δ and τ) were not significant and might be masked by the variability of the measured data.

Our preliminary data indicate that the proposed stiffness parameter η can be used for the characterization of human uterine cervices since the experimental procedure enables detection of changes of 30% with respect to a reference stiffness value.

References

Aoki, T., Ohashi, T., Matsumoto, T., and Sato, M. (1997). The pipette aspiration applied to the local stiffness measurement of soft tissues. *Ann. Biomed. Eng.* 25:581–587.

Bishop, E. H. (1964). Pelvic scoring for elective induction. *Obstet. Gynecol.* 24:266–268.

Brown, J. D., Rosen, J., Kim, Y. S., Chang, L., Sinananand, M., and Hannaford, B. (2003). In-vivo and in-situ compressive properties of porcine abdominal soft tissues. In Westwood, J. D., Hoffman, H. M., Mogel, G. T., Phillips, R., Robb, R. A., and Stredney, D., eds., *Medicine Meets Virtual Reality 11, Studies in Health Technology and Informatics*, volume 94, 26–32.

Carter, F. J., Frank, T. G., Davies, P. J., McLean, D., and Cuschieri, A. (2001). Measurement and modelling of the compliance of human and porcine organs. *Med. Image Anal.* 5:231–236.

Fung, Y. C. (1993). *Biomechanics. Mechanical Properties of Living Tissues.* New York: Springer-Verlag, 2nd edition.

Gefen, A., and Margulies, S. S. (2004). Are in vivo and in situ brain tissues mechanically similar? *J. Biomech.* 37:1339–1352.

Hendriks, F. M., Brokken, D., van Eemeren, J., Oomens, C. W. J., Baaijens, F. P. T., and Horsten, J. B. A. M. (2004). A numerical-experimental method to characterize the non-linear mechanical behaviour of human skin. *Skin Res. Technol.* 9:274–283.

Holzapfel, G. A., and Ogden, R. W., eds. (2003). *Biomechanics of Soft Tissue in Cardiovascular Systems.* Wien – New York: Springer-Verlag.

Humphrey, J. D. (2002). *Cardiovascular Solid Mechanics. Cells, Tissues, and Organs.* New York: Springer-Verlag.

Kauer, M., Vuskovic, V., Dual, J., Szekely, G., and Bajka, M. (2002). Inverse finite element characterization of soft tissues. *Med. Image Anal.* 6:275–287.

Manduca, A., Oliphant, T. E., Dresner, M. A., Mahowald, J. L., Kruse, S. A., Amromin, E., Felmlee, J. P., Greenleaf, J. F., and Ehman, R. L. (2001). Magnetic resonance elastography: Non-invasive mapping of tissue elasticity. *Med. Image Anal.* 5:237–254.

Mazza, E., Nava, A., Bauer, M., Winter, R., Bajka, M., and Holzapfel, G. A. (2005). Mechanical properties of the human uterine cervix: an in vivo study. *Med. Image Anal.* in press.

Miller, K., Chinzei, K., Orssengo, G., and Bednarz, P. (2000). Mechanical properties of brain tissue in vivo: experiment and computer simulation. *J. Biomech.* 33:1369–1376.

Nasseri, S., Bilston, L. E., and Phan-Thien, N. (2002). Viscoelastic properties of pig kidney in shear, experimental results and modeling. *Rheol. Acta* 41:180–192.

Nava, A., Mazza, E., Häfner, O., and Bajka, M. (2004a). Experimental observation and modelling of preconditioning in soft biological tissues. In Metaxas, D., and Cotin, S., eds., *Lecture Notes in Computer Science*, volume 3078, 1–8.

Nava, A., Mazza, E., Kleinermann, F., Avis, N. J., McClure, J., and Bajka, M. (2004b). Evaluation of the mechanical properties of human liver and kidney through aspiration experiments. *Technol. Health Care* 12:269–280.

Ophir, J., Cespedes, I., Ponnekanti, H., Yazdi, Y., and Li, X. (1991). Elastography - a quantitative method for imaging elasticity of biological tissues. *Ultrason. Imaging* 13:111–134.

Ottensmeyer, M. P. (2002). TeMPeST I-D: An instrument for measuring solid organ soft tissue properties. *Exp. Techn.* 26:48–50.

Rechberger, T., Uldbjerg, N., and Oxlund, H. (1988). Connective tissue changes in the cervix during normal pregnancy and pregnancy complicated by cervical incompetence. *Obstet. Gynecol.* 71:563–567.

Tonuk, E., and Silver-Thorn, M. B. (2004). Nonlinear viscoelastic material property estimation of lower extremity residual limb tissues. *J. Biomech. Eng.* 126:289–300.

Vuskovic, V. (2001). *Device for in vivo measurement of mechanical properties of internal human soft tissues.* Dissertation, Swiss Federal Institute of Technology Zürich. No. 14222.

Zheng, Y., and Mak, A. F. T. (1996). An ultrasound indentation system for biomechanical properties assessment of soft tissues in vivo. *IEEE Trans. Biomed. Eng.* 43:912–918.

Viscoelastic Response of Vocal Fold Tissues and Scaffolds at High Frequencies

R.J. Clifton[1], X. Jia[2], T. Jiao[1], C. Bull[1], M.S. Hahn[3]

[1] Brown University, Division of Engineering, USA
clifton@engin.brown.edu, Tong_Jiao@brown.edu,
Christopher_Bull@brown.edu
[2] Massachusetts Institute of Technology, Department of Chemical Engineering, USA xinqiao@mit.edu
[3] Texas A& M University, Department of Chemical Engineering, USA
mariah.hahn@chemail.tamu.edu

A new experimental capability is presented for measuring the viscoelastic properties of vocal folds and candidate scaffold materials at phonation frequencies. Thin cylindrical samples are subjected to torsional wave loading by using a high frequency galvanometer to oscillate a plate at the base of the sample. The rotation of the base plate and a top plate are recorded using optical levers. Viscoelastic properties of the sample are determined by fitting the amplification of the motion of the top of the sample to that predicted by a linear viscoelastic wave analysis. Results are reported for canine and ferret vocal folds as well as for hyaluronic acid (HA)-based hydrogels.

1 Introduction

During normal speech, human vocal folds sustain more than 100 high impact collisions each second. Voice overuse may generate nodules on the outer layer of the vocal folds – the so-called superficial lamina propria (SLP). In other cases, pathological conditions may render part of the tissue cancerous. In any case, when damaged tissue is removed by surgery, the resulting scar tissue lacks the pliability of the original tissue and voice quality is often seriously reduced. Over the past few years, significant research effort has been directed toward using tissue engineering approaches to regenerate vocal fold vibratory tissue that responds as normal LP. As a first step toward developing suitable replacement materials it is important to understand the mechanical response of the natural tissue that they will replace, as well as their own mechanical response at frequencies of human phonation – approximately 100–900 Hz. Chan and Titze (1999, 2000) measured the viscoelastic properties of human vocal folds by subjecting LP tissue to torsional oscillations in a parallel-plate rheometer. From these experiments they obtained the frequency-dependent storage (shear) modulus and viscosity of the LP over frequencies ranging from 0.01 to 10 Hz. Measurements at 15 Hz did not follow the trend lines of the results at lower frequencies and were deemed, by the authors, to be 'marginally acceptable'.

An important conclusion from the Chan and Titze (1999, 2000) experiments is that the shear modulus of human vocal folds is very low. For most subjects the shear modulus G_1 at a frequency of 10 Hz ranges from approximately 10 to 100 Pa. The elastic shear wave speed $c_s = \sqrt{G_1/\rho}$ for a material with a shear modulus $G_1 = 100\,\mathrm{Pa}$ and a density $\rho = 1000\,\mathrm{kg/m^3}$ is approximately 30 cm/s. Because this wave speed is so small relative to wave speeds in most solid materials, one can expect that strong limitations will be imposed on the maximum frequency for which the mechanical properties of the sample can be inferred by means of the usual interpretation of rheometric tests — based on assuming that the stress is nominally uniform through the thickness of the sample. For solid samples the latter assumption holds when the time required for roundtrip transit of stress waves through the thickness of the sample is much less than the period of a single oscillation, i.e. for

$$f \ll c_s/2h, \tag{1}$$

where h is the sample thickness and f is the driving frequency. For human LP with a thickness $h = 0.03\,\mathrm{cm}$, as used by Chan and Titze (1999, 2000), the limitation (1) becomes $f \ll 500\,\mathrm{Hz}$ or, say, $f < 30\,\mathrm{Hz}$. Even lower limits on allowable frequencies are obtained for samples with shear moduli near the lower limits of the range of measured values. Analogously, for fluid samples, a requirement that stresses due to sample inertia are small relative to those due to sample viscosity leads to the limitation (e.g., Schlichting, 1960)

$$f \ll \eta/\pi\rho h^2, \tag{2}$$

where η is the viscosity of the sample. From viscosity measurements of Chan and Titze (1999, 2000) the frequency dependent viscosity of human LP can be described approximately by $\eta = \eta_0 f^{-0.85}$ with $\eta_0 \approx 1.0\,\mathrm{Pa\,s^{0.15}}$. Substitution of this expression for η into (2) gives the limitation $f \ll 83\,\mathrm{Hz}$ or, say, $f < 10\,\mathrm{Hz}$.

Whether the frequency limitation is obtained from a constraint of type (1) or (2) it appears that the frequencies for which the viscoelastic properties of human LP can be measured by standard rheometric methods are likely to be below the frequencies of phonation. Other methods for measuring viscoelastic properties at high frequencies include electromagnetic torsion methods (Brodt et al., 1995), electromechanical tensile test methods (Hemler et al., 2001), and stress-controlled rheometer methods at low audio frequencies (Titze et al., 2004). While all of these methods have attractive features they all neglect wave propagation in the sample. Consequently, they all have frequency limitations similar to those of (1) and (2), although the upper limit on allowable frequencies may be extended by using smaller samples.

An alternative approach, based on the analysis of longitudinal waves in viscoelastic cylindrical rods, has been introduced by Jia et al. (2004) to determine the viscoelastic properties of photo-cross-linked hydrogels. An acoustic shaker was used to subject the base of the rod to an oscillatory vertical motion. The motion of the free, top end of the rod was monitored using a laser-Doppler vibrometer. From the measured amplification factor (the amplitude of the velocity at the top of the rod divided by the amplitude of the velocity at the shaker surface), and the phase shift between the top and bottom ends of the rod, the wave propagation solution was used to determine the frequency dependent viscoelastic moduli for the hydrogel. While this method could extend the measurement of viscoelastic properties into the range of phonation frequencies, it could not be used for LP because the LP geometry does not allow the preparation of slender cylindrical specimens.

In the present paper we retain the basic ideas of the method of Jia et al. (2004), but modify the configuration to be that of torsional waves in a cylindrical sample. With this modification it becomes possible to measure viscoelastic properties of LP and of candidate replacement materials at phonation frequencies. Furthermore, because small amplitude torsional waves are described exactly by a one dimensional wave theory the interpretation of the experiments is free of such complications as geometric dispersion and lateral inertia effects that encumber the interpretation of experiments involving longitudinal wave propagation in rods.

In the pilot experiments described here, results are reported for canine LP, ferret LP, and a number of hyaluronic acid (HA)-based hydrogels. These hydrogel networks exhibit prolonged *in vivo* residence time, while providing a range of viscoelastic properties that span the range of responses observed for LP tissue. These hydrogels are expected to be biocompatible and to be favorable for promoting cell growth.

2 Torsional Wave Analysis

Consider the propagation of torsional waves in a cylindrical, viscoelastic sample with its axis lying along the z-direction. The torsional moment or torque at position z at time t can be represented as

$$T(z,t) = 2\pi \int_0^a r^2 \tau(z,r,t)\mathrm{d}r, \tag{3}$$

where a is the radius of the sample and $\tau(z,r,t)$ is the shear stress corresponding to the strain rate history $\partial\gamma(z,r,t)/\partial t$ according to (e.g., Ferry, 1980)

$$\tau(z,r,t) = \int_0^t G(t-t')\frac{\partial\gamma(z,r,t')}{\partial t'}\mathrm{d}t', \tag{4}$$

where $G(t)$ is the viscoelastic relaxation modulus in shear. Assume that the deformation within the sample can be described adequately as pure torsion in

From the boundary condition (15) and the relation (19), the coefficient B is

$$B(s) = -\frac{J\tilde{G}(s)\lambda(s)\sinh[\lambda(s)h] + \rho_0 I_0 s\cosh[\lambda(s)h]}{J\tilde{G}(s)\lambda(s)\cosh[\lambda(s)h] + \rho_0 I_0 s\sinh[\lambda(s)h]}\frac{\theta_0}{(s - \mathrm{i}\omega_0)}. \tag{20}$$

To invert the Laplace transform $\tilde{\theta}$ we use the definition of the inverse transform:

$$\theta(h,t) = \frac{1}{2\pi\mathrm{i}}\int_{\varepsilon-\mathrm{i}\infty}^{\varepsilon+\mathrm{i}\infty} \exp(st)\tilde{\theta}(h,s)\mathrm{d}s. \tag{21}$$

The parameter ε is a positive real number so the integration path is to the right of the imaginary axis in the s-plane. The integration contour can be closed in the left half plane. No contribution is obtained from the contour at ∞. Therefore, by the Residue Theorem, the integral (21) is simply equal to $2\pi\mathrm{i}$ times the sum of the residues of the integrand at poles to the left of $s = \varepsilon$ plus possible contributions from integrals along branch cuts. The only pole is at $s = \mathrm{i}\omega_0$. Here

$$\lambda(\mathrm{i}\omega_0) = \sqrt{\frac{\rho\mathrm{i}\omega_0}{\tilde{G}(\mathrm{i}\omega_0)}} = \mathrm{i}\sqrt{\frac{\rho\omega_0^2}{G^*(\omega_0)}}, \tag{22}$$

in which (e.g., Ferry, 1980)

$$G^*(\omega_0) \equiv \mathrm{i}\omega_0\tilde{G}(\mathrm{i}\omega_0) \tag{23}$$

is the complex modulus for the viscoelastic response of the sample material in simple shear at circular frequency ω_0. Characterization of the viscoelastic response can be put into more readily interpretable form by representing the complex modulus in terms of its magnitude $|G^*(\omega)|$ and its phase angle $\delta(\omega)$:

$$G^*(\omega) = |G^*(\omega)|\mathrm{e}^{\mathrm{i}\delta(\omega)}. \tag{24}$$

In writing the solution it is helpful to simplify the argument to

$$\lambda h = \mathrm{i}(\xi\alpha - \mathrm{i}\xi\beta), \tag{25}$$

where

$$\xi = \omega_0 h\sqrt{\frac{\rho}{|G^*(\omega_0)|}}, \quad \alpha = \cos\left(\frac{\delta(\omega_0)}{2}\right), \quad \beta = \sin\left(\frac{\delta(\omega_0)}{2}\right).$$

Then, the angular velocity at the end $z = h$ can be written in the form

$$\theta(h,t) = M(\omega_0)\theta_0\cos[\omega_0 t - \phi(\omega_0)], \tag{26}$$

in which $M(\omega_0)$ is the amplification factor:

$$M = \frac{c\sqrt{[c\cos(\xi\alpha) - d\sin(\xi\alpha)]^2 + (c^2 + d^2)\sinh^2(\xi\beta)}}{[c\cos(\xi\alpha) + d\sin(\xi\alpha)]^2 + (c^2 + d^2)\sinh^2(\xi\beta)}$$

An alternative approach, based on the analysis of longitudinal waves in viscoelastic cylindrical rods, has been introduced by Jia et al. (2004) to determine the viscoelastic properties of photo-cross-linked hydrogels. An acoustic shaker was used to subject the base of the rod to an oscillatory vertical motion. The motion of the free, top end of the rod was monitored using a laser-Doppler vibrometer. From the measured amplification factor (the amplitude of the velocity at the top of the rod divided by the amplitude of the velocity at the shaker surface), and the phase shift between the top and bottom ends of the rod, the wave propagation solution was used to determine the frequency dependent viscoelastic moduli for the hydrogel. While this method could extend the measurement of viscoelastic properties into the range of phonation frequencies, it could not be used for LP because the LP geometry does not allow the preparation of slender cylindrical specimens.

In the present paper we retain the basic ideas of the method of Jia et al. (2004), but modify the configuration to be that of torsional waves in a cylindrical sample. With this modification it becomes possible to measure viscoelastic properties of LP and of candidate replacement materials at phonation frequencies. Furthermore, because small amplitude torsional waves are described exactly by a one dimensional wave theory the interpretation of the experiments is free of such complications as geometric dispersion and lateral inertia effects that encumber the interpretation of experiments involving longitudinal wave propagation in rods.

In the pilot experiments described here, results are reported for canine LP, ferret LP, and a number of hyaluronic acid (HA)-based hydrogels. These hydrogel networks exhibit prolonged *in vivo* residence time, while providing a range of viscoelastic properties that span the range of responses observed for LP tissue. These hydrogels are expected to be biocompatible and to be favorable for promoting cell growth.

2 Torsional Wave Analysis

Consider the propagation of torsional waves in a cylindrical, viscoelastic sample with its axis lying along the z-direction. The torsional moment or torque at position z at time t can be represented as

$$T(z,t) = 2\pi \int_0^a r^2 \tau(z,r,t)\mathrm{d}r, \tag{3}$$

where a is the radius of the sample and $\tau(z,r,t)$ is the shear stress corresponding to the strain rate history $\partial\gamma(z,r,t)/\partial t$ according to (e.g., Ferry, 1980)

$$\tau(z,r,t) = \int_0^t G(t-t')\frac{\partial\gamma(z,r,t')}{\partial t'}\mathrm{d}t', \tag{4}$$

where $G(t)$ is the viscoelastic relaxation modulus in shear. Assume that the deformation within the sample can be described adequately as pure torsion in which planes of constant z rotate about the axis of symmetry by an amount $\theta(z,t)$ so that the condition of kinematical compatibility becomes

$$\frac{\partial\gamma(z,r,t')}{\partial t'} = \frac{\partial\nu(z,r,t')}{\partial z} = \frac{\partial\Omega(z,t')}{\partial z}r, \tag{5}$$

where $\nu(z,r,t')$ is the (circumferential) particle velocity and

$$\Omega(z,t) = \frac{\partial\Theta(z,t)}{\partial t} \tag{6}$$

is the angular velocity of the cross-section. Substitution of (5) and (4) into (3) gives

$$T(z,t) = J\int_0^t G(t-t')\frac{\partial\Omega(z,t')}{\partial z}\mathrm{d}t', \tag{7}$$

where

$$J = \frac{\pi a^4}{2} \tag{8}$$

is the polar moment of inertia of the cross-sectional area. Consideration of the rate of change of angular momentum for a circular slice of the sample gives the dynamical equation

$$\rho J\frac{\partial\Omega(z,t)}{\partial t} = \frac{\partial T(z,t)}{\partial z}, \tag{9}$$

where ρ is the mass density of the sample material. Substitution of (7) into (9) gives the following integral-partial-differential equation for the angular velocity $\Omega(z,t)$:

$$\rho\frac{\partial\Omega(z,t)}{\partial t} = \int_0^t G(t-t')\frac{\partial^2\Omega(z,t')}{\partial z^2}\mathrm{d}t'. \tag{10}$$

2.1 Solution

We seek a solution of (10) that vanishes for $t < 0$ and matches measured angular velocities at the two ends of the sample, say $z = 0$ and $z = h$. More specifically, the objective here is to find the relaxation function $G(t)$ that corresponds to measured angular velocities $\Omega_0(t)$ at $z = 0$ and $\Omega_h(t)$ at $z = h$. Just as the direct problem requires an additional boundary condition at $z = h$ to predict the motion $\Omega_h(t)$ from the motion $\Omega_0(t)$ and the function $G(t)$, the determination of $G(t)$ from $\Omega_0(t)$ and $\Omega_h(t)$ requires an additional boundary condition at $z = h$. This additional boundary condition expresses the nature of the restraint at $z = h$.

For interpretation of experimental results on a cylindrical sample sandwiched between two rigid plates we consider the cylindrical sample to be attached to a rigid plate at $z = h$. The rigid plate is taken to have density

ρ_0 and polar moment of inertia I_0. The plate is assumed to be bonded to the cylindrical viscoelastic sample at $z = h$, but is otherwise free. The dynamical equation for the angular acceleration of the rigid plate gives the following boundary condition at $z = h$:

$$-J \int_0^t G(t - t') \frac{\partial \Omega(z, t')}{\partial z}\bigg|_{z=h} \mathrm{d}t' = \rho_0 I_0 \frac{\partial \Omega(h, t)}{\partial t}. \tag{11}$$

To solve equations (10) and (11) we take the Laplace transform of the field equation (10) and the boundary condition (11) to obtain

$$\rho s \tilde{\Omega}(z, s) = \tilde{G}(s) \frac{\partial^2 \tilde{\Omega}(z, s)}{\partial z^2} \tag{12}$$

and

$$-J \tilde{G}(s) \frac{\partial \tilde{\Omega}(h, s)}{\partial z} = \rho_0 I_0 s^2 \tilde{\theta}(h, s), \tag{13}$$

respectively, where a superposed ˜ denotes the Laplace transform of the underlying function. In taking the Laplace transforms of terms involving time derivatives the initial values of Ω and θ are assumed to vanish. Replacement of $\tilde{\Omega}$ by $s\tilde{\theta}$ in (12) and (13) gives the ordinary differential equation

$$\rho s \tilde{\theta}(z, s) = \tilde{G}(s) \frac{\partial^2 \tilde{\theta}(z, s)}{\partial z^2} \tag{14}$$

for $\tilde{\theta}(z, s)$ and the boundary condition

$$-J \tilde{G}(s) \frac{\partial \tilde{\theta}(h, s)}{\partial z} = \rho_0 I_0 s \tilde{\theta}(h, s). \tag{15}$$

Equation (14) has a solution of the form

$$\tilde{\theta}(z, s) = A \cosh(\lambda z) + B \sinh(\lambda z), \tag{16}$$

where

$$\lambda(s) = \sqrt{\rho s / \tilde{G}(s)} \tag{17}$$

and the coefficients A and B are to be determined from the boundary condition (15) and the imposed motion at $z = 0$. We assume that the imposed motion is the harmonic motion

$$\theta(0, t) = \theta_0 \exp(\mathrm{i}\omega_0 t), \tag{18}$$

where θ_0 and ω_0 are constants.

The coefficient A obtained by the substitution of (16) into the Laplace transform of (18) is

$$A(s) = \frac{\theta_0}{s - \mathrm{i}\omega_0}. \tag{19}$$

From the boundary condition (15) and the relation (19), the coefficient B is

$$B(s) = -\frac{J\tilde{G}(s)\lambda(s)\sinh[\lambda(s)h] + \rho_0 I_0 s\cosh[\lambda(s)h]}{J\tilde{G}(s)\lambda(s)\cosh[\lambda(s)h] + \rho_0 I_0 s\sinh[\lambda(s)h]}\frac{\theta_0}{(s-\mathrm{i}\omega_0)}. \tag{20}$$

To invert the Laplace transform $\tilde{\theta}$ we use the definition of the inverse transform:

$$\theta(h,t) = \frac{1}{2\pi\mathrm{i}}\int_{\varepsilon-\mathrm{i}\infty}^{\varepsilon+\mathrm{i}\infty} \exp(st)\tilde{\theta}(h,s)\mathrm{d}s. \tag{21}$$

The parameter ε is a positive real number so the integration path is to the right of the imaginary axis in the s-plane. The integration contour can be closed in the left half plane. No contribution is obtained from the contour at ∞. Therefore, by the Residue Theorem, the integral (21) is simply equal to $2\pi\mathrm{i}$ times the sum of the residues of the integrand at poles to the left of $s=\varepsilon$ plus possible contributions from integrals along branch cuts. The only pole is at $s=\mathrm{i}\omega_0$. Here

$$\lambda(\mathrm{i}\omega_0) = \sqrt{\frac{\rho\mathrm{i}\omega_0}{\tilde{G}(\mathrm{i}\omega_0)}} = \mathrm{i}\sqrt{\frac{\rho\omega_0^2}{G^*(\omega_0)}}, \tag{22}$$

in which (e.g., Ferry, 1980)

$$G^*(\omega_0) \equiv \mathrm{i}\omega_0\tilde{G}(\mathrm{i}\omega_0) \tag{23}$$

is the complex modulus for the viscoelastic response of the sample material in simple shear at circular frequency ω_0. Characterization of the viscoelastic response can be put into more readily interpretable form by representing the complex modulus in terms of its magnitude $|G^*(\omega)|$ and its phase angle $\delta(\omega)$:

$$G^*(\omega) = |G^*(\omega)|\mathrm{e}^{\mathrm{i}\delta(\omega)}. \tag{24}$$

In writing the solution it is helpful to simplify the argument to

$$\lambda h = \mathrm{i}(\xi\alpha - \mathrm{i}\xi\beta), \tag{25}$$

where

$$\xi = \omega_0 h\sqrt{\frac{\rho}{|G^*(\omega_0)|}}, \quad \alpha = \cos\left(\frac{\delta(\omega_0)}{2}\right), \quad \beta = \sin\left(\frac{\delta(\omega_0)}{2}\right).$$

Then, the angular velocity at the end $z=h$ can be written in the form

$$\theta(h,t) = M(\omega_0)\theta_0\cos[\omega_0 t - \phi(\omega_0)], \tag{26}$$

in which $M(\omega_0)$ is the amplification factor:

$$M = \frac{c\sqrt{[c\cos(\xi\alpha) - d\sin(\xi\alpha)]^2 + (c^2+d^2)\sinh^2(\xi\beta)}}{[c\cos(\xi\alpha) + d\sin(\xi\alpha)]^2 + (c^2+d^2)\sinh^2(\xi\beta)}$$

where

$$c = c(\omega_0) \equiv J\sqrt{|G^*(\omega_0)|\rho}, \quad d = d(\omega_0) \equiv \rho_0 I_0 \omega_0,$$

and $\phi(\omega_0)$ is the phase shift characterized by

$$\tan[\phi(\omega_0)] = \frac{[c\sin(\xi\alpha) + d\cos(\xi\alpha)]\sinh(\xi\beta)}{[c\cos(\xi\alpha) - d\sin(\xi\alpha)]\cosh(\xi\beta)}.$$

If the response $M(\omega_0), \phi(\omega_0)$ is measured over a range of frequencies, then the moduli $|G^*(\omega_0)|, \delta(\omega_0)$ can be estimated by means of a regression analysis. Parameter values in the regression models are readily obtained using an Excel spreadsheet and minimizing the differences between calculated and measured values by using the Solver routine.

3 Experimental Setup

The experimental setup is shown in Fig. 1. A thin cylindrical sample of a soft material is placed between two hexagonal plates. The bottom plate is rotated back and forth through angles up to ±6° by means of a galvanometer (obtained from a commercially available optical scanner: GSI Lumonics, Model 000-G112) at frequencies up to 2500 Hz. To ensure that the shear strains in the sample are sufficiently small for linear viscoelasticity to be a satisfactory

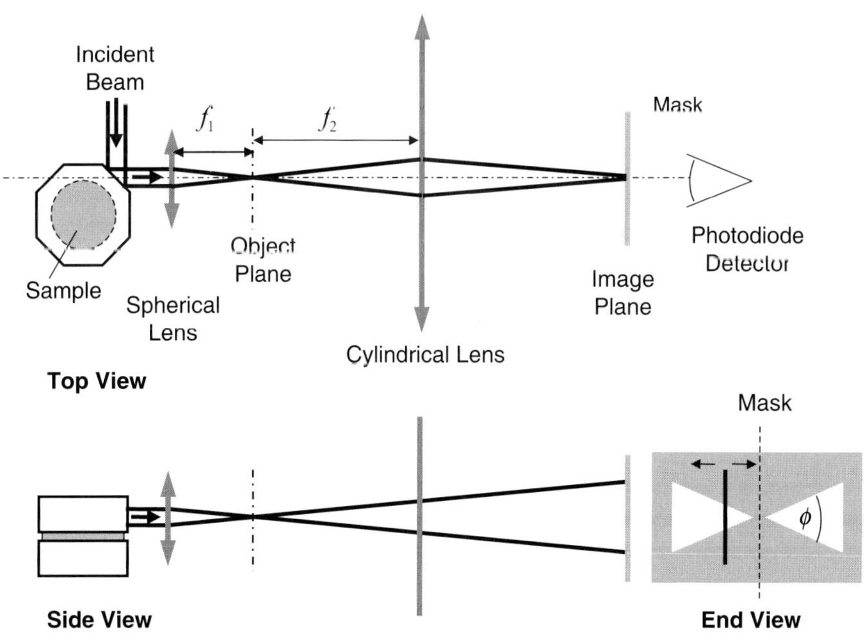

Fig. 1. Optical layout for torsional wave experiments.

approximation, the imposed rotations are kept less than $\pm 0.2°$. The rotations of the top and bottom plates are monitored by an optical lever technique in which laser beams, reflected off aluminized faces of the hexagonal plates, are passed through a spherical lens, a cylindrical lens and a butterfly-shaped aperture before being captured by photodiode detectors. The spherical lens expands the beam and the cylindrical lens focuses the beam into a vertical sheet of light at the aperture. The spherical lens is selected to ensure that the height of the sheet of light at the aperture is greater than the maximum height of the aperture. Because of the shape of the aperture (see End View in Fig. 1) the light power collected, and consequently the output of each photodiode, is proportional to the angular rotation of the respective plate.

The rotation of the bottom plate is driven by a computer-controlled frequency generator that steps through a sequence of frequencies from a minimum frequency, f_{min}, to a maximum frequency, f_{max}, in steps Δf . At each frequency the amplitude of the rotation of each plate is obtained as the average amplitude over a fixed number of cycles, usually 10, after transients from the change in frequency have died out. The output of the photodiodes is monitored by an oscilloscope (Agilent 54622A) that provides directly the average amplitude of the signal, less its minimum value. The difference between these two quantities is proportional to the rotation of the respective plate. The experimentally determined amplification factor is obtained as the ratio of the amplitude of the rotation of the top plate divided by that of the bottom plate. Calibration differences between the recorded outputs for the two plates are accounted for by adjusting the experimentally determined amplification factor to approach the required value of unity as the frequency goes to zero.

Amplification factors for each frequency, determined from the experimental records as described above, are compared with those predicted by the viscoelasticity model described in the previous section. For that model the viscoelastic description of the material is expressed in terms of the amplitude of the complex shear modulus, $|G^*(\omega)|$, and the loss angle, $\delta(\omega)$. The height and diameter of the sample, required for the wave analysis, are obtained from digital images of the sample in place. The output signal shows a peak in the amplification factor at a frequency, f_{peak}. For each test, a constant modulus $|G^*|$ and phase shift δ are obtained that provide the best fit between the model and the experimental results over a range of frequencies spanning the frequency f_{peak}.

4 Results and Discussion

An example of results for HA-based hydrogels is shown in Fig. 2. The cylindrical samples were punched out of sheets, synthesized in a flat-bottomed container and cross-linked using a UV light source. Tests were conducted under conditions of ambient temperature and humidity. Close agreement between the results of the experiment and the predictions of the model for constant

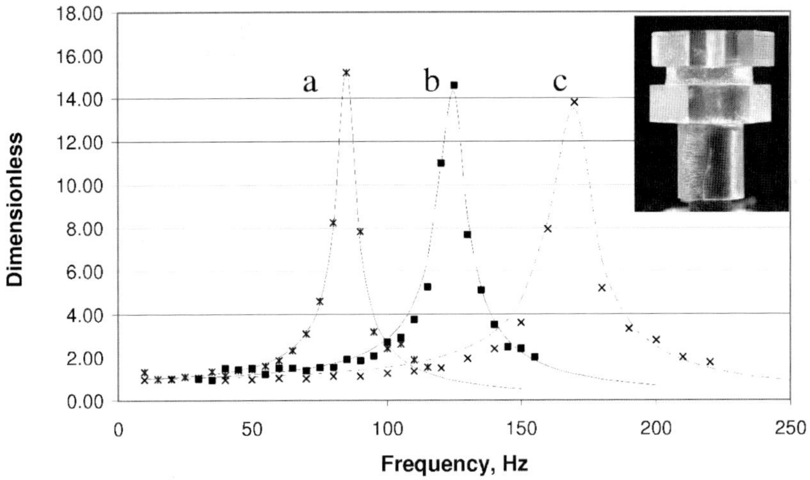

Fig. 2. Frequency dependence of amplification factors for three synthetic hydrogels; curves are model fits and symbols are experimental results. Hydrogel compositions and the viscoelastic moduli that provide the best fit between model predictions and experimental results are: (a) HAGMA (2%), $h = 0.969$ mm, $a = 2.52$ mm, $G_1 = 1514$ Pa, $\delta = 0.127$ radians; (b) HAGMA (2%)/PEGDA[MW = 575](6%), $h = 0.997$ mm, $a = 2.54$ mm, $G_1 = 3230$ Pa, $\delta = 0.132$ radians; (c) HAGMA (2%)/PEGDA[MW = 2000](12%), $h = 0.969$ mm, $a = 2.51$ mm, $G_1 = 6066$ Pa, $\delta = 0.14$ radians. Inset shows the sample (c) between the two plates of the loading system. PEGDA: polyethylene glycol diacrylate. HAGMA: Glycidyl methacrylate modified hyaluronic acid. [MW]: molecular weight.

values of the viscoelastic parameters (G_1, δ) suggests the validity of the model and the relatively weak dependence of these parameters on frequency over a range of frequencies comparable to the widths of the peaks, say 50 Hz. The increase in modulus with increasing fractions of PEGDA is consistent with expectations of the stiffening effect of this additive. Results are repeatable for successive tests on the same sample as long as the time between tests is not so long that the moisture content of the sample changes significantly. Results are also reproducible for separate samples prepared in the same way.

Examples of results for lamina propria samples taken from a dog larynx and from a ferret larynx are shown in Fig. 3. Samples were prepared by separating the LP from the vocalis muscle and then punching out a cylindrical sample with a thickness equal to the full thickness of the LP. Again the tests were conducted under conditions of ambient temperature and humidity. The fit between model and experiment is not as good as that shown in Fig. 2 for hydrogels. Better agreement can be obtained by including weak dependence of the viscoelastic moduli on frequency. Nevertheless, the agreement is sufficiently good to give confidence in the approximate levels of the complex

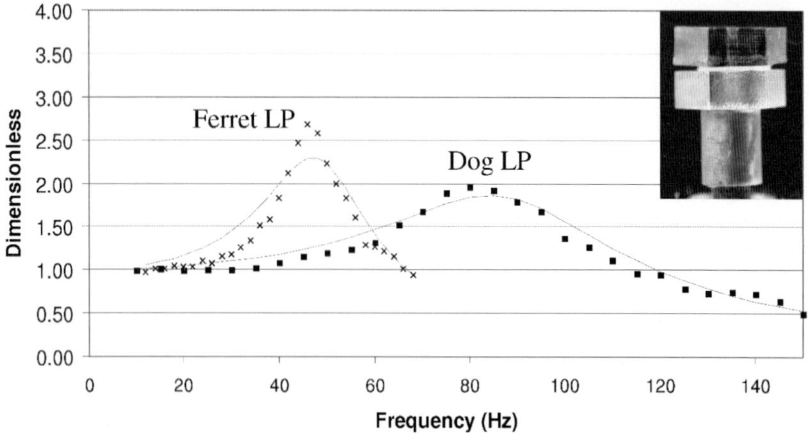

Fig. 3. Amplification factors for lamina propria samples obtained from a dog and a ferret; curves are model fits and symbols are experimental results. Sample dimensions and the values of the viscoelastic moduli that provide the best fit to the experimental measurements are as follows. Dog LP: $h = 0.371\,\text{mm}$, $a = 1.74\,\text{mm}$, $G_1 = 1423\,\text{Pa}$, $\delta = 0.898$ radians. Ferret LP: $h = 0.27\,\text{mm}$, $a = 1.43\,\text{mm}$, $G_1 = 1944\,\text{Pa}$, $\delta = 1.131$ radians. Inset shows the dog LP between the two plates of the loading system.

modulus characterized by the parameters (G_1, δ). Much lower peak values for the amplification factors result in much larger values for the loss angles δ.

Acknowledgements. RJC is grateful to Brown University for Sabbatical Leave support that made this research possible as well as to Bob Langer at MIT and Steven Zeitels at Massachusetts General Hospital for giving him the opportunity to join their research effort on vocal fold regeneration. We thank James Kobler for his help with tissue processing. XJ and MSH are grateful to the Eugene B. Casey Foundation and the Advisory Board Foundation for their financial support.

References

Brodt, M., Cook, L. S., and Lakes, R. S. (1995). Apparatus for measuring viscoelastic properties over ten decades: refinements. *Rev. Scientific Instruments* 66:5292–5297.

Chan, R. W., and Titze, I. R. (1999). Viscoelastic shear properties of human vocal fold mucosa: Measurement methodology and empirical results. *J. Acoust. Soc. Am.* 106:2008–2021.

Chan, R. W., and Titze, I. R. (2000). Viscoelastic shear properties of human vocal fold mucosa: theoretical characterization based on constitutive modeling. *J. Acoust. Soc. Am.* 107:565–580.

Ferry, J. D. (1980). *Viscoelastic Properties of Polymers.* New York: Wiley and Jones, 3rd edition.

Hemler, R. J., Wieneke, G. H., van Riel, A. M., Lebacq, J., and Dejonckere, P. H. (2001). A new method for measuring mechanical properties of laryngeal mucosa. *Eur. Arch. Otorhinolaryngol.* 258:130–136.

Jia, X., Burdick, J., Kobler, J., Clifton, R. J., Rosowski, J. J., Zeitels, S. M., and Langer, R. (2004). Synthesis and characterization of *in situ* crosslinkable hyaluronic acid-based hydrogels with potential application for vocal fold regeneration. *Macromolecules* 37:3239–3248.

Schlichting, H. (1960). *Boundary Layer Theory.* New York: McGraw-Hill, 4th edition. (Translated by J. Kestin).

Titze, I. R., Klemuk, S. A., and Gray, S. (2004). Methodology for rheological testing of engineering biomaterials at low audio frequencies. *J. Acoust. Soc. Am.* 115:392–401.

An Alternative Fabric-based Yield and Failure Criterion for Trabecular Bone

Ph. Zysset[1], L. Rincón[2]

[1] TU-Wien, Institute for Lightweight Design and Structural Biomechanics, Austria
philippe.zysset@ilsb.tuwien.ac.at
[2] Stryker Trauma, Switzerland
liliana.rincon@stryker.com

Computational analysis of bone and bone-implant systems relies on appropriate knowledge of the mechanical properties of trabecular bone. Architecture of trabecular bone can be quantified with increasing accuracy using quantitative computer tomography or magnetic resonance imaging, which opens the perspective of estimating the associated mechanical properties from 3D image reconstructions. For this purpose, the most promising variables are volume fraction and fabric, which is a second rank tensor describing the orientation and degree of architectural anisotropy. Motivated by previous successful estimates of elastic properties, the objective of this work is to develop an adequate model for describing a multi-axial yield and/or failure criterion for trabecular bone based on volume fraction and fabric. Due to linearity of the elastic behaviour of trabecular bone, the complementary elastic free energy of a previous model relating architectural and elastic properties is used for formulation of a generalized Hill criterion, which is quadratic with respect to the stress tensor. To model proper material symmetry, the criterion is restricted to an isotropic scalar function of volume fraction, fabric tensor and stress. In order to account for the distinct yield behaviour in compression and tension, a half-space-wise definition of this generalized Hill criterion is proposed. In order for the resulting global criterion to conform with the adopted material symmetry, the hyperplane separating the two half-spaces is also characterized by an isotropic scalar function of volume fraction, fabric and stress. A differentiability condition requires smoothness of the criterion at the interface between the half-spaces for arbitrary architectural parameters. A minimal set of four constants and two exponents is identified for description of the yield or failure surface. These constants coincide with the parameters of a constant half-space-wise criterion with cubic or higher symmetry when volume fraction and all fabric eigenvalues are equal to 1. Finally, the criterion developed is successfully applied to the analysis of a large experimental data set of human trabecular bone from several anatomical locations.

1 Introduction

The yield and failure properties of trabecular bone are of key interest in understanding and predicting fractures of bones and bone-implant systems. Heterogeneity and anisotropy of trabecular bone lead to variations in uniaxial compressive strength beyond an order of magnitude and a factor of two, respectively. Volume fraction (BV/TV or structural density ρ) and fabric tensors are the architectural variables characterizing heterogeneity and anisotropy, respectively, and need therefore to be accounted for in the elaboration of a meaningful yield or failure prediction.

The uniaxial yielding and strength properties were studied extensively in the past century and the dependence of the results with respect to volume fraction were reviewed by Rice et al. (1988). They concluded that uniaxial strength of cancellous bone is proportional to the square of apparent density and depends on the anatomical loading direction. Compressive strength of bovine tibial trabecular bone was found to be higher than tensile strength, and the difference seems to increase with apparent density (Keaveny et al. (1994b)). Ultimate strain was observed to be approximately constant but significantly higher in compression. Tensile yield strains of human vertebral trabecular bone were found to be independent of apparent density, while compressive yield strains were weakly and positively correlated with apparent density (Kopperdahl and Keaveny (1998)). Yield strains were found to be independent of the orientation of trabecular architecture (Turner (1989)) but dependent on anatomical site (Morgan et al. (1999)). A major drawback of these various studies is the lack of quantitative characterization of the architecture of the tested specimens. In particular, it remains unclear from these studies whether the differences in mechanical properties observed across anatomical locations are explained by volume fraction and the degree of anisotropy. The physiological loading mode of trabecular bone located at the end of long bones or in the core of short or flat bones is essentially multi-axial because of the confining presence of the cortical shell.

The availability of 3D reconstructions and morphological analysis software for trabecular architecture allows both the magnitude and the principal directions of trabecular anisotropy to be quantified using a second-order fabric tensor (Harrigan and Mann (1984)). Several authors reported correlation of strengths with mean intercept length (Goulet et al. (1994)) without providing a theoretical framework for general analysis. Fabric-elasticity models were formulated that describe the dependence of the elasticity tensor on volume fraction and fabric (Cowin (1985), Zysset and Curnier (1995)). Both experimental and numerical studies confirmed the validity of these models (Zysset (2003)) for predicting the multi-axial elastic properties of human trabecular bone from volume fraction and a second-order fabric tensor. Unfortunately, little experimental data are available for yielding and failure in the multi-axial loading cases. A preliminary attempt to identify multi-axial strength characteristics of trabecular bone is due to Stone et al. (1983). They used a device for applying plane stress states developed by Arcan et al. (1978) to measure coupled shear and tensile strengths and fitted their data to an isotropic version of Hoffmann's criterion for brittle orthotropic materials (Hoffman (1967)). A normalized cellular solid criterion was reported by Fenech and Keaveny (1999) that describes the failure envelope for the combination of torsion and uniaxial loading of bovine trabecular bone. In an effort to identify a complete multi-axial failure criterion for bovine tissue, they performed additional triaxial tests and concluded that the Tsai-Wu criterion was not the best choice to match their experimental results. However, no theoretical approach was proposed to account systematically for the variability in volume fraction and architecture. A first formulation of a fabric-based yield criterion for trabecular bone was proposed by Cowin and Van Buskirk (1986) using the Malmeister-Tsai-Wu phenomenological approach. The criterion is defined by a scalar function of stress, volume fraction and a second-order fabric tensor. It includes quadratic and linear stress components that account for distinct thresholds under tensile and compressive loading. Qualitatively, it describes an ellipsoid in stress space whose center is shifted from the origin. An experimental program for quantifying the 13 material functions of volume fraction of the model has not been undertaken. An anisotropic fracture criterion inspired from soil mechanics was suggested for bone that includes the third invariant of the stress tensor (Pietruszak et al. (1999)). Again, no experimental program was reported to verify the validity of the model.

Consequently, the objective of the present work is to propose an alternative yield and failure criterion that accounts for volume fraction and architecture of human trabecular bone. The predictions of the model are then compared with the results of a recent experimental program that includes the relevant morphological and mechanical data.

2 Model

Trabecular architecture is characterized by volume fraction ρ and fabric tensors that emerge from the approximation of a square integrable orientation distribution function $a(\mathbf{N})$ using spherical harmonics (Kanatani (1984)). This is given by

$$a(\mathbf{N}) = GA + \mathbf{G} : \mathbf{A}(\mathbf{N}) + \mathbb{G} :: \mathbb{A}(\mathbf{N}) + \ldots, \quad \forall \mathbf{N} = \mathbf{n} \otimes \mathbf{n}, \tag{1}$$

where the first tensorial basis functions are

$$A = 1, \quad \mathbf{A} = \mathbf{N} - \frac{1}{3}\mathbf{I},$$
$$\mathbb{A} = \mathbf{N} \otimes \mathbf{N} - \tfrac{1}{7}(\mathbf{I} \otimes \mathbf{N} + \mathbf{N} \otimes \mathbf{I}) - \tfrac{2}{7}(\mathbf{I} \underline{\overline{\otimes}} \mathbf{N} + \mathbf{N} \underline{\overline{\otimes}} \mathbf{I}) + \tfrac{1}{35}(\mathbf{I} \otimes \mathbf{I}) + \tfrac{2}{35}(\mathbf{I} \underline{\overline{\otimes}} \mathbf{I}), \tag{2}$$

and the fabric tensors are the traceless tensorial coefficients

$$G = \frac{1}{4\pi}\int_S a(\mathbf{N})\mathrm{d}A, \quad \mathbf{G} = \frac{15}{8\pi}\int_S a(\mathbf{N})\mathbf{A}(\mathbf{N})\mathrm{d}A, \quad \mathbb{G} = \frac{315}{32\pi}\int_S a(\mathbf{N})\mathbb{A}(\mathbf{N})\mathrm{d}A, \tag{3}$$

S being the unit sphere. When the orientation distribution function matches an ellipsoid, like the commonly used mean intercept length (Harrigan and Mann (1984)), a fabric tensor $\mathbf{H}$ can be defined through

$$a(\mathbf{N}) = \frac{1}{\sqrt{\mathbf{N}:\mathbf{H}}}, \tag{4}$$

where $\mathbf{H}$ is symmetric and positive definite. In an attempt to include both approaches in this work, a single symmetric positive definite fabric tensor $\mathbf{M}$ may be defined by

$$\mathbf{M} = G\mathbf{I} + \mathbf{G}, \quad \mathbf{M} = \mathbf{H}^{-\frac{1}{2}} \tag{5}$$

where a second-order approximation is retained in the spherical harmonics approach. The fabric tensor is normalized by

$$\det \mathbf{M} = 1, \tag{6}$$

which ensures that $\mathbf{M} = \mathbf{I}$ when its three eigenvalues degenerate.

A phenomenological yield or strength model consists of a scalar function of stress $f(\mathbf{S})$ representing a surface that separates the intact elastic states of stress from the damaged or failed ones. To account for volume fraction and to respect the symmetry of the trabecular architecture, this scalar function of stress is required to be an isotropic function of volume fraction and the spectral elements of the fabric tensor $\mathbf{M}$, i.e.

$$f(\mathbf{S}, \rho, m_1, m_2, m_3, \mathbf{M}_1, \mathbf{M}_2, \mathbf{M}_3) = 1, \tag{7}$$

with

$$\mathbf{M} = \sum_{i=1}^{3} m_i \mathbf{M}_i = \sum_{i=1}^{3} m_i (\mathbf{m}_i \otimes \mathbf{m}_i). \tag{8}$$

The isotropy condition (Boehler (1987)) is

$$f\left(\mathbf{S}, \rho, m_1, m_2, m_3, \mathbf{M}_1, \mathbf{M}_2, \mathbf{M}_3\right) =$$
$$f\left(\mathbf{Q}\mathbf{S}\mathbf{Q}^T, \rho, m_1, m_2, m_3, \mathbf{Q}\mathbf{M}_1\mathbf{Q}^T, \mathbf{Q}\mathbf{M}_2\mathbf{Q}^T, \mathbf{Q}\mathbf{M}_3\mathbf{Q}^T\right) \quad \forall \mathbf{Q} \in \Theta, \tag{9}$$

where Θ is the full orthogonal group of second-order tensors.

Since the indices of the eigenvalues m_i are associated with the indices of the structural tensors $\mathbf{M}_i$, an additional requirement for isotropy is the invariance of the function f under any index permutation.

In order to exploit the expression of the quadratic free energy potential in fabric-based linear elasticity (Zysset and Curnier (1995)) and the development of cone-wise linear elasticity (Curnier et al. (1995)), a half-space-wise quadratic function is sought in the form

$$f = \begin{cases} \mathbf{S} : \mathbb{F}^{-}\mathbf{S} & h(\mathbf{S}, \rho, m_1, m_2, m_3, \mathbf{M}_1, \mathbf{M}_2, \mathbf{M}_3) < 0 \\ \mathbf{S} : \mathbb{F}^{+}\mathbf{S} & h(\mathbf{S}, \rho, m_1, m_2, m_3, \mathbf{M}_1, \mathbf{M}_2, \mathbf{M}_3) \geq 0, \end{cases} \tag{10}$$

where the arguments of the fourth-order tensor functions $\mathbb{F}^{\mp}$ were omitted for conciseness and h is the parametrization of the hyperplane separating the compressive and tensile half-spaces. In order to respect again the selected material symmetry, the hyperplane must satisfy

$$\begin{aligned} &h\left(\mathbf{S}, \rho, m_1, m_2, m_3, \mathbf{M}_1, \mathbf{M}_2, \mathbf{M}_3\right) = \\ &h\left(\mathbf{Q}\mathbf{S}\mathbf{Q}^T, \rho, m_1, m_2, m_3, \mathbf{Q}\mathbf{M}_1\mathbf{Q}^T, \mathbf{Q}\mathbf{M}_2\mathbf{Q}^T, \mathbf{Q}\mathbf{M}_3\mathbf{Q}^T\right) \quad \forall \mathbf{Q} \in \Theta, \end{aligned} \tag{11}$$

with the same invariance under a permutation of the indices as the function f. Accordingly, the hyperplane can be expressed as a linear combination of the three mixed linear invariants of the fabric and stress tensors, i.e.

$$h = \sum_{i=1}^{3} \alpha_i(\rho, m_1, m_2, m_3)\mathrm{tr}\left(\mathbf{M}_i\mathbf{S}\right), \quad \sum_{i=1}^{3} \alpha_i^2 = 1. \tag{12}$$

The general forms of the fourth-order tensors $\mathbb{F}^{\mp}$ are

$$\mathbb{F}^{-} = \sum_{i=1}^{3} \frac{1}{(\sigma_{ii}^{-})^2}\mathbf{M}_i \otimes \mathbf{M}_i - \sum_{i,j=1;i\neq j}^{3} \frac{\chi_{ij}^{-}}{(\sigma_{ii}^{-})^2}\mathbf{M}_i \otimes \mathbf{M}_j + \sum_{i,j=1;i\neq j}^{3} \frac{1}{2\tau_{ij}^2}\mathbf{M}_i \underline{\overline{\otimes}} \mathbf{M}_j, \tag{13}$$

$$\mathbb{F}^{+} = \sum_{i=1}^{3} \frac{1}{(\sigma_{ii}^{+})^2}\mathbf{M}_i \otimes \mathbf{M}_i - \sum_{i,j=1;i\neq j}^{3} \frac{\chi_{ij}^{+}}{(\sigma_{ii}^{+})^2}\mathbf{M}_i \otimes \mathbf{M}_j + \sum_{i,j=1;i\neq j}^{3} \frac{1}{2\tau_{ij}^2}\mathbf{M}_i \underline{\overline{\otimes}} \mathbf{M}_j, \tag{14}$$

where $\sigma_{ii}^{-}(\rho, m_1, m_2, m_3)$ and $\sigma_{ii}^{+}(\rho, m_1, m_2, m_3)$ are the uniaxial compressive and tensile strengths along the axis of index $i = 1, 2, 3$, $\tau_{ij}(\rho, m_1, m_2, m_3)$ are the shear strengths in the plane of index $i, j = 1, 2, 3$, $i \neq j$, and $\chi_{ij}^{-}(\rho, m_1, m_2, m_3)$ and $\chi_{ij}^{+}(\rho, m_1, m_2, m_3)$ are stress interaction coefficients. Following the approach used in fabric-based elasticity (Zysset (2003)), power functions are selected for the dependence of the material properties on volume fraction and fabric eigenvalues. These are

$$\begin{aligned} &\sigma_{ii}^{-} = \sigma_0^{-}\rho^p m_i^{2q}, \quad \sigma_{ii}^{+} = \sigma_0^{+}\rho^p m_i^{2q}, \quad \tau_{ij} = \tau_0\rho^p m_i^q m_j^q, \\ &\chi_{ij}^{-} = \chi_0^{-}\frac{m_i^{2q}}{m_j^{2q}}, \quad \chi_{ij}^{+} = \chi_0^{+}\frac{m_i^{2q}}{m_j^{2q}}, \end{aligned} \tag{15}$$

where σ_0^- and σ_0^+ are the uniaxial compressive and tensile strengths, τ_0 is the shear strength, and χ_0^- and χ_0^+ are interaction coefficients for a pore-less ($\rho = 1$) bone material with at least cubic symmetry ($m_1 = m_2 = m_3 = 1$).

Continuity and differentiability of the yield or failure criterion f across the hyperplane can be expressed by

$$\mathbf{S} : \mathbb{F}^- \mathbf{S} = \mathbf{S} : \mathbb{F}^+ \mathbf{S} \quad \forall \mathbf{S} \mid h(\mathbf{S}, \rho, m_1, m_2, m_3, \mathbf{M}_1, \mathbf{M}_2, \mathbf{M}_3) = 0, \tag{16}$$

$$\mathbb{F}^- \mathbf{S} = \mathbb{F}^+ \mathbf{S} \quad \forall \mathbf{S} \mid h(\mathbf{S}, \rho, m_1, m_2, m_3, \mathbf{M}_1, \mathbf{M}_2, \mathbf{M}_3) = 0. \tag{17}$$

Differentiability is not a necessary condition for the stated purpose but offers a significant advantage for computational analysis, in particular for projection algorithms.

Characterization of differentiability can be reformulated through

$$\mathbb{F}_+ - \mathbb{F}_- = C \nabla_{\mathbf{S}} h \otimes \nabla_{\mathbf{S}} h \quad \forall \mathbf{S} \mid h(\mathbf{S}, \rho, m_1, m_2, m_3, \mathbf{M}_1, \mathbf{M}_2, \mathbf{M}_3) = 0, \tag{18}$$

which expresses that the jump between the fourth-order tensors must be normal to the separating hyperplane.

Using the result

$$\nabla_{\mathbf{S}} h = \sum_{i=1}^{3} \alpha_i \mathbf{M}_i, \tag{19}$$

we obtain

$$\alpha_i = \sqrt{\frac{1/(\sigma_{ii}^+)^2 - 1/(\sigma_{ii}^-)^2}{C}} = \sqrt{\frac{1/m_i^{4q}}{\sum_{i=1}^{3} 1/m_i^{4q}}}, \quad C = \sum_{k=1}^{n} \left(\frac{1}{(\sigma_{kk}^+)^2} - \frac{1}{(\sigma_{kk}^-)^2} \right). \tag{20}$$

The orientation of the half-space is therefore specified by the jumps of the yield or ultimate stress σ_0 in uniaxial compression and tension along each material axis. For the jump of the coupling terms $\chi_{ij}^{\mp}$ to be consistent, the following relation must hold:

$$\frac{\chi_0^- + 1}{(\sigma_0^-)^2} = \frac{\chi_0^+ + 1}{(\sigma_0^+)^2}. \tag{21}$$

Interestingly, the eigenvalues of the fabric tensor determine the orientation of the hyperplane. For instance, when the three fabric eigenvalues are equal, the hyperplane is characterized by $\operatorname{tr} \mathbf{S} = 0$ and is normal to the direction of hydrostatic pressure (see Fig. 1). The continuity and differentiability conditions are valid for an arbitrary volume fraction ρ and fabric tensor $\mathbf{M}$.

Convexity of the global yield criterion requires continuity and convexity of both parts of the criterion (Curnier et al. (1995)). Given the above required continuity and differentiability conditions, the condition for global convexity reduces to positive definiteness of both fourth-order tensors $\mathbb{F}_-$ and $\mathbb{F}_+$.

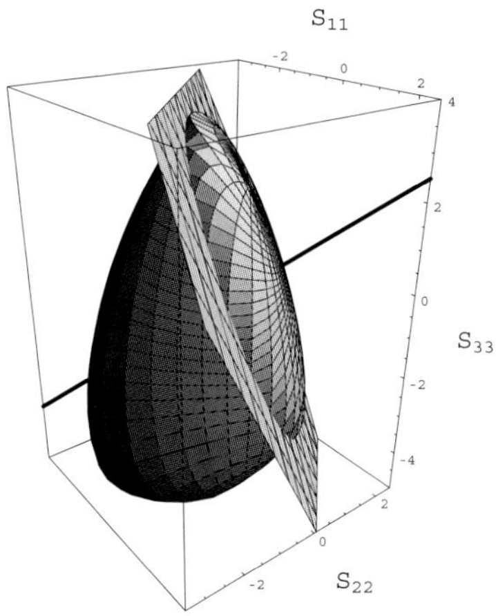

Fig. 1. Shape of the yield or failure criterion in the space of normal stresses with respect to the orthotropic planes of symmetry of the material. The hyperplane divides the stress space into compressive and tensile half-spaces. The black line indicates the states of hydrostatic stress.

Positive definiteness of these tensors is ensured by the following constraints on the material constants:

$$\begin{gathered}\sigma_0^+ \geq 0, \quad \sigma_0^- \geq 0, \quad \tau_0 \geq 0, \\ \tfrac{1}{2} \geq \chi_0^+ \geq -1, \quad \tfrac{1}{2} \geq \chi_0^- \geq -1. \end{gathered} \tag{22}$$

It should be emphasized that theses constraints are independent of volume fraction and fabric.

In fact, convexity of the elastic domain is not necessary, but in conjunction with a rate-independent associated flow rule for internal variables and the inclusion of the stress origin, it is sufficient to satisfy the entropy inequality *a priori*.

The experimental identification of the material constants requires uniaxial strength tests in tension and compression, shear tests and some biaxial or triaxial tests. Due to the homogeneity property of the model, the dependence on volume fraction and fabric may be determined from uniaxial data. However, further tests are then needed to check the hypotheses of the model in predicting yield or failure along new loading paths.

3 Results

The above model was applied to the identification of the yield and strength criteria of human cancellous bone specimens from several anatomical locations, including the proximal femur, proximal tibia, distal radius and the vertebral body. A total of 128 specimens were tested monotonically by Rincon-Kohli (2003) in axial compression, tension, torsion and multi-axial mode. The multi-axial mode consisted of proportional application of an axial compression and lateral pressure and was performed with three different axial to radial stress ratios. The details of the testing protocols can be found in Rincon-Kohli (2003). Yield was considered as a threshold for damage and was arbitrarily defined as a 20 % reduction in the slope of the axial or shear stress-strain curves. Strength was defined by the maximal axial or shear stress on these same curves.

In the model presented, material heterogeneity and anisotropy can be viewed as a stretch of the original space with stretch eigenvectors $\mathbf{m}_i$ and eigenvalues $\rho^p m_i^q$. For collective comparison of the results, the experimental stress data are therefore normalized with respect to the individual morphology by inverting this stretch operation of the original space, which leads to the effective stress

$$\hat{\mathbf{S}} = \frac{\mathbf{M}^{-q}\mathbf{S}\mathbf{M}^{-q}}{\rho^p}. \tag{23}$$

Using this transformation, the proposed half-space-wise criterion can be rewritten as

$$f(\mathbf{S}, \rho, m_1, m_2, m_3, \mathbf{M}_1, \mathbf{M}_2, \mathbf{M}_3) = f(\hat{\mathbf{S}}, \mathbf{M}_1, \mathbf{M}_2, \mathbf{M}_3) = \begin{cases} \hat{\mathbf{S}} : \hat{\mathbb{F}}^{-}\hat{\mathbf{S}} & \mathrm{tr}\,(\hat{\mathbf{S}}) < 0 \\ \hat{\mathbf{S}} : \hat{\mathbb{F}}^{+}\hat{\mathbf{S}} & \mathrm{tr}\,(\hat{\mathbf{S}}) \geq 0, \end{cases} \tag{24}$$

with

$$\hat{\mathbb{F}}^{\mp} = \sum_{i=1}^{3} \frac{1}{(\sigma_0^{\mp})^2}\mathbf{M}_i \otimes \mathbf{M}_i - \sum_{i,j=1;i\neq j}^{3} \frac{\chi_0^{\mp}}{(\sigma_0^{\mp})^2}\mathbf{M}_i \otimes \mathbf{M}_j + \sum_{i,j=1;i\neq j}^{3} \frac{1}{2(\tau_0)^2}\mathbf{M}_i \underline{\overline{\otimes}} \mathbf{M}_j. \tag{25}$$

A least-squares method was used to determine the model parameters for both the yield and strength criteria. This is stated as

$$\mathrm{Min} \sum_{k=1}^{n} \left[f(\mathbf{S}^k, \rho^k, m_1^k, m_2^k, m_3^k, \mathbf{M}_1^k, \mathbf{M}_2^k, \mathbf{M}_3^k) - 1 \right]^2, \tag{26}$$

where $\mathbf{S}^k$ represents the stress state at yield or ultimate stress and ρ^k, m_i^k, $\mathbf{M}_i^k$ characterize the morphology of specimen k.

The yield data provided the model constants

σ_0^-	σ_0^+	τ_0	χ_0^-	χ_0^+	p	q
43.6 MPa	19.2 MPa	7.5 MPa	0.33	−0.74	1.35	0.57

with $N = 112$ the number of observations, SEE $= 0.318$ the standard error of the estimate and $R^2_{\mathrm{adj}} = 0.933$ the correlation coefficient adjusted for the number of estimated parameters (see Fig. 2).

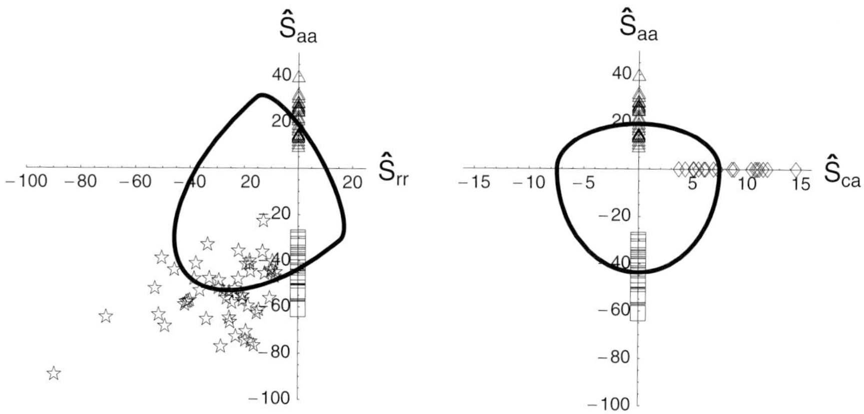

Fig. 2. Shape of the yield criterion together with the experimental data points in the normalized space of axial versus radial normal stresses (left) as well as the axial normal versus circumferential-axial shear stresses (right).

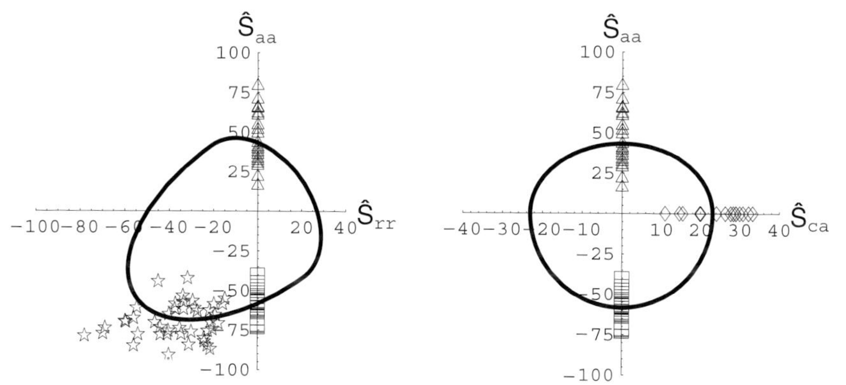

Fig. 3. Shape of the strength criterion together with the experimental data points in the space of axial and radial normal stresses (left) as well as the axial normal and circumferential-axial shear stresses (right).

The strength data lead to the alternative model constants

σ_0^-	σ_0^+	τ_0	χ_0	χ_0^+	p	q
58.4 MPa	43.3 MPa	23.1 MPa	0.31	−0.28	1.32	0.56

with $N = 96$, SEE $= 0.274$ and $R^2_{\text{adj}} = 0.932$ (see Fig. 3).

4 Discussion

The aim of this work was to develop and check the applicability of an alternative fabric-based yield or strength criterion for trabecular bone that accounts for distinct behaviour in compression and tension.

Trabecular bone is an anisotropic open-cell structure filled with marrow and made of a moderately plastic extracellular matrix. Under quasi-static loading conditions, the effect of the marrow can be neglected and the mechanical behaviour is determined by volume fraction, the degree of architectural anisotropy and the material properties of the matrix.

As a first approximation, the material properties of the healthy matrix are relatively constant and the observed variability in mechanical properties of trabecular bone is dominated by volume fraction and degree of anisotropy (Rincon-Kohli (2003)). However, the moderately plastic material properties of the matrix determine the mode of yield and failure of the trabecular structure. The high elastic modulus of the matrix almost excludes elastic buckling in the range of physiological volume fractions (0.04–0.44) and the commonly reported mode of failure is plastic yielding with subsequent formation of plastic hinges or failure of single struts (Gibson and Ashby (1988)).

Fabric tensors of increasing order can be used to describe the orientation distribution function of the trabecular structure and a second-order approximation implies orthotropic material symmetry. In the case of isotropic material symmetry, Gioux et al. (2000) used an analytical yield surface for open foams, specifically

$$f = \frac{\sqrt{\frac{3}{2}\mathbf{S}' : \mathbf{S}'}}{S^*_{pl}} + 0.81\rho \left(\frac{\operatorname{tr}\mathbf{S}}{3S^*_{pl}} \right)^2 , \tag{27}$$

where $\operatorname{tr}\mathbf{S}$ is the hydrostatic part, $\mathbf{S}'$ the deviatoric part of the stress tensor, ρ is volume fraction and S^*_{pl} is the uniaxial plastic collapse strength.

Alternative, phenomenological yield surfaces were proposed for metal foams by, e.g., Miller (2000) and Deshpande and Fleck (2000). These criteria demonstrate similar yield conditions in tension and compression that seem to be verified experimentally, but are not valid for trabecular bone (Keaveny et al. (1994a)). Due to the increase in the number of independent stress invariants, the extension of these yield surfaces to anisotropic foams is not straightforward. An attractive approach was presented by Arramon et al. (2000) that extends the basic idea of von Mises to anisotropic materials, namely to decompose the stress tensor along the spectral decomposition of the elasticity tensor.

Unfortunately, the different behaviour in tension and compression leads to gaps in the yield surface, and complementary failure modes must be defined. In the case of orthotropy, it remains unclear if the resulting polyhedral decomposition of the yield surface corresponds to specific failure modes. In this respect, phenomenological criteria such as Tsai-Wu have the mathematical and computational advantage of smoothness and uniqueness of the normal at each point.

In fact, the criterion proposed in this study has strong similarities with the orthotropic Tsai-Wu criterion (Tsai and Wu (1971)) and, in particular, with the fabric-based Tsai-Wu criterion of Cowin and Van Buskirk (1986): the major stress terms are quadratic and lead to an ellipsoid in the normal material stress space. Instead of translating the ellipsoidal criterion with additional linear stress terms, to distinguish the behaviour in tension and compression our model combines smoothly two distinct ellipsoids across a given hyperplane to achieve the same goal.

The key advantages of the model formulated here are simplicity and convexity. First, the full dependence of the model on volume fraction is contained in the exponent p. Second, the anisotropy of the model is entirely contained in the fabric tensor and the exponent q. Then, the yield surface for a plane and at least cubic material $\rho = 1$ and $\mathbf{M} = \mathbf{I}$ involves only four constants: uniaxial strength in compression and tension, shear strength and one interaction term that can be related to the resistance to hydrostatic pressure. Finally, convexity of the model and therefore closure of the yield surface is ensured under the simple conditions (22) on these four constants for an arbitrary positive volume fraction and positive definite fabric tensor. Comparison of our model with the available experimental data set is encouraging. It represents the first attempt to predict the multi-axial yield and failure characteristics for a broad spectrum of volume fraction and degree of anisotropy. The scatter of the reported data reflects primarily the boundary/size artifacts that are associated with the limited specimen dimensions and homogeneity associated with human tissue (Rincon-Kohli (2003)). Interestingly, the transformation of the identified yield criterion from the stress space into the strain space using the fabric-elasticity relationships (Zysset (2003)) suggests that uniaxial yield strains in tension and compression are not isotropic as believed by some authors (Turner et al. (1999), Chang et al. (1999), Cowin and He (2005)). Accordingly, the use of a piecewise isotropic criterion in strain space with distinct behaviour in tension and compression leads to inferior correlations $N = 112, \mathrm{SEE} = 2.03E-5, R^2_{\mathrm{adj}} = 0.626$ for the model formulated in stress space.

As pointed out by previous investigators, quadratic criteria such as Tsai-Wu, due to their ellipsoidal shape, do not accurately describe the observed interaction of axial loading with shear (Fenech and Keaveny (1999)) and the multi-axial failure in bovine trabecular bone (Keaveny et al. (1999)). Unfortunately, no attempt was made by these investigators to account for architectural anisotropy and no alternative model was proposed to improve the predictions. Nevertheless, refinements of our model may be considered in the future with the help of combined experimental and numerical approaches.

To conclude, an alternative yield and failure criterion for trabecular bone was formulated that is based on volume fraction and fabric with a minimal set of four constants. The parameters of the model were successfully identified using a large data set from uniaxial, torsional and multi-axial tests on human trabecular bone specimens with a broad range of volume fraction and degree of architectural anisotropy. This encouraging result provides the basis for improved computational analysis of the mechanical behaviour of bone and bone-implant structures using morphological information from state-of-the-art medical imaging techniques.

Acknowledgements. This research was supported by Swiss National Fund grant no 32-52821.97, a Fellowship of the Hôpital de la Suisse Romande and a Fellowship of the Ecole Polytechnique Fédérale de Lausanne. Prof. Peter Rüegsegger from ETHZ and Prof. L. Vulliet from EPFL are gratefully acknowledged for their experimental support. We owe special thanks to Prof. Alain Curnier for his careful review and relevant comments.

Notations

We use intrinsic notation and the tensor products

$$
\begin{aligned}
&[\mathbf{a} \otimes \mathbf{b}]\,\mathbf{x} = (\mathbf{x} \cdot \mathbf{b})\,\mathbf{a}, \quad \forall \mathbf{x}, && [\mathbf{e}_i \otimes \mathbf{e}_i]\,\mathbf{x} = \mathrm{x}_i\,\mathbf{e}_i,\\
&[\mathbf{A} \otimes \mathbf{B}]\,\mathbf{X} = (\mathbf{X} : \mathbf{B})\,\mathbf{A}, \quad \forall \mathbf{X}, && [\mathbf{I} \otimes \mathbf{I}]\,\mathbf{X} = (\mathrm{tr}\,\mathbf{X})\mathbf{I},\\
&[\mathbf{A} \underline{\otimes} \mathbf{B}]\,\mathbf{X} = \mathbf{A}\,\mathbf{X}\,\mathbf{B}^T, \quad \forall \mathbf{X}, && [\mathbf{I} \underline{\otimes} \mathbf{I}]\,\mathbf{X} = \mathbf{X},\\
&[\mathbf{A} \overline{\otimes} \mathbf{B}]\,\mathbf{X} = \mathbf{A}\,\mathbf{X}^T\,\mathbf{B}^T, \quad \forall \mathbf{X}, && [\mathbf{I} \overline{\otimes} \mathbf{I}]\,\mathbf{X} = \mathbf{X}^T,\\
&[\mathbf{A} \overline{\underline{\otimes}} \mathbf{B}]\,\mathbf{X} = \mathbf{A}\,\mathbf{X}\,\mathbf{B}^T, \quad \forall \mathbf{X} = \mathbf{X}^T, && [\mathbf{I} \overline{\underline{\otimes}} \mathbf{I}]\,\mathbf{X} = \mathbf{X},
\end{aligned}
$$

with $\mathbf{A} \overline{\underline{\otimes}} \mathbf{B} = \frac{1}{2}(\mathbf{A} \underline{\otimes} \mathbf{B} + \mathbf{A} \overline{\otimes} \mathbf{B})$.

References

Arcan, M., Hashin, Z., and Voloshin, A. (1978). A method to produce uniform plane-stress states with applications to fiber-reinforced materials. *Exp. Mech.* 18:141–146.

Arramon, Y. P., Mehrabadi, M. M., Martin, D. W., and Cowin, S. C. (2000). A multidimensional anisotropic strength criterion based on kelvin modes. *Int. J. Solids Structures* 37:2915–2935.

Boehler, J. P., ed. (1987). *Applications of Tensor Functions in Solid Mechanics.* Wien: Springer-Verlag. CISM Courses and Lectures No. 292, International Centre for Mechanical Sciences.

Chang, W. C. W., Christensen, T. M., Pinilla, T. P., and Keaveny, T. M. (1999). Isotropy of uniaxial yield strains for bovine trabecular bone. *J. Orthop. Res.* 17:582–585.

Cowin, S. C., and He, Q.-C. (2005). Tensile and compressive stress yield criteria for cancellous bone. *J. Biomech.* 38:141–144.

Cowin, S. C., and Van Buskirk, W. C. (1986). Thermodynamic restriction of the elastic constants of bone. *J. Biomech.* 19:85–87.

Cowin, S. C. (1985). The relationship between the elasticity tensor and the fabric tensor. *Mech. Mat.* 4:137–147.

Curnier, A., He, Q.-C., and Zysset, P. (1995). Conewise linear elastic materials. *J. Elasticity* 37:1–38.

Deshpande, V. S., and Fleck, N. A. (2000). Isotropic constitutive model for metallic foams. *J. Mech. Phys. Solids* 48:1253–1283.

Fenech, C. M., and Keaveny, T. M. (1999). A cellular solid criterion for predicting the axial-shear failure properties of bovine trabecular bone. *J. Biomech. Eng.* 121:414–422.

Gibson, L. J., and Ashby, M. F. (1988). *Cellular solids.* Oxford: Pergamon Press.

Gioux, G., McCormack, T. M., and Gibson, L. J. (2000). Failure of aluminium foams under multiaxial loads. *Int. J. Mech. Sci.* 42:1097–1117.

Goulet, R. W., Goldstein, S. A., Ciarelli, M. J., Kuhn, J. L., Brown, M. B., and Feldkamp, L. A. (1994). The relationship between the structural and orthogonal compressive properties of trabecular bone. *J. Biomech.* 27: 375–389.

Harrigan, T. P., and Mann, R. W. (1984). Characterization of microstructural anisotropy in orthotropic materials using a second rank tensor. *J. Mat. Sci.* 19:761–767.

Hoffman, O. (1967). The brittle strength of orthotropic materials. *J. Compos. Mater.* 1:200–206.

Kanatani, K.-I. (1984). Distribution of directional data and fabric tensors. *Int. J. Eng. Sci.* 22:149–164.

Keaveny, T. M., Wachtel, E. F., Ford, C. M., and Hayes, W. C. (1994a). Differences between the tensile and compressive strength of bovine tibial trabecular bone depend on modulus. *J. Biomech.* 27:1137–1146.

Keaveny, T. M., Wachtel, E. F., Guo, X. E., and Hayes, W. C. (1994b). Mechanical behavior of damaged trabecular bone. *J. Biomech.* 27:1309–1318.

Keaveny, T. M., Wachtel, E. F., Zadesky, S. P., and Arramon, Y. P. (1999). Application of the tsai-wu quadratic multiaxial failure criterion to bovine trabecular bone. *J. Biomech. Eng.* 121:99–107.

Kopperdahl, D. L., and Keaveny, T. M. (1998). Yield strain behavior of trabecular bone. *J. Biomech.* 31:601–608.

Miller, R. E. (2000). A continuum plasticity model for the constitutive and indentation behaviour of foamed metals. *Int. J. Mech. Sci.* 42:729–754.

Morgan, E. F., Arramon, Y. P., Kopperdahl, D. L., and Keaveny, T. M. (1999). Dependence of yield strain on anatomic site for human trabecular bone. In *Bioengineering Conference.* The American Society of Mechanical Engineers (ASME). BED-Vol. 43, 23–24.

Pietruszak, S., Inglis, D., and Pande, G. N. (1999). A fabric-dependent fracture criterion for bone. *J. Biomech.* 32:1071–1079.

Rice, J. C., Cowin, S. C., and Bowman, J. A. (1988). On the dependence of the elasticity and strength of cancellous bone on apparent density. *J. Biomech.* 21:155–168.

Rincon-Kohli, L. (2003). *Identification of a multiaxial failure criterion for human trabecular bone.* Ph.D. Dissertation, Swiss Federal Institute of Technology, Lausanne.

Stone, J. L., Beaupre, G. S., and Hayes, W. C. (1983). Multiaxial strength characteristics of trabecular bone. *J. Biomech.* 16:743–752.

Tsai, S. W., and Wu, E. M. (1971). A general theory of strength of anisotropic materials. *J. Compos. Mater.* 5:58–80.

Turner, C. H., Rho, J., Takano, Y., Tsui, T. Y., and Pharr, G. M. (1999). The elastic properties of trabecular and cortical bone tissues are similar: results from two microscopic measurement techniques. *J. Biomech.* 32:437–441.

Turner, C. H. (1989). Yield behavior of bovine cancellous bone. *J. Biomech. Eng.* 111:256–260.

Zysset, P. K., and Curnier, A. (1995). An alternative model for anisotropic elasticity based on fabric tensors. *Mech. Mat.* 21:243–250.

Zysset, P. (2003). A review of fabric-elasticity relationships for human trabecular bone: theories and experiments. *J. Biomech.* 36:1469–1485.

Part V

Image-based Analysis

Functional Micro-imaging at the Interface of Bone Mechanics and Biology

R. Müller, A. Nazarian, P. Schneider, M. Stauber, P. Thurner, G.H. van Lenthe, R. Voide

Swiss Federal Institute of Technology (ETH) and University of Zürich, Institute for Biomedical Engineering, Switzerland
ralph.mueller@ethz.ch

With recent advances in genetics and molecular medicine there is a strong need for quantitative image processing of three-dimensional (3D) biological structures. A number of new microstructural imaging modalities have been put forward recently, allowing phenotypic quantification with high precision and accuracy in humans and animals; especially in the mouse. Although biomedical imaging technology is now readily available, few attempts have been made to expand the capabilities of these systems by adding quantitative analysis tools as an integrative part of biomedical information technology. New strategies for 3D approaches for quantitative image processing and analysis of biological structures are presented. The focus will be on aspects of bioengineering and biomedical information technology in micro-imaging and image-guided biomechanics.

1 Introduction

Quantitative endpoints have become an important factor for success in basic research and in the development of novel therapeutic strategies in biology and biomedicine. Biomedical imaging of three-dimensional (3D) biological structures has therefore received increased attention for it is often the basis on which both qualitative (i.e. visualization) and quantitative (i.e. morphometry) image processing is performed, especially for the assessment of microstructural properties in small animals (i.e. mice). A number of new imaging modalities have been introduced in recent years. These techniques can typically be used to image a variety of different biological materials ranging from soft to hard tissues. Whereas soft tissue imaging is actually a much larger market and is nowadays regularly used as a standard procedure to image and visualize biological structures, hard tissue (i.e. bone) imaging has made large progress with respect to not only qualitative imaging but also direct 3D quantitative analysis of microstructural images in both human and animal bone.

In this article, strategies for new 3D approaches of quantitative image processing in the study and treatment of osteoporosis and bone loss will be presented. The focus will be on the bioengineering and imaging aspects of osteoporosis research. With the introduction of microstructural imaging systems such as desktop micro-computed tomography (μCT), a new generation of imaging instruments has entered the arena, allowing easy and relatively inexpensive access to the 3D microstructure of bone, and thereby giving researchers a powerful tool for the exploration of age-related bone loss and osteoporosis.

2 Motivation

Osteoporosis, which occurs most frequently in post-menopausal women and the aged, is defined as a systemic skeletal disease characterized by low bone mass and micro-architectural deterioration, with a concomitant increase in bone fragility and fracture risk (Conference (2000)). It is increasingly recognized that osteoporosis is an important public health problem because of the large size of the affected population and the devastating impact of osteoporotic fractures on morbidity and mortality of patients as well as the associated social costs. Expenditure in the United States attributable to osteoporotic fractures in 1995 reached 14 billion dollars (Ray et al. (1997)).

Although many older people may lose bone, not all develop fractures. Bone mineral density, geometry, micro-architecture and quality of the bone material are all components that determine bone strength as defined by the bone's ability to withstand loading. Neuromuscular function and environmental hazards influencing the risk of failing are important factors in determining the fracture risk. As a result, it has been found that, on an individual basis, 10–90% of the variation in the strength of trabecular bone cannot be explained by bone density. Preliminary data have shown that predicting bone strength can be greatly improved by including micro-architectural parameters in the analysis (Turner et al. (1990), Goldstein (1987)).

3 Microstructural Imaging

Quantitative bone morphometry is a method for assessing structural properties of the trabecular bone. Trabecular morphometry has traditionally been assessed in two dimensions (2D), where the structural parameters are either inspected visually or measured from sections, and the third dimension is added on the basis of stereology (Parfitt et al. (1983)). To overcome some of the limitations of 2D histology, several 3D measurement and analysis techniques have been developed over the last two decades. The most common is the use of stereo- or scanning-microscopy to qualitatively assess 3D microstructure. In using these methods, researchers were able to demonstrate the loss of 3D connectivity with age by visual observation (Mosekilde (1990)). Serial sectioning has been employed to explore the third dimension also quantitatively allowing true measurements of connectivity and other 3D structural properties like volume fraction and surface area (Odgaard (1997)). Nevertheless, being truly destructive, serial sectioning techniques will not allow secondary measurements such as mechanical testing or dynamic histomorphometry.

If we want to define an ideal imaging approach, we would like this imaging modality to be hierarchical, volumetric (3D), multi-contrast (hard vs. soft tissue), and above all fully non-invasive. Hierarchical imaging denotes the ability to resolve anatomical features at a variety of resolutions and size scales using basically the same imaging modality and ideally covering a few orders of magnitude in resolution. This ability will allow measurements starting at the organ level (500 μm resolution; 1000 mm object size), and going down from the structural level (50 μm resolution; 100 mm object size), to tissue level (5 μm resolution; 10 mm object size), and even down to the cell level (0.5 μm resolution; 1 mm object size) using one technology.

Computed tomography is such an approach to image and quantify trabecular bone in three dimensions, providing multi-scale biological imaging capabilities (Fig. 1) with isotropic resolutions ranging from a few millimeters (clinical CT), to few tens of micrometers (μCT) down to one hundred nanometers (synchrotron radiation nanoCT).

The field was pioneered by Feldkamp et al. (1989), who used a microfocus X-ray tube as a source, an image intensifier as a two-dimensional detector, and a cone-beam reconstruction algorithm to create a 3D object with a typical resolution of 50 μm. Others have used synchrotron radiation to get spatial resolutions on the order of micrometers (Bonse et al. (1994)). Where early implementations of microtomography focused more on methodological aspects of the systems and required equipment not normally available to a large public, a more recent development (Rüegsegger et al. (1996)) emphasized the practical aspects of 3D micro-tomographic imaging. The project aimed to enlarge the availability of the technology in basic research and clinical laboratories. This and other similar types of systems, now also commercially available, can be used routinely in basic research and clinical laboratories Also referred to as desktop μCT, they provide nominal resolutions ranging from roughly 5 to 100 μm. Specimens with diameters ranging from a few millimeters to 100 mm can be measured. Desktop μCT is a precise and validated technique (Balto et al. (2000), Kapadia et al. (1998), Müller et al. (1998b)), and has been used extensively for different research projects involving micro-architectural bone (von Stechow et al. (2003), Alexander et al. (2001), Dempster et al. (2001), Turner et al. (2000), Müller and Rüegsegger (1997)) and biomaterials (Lutolf et al. (2003), Zeltinger et al. (2001)). Since the introduction of these systems, there has been an increasing demand for micro-tomographic technology throughout the world.

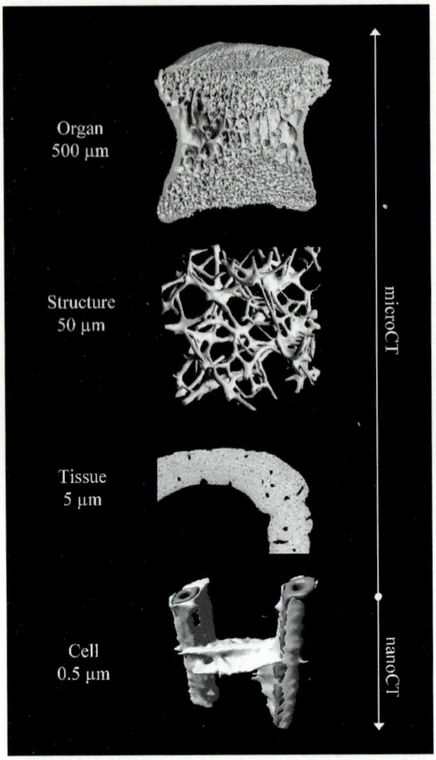

Fig. 1. Schematic overview of hierarchical imaging. From top to bottom: organ, cut through vertebral body demonstrating the large heterogeneity in the vertebra (desktop μCT); structure, three-dimensional representation of the trabecular micro-architecture in a vertebral sample (desktop μCT); tissue, cortical bone sample of a mouse femur illustrating vascular channels (large holes) and cell lacunae (synchrotron radiation μCT, SLS/PSI); osteoblast-like cell stretching between two polymer yarn filaments (synchrotron radiation μCT, SLS/PSI).

4 Three-dimensional Image Processing

Where micro-tomographic imaging provides truly 3D digitizations of the objects of interest, different image processing techniques are actually used for the modeling and analysis of that data. In the following, a few techniques available for the quantification of mineralized bone structures will be discussed.

Surface modeling and reconstruction: for surface modeling and visualization of 3D binary objects, the method described by Lorensen and Cline (1987) is often followed. They developed an algorithm, called marching cubes, which makes it possible to triangulate the surface of any given voxel array very quickly. The marching cubes algorithm is directly applicable to 3D digital data. Using this divide-and-conquer approach, it is not only possible to obtain a 3D triangular surface representation, but also to effectively smooth the surface. In order to demonstrate the capability of the surface reconstruction algorithm, a series of 3D visualizations are given in Fig. 2, illustrating the large variability in mouse and human bone architectures.

In terms of understanding the basic structure types that exist in trabecular bone, simple models have been proposed. Today, we usually distinguish between rod-like and plate-like structures (Fig. 2(d)–(f). These highly idealized models can be considered as two ends of a spectrum, where the architecture of a real bone specimen will be a mixture of both rods and plates. The prevalence of these types is dependent on the anatomical site as well as the bone age, with typical progression from a plate-like to a rod-like structure. Figure 3 displays three-dimensional, microstructural reconstructions of trabecular bone specimens from four anatomical sites, illustrating very nicely the differences between sites but also the large variance within a given site (Nägele et al. (2004)). Although such 3D visualizations are very helpful and illustrative in describing effects of age and disease on bone architecture, they cannot provide statistically analyzable data.

One method of quantitatively describing bone architecture and the changes associated with age or stage of a disease is the calculation of morphometric indices, also referred to as quantitative bone morphometry. In the past,

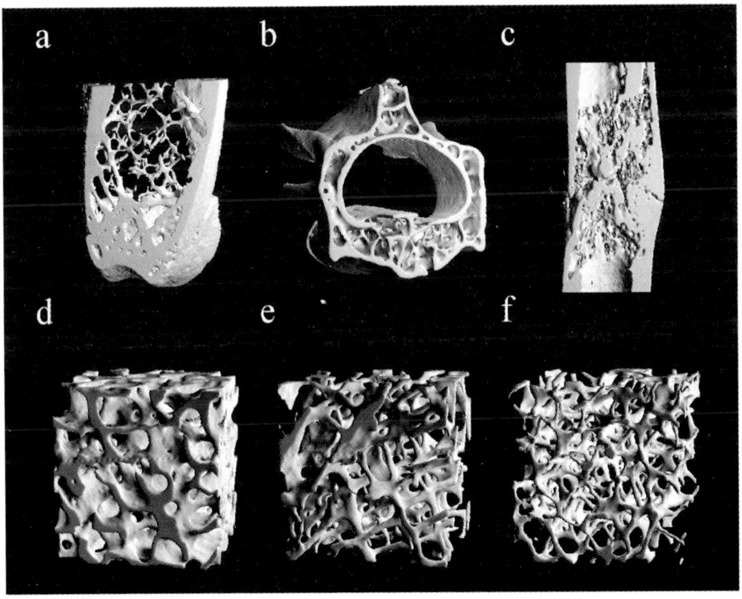

Fig. 2. High-resolution micro-tomographic images from (a) mouse distal femur, (b) mouse lumbar vertebra, (c) mouse femur diaphyseal fracture, (d) human proximal femur (plate-like structure), (e) human iliac crest (hybrid structure), (f) human lumbar spine (rod-like structure).

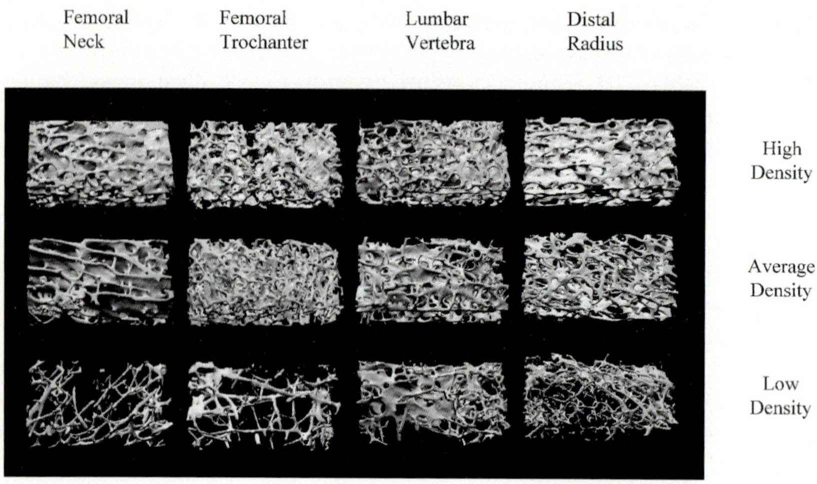

Fig. 3. Three-dimensional micro-tomographic reconstructions of bone autopsies at four different anatomical sites for low, average, and high volume density (data courtesy of Felix Eckstein, Ludwig-Maximilians-University Munich, Germany).

structural properties of trabecular bone have been investigated by the examination of 2D sections of bone biopsies. 3D morphometric parameters were then derived from 2D images using stereological methods (Parfitt et al. (1983)). Highly significant correlations between 2D histology and 3D μCT have been found for bone volume density (BV/TV) and bone surface density (BS/TV) (Müller et al. (1998b)). While measurements like BV/TV and BS/TV can be directly obtained from 2D images, a range of important parameters such as trabecular thickness (Tb.Th), trabecular separation (Tb.Sp), and trabecular number (Tb.N) are derived indirectly assuming a fixed-structure model. Typically, an ideal plate or ideal rod model is used; however, such assumptions are critical due to the fact that trabecular bone architecture differs for different sites and that it continuously changes its structure as a result of remodeling. This has been demonstrated clearly in a large study on 260 human bone biopsies taken from five different skeletal sites, and evaluated with both traditional 2D histomorphometry and newly developed 3D methods (Hildebrand et al. (1999)). For Tb.Th, Tb.Sp and Tb.N, marked differences between the two methods were found, and correlations were only moderate. Furthermore, correlations depended on anatomical site. The latter is caused by anatomical differences in bone architecture; for one site the discrepancy between the true trabecular architecture and the assumed structure is larger than for another site. Hence, deviation from the assumed model will lead to an unpredictable error of the indirectly derived parameters. This is particularly true in studies that follow the changes in bone structure in the course of age-related bone loss, and in evaluating drug therapy. In such cases, a predefined model assumption could easily overestimate or underestimate the effects of the bone atrophy depending on the assessed index.

For these reasons, and in order to take full advantage of the volumetric measurements, several new 3D image processing methods have recently been presented, allowing direct quantification of bone micro-architecture (Hildebrand et al. (1999), Odgaard (1997)). These techniques calculate actual distances in 3D space and therefore do not rely on an assumed model type and are not biased by possible deviations. In addition to the computation of direct metric parameters, nonmetric parameters can be calculated to describe the 3D nature of a bone structure. An estimate of the plate-rod characteristics can be achieved using the structure model index (SMI). For an ideal plate and ideal rod structure, the SMI is 0 and 3, respectively. For a structure with both plates and rods, the value will be between 0 and 3. Another parameter often used as an architectural index is structural anisotropy (Whitehouse (1974)), a measure of the primary orientation of the trabeculae also often referred to as degree of anisotropy (DA).

The introduction of such direct 3D measures of structural morphometry has added tremendously to the quantitativeness of microstructural imaging. State-of-the-art 3D morphometry is nowadays widely accepted as a gold standard for the characterization of trabecular bone. Nevertheless, new algorithms to better describe the complex nature of bone morphology are currently under development. They typically aim at the better characterization of local bone topology and the assessment of dynamic bone properties as assessed from *in vivo* investigations in humans and animal models of human disease.

5 Experimental Assessment of Bone Competence

Although quantification of bone micro-architecture is of great value in assessing age-related bone loss and the effects of interventions on bone micro-architecture, ultimately the aim of any bone measurement in patients is to assess bone strength. The gold standard to determine bone competence is direct mechanical testing of bone. Although in itself, it is a straightforward procedure, care must be taken in interpreting the results, as they are influenced by anatomical site and loading direction and can be influenced to a large extent by end-artifacts (Keaveny et al. (1997)). Mechanical testing has shown huge heterogeneity in bone mechanical properties, not only across sites and specimens, but even within the same bone these properties can differ 50-fold. Testing also showed that bone is not equally strong in all directions. This mechanical anisotropy is expressed as the ratio of the stiffness in the strongest direction to the stiffness in the weakest direction; it can range from basically 1 (no preferential orientation) to over 10 for both stiffness and strength (Keaveny et al. (2001), Ciarelli et al. (1991)). Because the mechanical behavior of trabecular bone is largely determined by its architecture, many investigators have correlated structural parameters with mechanical properties. Several studies have shown that where bone density alone explains around 70% of bone elastic properties, by including structural anisotropy the predictive power increases to over 90% (Van Rietbergen et al. (1998), Turner et al. (1990)).

The afore-mentioned studies were all performed on excised bone specimens. In a recent study (Müller et al. (2004)) we showed for the first time that the inclusion of bone architectural indices was also beneficial for the prediction of mechanical competence of whole bones. In that study we evaluated the effect of ibandronate on bone mass, architecture and strength in a large animal study of ovariectomized macaques. Sixty-one adult female macaques were divided into 5 groups ($N = 11$–15): sham control, OVX control, and OVX low-, medium- and high-dose ibandronate. We showed that, in pooled populations, bone mass, as assessed by Bone Mineral Content (BMC), is the single most important predictor for ultimate load ($r^2 = 67\%$). In step-wise multiple regression analysis, Tb.Sp, SMI and BS/BV contributed an additional 21% independently of BMC, such that a total of 88% of the mechano-structure relationship was explained.

6 Computational Assessment of Bone Competence

Although the inclusion of architectural parameters has strongly improved the prediction of strength for bone specimens, they only do so in a statistical sense. They do not explain the real physical contribution of the micro-architecture to the mechanical failure behavior of bone. To understand how differences in bone micro-architecture influence bone strength, insight into load transfer through the bone architecture is needed. With the advent of fast and powerful computers, simulation techniques are becoming popular for investigating the mechanical properties of bone.

Using microstructural finite element (μFE) models generated directly from computer reconstructions of trabecular bone it is now possible to perform a 'virtual experiment', i.e. to simulate a mechanical test in great detail and with high precision. Detailed FE models of trabecular bone can be created using 3D microstructural images, as described before. They are often denoted as 'high-resolution', 'large scale' or 'microstructural' FE models. These models typically represent small trabecular regions in the order of 5 to 10 mm cubes, a scale at which the bone behaves as a continuum. After assigning appropriate material properties to the elements defining the structure, these computer models provide realistic response characteristics to simulated loading. For linear deformation conditions, comparison between biomechanical compression tests and μFE show very good agreement when a homogeneous, isotropic tissue modulus is applied (Kabel et al. (1999), Ladd et al. (1998)). This holds true for normal as well as osteoporotic bone (Homminga et al. (2003)). These computer models allow calculation of loads at the microstructural or even the tissue level (Ladd et al. (1998), Van Rietbergen et al. (1995)) and have been used extensively to accurately determine the apparent mechanical properties of bone specimens.

Recently, it has been shown that non-linear μFE models can accurately predict trabecular bone failure for bovine (Niebur et al. (2000)) as well as human trabecular bone (Bayraktar et al. (2004)) when a bilinear constitutive material model is implemented. After assigning appropriate values for tensile and compressive yield strains, the μFE predicted apparent stresses and strains at failure were equal to experimentally measured values for the same bone specimens, thereby demonstrating that the quality of μFE analyses has reached such accuracy that the use of such simulation techniques can be an effective way to reduce experimental errors (Van Rietbergen et al. (1998)), and can be used as an alternative to destructive mechanical tests (Niebur et al. (2000)). A great advantage of μFE analysis is that the models can be analyzed multiple times under different conditions to simulate various types of loading. Furthermore, bone μFE models provide better insight into the relationship of structure and strength by allowing us to look inside the bone to see where stresses are localizing, and therefore where they may cause fracture.

The μFE models have shown that predicted tissue stresses and strains in the individual trabeculae can differ considerably from those estimated from apparent stresses and strains for the material as a continuum. However, since the *in situ* loading conditions for bone specimens are not precisely known, these models are inadequate for the determination of realistic loading conditions in bone tissue. For the calculation of physiological tissue loading, natural boundary conditions must be applied, which is possible only when μFE models can represent whole bones. Thanks to recent advances in scanning abilities and computer resources, this is now possible. As an example for this approach, the effect of osteoporosis on load transfer in the proximal human femur could be simulated, although at the expense of large computational efforts (Van Rietbergen et al. (2003)). The advantages of these types of analyses are that the complete geometrical and structural organization of a particular bone is taken into account and realistic boundary conditions in modeling the musculoskeletal interface can be applied, such that its outcomes can be directly related to bone failure as a better surrogate for fracture risk. In fact, when the failure characteristics of bone tissue are added to μFE models, the failure process can be followed over time, from local trabecular bone failure to complete fracture (Niebur et al. (2000)).

7 Image-guided Failure Assessment

Thus far, μFE models have mainly assessed bone loading in the elastic range, based on the notion that bone strength is highly correlated with its elastic properties. However, bone fracture is a time-dependent, non-linear event including high local deformations and local trabecular fractures. Although some work on bone failure characteristics has been done, basic knowledge of local trabecular failure is still lacking. In estimating the risk of spontaneous fractures, an extended understanding of the failure behavior of trabecular bone is essential. For this reason, our group has developed an image-guided failure assessment technique, allowing direct time-lapsed 3D visualization and quantification of fracture progression on the microscopic level (Nazarian and Müller (2004), Müller et al. (1998a)). This technique has recently been validated as compared to classical continuous mechanical testing (Nazarian and Müller (2004)). In order to compute local displacements and strains in the compressed structures a novel image analysis approach has been developed (Müller et al. (2002)). This new method uses sequential structural images from step-wise micro-compression testing to: first, align all images from the different steps; second, identify local anatomical features (nodes) in the structure that can be followed directly in the structure throughout all the steps (Fig. 4(a)); third, create a network of connections between those nodes by image decomposition (Fig. 4(b)); fourth, to compute nodal displacements and local strains between the nodes. In an initial study, trabecular bone specimens were compressed in steps of 0%, 1%, 2%, 4%, 8%, and 12% and local strains were determined experimentally. The method was verified using images, which contained only translations or a uniform stretch in the direction of compression. For the translation, our method predicted the correct translations and the computed strains were zero, as expected. For the uniformly stretched images, we computed strain averages close to the stretch factor (< 1% error) with a standard deviation of about 10%. For the actual bone micro-compression tests, the results showed that average strains were much smaller than the externally applied strain, but maximum local strain values were 5 to 8 times greater than the externally applied strain, thus providing further evidence for a band-like, local failure behavior of trabecular bone (Fig. 5).

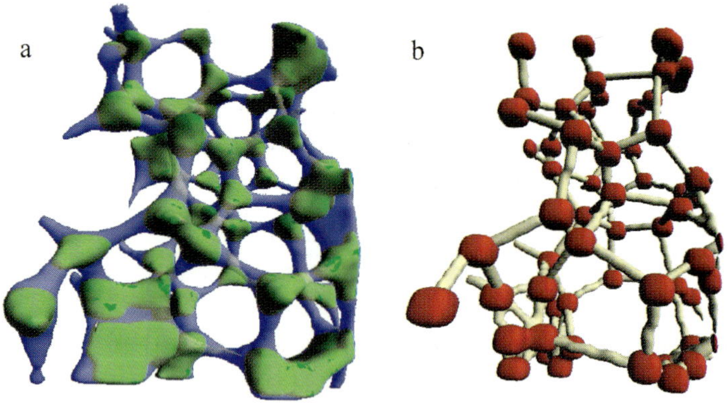

Fig. 4. Three-dimensional image decomposition: (a) thresholded reconstruction of trabecular bone structure with volumetric nodes already identified; (b) topological classification algorithm allows identification of each individual trabecula providing an element node and connection list.

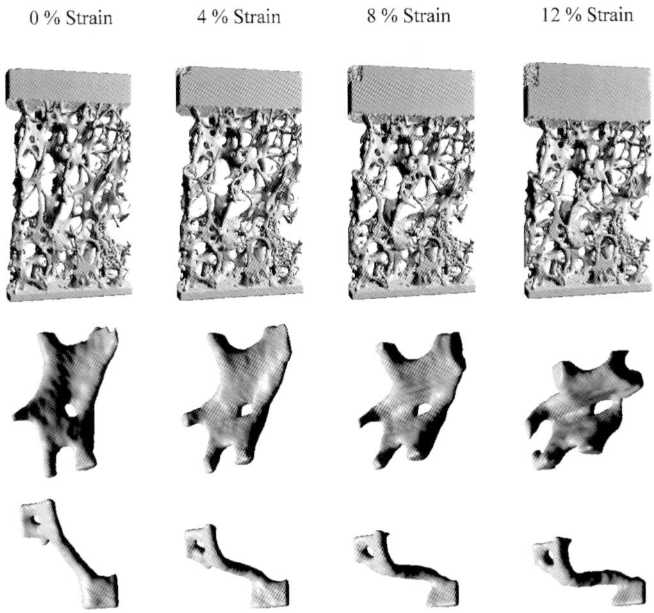

Fig. 5. Failure assessment in a human spine sample using time-lapsed tomographic imaging. Upper row shows a compressed specimen, imaged in steps of 4% strain. Middle and lower rows show how micro-compression can be used to non-invasively monitor the deformation of individual plates and rods, respectively.

These strains were found in rod-like elements that were aligned with the main strain axis. Some of these elements bend, others buckle, and some are compressed. Although inter-node strains can indicate active structure elements, they cannot distinguish the different behavior of these elements and how much energy is absorbed. They also do not correlate well with the amount of deformation in an element. Therefore, inter-node strains alone cannot fully explain the failure mechanisms of trabecular bone. Nevertheless, it is also important to note that failure behavior might be quite different for bone in an *in vivo* environment where boundary conditions differ significantly from *in vitro* setups. Therefore, extrapolation of the results on the material level to the *in vivo* situation in whole bones might prove to be difficult and will have to be addressed in future studies of whole bone failure behavior including not only trabecular micro-architecture but also cortical bone.

Image-guided failure assessment has shown that failure of an individual trabecula can lead to global bone failure; hence, bone failure can be best predicted using a 'weakest link of the chain' approach (Müller et al. (1998a)). Local bone morphometry allows identification of 'weak' trabeculae and therefore improves the predictive ability to determine bone strength and failure behavior. A 10% change in local thickness (in the rods only) was found to be responsible for a three-fold increase in the mechanical strength of osteoporotic bone, where changes in bone density were only linearly related to bone strength. This might explain a number of findings where only small changes in bone density can achieve up to a 70% reduction in the incidence of fracture. If it were possible to preferentially treat the weakest links of the structure, a minor increase in individual element quality may be enough to prevent a fracture rather than large quantities of new bone. However, this idea is momentarily purely speculative since there are no data from treated patients yet, although it will be possible to obtain such data from iliac bone samples of real patients undergoing treatment. The ability to analyze the same bone sample using both non-invasive micro-architectural imaging and mechanical testing is likely to cause a revival in the use of iliac bone biopsies in the initial assessment of therapeutic success and the regulatory approval of new drugs.

Alternatively, high-resolution *in vivo* imaging modalities are currently under investigation allowing micro-architectural measurements in live patients with resolutions better than 100 μm (Müller (2003)). Although currently only peripheral imaging systems (i.e. measures at the forearm) can achieve this kind of resolution in patients, there is hope that in the future also more central sites such as spine and hip might become accessible to such investigations. This will allow following transient changes in bone micro-architecture in individual patients, giving the clinician a powerful tool for better diagnosis of diseases and online monitoring of treatment regimens.

8 Conclusions

Micro-architectural bone imaging is a nondestructive, non-invasive, and precise procedure that allows the measurement of trabecular and compact bone as well as the repetitive 3D assessment and computation of microstructural and micromechanical properties in patients. The procedure can help improve predictions of fracture risk, clarify the pathophysiology of skeletal diseases, and define the response to therapy. Hierarchical bioimaging in combination with biocomputational approaches are well suited for investigating structure-function relationships as well as failure mechanisms in normal, osteoporotic and treated bone. We expect these findings to improve our understanding of the influence of densitometric, morphological but also loading factors in the etiology of spontaneous fractures of the hip and the spine. Eventually, this improved understanding may lead to more successful approaches in the prevention of such age- and disease-related fractures.

Acknowledgements. These studies were partly funded by a Biomedical Engineering Research Grant of the Whitaker Foundation and the SNF Professorship in Bioengineering of the Swiss National Science Foundation (FP

620-58097.99). We would like to thank Dr. Marco Stampanoni for his support in setting up the tomography station at the MS beamline, Swiss Light Source, PSI, Switzerland.

References

Alexander, J. M., Bab, I., Fish, S., Müller, R., Uchiyama, T., Gronowicz, G., Nahounou, M., Zhao, Q., Chorev, M., Gazit, D., and Rosenblatt, M. (2001). Human parathyroid hormone 1-34 reverses bone loss in ovariectomized mice. *J. Bone Miner. Res.* 16:1665–1673.

Balto, K., Müller, R., Carrington, D. C., Dobeck, J., and Stashenko, P. (2000). Quantification of periapical bone destruction in mice by micro-computed tomography. *J. Dent. Res.* 79:35–40.

Bayraktar, H. H., Morgan, E. F., Niebur, G. L., Morris, G. E., Wong, E. K., and Keaveny, T. M. (2004). Comparison of the elastic and yield properties of human femoral trabecular and cortical bone tissue. *J. Biomech.* 37:27–35.

Bonse, U., Busch, F., Gunnewig, O., Beckmann, F., Pahl, R., Delling, G., Hahn, M., and Graeff, W. (1994). 3D computed X-ray tomography of human cancellous bone at 8 microns spatial and $10^{(-4)}$ energy resolution. *Bone Miner.* 25:25–38.

Ciarelli, M. J., Goldstein, S. A., Kuhn, J. L., Cody, D. D., and Brown, M. B. (1991). Evaluation of orthogonal mechanical properties and density of human trabecular bone from the major metaphyseal regions with materials testing and computed tomography. *J. Orthop. Res.* 9:674–682.

Conference, C. D. (2000). Osteoporosis prevention, diagnosis, and therapy. In *NIH consensus statement*, volume 17, 1–45.

Dempster, D. W., Cosman, F., Kurland, E. S., Zhou, H., Nieves, J., Woelfert, L., Shane, E., Plavetic, K., Müller, R., Bilezikian, J., and Lindsay, R. (2001). Effects of daily treatment with parathyroid hormone on bone microarchitecture and turnover in patients with osteoporosis: a paired biopsy study. *J. Bone Miner. Res.* 16:1846–1853.

Feldkamp, L. A., Goldstein, S. A., Parfitt, A. M., Jesion, G., and Kleerekoper, M. (1989). The direct examination of three-dimensional bone architecture in vitro by computed tomography. *J. Bone Miner. Res.* 4:3–11.

Goldstein, S. A. (1987). The mechanical properties of trabecular bone: dependence on anatomic location and function. *J. Biomech.* 20:1055–1061.

Hildebrand, T., Laib, A., Müller, R., Dequeker, J., and Rüegsegger, P. (1999). Direct three-dimensional morphometric analysis of human cancellous bone: microstructural data from spine, femur, iliac crest, and calcaneus. *J. Bone Miner. Res.* 14:1167–1174.

Homminga, J., McCreadie, B. R., Weinans, H., and Huiskes, R. (2003). The dependence of the elastic properties of osteoporotic cancellous bone on volume fraction and fabric. *J. Biomech.* 36:1461–1467.

Kabel, J., Van Rietbergen, B., Dalstra, M., Odgaard, A., and Huiskes, R. (1999). The role of an effective isotropic tissue modulus in the elastic properties of cancellous bone. *J. Biomech.* 32:673–680.

Kapadia, R. D., Stroup, G. B., Badger, A. M., Koller, B., Levin, J. M., Coatney, R. W., Dodds, R. A., Liang, X., Lark, M. W., and Gowen, M. (1998). Applications of micro-CT and MR microscopy to study pre-clinical models of osteoporosis and osteoarthritis. *Technol. Health Care* 6:361–372.

Keaveny, T. M., Pinilla, T. P., Crawford, R. P., Kopperdahl, D. L., and Lou, A. (1997). Systematic and random errors in compression testing of trabecular bone. *J. Orthop. Res.* 15:101–110.

Keaveny, T. M., Morgan, E. F., Niebur, G. L., and Yeh, O. C. (2001). Biomechanics of trabecular bone. *Ann. Rev. Biomed. Eng.* 3:307–333.

Ladd, A. J., Kinney, J. H., Haupt, D. L., and Goldstein, S. A. (1998). Finite-element modeling of trabecular bone: comparison with mechanical testing and determination of tissue modulus. *J. Orthop. Res.* 16:622–628.

Lorensen, W. E., and Cline, H. E. (1987). Marching cubes: A high resolution 3D surface construction algorithm. *Computer Graphics* 21:163–169.

Lutolf, M. P., Weber, F. E., Schmoekel, H. G., Schense, J. C., Kohler, T., Müller, R., and Hubbell, J. A. (2003). Repair of bone defects using synthetic mimetics of collagenous extracellular matrices. *Nat. Biotechnol.* 21:513–518.

Van Rietbergen, B., Weinans, H., Huiskes, R., and Odgaard, A. (1995). A new method to determine trabecular bone elastic properties and loading using micromechanical finite-element models. *J. Biomech.* 28:69–81.

Van Rietbergen, B., Odgaard, A., Kabel, J., and Huiskes, R. (1998). Relationships between bone morphology and bone elastic properties can be accurately quantified using high-resolution computer reconstructions. *J. Orthop. Res.* 16:23–28.

Van Rietbergen, B., Huiskes, R., Eckstein, F., and Rüegsegger, P. (2003). Trabecular bone tissue strains in the healthy and osteoporotic human femur. *J. Bone Miner. Res.* 18:1781–1788.

von Stechow, D., Balto, K., Stashenko, P., and Müller, R. (2003). Three-dimensional quantitation of periradicular bone destruction by micro-computed tomography. *J. Endod.* 29:252–256.

Mosekilde, L. (1990). Consequences of the remodelling process for vertebral trabecular bone structure: a scanning electron microscopy study (uncoupling of unloaded structures). *Bone Miner.* 10:13–35.

Müller, R., and Rüegsegger, P. (1997). Micro-tomographic imaging for the nondestructive evaluation of trabecular bone architecture. *Stud. Health Technol. Inform.* 40:61–79.

Müller, R., Gerber, S. C., and Hayes, W. C. (1998a). Micro-compression: a novel technique for the nondestructive assessment of local bone failure. *Technol. Health Care* 6:433–444.

Müller, R., Van Campenhout, H., Van Damme, B., Van der Perre, G., Dequeker, J., Hildebrand, T., and Rüegsegger, P. (1998b). Morphometric analysis of human bone biopsies: a quantitative structural comparison of histological sections and micro-computed tomography. *Bone* 23:59–66.

Müller, R., Boesch, T., Jarak, D., Stauber, M., Nazarian, A., Tantillo, M., and Boyd, S. K. (2002). Micro-mechanical evaluation of bone microstructures under load. In Bonse, U., ed., *Developments in X-Ray Tomography III.* San Diego, CA: SPIE. 189–200.

Müller, R., Hannan, M., Smith, S. Y., and Bauss, F. (2004). Intermittent ibandronate preserves bone quality and bone strength in the lumbar spine after 16 months of treatment in the ovariectomized cynomolgus monkey. *J. Bone Miner. Res.* 19:1787–1796.

Müller, R. (2003). Bone microarchitecture assessment: current and future trends. *Osteoporos. Int.* 14 (Suppl. 5):89–99.

Nägele, E., Kuhn, V., Vogt, H., Linka, T. M., Müller, R., Lochmüller, E. M., and Eckstein, F. (2004). Technical considerations for microstructural analysis of human trabecular bone from specimens excised from various skeletal sites. *Calcif. Tissue Int.* 75:15–22.

Nazarian, A., and Müller, R. (2004). Time-lapsed microstructural imaging of bone failure behavior. *J. Biomech.* 37:55–65.

Niebur, G. L., Feldstein, M. J., Yuen, J. C., Chen, T. J., and Keaveny, T. M. (2000). High-resolution finite element models with tissue strength asymmetry accurately predict failure of trabecular bone. *J. Biomech.* 33:1575–1583.

Odgaard, A. (1997). Three-dimensional methods for quantification of cancellous bone architecture. *Bone* 20:315–328.

Parfitt, A. M., Mathews, C. H., Villanueva, A. R., Kleerekoper, M., Frame, B., and Rao, D. S. (1983). Relationships between surface, volume, and thickness of iliac trabecular bone in aging and in osteoporosis. implications for the microanatomic and cellular mechanisms of bone loss. *J. Clin. Invest.* 72: 1396–1409.

Ray, N. F., Chan, J. K., Thamer, M., and Melton 3rd, L. J. (1997). Medical expenditures for the treatment of osteoporotic fractures in the united states in 1995: report from the national osteoporosis foundation. *J. Bone Miner. Res.* 12:24–35.

Rüegsegger, P., Koller, B., and Müller, R. (1996). A microtomographic system for the nondestructive evaluation of bone architecture. *Calcif. Tissue Int.* 58:24–29.

Turner, C. H., Cowin, S. C., Rho, J. Y., Ashman, R. B., and Rice, J. C. (1990). The fabric dependence of the orthotropic elastic constants of cancellous bone. *J. Biomech.* 23:549–561.

Turner, C. H., Hsieh, Y. F., Müller, R., Bouxsein, M. L., Baylink, D. J., Rosen, C. J., Grynpas, M. D., Donahue, L. R., and Beamer, W. G. (2000). Genetic regulation of cortical and trabecular bone strength and microstructure in inbred strains of mice. *J. Bone Miner. Res.* 15:1126–1131.

Whitehouse, W. J. (1974). The quantitative morphology of anisotropic trabecular bone. *J. Microscopy* 101:153–168.

Zeltinger, J., Sherwood, J. K., Graham, D. A., Müller, R., and Griffith, L. G. (2001). Effect of pore size and void fraction on cellular adhesion, proliferation, and matrix deposition. *Tiss. Engrg.* 7:557–572.

Strain Measurement Using Deformable Image Registration

J.A. Weiss[1,2], A.I. Veress[1], G.T. Gullberg[3], N.S. Phatak[1,2], Q. Sun[1,2], D. Parker[4], R.D. Rabbitt[1]

[1] University of Utah, Department of Bioengineering, USA
jeff.weiss@utah.edu
[2] University of Utah, Scientific Computing and Imaging Institute, USA
[3] Lawrence Berkeley National Laboratory, Center for Functional Imaging, USA
[4] Utah Center for Advanced Imaging Research, Center for Advanced Medical Technologies, USA

The accurate determination of strain in deforming biological tissues is a necessary and important part of experimental investigations in biomechanics. We have developed a method, referred to as Hyperelastic Warping, to combine medical image data with a solid mechanics analysis approach to allow estimation of tissue strains in the absence of detailed information about boundary conditions and in some cases constitutive information. The method makes use of medical image data to provide information about the deforming tissue and thus the strain and stress fields. The mathematical problem is to search through all admissible configurations for the one that minimizes the difference between a transformed template image and a target image collected experimentally. The resulting deformation map is used to determine the strain field. This technique has been applied successfully to determine strain in several biological soft tissues. This paper describes the theory, implementation and validation of the technique for measurement of transmural strains in the left ventricle.

1 Introduction

Nonlinear strains in biological tissues and cells are usually measured using fiducial markers (e.g., Holmes et al. (1995) and McCulloch et al. (1989)). In some cases, the 3D strain can be estimated directly from changes in the distances between groups of markers making up tetrahedral sets. The inherent invasiveness, inhomogeneous strain fields and physical dimensions of the markers can limit the applicability of this technique. This paper describes a non-invasive method to estimate strain fields in biological soft tissues that may be applied in the absence of discrete fiducials or exact information regarding tissue constitutive properties. The method is based on an algorithm for deformable image registration termed *Hyperelastic Warping* (Rabbitt et al. (1995), Veress et al. (2004), Weiss et al. (1998)). The method makes use of the textural information in image data to determine the deformation map between pairs of images that represent distinct states of deformation. Validation of the method in the context of measuring transmural left ventricular strains is presented.

2 Methods

2.1 Finite Deformation Theory

A Lagrangian reference frame is assumed. A pair of image datasets, the *template* and *target*, represents some physical characteristic of the body before and after deformation, respectively. The images have spatially varying scalar intensity fields defined with respect to the reference configuration and denoted by T and S, respectively. The deformation map is $\varphi(\mathbf{X}) = \mathbf{x} = \mathbf{X} + \mathbf{u}(\mathbf{X})$, where $\mathbf{x}$ are deformed coordinates, $\mathbf{u}(\mathbf{X})$ is the displacement field, and $\mathbf{F} = \partial\varphi(\mathbf{X})/\partial\mathbf{X}$ is the deformation gradient (Spencer (1980)). The density ρ is related to the Jacobian, $J := \det \mathbf{F} = \rho_0/\rho$, where ρ_0 is the reference density. It is assumed that T and S have a general dependence on position $\mathbf{X}$ and deformation map $\varphi(\mathbf{X})$. The positive definite, symmetric right Cauchy-Green deformation tensor is $\mathbf{C} = \mathbf{F}^T\mathbf{F}$.

2.2 Variational Framework for Hyperelastic Warping

Most deformable image registration methods can be posed as the minimization of an energy functional E that consists of at least two terms. The combined energy functional can be defined with respect to the current (deformed) configuration as

$$E(\mathbf{X}, \varphi) = \int_{\mathcal{B}} W(\mathbf{X}, \varphi)\frac{dv}{J} + \int_{\mathcal{B}} U\left(T(\mathbf{X}, \varphi), S(\mathbf{X}, \varphi)\right)\frac{dv}{J}. \tag{1}$$

W is an energy term that provides regularization and/or some type of constraint on the deformation map (e.g., one-to-one mapping or no negative volumes admitted), while U represents an energy that depends on the registration of the image data in the template and target images. $\mathcal{B}$ represents the current volume of integration. In Hyperelastic Warping, a spatial discretization of the template image is deformed into alignment with the target, which remains fixed in the reference configuration. Coordinates of template material points are tracked. In the following presentation, the scalar intensity field of the template, T, is not changed directly by the deformation, and thus it is represented as $T(\mathbf{X})$. Since the values of S at material points associated with the deforming template change as the template deforms with respect to the target, it is written as $S(\varphi)$. The standard formulation uses a Gaussian sensor model:

$$U(\mathbf{X},\varphi) = \frac{\lambda}{2}\left(T(\mathbf{X}) - S(\varphi)\right)^2 . \tag{2}$$

λ is a penalty parameter (Reddy (2002)) that enforces the alignment of the images. As $\lambda \to \infty$, $\left(T(\mathbf{X}) - S(\varphi)\right)^2 \to 0$, and the image energy converges to a finite, minimized value.

It is assumed that W, a hyperelastic strain energy, defines the material behavior. It depends on $\mathbf{C}$, which is independent of rotation, and thus hyperelasticity provides an objective (invariant under rotation) constitutive framework, in contrast to linearized elasticity. Equation (1) becomes

$$E = \int_{\mathcal{B}} W(\mathbf{X},\mathbf{C})\frac{\mathrm{d}v}{J} + \int_{\mathcal{B}} U\left(T(\mathbf{X}), S(\varphi)\right)\frac{\mathrm{d}v}{J}. \tag{3}$$

The Euler-Lagrange equations are obtained by taking the first variation of $E(\mathbf{X},\varphi)$ with respect to φ. This can be thought of as a 'virtual displacement' – a small variation in the current coordinates $\mathbf{x}$, denoted $\varepsilon\boldsymbol{\eta}$, where ε is an infinitesimal scalar. Variations are calculated by taking the Gateaux derivative of U evaluated at with respect to $\varphi + \varepsilon\boldsymbol{\eta}$ and then letting $\varepsilon \to 0$. The weak form of the Euler-Lagrange equations for (3) is (Simo and Hughes (1998))

$$G(\varphi,\boldsymbol{\eta}) := DE(\varphi)\cdot\boldsymbol{\eta} = \int_{\mathcal{B}} \boldsymbol{\sigma} : \nabla\boldsymbol{\eta}\mathrm{d}v - \int_{\mathcal{B}} \lambda\left[(T-S)\frac{\partial S}{\partial\varphi}\cdot\boldsymbol{\eta}\right]\frac{\mathrm{d}v}{J} = 0. \tag{4}$$

Here, $\boldsymbol{\sigma}$ is the second-order symmetric Cauchy stress tensor

$$\boldsymbol{\sigma} = \frac{1}{J}\mathbf{F}\frac{\partial W}{\partial\mathbf{C}}\mathbf{F}^T. \tag{5}$$

Forces applied to the template physical model due to the image data are opposed by internal forces from the deformation of the material through the constitutive model. The form of W depends on the material (Spencer (1980), Weiss et al. (1996)).

Equation (4) is highly nonlinear and an incremental iterative solution method is necessary to obtain the configuration φ that satisfies this equation (Matthies and Strang (1979)). A common approach is based on linearization and iterative solution using Newton's method or some variant. Assuming a solution at configuration φ^* is known, a solution is sought at some small increment $\varphi^* + \Delta\mathbf{u}$. Linearization of (4) at φ^* in the direction $\Delta\mathbf{u}$ yields

$$\begin{aligned}\mathbf{L}\varphi\cdot G(\varphi,\boldsymbol{\eta}) &= \int_{\mathcal{B}} \boldsymbol{\sigma} : \nabla\boldsymbol{\eta}\mathrm{d}v - \int_{\mathcal{B}} \lambda\left[(T-S)\frac{\partial S}{\partial\varphi^*}\cdot\boldsymbol{\eta}\right]\frac{\mathrm{d}v}{J} \\ &+ \int_{\mathcal{B}} \nabla\boldsymbol{\eta} : \boldsymbol{\sigma} : \nabla(\Delta\mathbf{u})\mathrm{d}v + \int_{\mathcal{B}} \nabla^s\boldsymbol{\eta} : \mathbf{c} : \nabla^s(\Delta\mathbf{u})\mathrm{d}v + \int_{\mathcal{B}} \boldsymbol{\eta}\cdot\mathbf{k}\cdot\Delta\mathbf{u}\frac{\mathrm{d}v}{J},\end{aligned} \tag{6}$$

where $\mathbf{c}$ is the fourth order spatial elasticity tensor and $\nabla^s[\cdot]$ is the symmetric gradient (Bathe (1982)). The first two terms in the second line of equation (6) are referred to as the *geometric* and *material stiffnesses*, respectively (Bathe (1996)). The third term is the second-order image stiffness for Hyperelastic Warping:

$$\mathsf{k} = \frac{\partial^2 U}{\partial \varphi \partial \varphi} = \lambda \left[\left(\frac{\partial S}{\partial \varphi} \right) \otimes \left(\frac{\partial S}{\partial \varphi} \right) - (T - S) \left(\frac{\partial^2 S}{\partial \varphi \partial \varphi} \right) \right] . \tag{7}$$

These three terms form the basis for evaluating the relative influence of the image-derived forces and the forces due to internal stresses on the converged solution to the deformable image registration problem.

2.3 Finite Element Discretization and Solution Procedure

Hyperelastic Warping is based on a FE discretization of the template. A FE mesh is constructed to correspond to all or part of the template image. T is interpolated to the FE mesh and is convected with the FE mesh. As the FE mesh deforms, the values of S are queried at the current location of the FE mesh nodes. Using Voigt notation (Bathe (1996)), the linearized equations (6) can be written as

$$\sum_{i=1}^{N_{\text{nodes}}} \sum_{j=1}^{N_{\text{nodes}}} \left(\mathbf{K}^R (\varphi^*) + \mathbf{K}^I (\varphi^*) \right)_{ij} \Delta \mathbf{u}_j = \sum_{i=1}^{N_{\text{nodes}}} \left(\mathbf{F}^{\text{ext}} (\varphi^*) - \mathbf{F}^{\text{int}} (\varphi^*) \right)_i . \tag{8}$$

Equation (8) is a system of linear equations. The term on the left-hand side is the (symmetric) tangent stiffness matrix. $\Delta \mathbf{u}$ is the vector of unknown incremental nodal displacements – for a FE mesh of 8-noded hexahedral elements in three dimensions, $\Delta \mathbf{u}$ has length $8 \times 3 \times N_{\text{el}}$, where N_{el} is the number of elements in the mesh. $\mathbf{F}^{\text{ext}}$ is the vector of external forces arising from the differences in the image intensity fields, and $\mathbf{F}^{\text{int}}$ is the vector of internal forces resulting from the stress divergence. The material and geometric stiffnesses combine to give the mechanics regularization stiffness $\mathbf{K}^R$, and the FE discretization of (7) gives the image-based tangent stiffness $\mathbf{K}^I$. An initial estimate of the incremental nodal displacements is obtained by solving (8) for $\Delta \mathbf{u}$ and this solution is improved iteratively using a quasi-Newton method.

In the applications described below, U is treated as a hard constraint, with W providing a regularization. Indeed, the U in equation (2) is essentially a penalty function. The main problem with the penalty method is that as the penalty parameter λ is increased, some of the diagonal terms in $\mathbf{K}^I$ become very large with respect to others, leading to numerical ill-conditioning. To circumvent this problem, the augmented Lagrangian method is used (Fletcher (1989)). A solution to the governing equations at a particular computational timestep is first obtained with a relatively small penalty parameter λ. Then the total image-based body forces $\partial U / \partial \varphi$ are incrementally increased in a second iterative loop, resulting in better satisfaction of the constraint. This leads to a stable algorithm that allows the constraint to be satisfied to a user-defined tolerance. Ill conditioning of the stiffness matrix is entirely avoided.

3 Application – Strain Measurement in the Left Ventricle

Assessment of regional heart wall motion (wall motion, thickening, strain, etc.) can identify impairment of cardiac function due to a variety of cardiomyopathies and ischemic myocardial disease. The objectives of this study were to validate the use of Hyperelastic Warping for the extraction of high-resolution strain maps of the left ventricle from cine-MRI images. The sensitivity of predictions to errors in material model selection, material parameter estimation, and simulated noise in the image data was determined. The hypotheses were that (1) Hyperelastic Warping could accurately predict the fiber stretch (final length/initial length along the local fiber direction) and in-plane strain distributions during diastolic (passive) filling from cine-MRI image datasets, (2) variations in the assumed material properties and constitutive model would have a minimal effect on the predicted fiber stretch distribution, and (3) the results of Warping cardiac cine-MRI images would be relatively insensitive to noise in the image data.

3.1 Methods

Cardiac Image Acquisition. To mimic the typical imaging procedure used in clinical MR acquisition for patients with cardiac pathologies, gated cine-MRI images of a normal male volunteer's heart (35 years old) were acquired on a 1.5T Siemens scanner using standard clinical resolution and settings (256×256 image matrix, 378 mm FOV, 10 mm slice thickness, 10 slices) (Fig. 1). The volumetric MRI dataset corresponding to beginning of diastole was designated as the template image. The template image was cropped to 64×64 image matrix by 8 slices to focus on the heart.

FE Mesh Generation and Boundary Conditions. A conforming FE mesh was used. The boundaries of the LV were obtained by manual segmentation of the epi and endocardium from the template image (beginning of diastole). The

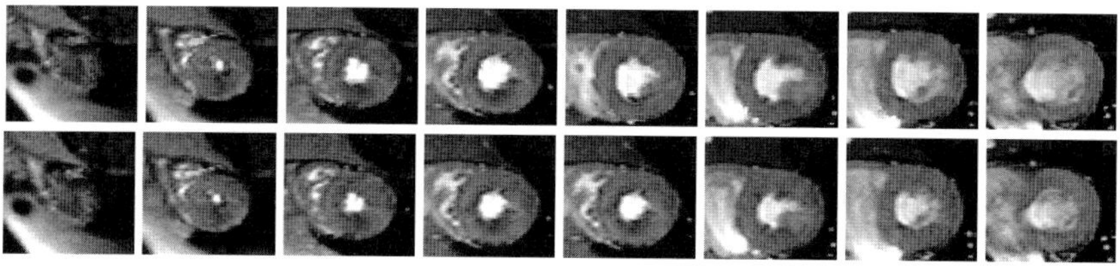

Fig. 1. Target (top) and template (bottom) image datasets used in the Warping analysis. The Target image dataset was created by mapping the template image dataset with displacements determined from a forward FE simulation of passive diastolic filling.

3D FE model was constructed to include the entire image domain, with the lumen and the tissue surrounding the myocardium represented by a compressible neo-Hookean isotropic hyperelastic constitutive model with relatively soft properties (equivalent linear material coefficients of $E = 0.3\,\text{kPa}$ and $n = 0.3$) so that the entire template image could be mapped by the resulting deformation. The edges of the FE mesh were fixed, eliminating rigid body modes.

Constitutive Model and Material Coefficients. The myocardium was represented as a transversely isotropic material with fiber angles varying from $-90°$ at the epicardial surface, through $0°$ at the mid-wall, to $+90°$ at the endocardial surface. The transversely isotropic material model represented fibers embedded in a neo-Hookean matrix with

$$W = \mu(\tilde{I}_1 - 3) + F_2(\tilde{\lambda}) + \frac{\kappa}{2}[\ln(J)]^2, \tag{9}$$

where $\tilde{I}_1$ is the first deviatoric invariant of the right Cauchy deformation tensor (Spencer (1980)), $\tilde{\lambda} = \sqrt{\mathbf{a}_0 \cdot \tilde{\mathbf{C}} \cdot \mathbf{a}_0}$ is the deviatoric fiber stretch along the local direction $\mathbf{a}_0$, μ is the shear modulus of the matrix and κ is the bulk modulus. The stress-stretch behavior for the fibers was represented as exponential, with no resistance to compressive load:

$$\begin{aligned} \tilde{\lambda}\frac{\partial F_2}{\partial \tilde{\lambda}} &= 0, && \tilde{\lambda} < 1, \\ \tilde{\lambda}\frac{\partial F_2}{\partial \tilde{\lambda}} &= C_3\left[\exp\left(C_4(\tilde{\lambda} - 1)\right) - 1\right], && \tilde{\lambda} \geq 1. \end{aligned} \tag{10}$$

Here, C_3 scales the stresses and C_4 defines the rate of stiffening along the preferred direction. A description of the constitutive model and its FE implementation can be found in Weiss et al. (1996).

Material coefficients were determined by a nonlinear least-squares fit of the constitutive equation to published equibiaxial stress/strain curves (Humphrey et al. (1990a,b)) ($\mu = 2.10\,\text{kPa}$, $C_3 = 0.14\,\text{kPa}$, and $C_4 = 22.0$). A bulk modulus κ of $160.00\,\text{kPa}$ was chosen such that the change in relative volume would remain under 5%, in other words the material would be nearly incompressible. The LV material properties do not need to be exact for this validation analysis because the stretch and strain results from the forward FE model were used as the 'gold standard' for comparison to the Warping results. A physiological internal pressure load was applied to the endocardial surface and a forward nonlinear FE analysis was performed using NIKE3D (Maker et al. (1990)). The forward FE model was validated by comparing predictions of fiber stretch, circumferential, radial and in-plane (radial-circumferential) shear strains, wall thickness, ejection fraction and end-diastolic diameter to values reported in the literature (data not shown).

Creation of Synthetic Target Image. Warping strain predictions were validated by comparing various strain measures against a forward FE simulation in which the FE model (based on the template image) was used to generate a synthetic target image. The template and synthetic target image were then input to the warping analysis but with no boundary conditions applied. A synthetic target image was created by applying the displacement map of the forward FE model to the original template MRI image (Fig. 1).

The Warping FE model used the same geometry and material properties as the forward FE model. The Warping analysis was performed using the original template image and the synthetic target image. The nodal values of fiber stretch, circumferential strain, radial strain and shear strain were averaged for each image slice and compared with the values from the forward FE model on a slice-by-slice basis. To assess transmural patterns of deformation, these same measures of deformation were computed as functions of wall position and were then compared with the forward FE predictions. Finally, scatter plots of all nodal values of the four measures of deformation were generated to determine coefficients of determination (R^2) between the Warping and forward FE model predictions.

Sensitivity to Material Coefficients and Constitutive Model. To determine the sensitivity of the Warping analysis to changes in material coefficients, μ and C_3 were increased and decreased by 24% of the baseline values, corresponding to the 95% confidence interval of the material coefficients reported in Humphrey et al. (1990a,b)). To assess the importance of the fiber reinforcement in the constitutive model, the constitutive model was changed to an isotropic neo-Hookean model for the Warping simulations. Finally, the effect of the material bulk modulus was assessed by increasing and decreasing κ by a factor of 10.0 in the Warping analysis. R^2 values were determined for the four measures of local deformation between the Warping and forward FE model predictions.

Sensitivity to Image Noise. To assess the effects of noise on the Warping predictions of deformation, an additive noise model was used to modify the template and target images. Random noise $n(i,j)$ was added to the images $s(i,j)$, where i and j represent pixel coordinates, to create a noisy image $i(i,j)$:

$$i(i,j) = s(i,j) + n(i,j). \tag{11}$$

$n(i,j)$ was defined as the standard deviation σ_N of a zero mean normal probability distribution for noise image intensities (Gonzalez and Woods (1992), pp. 187-213). The signal to noise ratio (SNR) was defined as

$$\text{SNR} = \frac{\sigma_I}{\sigma_N}. \tag{12}$$

For the images used, $\sigma_I = 42$ gray levels. SNRs of $16, 8, 4, 1$ and 0.5 were examined. Local measures of deformation obtained from Warping with the noisy images were compared with results from forward FE predictions to determine the effect of SNR on the R^2 values.

3.2 Results

Comparison of Forward FE and Warping Predictions. There was good qualitative and quantitative agreement between forward and Warping predictions of fiber stretch (Fig. 2A). The highest values of fiber stretch were near the mid-wall (40–50% through the wall), decreasing toward the endo- and epicardial surfaces. Fiber stretch was slightly higher at the endocardial surface than the epicardial surface. The magnitudes of circumferential, radial and shear strain were highest at the endocardial wall and decreased toward the epicardial surface (Fig. 2B–D). There was very good agreement between the forward FE and Warping predictions in terms of the magnitudes of strains and their transmural variation for all four measures of local deformation. The average fiber stretch and strains for both the Warping and forward FE results had generally higher standard deviations than the results of each of the individual

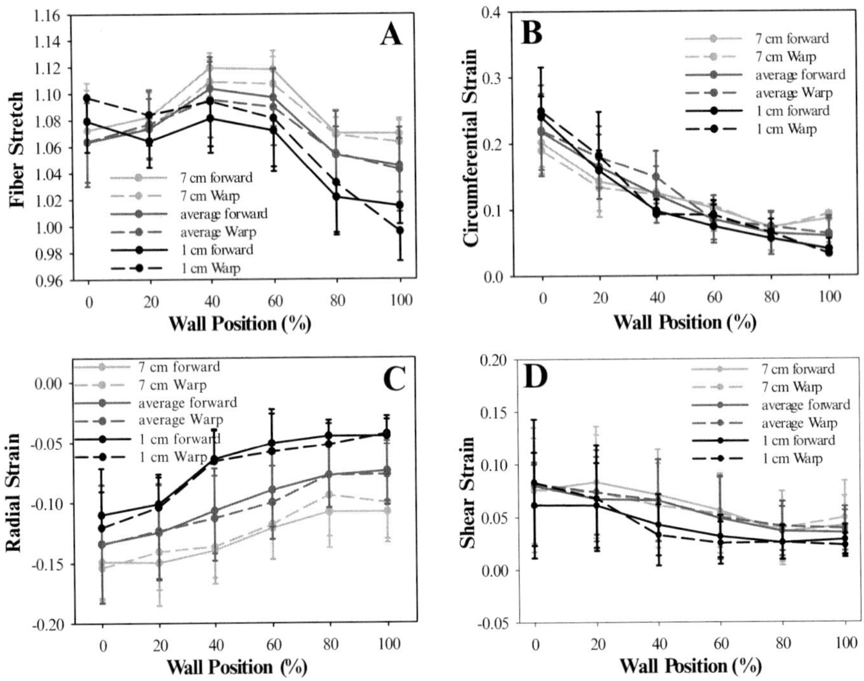

Fig. 2. Forward and FE predictions of local wall deformation at end-diastole as a function of distance through myocardial wall (mean ± SD): A – fiber stretch; B – circumferential Green-Lagrange strain; C – radial Green-Lagrange strain; D – in-plane Green-Lagrange shear strain (circumferential/radial). 0% denotes endocardial and 100% denotes epicardial. Results are presented for image cross-sectional slices at 1 cm (black), 7 cm (light gray) and as an average over all slices (dark grey). 7 cm corresponds to the base of the LV and 1 cm is near the apex of the heart. Solid lines are results from forward FE model and dashed lines are results for Hyperelastic Warping. All values are referenced to the undeformed geometry (beginning-diastole).

image planes (Fig. 2). Thus, the variability depicted in the overall averages is partially due to variability between the results for each image plane. These differences between axial locations are expected because of the changes in LV geometry from apex to base.

Correlation analyses between the forward FE and Warping stretch/strain distributions for each image plane indicated that good agreement was obtained (Fig. 3A–D). There was a significant correlation between the predictions for all four measures of strain ($p < 0.001$ for all cases). The circumferential strain predictions had the best agreement with the forward FE results ($R^2 = 0.76$), while the fiber stretch predictions had the lowest coefficient of determination ($R^2 = 0.67$).

Sensitivity to Changes in Material Coefficients and Constitutive Model. Warping predictions of fiber stretch and strain were insensitive to changes in material parameters μ and C_3 (Table 1). The predictions appear to be somewhat sensitive to material model, with the R^2 values dropping from 1 to 10 points depending on the particular case (Table 1). The technique also appears to be somewhat sensitive to changes in the bulk modulus κ of the model depending on whether the material is being made more or less compressible. Increasing the bulk modulus by an order of magnitude resulted in little in-plane strain degradation; however, the stretch results did show a decrease in the R^2 value.

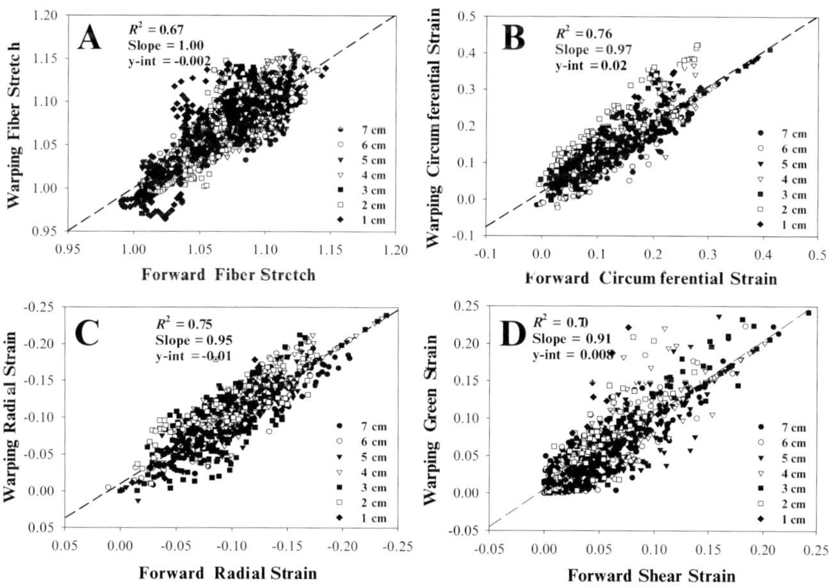

Fig. 3. Scatter plots of forward FE versus Warping stretch/strains: A – fiber stretch; B – circumferential strain; C – radial strain; D – in-plane shear strain. 7 cm corresponds to the base of the LV and 1 cm is near the apex of the heart.

Table 1. Effect of changes in material coefficients and constitutive model on (R^2) between the Warping and forward FE predictions for the four measures of strain.

	Fiber Stretch	Circ. Strain	Radial Strain	Shear Strain
Baseline Warping	0.67	0.72	0.66	0.70
$\mu + 24\%$	0.65	0.69	0.65	0.71
$\mu - 24\%$	0.64	0.70	0.65	0.70
$C_3 + 24\%$	0.65	0.71	0.66	0.72
$C_3 - 24\%$	0.65	0.70	0.66	0.72
neo-Hookean	0.57	0.65	0.58	0.69
$\kappa \times 10$	0.50	0.84	0.65	0.63
$\kappa/10$	0.55	0.40	0.30	0.27

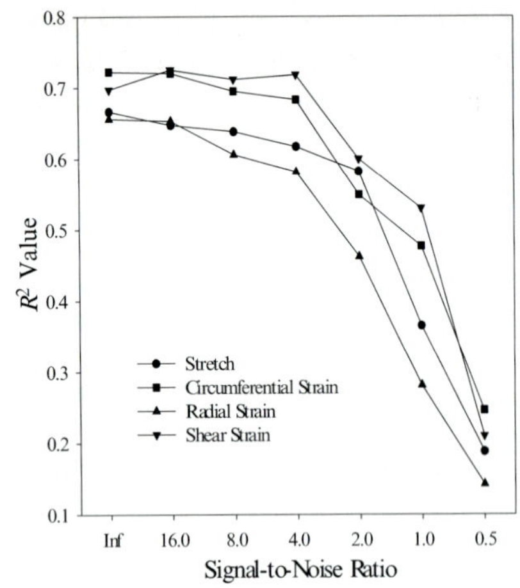

Fig. 4. Effect of signal-to-noise ratio on coefficient of determination.

Decreasing the bulk modulus by an order of magnitude resulted in severe degradation of the results for all measures of deformation.

Sensitivity to Image Noise. There was little change in R^2 values between the forward FE and Warping predictions down to a SNR of 4.0. Degradation in R^2 values was progressive for SNRs below 4.0 (Fig. 4).

4 Discussion

The results of this study indicate that Hyperelastic Warping can predict strain and fiber stretch distributions of the left ventricle during diastole from clinical cine-MRI images. The material sensitivity studies demonstrated that strain predictions are not highly dependent on the material coefficients used to regularize the registration problem, with the exception that a reasonable estimate of the bulk modulus is needed. Warping analysis can accurately determine fiber stretch distribution in relatively noisy images, down to a SNR of 4. Hyperelastic Warping has the inherent flexibility to analyze images acquired using a variety of modalities (Veress et al. (2003)).

Hyperelastic Warping minimizes the image-based energy in a 'hard' sense using an augmented Lagrangian, while a hyperelastic strain energy regularizes the image registration. This approach is considered to be a major strength of the method. First and foremost, the use of a hyperelastic strain energy in combination with a FE discretization ensures that deformations will be diffeomorphic (one-to-one, onto, and differentiable with a differentiable inverse; see, e.g., Miller et al. (2002). Other image-based techniques that have been used to estimate strains via deformable registration, such as optical flow (Klein et al. (1997)) and texture correlation (Gilchrist et al. (2004)), use only an image-based minimization term. Thus, these techniques do not ensure physically reasonable deformations in regions that lack image contrast or texture and they are sensitive to noise. Further, there is no guarantee that deformation maps will be diffeomorphic. Second, hyperelasticity is objective for large strains and rotations, while previous solid mechanics-based regularizations were based on linear elasticity (Bajcsy et al. (1983), Davatzikos (1996), Gee et al. (1993)), which is not objective and penalizes finite strains and rotations. Finally, the use of a hyperelastic constitutive model ensures that deformation maps reflect the behavior of an elastic material under finite deformation. In regions of the template model that have large intensity gradients, large image-based forces will be generated and the solution will be primarily determined by the image data. In regions that lack image texture or gradients, image-based forces will be smaller and the predicted deformation will be more dependent on the regularization (material model). A reasonable representation of the material behavior helps to improve predictions in these areas. In the future, more accurate constitutive models should be incorporated to better represent the material properties of the myocardium. Further, more accurate estimates of local fiber direction may be used (Dou et al. (2003)) to further improve the predictions of myocardial strains based on Hyperelastic Warping.

Acknowledgements. Financial support from NSF grant #BES-0134503, NIH grant #R01-EB00121, and by the U.S. Department of Energy under contract #DE-AC03-76SF00098 is gratefully acknowledged. An allocation of computer time was provided by the Center for High Performance Computing at the University of Utah.

References

Bajcsy, R., Lieberson, R., and Reivich, M. (1983). A computerized system for the elastic matching of deformed radiographic images to idealized atlas images. *J. Comput. Ass. Tomography* 7:618–625.

Bathe, K.-J. (1982). *Finite Element Procedures in Engineering Analysis.* Englewood Cliffs: Prentice Hall.

Bathe, K.-J. (1996). *Finite Element Procedures.* New Jersey: Prentice Hall.

Davatzikos, C. (1996). Spatial normalization of 3D brain images using deformable models. *J. Comput. Ass. Tomography* 20:656–665.

Dou, J., Tseng, W. Y., Reese, T. G., and Wedeen, V. J. (2003). Combined diffusion and strain MRI reveals structure and function of human myocardial laminar sheets in vivo. *Magn. Reson. Med.* 50:107–113.

Fletcher, R. (1989). *Practical Methods of Optimization.* New Delhi: John Wiley & Sons.

Gee, J., Reivich, M., and Bajcsy, R. (1993). Elastically deforming 3D atlas to match anatomical brain images. *J. Comput. Ass. Tomography* 17:225–236.

Gilchrist, C. L., Xia, J. Q., Setton, L. A., and Hsu, E. W. (2004). High-resolution determination of soft tissue deformations using MRI and first-order texture correlation. *IEEE Trans. Med. Imaging* 23:546–553.

Gonzalez, R. C., and Woods, R. E. (1992). *Digital image processing.* Reading, Massachusetts: Addison-Wesley.

Holmes, J. W., Takayama, Y., LeGrice, I., and Covell, J. W. (1995). Depressed regional deformation near anterior papillary muscle. *Am. J. Physiol.* 269:H262–H270.

Humphrey, J. D., Strumpf, R. K., and Yin, F. C. P. (1990a). Determination of a constitutive relation for passive myocardium: I. A new functional form. *J. Biomech. Eng.* 112:333–339.

Humphrey, J. D., Strumpf, R. K., and Yin, F. C. P. (1990b). Determination of a constitutive relation for passive myocardium: II. Parameter estimation. *J. Biomech. Eng.* 112:340–346.

Klein, G. J., Reutter, B. W., and Huesman, R. H. (1997). Non-rigid summing of gated PET via optical flow. *IEEE Trans. Nucl. Sci.* 44:1509–1512.

Maker, B. N., Ferencz, R. M., and Hallquist, J. O. (1990). NIKE3D: A nonlinear, implicit, three-dimensional finite element code for solid and structural mechanics. Technical report UCRL-MA, Lawrence Livermore National Laboratory.

Matthies, H., and Strang, G. (1979). The solution of nonlinear finite element equations. *Int. J. Numer. Meth. Engr.* 14:1613–1626.

McCulloch, A. D., Smaill, B. H., and Hunter, P. J. (1989). Regional left ventricular epicardial deformation in the passive dog heart. *Circ. Res.* 64:721–733.

Miller, M. I., Trouvé, A., and Younes, L. (2002). On the metrics and Euler-Lagrange equations of computational anatomy. *Ann. Rev. Biomed. Eng.* 4:375–405.

Rabbitt, R. D., Weiss, J. A., Christensen, G. E., and Miller, M. I. (1995). Mapping of hyperelastic deformable templates using the finite element method. *SPIE* 2573:252–265.

Reddy, J. N. (2002). *Energy Principles and Variational Methods in Applied Mechanics.* 2nd edition.

Simo, J. C., and Hughes, T. J. R. (1998). *Computational Inelasticity.* New York: Springer.

Spencer, A. (1980). *Continuum Mechanics.* New York: Longman.

Veress, A. I., Weiss, J. A., Gullberg, G. T., Vince, D. G., and Rabbitt, R. D. (2003). Strain measurement in coronary arteries using intravascular ultrasound and deformable images. *J. Biomech. Eng.* 124:734–741.

Veress, A. I., Phatak, N., and Weiss, J. A. (2004). Deformable image registration with hyperelastic warping. In *The Handbook of Medical Image Analysis: Segmentation and Registration Models.* New York: Marcel Dekker.

Weiss, J. A., Maker, B. N., and Govindjee, S. (1996). Finite element implementation of incompressible, transversely isotropic hyperelasticity. *Comput. Meth. Appl. Mech. Engr.* 135:107–128.

Weiss, J. A., Rabbitt, R. D., and Bowden, A. E. (1998). Incorporation of medical image data in finite element models to track strain in soft tissues. *SPIE* 3254:477–484.

Image-based Hierarchical Analysis and Design of Tissue Engineering Scaffolds

S.J. Hollister, C.Y. Lin, C.Y. Lin

University of Michigan, Department of Biomedical Engineering, USA
scottho@umich.edu, lincy@umich.edu, cylin@umich.edu

Tissue engineering scaffolds must be able to replicate complex anatomic defects, provide temporary mechanical function, and provide biofactor (cells, genes, and/or proteins) delivery from a porous microstructure. Fulfilling these requirements necessitates a computational design approach that can predict microstructure influence on effective properties as well as integrate this microstructure within complex 3D anatomic defects. We present an image-based hierarchical homogenization approach for designing porous microstructured tissue engineering scaffolds. This method utilizes image databases for computational models at two scales of structure: (i) the anatomic global scale and (ii) the microstructural scale. Microstructural models can be analyzed using homogenization theory to estimate effective elastic and permeability coefficients. Furthermore, the image-based homogenization method is used in a topology optimization approach to predict *de novo* microstructures that attain desired elastic and permeability properties. Finally, the designed microstructure database is integrated with image database representations of an anatomic defect to create the final tissue engineering scaffold design that may be built using solid free-form fabrication (SFF) techniques.

1 Requirements for Tissue Engineering Scaffolds

Tissue engineering utilizes biomaterial scaffolds to deliver cells, genes and/or proteins (collectively referred to as 'biofactors') to regenerate normal tissue. The biofactors may include stem cells pushed to differentiate into specific cell types or proteins such as Bone Morphogenetic Proteins (BMP). These biofactors, however, cannot in general be injected directly into a tissue defect, since the biofactors will not be able to maintain shape or provide mechanical support for mechanically functioning tissue. Thus, the scaffold must provide temporary load bearing support, but at the same time deliver biofactors. Ideally, there must a sequential trade off between the scaffold bearing load and the regenerate tissue bearing load (Hutmacher (2000)). These two requirements generally conflict, since a more porous microstructure is generally believed to enhance biofactor delivery, while a denser microstructure will better attain load bearing requirements, especially when the base modulus of biopolymers is significantly less than the base modulus of the tissue to be replaced. Thus, the microstructure must be designed such that it provides sufficient load bearing while enhancing biofactor delivery. Finally, this heterogeneous microstructure must be defined within an arbitrarily complex three-dimensional (3D) volume. The three primary scaffold requirements are (Hollister et al. (2002)):

(i) have external shape to fill complex 3D tissue defects;
(ii) provide temporary mechanical load bearing;
(iii) enhance biofactor delivery for accelerated tissue regeneration.

These three requirements are very general and as such may be applied to most tissues. However, for design purposes it will be desirable to associate a physical or chemical property with each requirement. The first requirement is straightforward, namely that the 3D scaffold shape matches the 3D anatomic tissue defect volume. The second requirement can be associated with tissue mechanical properties. For hard tissues, mechanical load bearing may be associated with linear elastic modulus and ultimate strength. For soft tissues, mechanical load bearing may be associated with nonlinear elastic properties defined by an appropriate strain-energy function (Holzapfel (2000)). Thus, for a hard tissue scaffold, it may be desirable to match the native bone anisotropic elastic coefficients as closely as possible. For soft tissues, it may be desirable to match the entire nonlinear stress-strain curve as closely as possible. Furthermore, since the scaffold in most cases should degrade, the reduction in scaffold load bearing characteristics should match the uptake in load bearing by the regenerated tissue.

As daunting as these tasks may appear, the third requirement is even more difficult to quantify and fulfill given the myriad of available biomaterials that interact with innumerable biofactors. Therefore, little is known about what potential physical or chemical scaffold property to link with enhanced tissue regeneration, let alone the desirable value of such a property. One physical scaffold property that is linked to tissue regeneration is porosity, with increased porosity being linked to enhanced tissue regeneration. However, porosity is not uniquely defined since its scalar value indicates little as to whether the porosity is connected or what size pores are present. In fact, some researchers have found that level of connectedness is a better indicator of potential tissue ingrowth than the value of porosity (Hui et al. (1996), Hing et al. (2004)). Thus, a better measure of a scaffold property related to tissue regeneration and ingrowth may be permeability, the ability of fluid to flow through a porous media. The better connected the porosity, the higher the permeability. Therefore, permeability may be one targeted quantity, albeit a simplistic one, for scaffold designs.

There are two available design variables that may be manipulated to fulfill the three primary scaffold requirements: (i) base scaffold material, and (ii) hierarchical design of material microstructure. Base scaffold material has an intrinsic stiffness that obviously defines the upper bound of attainable scaffold effective stiffness. The base scaffold material inherently defines cell interaction and differentiation. Thus, the base scaffold material can be thought of as defining the bounds of scaffold function and cell interaction. Once the base material is given, design of the porous architecture further determines scaffold function and mass transport properties. Specifically, the architecture will determine where between the upper bound of base material stiffness and the lower bound of zero effective stiffness a particular scaffold effective stiffness will fall. The architecture will also determine achievable effective scaffold permeability. Thus, by combining base material synthesis with porous microstructure design, the scaffold properties affecting load bearing (namely elastic properties) and tissue regeneration via biofactor delivery (cell-material interactions and effective scaffold permeability) may be controlled. The purpose of this paper is to present an image-based computational approach for designing tissue engineering scaffolds incorporating hierarchical structure through homogenization theory.

2 Image-based Hierarchical Design

There are two important considerations for scaffold design. First, information on tissue defects, which can have extremely complex shapes, is presented in a structured, voxel grid from imaging modalities like Computed Tomography (CT) or Magnetic Resonance Imaging (MRI). Second, the need for a porous scaffold microstructure inherently requires a hierarchical approach for design and analysis. These considerations have led us to the use of a hierarchical image-based method (Hollister et al. (2000, 2002)) for designing tissue engineering scaffolds. Specifically, this method utilizes a density distribution within a voxel dataset to define shape and topology. A separate image database is created for each structure at every hierarchical level. Thus, a separate database is created for the external scaffold anatomic shape, and for each different porous scaffold microstructure that will exist within the anatomic shape. The density at the more global or macroscopic level will serve as a flag or indicator for the location of voxel density from the next more microscopic level. Formally, we may define the indicator function at a given level k for a specific voxel indexed as i counted sequentially in the x, y and z three-dimensional voxel directions:

$$\begin{aligned} \psi_i^k &= 1 \quad \text{if upper indicator } \rho \geq \rho_i^k \geq \text{lower indicator } \rho, \\ \psi_i^k &= 0 \quad \text{otherwise.} \end{aligned} \tag{1}$$

The density ρ_i, defined as a voxel image intensity ranging from 0 to 255 for a single byte representation, may be defined at specific points for each hierarchical level. At each level the density can depend on a number of factors. For the global anatomic level, the density is obviously a function of anatomic density data derived from the CT or MRI scan, termed anatomic. However, if we are applying topology optimization or computational remodeling schemes, then the global density may also be a function of global stress or strain. Thus, density at the 0th or global anatomic level is given by

$$\rho_i^0 = \rho_i^0(\text{anatomy}, \sigma^0, \ldots), \tag{2}$$

where the superscript 0 denotes a quantity at the global level. The stress σ^0 indicates that the density may be predicted by a topology optimization or computational remodeling algorithm.

The density at subsequent scales is represented by $k = 1, 2, 3$, etc. denoting increasingly smaller scales. Thus,

$$\begin{aligned} \rho_i^1 &= \psi_i^0 \rho_i^1 (C_{pqrs}^0 \text{ or } W^0, K_{pq}^0, \ldots), \\ \rho_i^2 &= \psi_i^0 \psi_i^1 \rho_i^2 (C_{pqrs}^1 \text{ or } W^1, K_{pq}^1, \text{cell attachment}, \ldots), \end{aligned} \tag{3}$$

where $\mathbf{C}^0$ is the elasticity tensor (for hard tissue) at the 0th level, W^0 is a strain-energy function at the 0th level (for soft tissue), and $\mathbf{K}^0$ is the permeability tensor at the 0th level. The 0th level properties at the 1st level indicate that the density arrangement at the 1st scale will determine these properties. For example, the arrangement of material at the 1st level will determine both the 0th level permeability and elasticity. Indeed, achieving specified values of these effective properties will be a scaffold design goal, with the arrangement of density in 3D space a design variable. The presence of an indicator function at a given level implies that if a pore exists ($\psi_i = 0$) at the most global level, a pore must exist at all subsequent microstructure levels. However, existence of a solid material at the most global level ($\psi_i = 1$) does not necessarily imply solid material at subsequent microstructural levels since the global solid material may consist of a solid-pore composite at more microstructural levels.

A significant challenge in scaffold design is timing the trade off between loss of scaffold effective stiffness and gain of regenerate tissue stiffness. As such, it may be desirable to design scaffold topology such that tissue stiffness can be prolonged through degradation. This would involve designing reinforcement of scaffold topology without significantly increasing scaffold mass and decreasing permeability. We (Lin (2004)) have developed a method for designing global density layouts using topology optimization that includes degradation kinetics. In this approach, the degraded stiffness to initial stiffness ratio and time past at different time points is used to weight the density predicted at discrete time points during the degradation cycle. This weighting constant and weighted global density are defined as

$$\begin{aligned} &\text{Time Weighting Factor:} \quad T_{wt} = \frac{T_{\text{tot}} - T_{\text{cur}}}{T_{\text{tot}}}, \\ &\text{Modulus Weighting Factor:} \quad E_{wt} = \frac{E_{ijkl}(T_{\text{cur}})}{E_{ijkl}(T_{\text{tot}})}, \end{aligned} \tag{4}$$

where T_{wt} is the time weighting factor, E_{wt} is the modulus weighting factor, T_{tot} is the total time until the material is degraded, T_{cur} is the current time in the degradation period, and $E_{ijkl}(T_i)$ are the components of the elasticity tensor at time T_i. These are combined to weigh the density predicted at a particular degradation point to achieve the final predicted density:

$$\rho^{\mathrm{deg}} = \sum_{n=1}^{\substack{\sharp\text{points during}\\ \text{degradation}}} \rho^n T_{wt}^n E_{wt}^n. \tag{5}$$

Note that ρ^{deg}, the density from degradation design, will always be greater than the density predicted based only at time T_0 considerations, since the weighting functions at T_0 will be one. Also, the final contribution from the end time will be zero, since T_{cur} will equal T_{tot} and $E_{ijkl}(T_i)$ will also be zero. Furthermore, topology optimization is done at each degradation time point to predict material distribution for the degraded base scaffold modulus. Figure 1 illustrates the degradation design procedure applied to a spinal cage design, showing a design with and without degradation weighting.

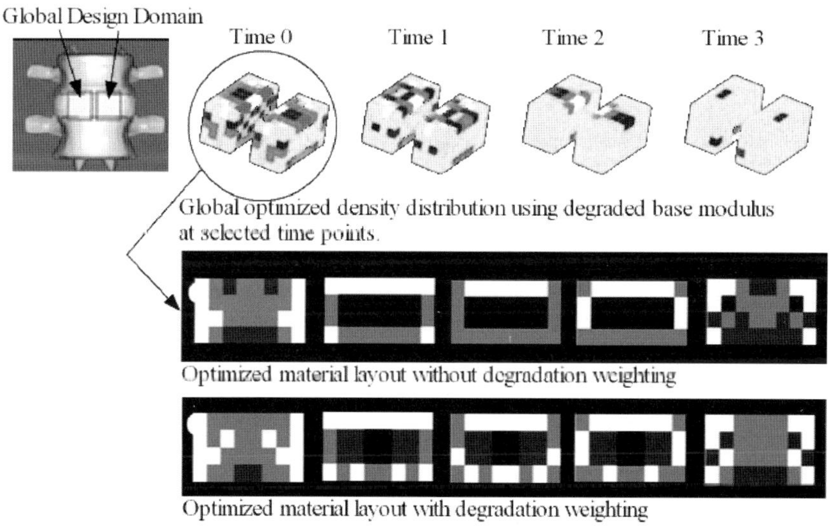

Fig. 1. Example of degradation design scheme applied to a spinal cage design. Top shows topology optimization prediction of density distribution for four different times during degradation. Light grey indicates material, dark grey indicates void. Results show that the volume is almost completely occupied for the last degraded density. Greyscale slice pictures (white is solid, black is void, grey is porous material) show density distribution on slices through the structure for topology design without degradation weighting (top) and with degradation weighting (bottom).

3 Homogenization Theory for Linear Elasticity and Permeability

We utilize image-based homogenization theory (Hollister et al. (1994)) to compute effective elasticity and permeability for scaffold architecture. Homogenization theory is based on asymptotic expansion of the relevant field variables in the governing field equations. For linear elasticity, we expand the displacement components as

$$u_i^{\varepsilon} = u_i^0 + \varepsilon u_i^1 + \dots, \tag{6}$$

where u_i^{ε} is the total displacement, u_i^0 is the global displacement, u_i^1 is the local fluctuation in displacement due to the presence of a microstructure, and ε represents the ratio of one unit on the microstructure scale divided by the global scale

$$\varepsilon = \frac{x_i^1}{x_i^0} \quad \Rightarrow \quad x_i^0 = \frac{x_i^1}{\varepsilon}, \tag{7}$$

where x_i^0 are the global scale coordinates and x_i^1 are the microscale coordinates. The total gradient may be defined using eq. (7) and the chain rule as

$$\frac{\partial}{\partial x_i^{\varepsilon}} = \frac{\partial}{\partial x_i^0} + \frac{\partial x_i^1}{\partial x_i^0}\frac{\partial}{\partial x_i^1} = \frac{\partial}{\partial x_i^0} + \frac{1}{\varepsilon}\frac{\partial}{\partial x_i^1}. \tag{8}$$

If we apply eqs (6) and (8) in the governing elastic equilibrium equation, we obtain a microstructure scale equilibrium equation, namely

$$\frac{\partial}{\partial x_i^1}\left(E_{ijkl}\frac{\partial \chi_k^{pq}}{\partial x_l^1}\right) = \frac{\partial}{\partial x_i^1}\left(E_{pqkl}\right), \tag{9}$$

where E_{ijkl} are the components of the microstructure material elasticity tensor and χ_k^{pq} are the characteristic displacements at the microstructural level. Note that eq. (9) indicates that six unit strains (due to symmetry of the strain tensor) are applied to the microstructure and χ_k^{pq} are the components of the resulting displacement vectors for each of these six unit strains. The effective elasticity tensor then depends on the microstructure elasticity and the characteristic displacements, which, in turn, depend on the 3D arrangement of microstructure:

$$E_{ijkl}^H = \frac{1}{|V_{\text{unit cell}}|}\int_{V_{\text{unit cell}}} E_{ijkl}\left(\delta_{pk}\delta_{ql} - \frac{\partial \chi_k^{pq}}{\partial x_l^1}\right)\mathrm{d}V_{\text{unit cell}}. \tag{10}$$

For permeability analysis, we expand the pressure p and fluid velocity components v_i as

$$p^{\varepsilon} = p^0 + \varepsilon p^1 + \dots, \quad v_i^{\varepsilon} = v_i^0 + \varepsilon v_i^1 + \dots, \tag{11}$$

where $p^\varepsilon/v_i^\varepsilon$ is the total pressure/velocity, p^0/v_i^0 is the average pressure/velocity and p^1/v_i^1 is the perturbation in pressure/velocity due to the presence of a microstructure. Substituting the expansions from eq. (11) into the equations for Stokes flow gives the governing equation for fluid velocity and pressure under unit pressure gradients. Thus,

$$\frac{\partial p^{1k}}{\partial x_i^1} - \frac{\partial}{\partial x_j^1}\left(\frac{\partial v_i^{0k}}{\partial x_j^1}\right) = e^k, \tag{12}$$

where e^k represents a unit pressure gradient for each of the x_1, x_2, and x_3 directions. Solving the above equation for p^1 and v^0 allows calculation of the effective permeability tensor through its components

$$K_{ik} = \frac{1}{\mu\left|V_{\text{unit cell}}^{\text{fluid}}\right|} \int\limits_{V_{\text{unit cell}}^{\text{fluid}}} v_i^{0k} \mathrm{d}V_{\text{unit cell}}^{\text{fluid}}. \tag{13}$$

Thus, by solving eqs (9) and (12) respectively on a structured voxel grid, the effective elasticity and permeability may be computed using eqs (10) and (13), respectively.

4 Scaffold Microstructure Design

Homogenization theory and image-based hierarchical design are used in two ways to design scaffolds. First, scaffold microstructure may be generated based on interconnecting pores in the shape of geometric entities like cylinders and spheres, with the effective elastic properties calculated after the fact. Second, algorithms for solving image-based voxel computational models may be incorporated with optimization schemes to generate *de novo* microstructures to attain desired elasticity coefficients with voxel density as a design variable. In the first approach, the geometric microstructures are shown in Fig. 2.

The effective elasticity and effective permeability were calculated for each of the two microstructures. The results showed that the spherical pore microstructure was stiffer but less permeable than the cylindrical pore microstructure at any given porosity (Fig. 3).

Although generating microstructures using geometric shapes has proved useful for tissue engineering, one would ultimately like greater control over the final effective elasticity and permeability of scaffold designs. Furthermore, the geometric design approach has inherent limitations with respect to effective properties that may be achieved. Therefore, one would like to generate *de novo* microstructure designs without inherent limitations. This may be accomplished by coupling algorithms used to solve eqs (9), (11) and (12) with (10) and (13) using optimization algorithms that update density within a structured grid. In this case, the optimization objective function is to match or

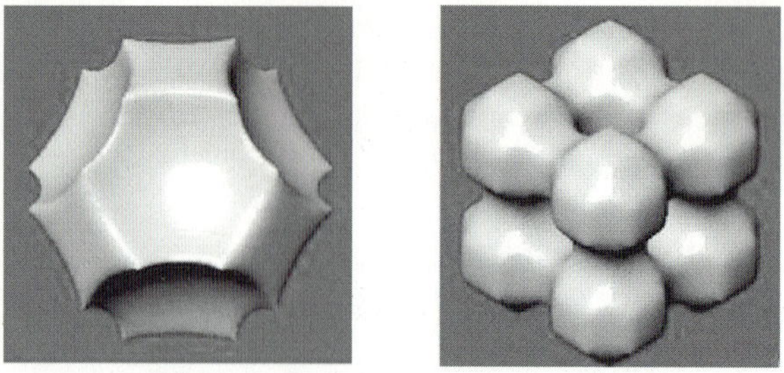

(a) (b)

Fig. 2. Examples of designed architecture based on geometric equations: (a) interconnecting porous spheres; (b) interconnecting porous cylinders.

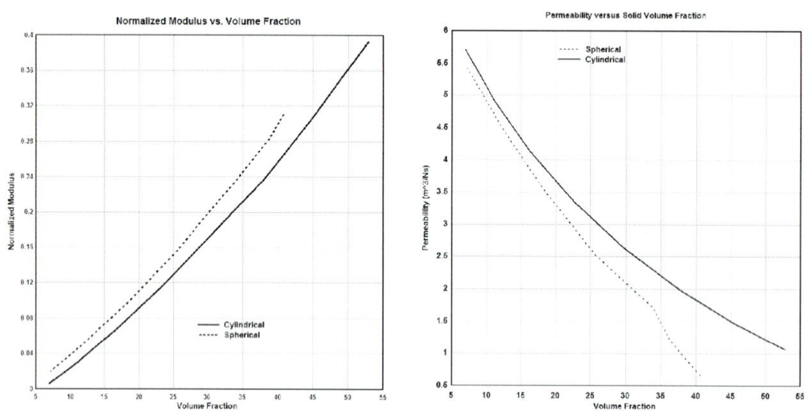

(a) (b)

Fig. 3. (a) plot of modulus versus volume fraction for the cylindrical pore microstructure and spherical pore microstructure. Modulus normalized by base material modulus (values range from 0.0 to 0.4); (b) plot of permeability versus volume fraction for the cylindrical pore microstructure and spherical pore microstructure (values range from 0.5 to 6.0 m^3/Ns). Note that spherical pore microstructures above 45% volume fraction are closed and not included in the calculations, as the permeability is obviously zero.

maximize effective elasticity or permeability, with a constraint on porosity to avoid trivial solutions. The formal optimization statement to match effective elasticity with a constraint on porosity is written as

$$\min_{v} \sum_{i,j,k=1}^{3} w_m \left\| C_{ijkl}^{H} - C_{ijkl}^{\text{target}} \right\|_{L_2}, \tag{14}$$

subject to

$$\frac{\int\limits_{V_{\text{unit cell}}} v \mathrm{d}V_{\text{unit cell}}}{V_{\text{unit cell}}} \leq \frac{V^{\text{target}}}{V_{\text{unit cell}}} \quad \Rightarrow \text{porosity constraint,} \tag{15}$$

where C^H_{ijkl} are the components of the effective elasticity tensor for the designed microstructure, C^{target}_{ijkl} those of the target elasticity tensor, $\left\| C^H_{ijkl} - C^{\text{target}}_{ijkl} \right\|_{L_2}$ represents the L_2-norm of the difference between the components of the homogenized and target elasticity tensor, v is the volume fraction in each element, restricted so that $0 < v \leq 1$, w_m are weighting functions, V^{target} is the target volume and $V_{\text{unit cell}}$ is the total volume of the design unit cell. The optimization approach to maximize permeability with a constraint on effective elasticity and porosity is written as

$$\max_{v} \, w_1 K^H_{11} + w_2 K^H_{22} + w_3 K^H_{33}, \tag{16}$$

subject to

$$\sum_{i,j,k=1}^{3} \left\| C^H_{ijkl} - C^{\text{target}}_{ijkl} \right\|_{L_2} \leq \alpha, \tag{17}$$

$$\frac{\int\limits_{V_{\text{unit cell}}} v \mathrm{d}V_{\text{unit cell}}}{V_{\text{unit cell}}} \leq \frac{V^{\text{target}}}{V_{\text{unit cell}}} \quad \Rightarrow \text{porosity constraint,} \tag{18}$$

where K_{ij} denote components of the effective permeability tensor, α is a tolerance factor governing how closely the homogenized elasticity tensor matches the target elasticity tensor, and the other entities are as defined for eqs (14) and (15). The optimization formulations posed in eqs (14)–(18) are both solved using the Method of Moving Asymptotes, with the voxel-based homogenization iterative solver used to solve for the homogenized elasticity tensor and the Marker and Cell (MAC) finite difference approach used to solve for the homogenized permeability tensor. Specific details as to implementation of the algorithms may be found in Lin et al. (2004a,b) and Lin (2005).

Both optimization approaches were used to design scaffold microstructure for a scaffold to replace a minipig mandibular condyle. Results showed that both approaches could match the three Young's moduli well. However, the optimization formulation that maximized permeability (eqs (16)–(18)) produced a much more permeable structure than the elasticity only optimization (eqs (14) and (15)), as shown in Fig. 4. Furthermore, the microstructure designed using permeability had a much more open and simpler geometry (Fig. 5).

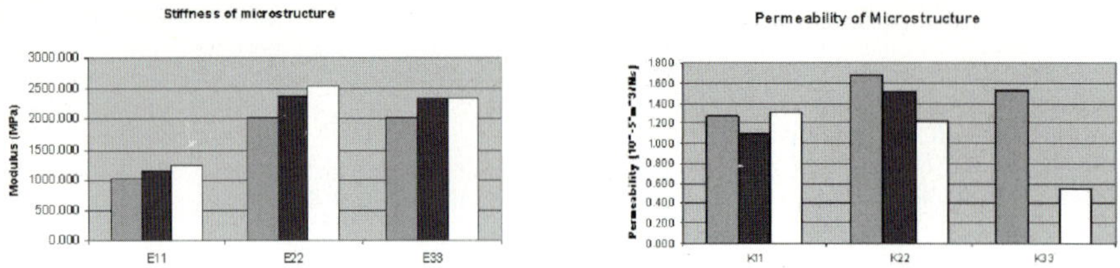

(a) (b)

Fig. 4. (a) Comparison of effective elastic moduli, and (b) effective permeability for actual trabecular bone structure (gray), elasticity only optimized (dark gray), and permeability plus elasticity optimized (white).

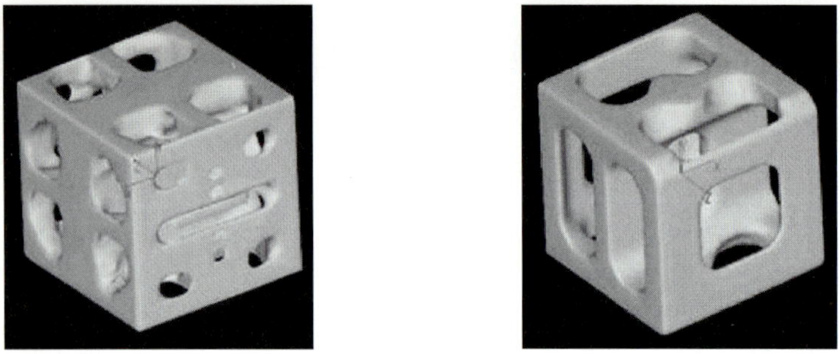

(a) (b)

Fig. 5. Example microstructure designed by (a) optimizing only for elasticity, and (b) maximizing permeability while constraining the elastic properties to match a target property.

5 Integrated Scaffold Design

Once the external shape voxel mapping database is created using methods from Section 2, and the scaffold microstructure is created using methods from Section 3, the final scaffold design may be created. To begin, the two levels of database are generated at the same resolution. Then, indicator function operations defined in eqs (1)–(3) are performed to place the microstructure designed databases into the correct location of the global anatomic shape database. The final outcome is a designed scaffold that matches the anatomic defect with a designed microstructure that attains desired elasticity and permeability properties. The entire sequence is illustrated in Fig. 6.

Once the final scaffold design is created in an image format, the image representation is readily converted to surface format called .STL using a marching cubes algorithm. The .STL format may be used to fabricate scaffolds using a SFF approach (Taboas et al. (2003)).

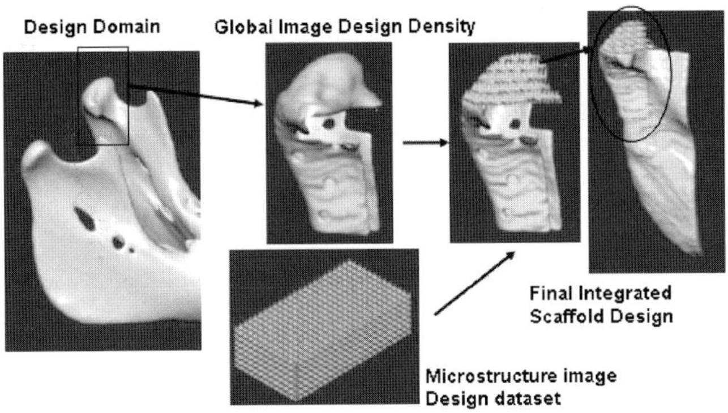

Fig. 6. Sequence of image-based hierarchical design (left to right). The tissue volume to be replaced is segmented from a CT or MRI scan. This creates the global image design density database that may be modified to include surgical attachments. A microstructure database is then designed. The global and microstructure databases are then combined using indicator functions to create the final scaffold design, which matches the original anatomic defect.

6 Discussion

We have proposed an image-based hierarchical scaffold design method based on homogenization theory. Multiple scales of image data are used to design the scaffold, with the topology at each level defined as a density distribution within a structured voxel database. This image based method can design hierarchical structures across multiple scales, limited by available computer memory. We have designed for only two structural scales due to the limitations in fabrication resolution, which is typically 300 to 700 μm. Thus, features at scales below this cannot be built in a controlled manner.

Results have shown that this approach can create microstructures that maximize effective permeability while attaining desired effective elastic properties. Of course, the ability to attain desired effective elastic properties will depend on the starting base properties of the scaffold material and the desired porosity constraint. A limitation in bone tissue engineering is that commonly used biomaterials, especially polymers, are much less stiff than natural bone tissue. Depending on material symmetry in the microstructure, we can expect that the effective properties will be defined within Hashin and Shtrikman (1963) bounds for an isotropic material, and at an absolute upper limit by Voigt bounds for a unidirectional fiber material in the direction of the fibers.

The global scale density distribution can be defined either by natural anatomic tissue density distribution, or by topology optimization schemes. In particular, one can also include the effect of degrading elasticity on the topology design. Degradation means that additional structural material should be included in the base scaffold for reinforcement. The key is to strategically place material so that the structure may be reinforced to keep elastic properties longer through the degradation period without decreasing porosity significantly. For the spinal cage design, the increase in volume fraction was 10%. Although the density weighting method we presented could increase the period for loss of stiffness, design for degrading materials is a new and unique field for structural design optimization. Further investigations will require more degradation kinetics from current materials incorporated in design algorithms and more *in vivo* degradation experiments to verify the results.

Finally, we have only utilized this approach to design linear elastic materials. Although this approach can be extended to the nonlinear behavior of soft tissues using strain-energy functions, this will be quite challenging. If the base material is linear, then the effective nonlinear behavior must be introduced through geometric or boundary condition nonlinear behavior such as buckling or contact at the microstructure level. Our group has recently (Saito et al. (2005)) designed and fabricated scaffolds that give rise to effective nonlinear behavior by introducing contact at the microstructural level.

In conclusion, we have introduced a hierarchical image-based approach for designing tissue engineering scaffolds that exhibit desired elastic and permeability properties at the global level through microstructure design. These scaffold microstructures may then be fit to fill any complex 3D anatomic tissue defect. However, it is important to recognize that these are hypothesized design targets. Verification of how fixed permeability and elasticity influence tissue regeneration will require *in vivo* experiments. Such experiments will provide additional data to update scaffold design targets.

Acknowledgements. This research was supported by NIH DE 13608 (A Bioengineering Research Partnership).

References

Hashin, Z., and Shtrikman, S. (1963). A variational approach to the theory of the elastic behaviour of multiphase materials. *J. Mech. Phys. Solids* 11: 127–140.

Hing, K. A., Best, S. M., Tanner, K. E., Bonfield, W., and Revell, P. A. (2004). Mediation of bone ingrowth in porous hydroxyapatite bone graft substitutes. *J. Biomed. Mater. Res. A* 68:187–200.

Hollister, S. J., Brennan, J. M., and Kikuchi, N. (1994). A homogenization sampling procedure for calculating trabecular bone effective stiffness and tissue level stress. *J. Biomech.* 27:433–444.

Hollister, S. J., Levy, R. A., Chu, T. M., Halloran, J. W., and Feinberg, S. E. (2000). An image-based approach for designing and manufacturing craniofacial scaffolds. *Int. J. Oral Maxillofac. Surg.* 29:67–71.

Hollister, S. J., Maddox, R. D., and Taboas, J. M. (2002). Optimal design and fabrication of scaffolds to mimic tissue properties and satisfy biological constraints. *Biomaterials* 23:4095–4103.

Holzapfel, G. A. (2000). *Nonlinear Solid Mechanics. A Continuum Approach for Engineering.* Chichester: John Wiley & Sons.

Hui, P. W., Leung, P. C., and Sher, A. (1996). Fluid conductance of cancellous bone graft as a predictor for graft-host healing. *J. Biomech.* 29:123–132.

Hutmacher, D. W. (2000). Scaffolds in tissue engineering bone and cartilage. *Biomaterials* 21:2529–2543.

Lin, C. Y., Kikuchi, N., and Hollister, S. J. (2004a). Computational analysis and optimization of bone tissue engineering scaffolds. In *6th International Symposium Computer Methods in Biomechanics and Biomedical Engineering.* Madrid, Spain.

Lin, C. Y., Kikuchi, N., and Hollister, S. J. (2004b). A novel method for biomaterial scaffold internal architecture design to match bone elastic properties with desired porosity. *J. Biomech.* 37:623–636.

Lin, C. Y. (2004). *A New Spine Interbody Fusion System Integrating Topology Optimized Biodegradable Interbody Fusion Cages with Cell-Based and Ex-Vivo Gene Therapy.* PhD dissertation, The University of Michigan, Ann Arbor.

Lin, C. Y. (2005). *Solid-Fluid Mixture Microstructure Design of Composite Materials with Application to Tissue Engineering Scaffold Design.* PhD dissertation, The University of Michigan, Ann Arbor.

Saito, E., Partee, B. I., Das, S., and Hollister, S. J. (2005). Engineered wavy fibered porous polycaprolactone soft tissue scaffolds: Design, fabrication and mechanical properties. In *Proceedings of the 51st Annual Meeting of the Orthopaedic Research Society.* Washington, D.C., to appear.

Taboas, J. M., Maddox, R. D., Krebsbach, P. H., and Hollister, S. J. (2003). Indirect solid free form fabrication of local and global porous, biomimetic and composite 3D polymer-ceramic scaffolds. *Biomaterials* 24:181–194.

Index

A

Adaptation, 3, 11, 17–19, 21, 24, 26, 29, 30, 34–37, 78, 91, 100, 242, 255, 256
Adenosin-Tri-Phosphate, 369, 372, 374, 375, 377
Adherent cell, 138, 177, 182
Adventitia, 6, 207, 211, 228, 229, 263, 285, 313
Aneurysm, 3, 307–311, 314–319, 323, 324, 328, 330, 332–334
  arotic, *see* Aortic aneurysm
  cerebral, *see* Cerebral aneurysm
Angiogenesis, 6, 216
Angioplasty, 208, 221, 225–228, 237, 242, 261
Anion, 422, 425, 426
Anulus fibrosus, 405, 407
Aorta, 140, 147–149, 209, 215, 270, 278, 283–285, 288, 289, 292, 293, 295, 296, 300, 301, 304, 324, 365, 375, 376
Aortic aneurysm, 13
Aortic sinus, 355, 356, 362, 364
Aortic valve, 355, 356, 361, 363, 366, 383
Arterial growth, 3, 7, 13
Artery, 3, 4, 6, 9–11, 18, 22, 27, 32, 40, 41, 48, 55, 56, 58, 60, 61, 138, 140, 147–150, 207–213, 215–217, 219, 220, 225, 228–230, 233–235, 238, 241–251, 255, 257–263, 265, 269, 271–276, 278, 279, 284, 285, 295–297, 299, 301, 302, 304, 307–310, 312–314, 316, 318, 324, 328–330, 332, 333
  carotid, 20, 26, 27, 140, 210, 216, 269, 272, 273, 275–278, 308, 329, 333
  coronary, 208, 211–213, 215, 219–221, 258, 259, 261, 271, 277, 278, 313
  femoral, 210, 257, 269, 272–278
  iliac, 147–150, 217, 225, 228–230, 301
Aspiration device, 432, 434, 439–441
Atheroma, 212, 269–272, 274, 277–279
Atherosclerosis, 4, 147–149, 207–212, 215, 216, 270, 318
Atomic force microscopy, 137, 138, 143–148, 287
Atrophy, 19, 479

B

Bioreactor, 64, 65, 67, 153, 154, 159–163
Blood cells, 137
Blood flow, 21, 22, 61, 210, 244, 245, 247, 250, 323, 324, 327, 329, 333, 334, 339, 376
Bone
  architecture, 477–480, 484
  cells, 6, 33, 35, 112–114
  healing, 105, 106, 112, 113, 117
Brain
  artery, 307, 309, 313, 314
  artery bifurcation, 307, 319

C

Cardiac mechanics, 370
Cartilage, 113, 119, 122, 130, 163, 349, 382, 383, 407, 429
  cells, 112–114
  growth, 119, 130, 131
Cation, 422, 425, 426
Cell
  adherent, *see* Adherent cells
  adhesion, 31, 163
  blood, *see* Blood cells
  bone, *see* Bone cells
  catrilage, *see* Catrilage cells
  concentration, 109, 260–263
  cultures, 24, 33
  debonding, 177, 183–185
  endothelial, *see* Endothelial cells
  epithelial, *see* Epithelial cells
  foam, *see* Foam cells
  geometry, 181–185
  hair, *see* Hair cells
  lacunae, 476
  level, 19, 31, 32, 284, 352, 371, 475
  mechanical properties, 137, 138, 143
  migration, 3, 243, 247, 250, 260–263
  muscle, *see* Muscle cells
  phenotype, 141, 142, 177
  proliferation, 113, 225, 241–244, 258, 260, 261, 263
  reorientation, 177, 182
  smooth muscle, *see* Smooth muscle cells
  stem, *see* Stem cells
  stiffness, 137, 138, 143–146, 149
  structure, 139, 147, 466
  turnover, 12
Cerebral
  aneurysm, 307, 309, 318, 323
  palsy, 339, 340, 349, 350
Cervix, 431, 432, 434–437, 439–441
Chain model, 70, 78, 79, 81, 82, 84–86, 88, 92, 194, 195
Chondrocytes, 42, 122, 153, 155, 156
Circumferential stress, 11, 18, 23, 48, 56–58, 60, 61, 213, 227, 234, 283, 284, 290, 292
Clausius-Duhem inequality, 52, 110, 111, 129
Cochlea, 165–167, 174, 175
Cohesive zone, 177, 180, 182
Collagen, 6, 8–10, 19–22, 29, 52, 57, 63–68, 70–74, 77–80, 86–88, 113, 119, 207, 213, 215, 218, 230, 307, 309, 310, 312–315, 317–319, 355, 356, 361, 363, 364, 367, 380–382, 392, 405–408, 412–415
  birefringence, 309, 313, 315
  turnover, 11
Computational fluid dynamics, 153, 156–159, 161, 334
Confocal laser microscopy, 139, 141, 142
Constitutive equation, *see* Material model
Constitutive model, *see* Material model
Contact
  algorithm, 233, 258
  area, 177, 183–185, 227
  pressure, 225, 227, 349, 350
Convexity, 299, 413, 462, 463, 467
Creep, 91, 178, 379, 380, 382–384, 386–393, 436, 439
Cytoskeleton, 33, 137, 142, 143, 146, 147, 149, 150, 163, 186

D

Distribution function, 12, 356, 414, 459, 460, 466
DNA, 78, 80, 143
Donnan osmosis, 423
Drug delivery, 241, 249, 251
Drug-eluting stent, 226, 241, 244, 249, 266

E

Elastic lamina, 19, 273, 283, 287, 289, 292, 313
Elastic properties, 92, 191, 210, 219, 270, 285, 290, 292, 369, 457, 458, 481, 504, 505, 509, 512–514
Elasticity, 4, 30, 34–37, 53, 77, 78, 80, 81, 123, 130, 178, 180, 217, 270, 310, 316, 319, 327, 346, 359, 360, 458, 461, 466, 491, 499, 506–512

Elastin, 3, 6, 9–11, 13, 17, 19, 22, 23, 29, 207, 283, 286, 292, 293, 318, 319, 380, 392
turnover, 11, 17, 22, 23
Electrochemical potential, 424, 426
Endothelial cells, 6, 137–140, 143, 144, 146, 148–150, 177, 178, 182, 271, 272, 275–277
Epithelial cells, 33
Equilibrium equation, 55, 56, 59, 61, 125, 131, 298, 326, 346, 508
Extracellular matrix, 6, 19, 21, 24, 29, 105, 108, 113, 117, 122, 130, 143, 309, 405–408, 412, 466

F

Fabric, 309–311, 316, 457, 458, 463, 467
Failure criterion, 457, 459, 462, 463, 467
Femur, 108, 340, 349, 350, 385, 476, 477, 481
Fiber reorientation, 78, 86, 87, 99
Fibroblasts, 6, 18, 63–65, 112, 113, 137–140, 143, 149, 177
Finite element method, 226, 375, 406
inverse, *see* Inverse finite element method
Fluid-structure interaction, 324, 328, 355
Fluorescence spectroscopy, 318
Foam cells, 213
Fourier law of heat conduction, 73

G

Glycosaminoglycans, 6, 10, 213, 391, 392, 407
Growth, 3, 4, 7–9, 11–13, 17–20, 29–35, 37–39, 41, 42, 47–56, 58–61, 63–67, 72–74, 91–96, 100, 105–108, 111–114, 117, 119–133, 153, 163, 261, 262, 264, 265, 278, 279, 314, 324, 339, 410, 447
arterial, *see* Arterial growth
cartilage, *see* Cartilage growth
energy, 51, 122, 123, 128
factor, 3, 6, 19, 117
law, 60, 113, 119, 120, 122–127, 130, 132
tensor, 39, 111, 120, 121, 124, 125, 127, 128
velocity gradient, 93, 94, 100, 121

H

Hair cells, 165, 166
Healing, 106, 112, 115, 116, 250, 311, 313, 348
Heart, 30, 41, 324, 372, 375, 392, 493, 496, 497
Histology, 26, 66, 228, 269, 272–275, 279, 307, 309, 316, 474, 478
Homeostatic state, 18, 49, 53–55, 243
Homogenization theory, 503, 505, 508, 509, 513
Hydrogel, 421, 422, 427–429, 445, 447, 452, 453
Hyerelasticity, *see* Elasticity
Hyperelastic warping, 490, 492, 493, 496, 498, 499
Hyperplasia, 3, 216, 225, 243, 249
Hypertension, 3, 213, 214
Hysteresis, 245, 382

I

Image
analysis, 286, 343, 433, 473, 474, 476, 479, 482
registration, 489, 490, 492, 499
Image-based design, 505, 509, 513
In vitro experiment, 166, 174, 229, 369
In vivo experiment, 247, 369, 431, 432, 434, 435, 438, 514
Incompressibility, 50, 53, 56, 59, 131, 230, 290, 297, 298, 345, 346, 356, 359, 403, 410, 422
Indentation test, 194, 288
Internal variable, 86, 88, 369–374, 463
Intervertebral disc, 382, 399, 404–409, 411, 413–417, 419
Intima, 6, 209, 211, 213, 214, 217, 218, 228, 229, 279
Intimal thickening, 219, 249, 261, 273

Invariant, 80, 81, 93, 113, 126, 230, 296, 298, 299, 343, 345, 347, 355–357, 361, 364, 366, 401, 412, 413, 422, 459, 461, 491, 494
Inverse finite element method, 364, 367, 432

L

Lamellar unit, 283, 285, 289
Lamina propria, 445, 453, 454
Langevin function, 195
Least squares, 228, 299, 341, 464, 494
Left ventricle, 284, 369, 370, 374–376, 489, 493, 498
Lesion, 147–149, 209, 210, 212, 215, 216, 228, 241, 248, 266, 269–272, 274–278, 308–310, 314–316, 318
Ligament, 200, 379–393, 395, 397, 399, 400, 403
Long-time memory, 400, 403

M

Mass balance, 8, 9, 36, 50–53, 58, 72
Material model, 36, 41, 48, 51, 122, 123, 125, 126, 129, 131, 165, 174, 178, 192, 194, 200, 216, 232, 257, 263, 348, 355, 356, 360–365, 371, 380, 386, 393, 399, 411, 433, 439, 481, 491, 493–495, 497–499
Material properties, 9, 23, 113, 120, 167, 180, 196, 230, 232, 247–249, 263, 269–272, 274, 277–279, 296, 339, 347, 348, 352, 358, 374, 461, 466, 480, 493–495, 499
Matrix metalloproteinases, 3, 6, 207, 213, 217
Mechanobiology, 230, 258
Mechanotransduction, 19, 24, 69
Media, 6, 25, 56, 65, 147, 148, 207, 210, 221, 228, 229, 273, 279, 283, 290, 292, 313, 314, 356
Microtomography, 156, 475
Mixture theory, 8, 9, 11, 20, 129, 131
Momentum balance, 8, 9, 50, 51, 410, 411
Mooney-Rivlin model, 257
Morphogenesis, 30, 39, 41, 42
Muscle, 372, 374
  cells, 137
Musculo-skeletal system, 339, 352
Myocardium, 41, 345, 347, 348, 355, 369, 375, 494, 499

N

Navier-Stokes equation, 158, 324
Neo-Hookean model, 57, 232, 361, 412, 494, 495
Network model, 77, 78, 80, 81, 83–85, 88
Non-invasive, 174, 300, 475, 483, 484, 489
Nucleus pulposus, 407

O

Objective function, 299, 303, 342, 348, 365, 509
Open system, 36, 48, 50, 92, 100, 105
Opening angle, 40, 57, 295, 301, 304
Optimization, 12, 19, 216, 231, 244, 248, 249, 269, 272–274, 277, 352, 439, 503, 506, 507, 509–511, 513, 514
Organ culture, 17, 24, 25
Orthotropy, 56, 125, 126, 131, 347, 348, 466
Osmotic pressure, 409, 411, 415, 424
Osteoporosis, 474, 481

P

Parameter identification, 295, 299
Patella, 339, 349
Penalty parameter, 299, 301, 491, 492
Permeability, 113, 114, 408, 411, 416, 503–506, 508–514
Persistence length, 79–81, 86
Phenotype, 6, 140–145, 149, 177
Plaque, 209, 212–215, 221, 230, 235, 242, 248
  cap, 209, 211–221, 228, 229, 235
  rupture, 209, 212–214, 220, 277, 280
Plasticity, 38, 100, 127, 232, 348
Platelet, 198, 201, 212, 242, 244, 246, 250
Polarized light microscopy, 307, 313–315

Porous media, 421, 423, 504
Preterm delivery, 431, 440
Proteoglycans, 9, 119, 380, 391, 392, 405, 407

R

Relaxation, 49, 54, 82, 86, 91, 97, 178, 179, 211, 221, 379–393, 399, 401–403, 438, 447, 448
Remodeling, 3, 4, 7–9, 11–13, 17–21, 29–34, 36, 37, 39, 41, 53, 63–65, 68–72, 74, 78, 82, 85–88, 91, 106–108, 117, 130, 208, 215, 216, 218, 241, 255, 256, 278, 307, 309, 318, 478, 506
Representative volume element, 189, 193–195, 200
Residual strain, 39, 289, 295, 296, 299, 301, 303, 304
Residual stress, 5, 6, 11, 12, 18, 37, 38, 40, 41, 47–49, 52, 55–61, 105, 117, 120, 121, 124, 131, 265, 283–285, 287–293, 295, 296, 303, 304, 375
Resorption, 29, 30, 34, 35, 41
Response function, 9, 20, 22, 121, 131, 249, 347, 412
Restenosis, 226, 227, 237, 241, 243–245, 249, 255, 256, 258–266
Reynolds number, 154, 158, 159, 162, 324, 329, 333
Risk factor, 4, 213, 244
Rupture, 189, 212–215, 242, 307, 308, 314, 323

S

Sarcomere, 372, 374, 375
Scaffold, 65, 86, 153–159, 161–163, 242, 250, 445, 503–506, 509, 511–514
Scanning acoustic microscopy, 138
Scanning electron microscopy, 155, 156, 193, 382
Short-time memory, 399–401, 403
Signal transduction, 352
Skeletal muscle, 339, 345, 348, 383
Smooth muscle, 5, 6, 9, 11, 18–20, 22, 220, 278, 279, 283
  cells, 6, 19, 137–147, 149, 210, 213–215, 228, 243, 262, 263, 269, 285, 292, 313, 356
Stem cells, 112–115, 503
Stenosis, 208, 212, 225, 226, 228, 233, 238, 242, 256, 260
Stent, 208, 225–227, 230–238, 241–244, 248–251, 255–258, 262–266, 279
  design, 225, 226, 235, 238, 241, 244, 245, 247–249, 256–258, 265, 266
Stress-free configuration, 8, 50, 52
Structure tensor, 81, 91, 92, 95, 96, 99, 100, 460
Swelling, 38, 39, 73, 406, 408, 412, 421, 424, 427–429

T

Tendon, 63–68, 70, 72, 77, 78, 86–88, 105, 139, 313, 382, 383, 392, 399, 403
  construct, 65–68, 73, 74
Theory of porous media, 405, 406, 409
Thermoelasticity, 52, 53
Thrombosis, 208, 212, 225, 235, 242, 243
Tissue
  damage, 105, 106, 108, 117, 241, 263, 445
  differentiation, 105, 106, 117
  engineering, 7, 64, 154, 163, 445, 503, 505, 509, 513, 514
  growth, 29–33, 37, 39, 64, 65, 106, 112, 113, 153, 163, 225, 261, 262
Trabecular
  bone, 114, 457–459, 465–467, 474, 475, 477–483
  failure, 481
Transmural pressure, 11, 18, 24, 210, 211, 270, 277, 279, 312, 375
Transverse isotropy, 77, 78, 80, 81, 83–86, 88, 91, 94, 95, 125, 126, 290, 347, 355, 357–359, 365, 494

U

Unloaded configuration, 11, 49, 58
Uterus, 138, 431, 432, 434, 436, 437, 440–442

V

Vascular growth, 3, 4, 13, 17, 18
Viscoelastic properties, 178, 381, 382, 399, 400, 406, 445–447
Viscoelasticity, 36, 178, 180, 182, 185, 348, 380, 382, 383, 385, 416, 451, 452
Vocal fold, 445, 446

Y

Yield criterion, 457, 459, 462, 463, 465, 467

Printing: Krips bv, Meppel
Binding: Stürtz, Würzburg